# Molecular Biology and Biotechnology of Plant Organelles

# Molecular Biology and Biotechnology of Plant Organelles

Edited by Kameron Gray

New York

Published by Syrawood Publishing House,
750 Third Avenue, 9th Floor,
New York, NY 10017, USA
www.syrawoodpublishinghouse.com

**Molecular Biology and Biotechnology of Plant Organelles**
Edited by Kameron Gray



International Standard Book Number: 978-1-64740-368-3 (Hardback)



**Cataloging-in-publication Data**

Molecular biology and biotechnology of plant organelles / edited by Kameron Gray.
p. cm.
Includes bibliographical references and index.
ISBN 978-1-64740-368-3
1. Plant organelles. 2. Plant molecular biology. 3. Plant genetic engineering. 4. Plant cell biotechnology.
5. Molecular biology. I. Gray, Kameron.

QK725 .M65 2023
581.072 4--dc23

# TABLE OF CONTENTS

# PREFACE

An organelle refers to a specialized subunit of a cell, which performs an important role. Some of the prominent organelles found in plants are mitochondria, nuclei, and plastids. Mitochondria are oblong shaped organelles, which are responsible for providing energy by breaking down carbohydrates and sugar molecules. The nucleus functions as the administrative and information processing center of the cell. Biotechnology has been used for genetically engineering plant organelles for agricultural purposes. Various studies are also being conducted for genome-wide analysis of organellar gene expression and technologies are being developed for selectively manipulating organellar genes and genomes. Such studies are used to study the functioning of organelles. Biotechnology of plant organelles can also be used for improving nutritional quality, crop yield, and resistance to abiotic and biotic stresses. This book contains some path-breaking studies on the molecular biology and biotechnology of plant organelles. Those in search of information to further their knowledge will be greatly assisted by it.

All of the data presented henceforth, was collaborated in the wake of recent advancements in the field. The aim of this book is to present the diversified developments from across the globe in a comprehensible manner. The opinions expressed in each chapter belong solely to the contributing authors. Their interpretations of the topics are the integral part of this book, which I have carefully compiled for a better understanding of the readers.

At the end, I would like to thank all those who dedicwated their time and efforts for the successful completion of this book. I also wish to convey my gratitude towards my friends and family who supported me at every step.

**Editor**

# 1

# Salinity Response in Chloroplasts: Insights from Gene Characterization

**Jinwei Suo [1], Qi Zhao [2], Lisa David [3], Sixue Chen [3] and Shaojun Dai [1,2,*]**

[1] Alkali Soil Natural Environmental Science Center, Northeast Forestry University, Key Laboratory of Saline-alkali Vegetation Ecology Restoration in Oil Field, Ministry of Education, Harbin 150040, China; suojinwei@nefu.edu.cn

[2] Development Center of Plant Germplasm Resources, College of Life and Environmental Sciences, Shanghai Normal University, Shanghai 200234, China; zhaoqizq@yeah.net

[3] Department of Biology, Genetics Institute, Plant Molecular and Cellular Biology Program, Interdisciplinary Center for Biotechnology Research, University of Florida, Gainesville, FL 32610, USA; lisaidavid@ufl.edu (L.D.); schen@ufl.edu (S.C.)

* Correspondence: daishaojun@hotmail.com

Academic Editor: Marcello Iriti

**Abstract:** Salinity is a severe abiotic stress limiting agricultural yield and productivity. Plants have evolved various strategies to cope with salt stress. Chloroplasts are important photosynthesis organelles, which are sensitive to salinity. An understanding of molecular mechanisms in chloroplast tolerance to salinity is of great importance for genetic modification and plant breeding. Previous studies have characterized more than 53 salt-responsive genes encoding important chloroplast-localized proteins, which imply multiple vital pathways in chloroplasts in response to salt stress, such as thylakoid membrane organization, the modulation of photosystem II (PS II) activity, carbon dioxide ($CO_2$) assimilation, photorespiration, reactive oxygen species (ROS) scavenging, osmotic and ion homeostasis, abscisic acid (ABA) biosynthesis and signaling, and gene expression regulation, as well as protein synthesis and turnover. This review presents an overview of salt response in chloroplasts revealed by gene characterization efforts.

**Keywords:** chloroplast; gene characterization; salinity response

---

## 1. Introduction

Soil salinity is one of the most severe abiotic stresses affecting agricultural yield and productivity worldwide [1]. In plants, salinity induces ion imbalance, hyperosmotic stress, oxidative damage, and other subsequent disturbances. Plants have evolved sophisticated salt-responsive signaling and metabolic processes, including photosynthetic adjustment, ion exclusion, the synthesis of compatible products, the enhancement of reactive oxygen species (ROS) scavenging, phytohormone regulation, and cell structure modulation [2].

Chloroplasts are organelles for photosynthesis. Chloroplasts also participate in many other important metabolic processes, including the biosynthesis of amino acids, vitamins, isoprenoids, fatty acids, and lipids, as well as the reduction of nitrites and sulfates [3,4]. About 3000 distinct proteins are estimated to be localized to the *Arabidopsis thaliana* chloroplasts [5,6]. Most of these proteins are encoded by the nuclear genome, and less than 150 proteins are predicted to be encoded by the plastome [7]. Previous gene characterization studies have revealed a number of genes/proteins involved in various signaling and metabolic processes in chloroplasts [7,8]. However, the characterization of salinity-responsive genes encoding chloroplast-localized proteins is limited, although many genes/proteins have been proposed to be involved in salt tolerance using transcriptomic and proteomic approaches [9–12].

Over the past twenty years, only about 53 salt-responsive genes have been characterized that encode chloroplast-localized proteins from Arabidopsis, rice (*Oryza sativa*), maize (*Zea mays*), wheat (*Triticum aestivum*), beet (*Beta vulgaris*), cotton (*Gossypium hirsutum*), mangrove (*Kandelia candel*), tobacco (*Nicotiana tabacum*), alfalfa (*Medicago sativa*), spinach (*Spinacia oleracea*), alkaligrass (*Puccinellia tenuiflora*), cowpea (*Vigna unguiculata*), pea (*Pisum sativum*), *Thellungiella halophila*, *Mesembryanthemum crystallinum*, *Porteresia coarctata*, *Suaeda salsa*, and *Synechocystis* sp. PCC 6803, respectively (Table 1). The proteins encoded by these salt-responsive genes are mainly involved in ROS scavenging, thylakoid membrane organization, photosystem II (PS II) activity, carbon dioxide ($CO_2$) assimilation, photorespiration, osmotic and ion homeostasis, abscisic acid (ABA) biosynthesis and signaling, and gene expression, as well as protein synthesis and turnover (Table 1). In this review, we summarize the genes encoding the chloroplast-localized proteins in response to salinity.

## 2. Salinity-Induced Diverse ROS Scavenging Pathways in Chloroplasts

Chloroplasts are unique organelles due to their highly oxidizing metabolic activity and increased rate of electron flow, being especially prone to generating ROS, such as superoxide anion ($O_2^-$), hydrogen peroxide ($H_2O_2$), singlet oxygen ($^1O_2$), and hydroxyl radical (OH•). The presence of ROS producing centers, such as triplet chlorophylls and the electron transport chain (ETC) in PS II and PS I, make chloroplasts a major site of ROS production [13]. Salinity stress enhances ROS production, leading to severe ROS-associated damage to chloroplasts. ROS seriously disrupt normal metabolism through oxidative damage to lipids, nucleic acids, and proteins, resulting in protein destruction and the peroxidation of membrane lipids [14]. Therefore, antioxidant enzyme systems and non-enzymatic antioxidants in chloroplasts offer greater protection from oxidative damage generated from salinity stress. Several salinity-responsive genes encoding antioxidant enzymes/proteins have been cloned (Table 1), which highlight specific ROS scavenging pathways in chloroplasts under salt stress.

### *2.1. The Water-Water Cycle Detoxifies $O_2^-$ and $H_2O_2$*

The water-water cycle operates as a ROS scavenging pathway in chloroplasts, which is essential for salinity tolerance (Figure 1A). Oxygen ($O_2$) generated in chloroplasts during photosynthesis can accept electrons passing through PS II and PS I, resulting in the formation of $O_2^-$ by the Mehler reaction [15]. The major site of $O_2^-$ production is the thylakoid membrane-bound primary electron acceptor of PS I. It has been revealed that the acceptor side of the ETC in PS II also provides sides ($Q_A$, $Q_B$) with electron leakage to $O_2$ producing $O_2^-$. Once produced on the internal "lumen" membrane surface, $O_2^-$ may be protonated to $HO_2^-$, which initiates lipid peroxidation. On the "stromal" membrane surface, a membrane attached copper/zinc superoxide dismutase (Cu/Zn SOD) in the vicinity of PS I catalyzes the dismutation of $O_2^-$ into $O_2$ and $H_2O_2$, and then the generated $H_2O_2$ is reduced to $H_2O$ by a membrane-bound thylakoid ascorbate peroxidase (tAPX), which is the commonly named water–water cycle [13].

The overexpression of *Cu/Zn SOD* in chloroplasts of Arabidopsis [16], tobacco [17,18], Chinese cabbage (*Brassica campestris* L. ssp. *pekinensis* cv. Tropical Pride) [19], and cotton [20] can enhance salinity tolerance through reducing ROS (Table 1). Similarly, tobacco plants with an overexpression of *APX* in chloroplasts showed a higher resistance to salt stress, and the isolated chloroplasts from the transgenic lines also showed higher APX activity than wild-type control plants [21]. These results indicated that the thylakoidal scavenging system of ROS is essential for salt tolerance.

Although catalase (CAT) has not been found in chloroplast stroma, PS II membranes associate with a heme CAT [22]. The CAT does not directly participate in the water–water cycle, but protects water oxidase in the lumen if the water–water cycle does not operate properly and $H_2O_2$ diffuses to the lumen [15]. An increased defense against oxidative damage induced by salt stress was conferred by targeting CAT to chloroplasts in both Chinese cabbage [19] and cotton plants [20].

**Table 1.** List of genes encoding chloroplast proteins in response to salinity.

| Gene Name | Plant Species | Encoding Protein | Salt Treatment Condition | Ref. |
|---|---|---|---|---|
| Stress and defense (22) | | | | |
| *Cu/Zn SOD* | Alkaligrass (*Puccinellia tenuiflora*) | Copper/zinc superoxide dismutase | NaCl (0, 50, 100 mM; 21 days); $NaHCO_3$ (0, 3, 5 mM; 21 days) | [16] |
| *Cu/Zn SOD* | Rice (*Oryza sativa*) | Copper/zinc superoxide dismutase | NaCl (300 mM; 2, 4, 6, 8, 10 days) | [17] |
| *Cu/Zn SOD* | Mangrove (*Kandelia candel*) | Copper/zinc superoxide dismutase | NaCl (100, 300 mM; 8, 24 h; 1, 2, 3 weeks) | [18] |
| *Cu/Zn SOD* | Maize (*Zea mays*) | Copper/zinc superoxide dismutase | NaCl (0, 50, 100, 150, 200 mM; 10 days; 4 weeks) | [19] |
| *Cu/Zn SOD* | Cotton (*Gossypium hirsutum*) | Copper/zinc superoxide dismutase | NaCl (50, 100, 150, 200 mM; 1, 2, 3, 4 weeks) | [20] |
| *APX* | *Arabidopsis thaliana* | Ascorbate peroxidase | NaCl (300 mM; 8 days) | [21] |
| *CAT* | Maize (*Z. mays*) | Catalase | NaCl (0, 50, 100, 150, 200 mM; 10 days; 4 weeks) | [19] |
| *CAT* | Cotton (*G. hirsutum*) | Catalase | NaCl (50, 100, 150, 200 mM; 1, 2, 3, 4 weeks) | [20] |
| *MDHAR* | Mangrove (*Avicennia marina*) | Monodehydroascorbate reductase | NaCl (200 mM; 4 days) | [23] |
| *DHAR* | Rice (*O. sativa*) | Dehydroascorbate reductase | NaCl (100, 150, 200 mM; 12, 14 days) | [24] |
| *GR3* | Rice (*O. sativa*) | Glutathione reductase | NaCl (200 mM; 0, 4, 8, 12, 16, 20, 24 days) | [25] |
| *GR3* | Rice (*O. sativa*) | Glutathione reductase | NaCl (100 mM; 7, 15 days) | [26] |
| *PrxQ* | *Suaeda salsa* | Peroxiredoxin Q | NaCl (0, 100, 150 mM; 3 weeks) | [27] |
| *NTRC* | Rice (*O. sativa*) | NADPH thioredoxin reductase | NaCl (170 mM; 1, 3 days) | [28] |
| *W69* | Wheat (*Triticum aestivum*) | Glutathione peroxidase | NaCl (150 mM; 7 days) | [29] |
| *W106* | Wheat (*T. aestivum*) | Glutathione peroxidase | NaCl (150 mM; 7 days) | [29] |
| *SIA1* | Arabidopsis | ABC1-like kinase | NaCl (200 mM; 3 days) | [30] |
| *γ-TMT* | Arabidopsis | γ-tocopherol methyltransferase | NaCl (0, 200, 300, 400 mM; 12, 24, 48 h; 4 weeks) | [31] |
| *WSL12* | Rice (*O. sativa*) | Nucleoside diphosphate kinase | NaCl (100, 150, 200 mM; 0, 1, 1.5, 2, 2.5, 3, 3.5, 4, 4.5, 5, 5.5, 6, 6.5 days) | [32] |
| *NDPK2* | Arabidopsis | Nucleoside diphosphate kinase 2 | NaCl (200 mM; 14 days) | [33] |
| *TSPO* | Arabidopsis | 18 kDa translocator protein | NaCl (150 mM; 1, 3, 6, 12, 24 h) | [34] |
| *MSRA4.1* | Rice (*O. sativa*) | Methionine sulfoxide reductase | NaCl (100 mM; 2 days) | [35] |
| Thylakoid membrane organization and PSII activity (5) | | | | |
| *Fad6* | Arabidopsis | ω-6 desaturase | NaCl (0, 75, 100, 125 mM; 8 days); (300 mM; 0, 1, 3, 6, 12, 24 h) | [36] |
| *GPAT* | Tomato (*Lycopersicon esculentum*) | Glycerol-3-phosphate acyltransferase | NaCl (200 mM; 1, 3, 5, 7 days); (150 mM; 30 days) | [37] |
| *MGD* | Rice (*O. sativa*) | Monogalactosyl-diacylglycerol synthase | NaCl (0, 200 mM; 10 days) | [38] |
| *RUB* | Alkaligrass (*P. tenuiflora*) | Rubredoxin family protein | NaCl (100, 125 mM; 10 days); $NaHCO_3$ (1.5, 3 mM; 10 days) | [39] |
| *RCI* | Wheat (*T. aestivum*) | Rare cold inducible protein | NaCl (150 mM; 2 weeks) | [40] |
| Photosynthesis and photorespiration (3) | | | | |
| *NADP-MDH* | *Mesembryanthemum crystallinum* | $NADP^+$-dependent malate dehydrogenase | NaCl (400 mM; 1, 6, 12, 30, 72, 126 h) | [41] |
| *GAPB* | *Thellungiella halophila* | Glyceraldehyde 3 phosphate dehydrogenase β subunit | NaCl (200 mM; 2 weeks) | [42] |
| *GS2* | Rice (*O. sativa*) | Glutamine synthetase | NaCl (150 mM, 12 days) | [43] |

**Table 1.** *Cont.*

| Gene Name | Plant Species | Encoding Protein | Salt Treatment Condition | Ref. |
|---|---|---|---|---|
| Osmotic and ion homoestasis (12) | | | | |
| *CMO* | Spinach (*Spinacia oleracea*) | Choline monooxygenase | NaCl (50, 100 mM; 0, 3, 6, 9, 12, 15 weeks) | [44] |
| *CMO* | Beet (*Beta vulgaris*) | Choline monooxygenase | NaCl (0, 100, 150 mM; 36 days) | [45] |
| *BADH* | Spinach (*S. oleracea*) | Betaine aldehyde dehydrogenase | NaCl (50, 100 mM; 0, 3, 6, 9, 12, 15 weeks) | [44] |
| *BADH* | Tobacco (*Nicotiana tabacum*) | Betaine aldehyde dehydrogenase | NaCl (100, 200, 300, 400, 500 mM; 1 month) | [46] |
| *BADH* | Spinach (*S. oleracea*) | Betaine aldehyde dehydrogenase | NaCl (0, 75, 150 mM; 3 weeks) | [47] |
| *TPPD* | Arabidopsis | Trehalose-6-phosphate phosphatase | NaCl (200 mM; 0, 1, 3, 8 h) | [48] |
| *INO1* | *Porteresia coarctata* | L-*myo*-inositol 1-phosphate synthase | NaCl (100, 200, 300, 400 mM; 96 h) | [49] |
| *IMT1* | *M. crystallinum* | Inositol methyl transferase | NaCl (100, 200, 300, 400 mM; 96 h) | [49] |
| *CHX23* | Arabidopsis | $Na^+(K^+)/H^+$ exchanger | NaCl (75 mM; 12 days) | [50] |
| *NHD1* | Arabidopsis | Sodium hydrogen antiporter | NaCl (150 mM; 72 h) | [51] |
| *KEA* | Arabidopsis | $K^+/H^+$ antiporter | NaCl (75 mM) | [52] |
| *YL1* | Arabidopsis | YqeH-type GTPase | NaCl (0, 150 mM; 2 days) | [53] |
| ABA and kinase signaling (5) | | | | |
| *ABA1* | Arabidopsis | Zeaxanthin epoxidase | NaCl (300 mM; 3 h) | [54] |
| *NCED1* | Cowpea (*Vigna unguiculata*) | 9-*cis*-epoxycarotenoid dioxygenase | NaCl (250 mM; 0, 1, 2, 5, 10, 24 h) | [55] |
| *MDA1* | Arabidopsis | Transcription termination factor | NaCl (100, 150, 200 mM; 4, 9, 10, 14 days) | [56] |
| *COR15* | Arabidopsis | 15 kDa protein | NaCl (150 mM; 3 days) | [57] |
| *MsK4* | Alfalfa (*Medicago sativa*) | Glycogen synthase kinase 3 like kinase | NaCl (100 mM; 4 weeks) | [58] |
| Gene expression and protein turnover (6) | | | | |
| *RH3* | Arabidopsis | DEAD-box RNA helicase | NaCl (100 mM; 0, 4, 12, 24 h) | [59] |
| *CRP1* | Arabidopsis | Chloroplast-targeted RNA-binding protein 1 | NaCl (150 mM, 7 days) | [60] |
| *S-RBP11* | Arabidopsis | RNA-binding group protein | NaCl (0, 130, 140, 150, 160 mM; 14 days); (300 mM; 0, 1, 2, 4, 8 h) | [61] |
| *TufA* | Pea (*Pisum sativum*) | Chloroplast translation elongation factor | NaCl (100, 200, 500 mM; 4, 6, 24 h) | [62] |
| *PsbA* | *Synechocystis* sp. PCC 6803 | Photosystem II D1 protein | NaCl (20 mM, 500 mM, 1000 mM; 0, 1, 2, 3, 4 h) | [63] |
| *DegP2* | Arabidopsis | Prokaryotic trypsin-type Deg/Htr serine protease | NaCl (400 mM; 2 h) | [64] |

Ref.: Reference.

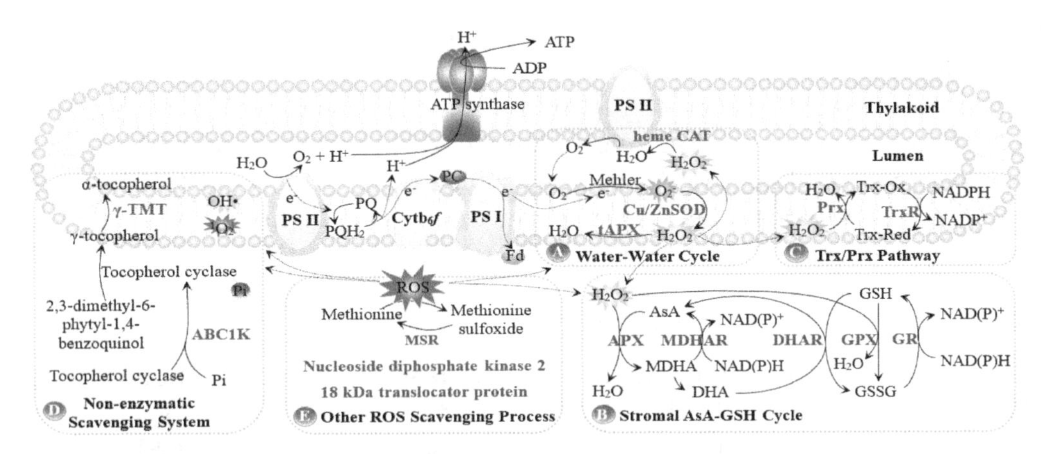


**Figure 1.** Schematic presentation of ROS scavenging pathway in chloroplasts. (**A**) Water–Water cycle; (**B**) Stromal AsA–GSH cycle; (**C**) Trx/Prx pathway; (**D**) Non-enzymatic scavenging system; (**E**) Other ROS scavenging process. The solid line indicates a single-step reaction, and a dotted line indicates the movement of molecules. Substrates and products are in black font, proteins are in blue bold font, and P in a blue circle indicates phosphorylated protein. Abbreviations: ABC1K, activity of bc1 complex-like kinase; APX/tAPX, ascorbate peroxidase/thylakoid ascorbate peroxidase; AsA, ascorbate; Cu/Zn SOD, copper/zinc superoxide dismutase; Cytb$_6$*f*, cytochrome b$_6$*f* complex; DHA, dehydroascorbate; DHAR, dehydroascorbate reductase; Fd, Ferredoxin; GPX, glutathione peroxidase; GR, glutathione reductase; GSH, reduced glutathione; GSSG, oxidized glutathione; $H_2O_2$, hydrogen peroxide; hemeCAT, heme catalase; MDHA, monodehydroascorbate; MDHAR, monodehydroascorbate reductase; MSR, sulfoxide reductase; $^1O_2$, singlet oxygen; $O_2$, oxygen; $O_2^-$, superoxide anion; OH•, hydroxyl radical; PC, plastocyanin; Prx, Trx-dependent peroxidase; PS II, photosystem II; ROS, reactive oxygen species; γ-TMT, γ-tocopherol methyltransferase; Trx-Ox, oxidized thioredoxin; Trx-Red, reduced thioredoxin; TrxR, thioredoxin reductase.

### 2.2. *Stromal Ascorbate (AsA)-Glutathione (GSH) Cycle*

Salinity-induced ROS generated in thylakoids and/or stroma undergo detoxification by the stromal AsA–GSH cycle. In this cycle, $H_2O_2$ is reduced to $H_2O$ catalyzed by stroma APX using AsA as the electron donor, and the oxidized AsA can be reduced back to AsA by monodehydroascorbate reductase (MDHAR), or be converted into dehydroascorbate (DHA) spontaneously. Then, DHA is reduced to AsA by dehydroascorbate reductase (DHAR) at the expense of GSH, generating oxidized glutathione (GSSG). Furthermore, GSSG is reduced by glutathione reductase (GR) using NADPH as an electron donor [14]. Genes encoding the aforementioned enzymes have been reported to be regulated by salinity (Figure 1B, Table 1). Tobacco plants overexpressing genes of *MDHAR* [23] and *DHAR* [24] showed significantly high enzyme activities of MDHAR and DHAR, as well as an increased level of reduced AsA and improved survival under salt stress. In addition, *OsGR3* was markedly induced in rice under salt treatment [25], and the salinity sensitivity of rice was increased when the *OsGR3* gene was knocked out [26]. These results indicate that stromal ROS scavenging in chloroplasts is crucial for redox homeostasis and supplying $NADP^+$, leading to the reduced loading of the ETC. Overall, this contributes to enhancing a plant's ability to withstand adverse environmental conditions [13].

### 2.3. *Thioredoxin/Peroxiredoxin (Trx/Prx) and Glutathione Peroxidase (GPX) Pathway*

Salinity-induced $H_2O_2$ is a potent oxidant for protein thiol groups, which are highly susceptible to oxidation. The thiol reduction is mainly controlled by the Trx/Prx pathway and the GPX pathway. Trx acts as an electron donor that couples with Trx-dependent peroxidase (Prx) to scavenge $H_2O_2$, and thioredoxin reductase (TrxR) utilizes NADPH to keep the Trx/Prx system in a reduced state (Figure 1C, Table 1) [65]. In this process, electrons are taken from NADPH via TrxR, and then transferred to the active site of Trx, which can reduce protein disulfides or other substrates. Transgenic Arabidopsis overexpressing a *S. salsa* thylakoid membrane-attached *SsPrxQ* gene showed enhanced salt tolerance [27]. Similarly, a NADPH thioredoxin reductase (OsNTRC) gene was cloned from rice, which encodes a chloroplast-localized bifunctional enzyme with both TrxR and thioredoxin activity [28]. An Arabidopsis *NTRC* knockout mutant showed growth inhibition and hypersensitivity to salt stress [28]. In addition, GPX is also involved in the reduction of $H_2O_2$, using GSH as the electron donor. Overexpressing wheat *W69* and *W106* genes, which encode chloroplast GPXs, can improve salt and $H_2O_2$ tolerance in Arabidopsis [29]. These results provided evidence that the Trx/Prx system and GPX pathway in chloroplasts are important for the stroma $H_2O_2$ removal in salt-stressed plants.

### 2.4. Non-Enzymatic OH• and $^1O_2$ Scavenging System

Salinity-induced $O_2^-$ generation may trigger more reactive OH• formation through the Fenton reaction. In PS II, $O_2$ of the ground (triplet) state is excited to $^1O_2$ by the reaction center chlorophyll of the triplet excited state [15]. Primarily, OH• and $^1O_2$ are scavenged by AsA, GSH, and tocopherol in chloroplasts [13].

Among these OH• and $^1O_2$ scavengers, tocopherol is a thylakoid membrane-localized lipid antioxidant, which can protect photosynthetic membranes from oxidative damage by scavenging ROS and prevent the propagation of lipid peroxidation under stress conditions [66]. Tocopherol biosynthesis is a finely balanced process in chloroplasts. Arabidopsis chloroplast plastoglobule-localized ABC1 (for activity of bc1 complex)-like kinase ABC1K3 phosphorylates tocopherol cyclase, possibly stabilizing it at plastoglobules and regulating tocopherol biosynthesis [67]. In addition, *AtSIA1* encodes a chloroplast-localized ABC1-like kinase, which is salinity-induced in Arabidopsis. Transgenic Arabidopsis seedlings that overexpress *AtSIA1* showed a higher tolerance to salt stress than Col-0 and the *AtSIA1* knockout mutant [30]. Additionally, $\gamma$-tocopherol methyltransferase ($\gamma$-TMT) is another important enzyme regulating tocopherol synthesis (Figure 1D, Table 1). Under NaCl stress, the overexpression of Arabidopsis *$\gamma$-TMT* in tobacco chloroplasts converted more $\gamma$-tocopherol to $\alpha$-tocopherol, enhanced sugar transport, and reduced ROS contents and ion leakage, which ultimately contributed to salt stress alleviation [31].

### 2.5. Other Genes Involved in Chloroplast ROS Scavenging

Several genes have been proposed to regulate the ROS scavenging process (Figure 1E, Table 1). Rice *WSL12* encodes a chloroplast nucleoside diphosphate kinase 2 (NDPK2), which plays an important role in chloroplast development and chlorophyll biosynthesis by regulating multiple gene expression levels [32]. The *WSL12* mutant showed high $O_2^-$ levels and sensitivity to salinity [32], probably due to the association of NDPK2 with ROS signaling and oxidative stress. Overexpressing *NDPK2* in Arabidopsis chloroplasts induces a higher expression of multiple antioxidant genes (e.g., *peroxidase*, *CAT*, *Trx*, *TrxR*, and *Prx*) [68]. Similarly, transgenic sweet potato (*Ipomoea batatas*) with an expression of *AtNDPK2* in chloroplasts showed an enhanced tolerance to salinity and increased activities of peroxidase, APX, and CAT [33].

As expected, the concentrations of photoreactive tetrapyrrole intermediates are tightly controlled as they can generate ROS under a variety of environmental stimuli. The Arabidopsis genome contains a single 18 kDa translocator protein (TSPO)-encoding gene *AtTSPO*, which is normally localized to the endoplasmic reticulum and vesicles, and is translocated to chloroplasts in the presence of 150 mM NaCl. *AtTSPO* is involved in transporting tetrapyrrole intermediates and protecting chloroplasts from oxidative damage [34].

In addition, methionine sulfoxide reductase (MSR) plays a role in the plant oxidative stress response. The methionine (Met) residue in this protein is especially sensitive to oxidation, leading to the formation of S- and R-epimers of methionine sulfoxide. The sulfoxide, in turn, can be enzymatically reduced back to Met by MSR. Thus, the oxidation and the enzymatically catalyzed reduction of Met is probably a critical molecular mechanism for cellular redox regulation under stress conditions. Transgenic rice plants overexpressing the chloroplast-localized *OsMSRA4.1* showed enhanced viability during salt stress, implying that MSR is important for Met reduction in chloroplasts [35].

## 3. Thylakoid Membrane Organization and Photosynthesis

Plants' photosynthetic machinery respond to salt stress by regulating thylakoid membrane fluidity and remodeling membrane lipid composition, thus maintaining an environment suitable for the function of critical integral proteins during stress [69].

The modification of membrane fluidity is mediated by the changes in unsaturated fatty acid levels of thylakoid membranes. *Fad6* encodes a chloroplast-localized ω-6 desaturase, which is a fatty acid desaturase that catalyzes the conversion of oleic acid (18:1) to linoleic acid (18:2) by inserting a double bond at the ω-6 position. The Arabidopsis *fad6* mutant has an increased sensitivity to salt stress, implying that Fad6 is required for salt tolerance during early seedling development [36]. In addition, a chloroplast-localized glycerol-3-phosphate acyltransferase (GPAT) is one of the main factors that determine the content of *cis*-unsaturated fatty acids in the phosphatidylglycerol of thylakoid membranes (Figure 2A, Table 1) [37]. The overexpression of *GPAT* in tomato increased the *cis*-unsaturated fatty acid content of thylakoid membranes, and the transgenic plants exhibited higher activities of chloroplastic antioxidant enzymes, lower ROS contents, and a better photosynthetic performance, as well as an increased efficiency in alleviating PS II photoinhibition [37]. The increased unsaturation of fatty acids seems to enhance structural flexibility to the thylakoid membranes, which is favorable to thylakoid membrane binding to antioxidant enzymes for excess ROS scavenging [37]. Importantly, the unsaturation of fatty acids might boost the tolerance of PS II to salt stress by accelerating the repair of photodamaged D1. The PS II reaction center subunits D1 and D2 are enclosed by a belt of 11 lipids, which provides a flexible environment and fosters a high mobility of subunits. This would be beneficial to the degradation of damaged D1, the acceleration of de novo synthesis, and the insertion of the D1 protein [70]. However, the specific steps of the PS II repair that are regulated by the unsaturation of fatty acids still remain to be clarified.

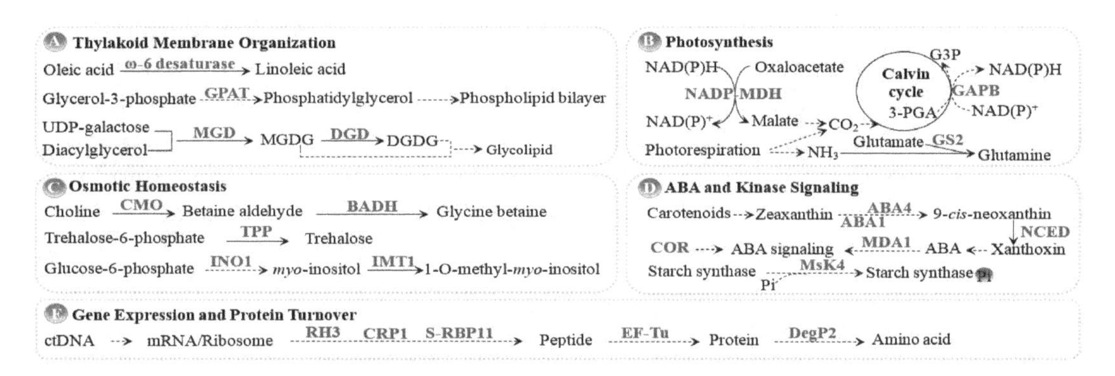


**Figure 2.** Salt tolerance pathways in chloroplasts. (**A**) Thylakoid membrane organization; (**B**) Photosynthesis; (**C**) Osmotic homeostasis; (**D**) ABA and kinase signaling; (**E**) Gene expression and protein turnover. The solid line indicates a single-step reaction, and the dashed line indicates a multistep reaction. Substrates and products are in black font, proteins are in blue bold font, and P in a blue circle indicates phosphorylated protein. Abbreviations: 3-PGA, glycerate-3-phosphate; ABA, abscisic acid; ABA1, zeaxanthin epoxidase; ABA4, neoxanthin synthase; BADH, betaine aldehyde dehydrogenase; CMO, choline monooxygenase; $CO_2$, carbon dioxide; COR, cold-responsive protein; CRP1/S-RBP11, RNA-recognition motif containing protein; DegP2, prokaryotic trypsin-type Deg/Htr serine protease; DGD, digalactodiacylglycerol synthase; DGDG, digalactodiacylglycerol; EF-Tu, translation elongation factor; G3P, glyceraldehyde-3-phosphate; GAPB, glyceraldehyde 3-phosphate dehydrogenase beta subunit; GPAT, glycerol-3-phosphate acyltransferase; GS2, glutamine synthetase; IMTI, inositol methyl transferase; INO1, L-*myo*-inositol 1-phosphate synthase; MDA1, transcription termination factor; MGD, monogalactosyldiacylglycerol synthase; MGDG, monogalactosyldiacylglycerol; MsK4, glycogen synthase kinase 3 like kinase; NADP-MDH, $NADP^+$-dependent malate dehydrogenase; NCED, 9-*cis*-epoxycarotenoid dioxygenase; RH3, DEAD-box RNA helicase; TPP, trehalose-6-phosphate phosphatase.

Two galactolipids, monogalactosyldiacylglycerol (MGDG) and digalactodiacylglycerol (DGDG), are major constituents of photosynthetic membranes in chloroplasts (Figure 2A, Table 1) [71]. MGDG has a conical shape with nonbilayer-forming characteristics, providing a high lateral pressure on the proteins embedded in the membrane. This permits the dense packing of membrane proteins, which facilitates the maintenance of the stability of the membrane structure. Furthermore, the non-bilayer lipid can easily segregate from the membrane and is crucial for the self-regulation of lipid content in the thylakoid membrane through the formation of osmiophilic lipid droplets in the chloroplasts. In addition, DGDG, a bilayer-prone lipid, is involved in lipid-mediated contact between adjacent trimers of the light harvesting complex II [71]. This is important for the stability of the lamellar structure of chloroplast membranes. Crystallization studies have revealed that MGDG was associated with the core of the reaction center of PS I, PS II, and the cytochrome $b_6f$ complex, while DGDG was a component of PS I, PS II, and the light-harvesting complex of PS II [38,72,73]. Thus, they can provide a membrane environment to physically support the photosynthetic complex, and also contribute directly to various photosynthesis-related processes. The biosynthesis of the two galactoglycerolipids is catalyzed by MGDG synthase (MGD) and DGDG synthase, respectively [74]. A chloroplast outer envelope membrane-localized MGD encoding gene has been cloned in rice, and transgenic tobacco plants that overexpress *OsMGD* exhibited significantly higher levels of MGDG and DGDG, as well as higher DGDG/MGDG ratios than wild-type plants. Chloroplasts from salt-stressed *OsMGD* transgenic tobacco had well-developed thylakoid membranes and properly stacked grana lamellae, whereas the chloroplasts from salt-stressed wild-type plants were fairly disorganized and had large membrane-free areas [38]. Therefore, the increased levels of MGDG and DGDG could contribute to the organization of the plant photosynthetic membrane structure for the enhancement of salt resistance.

Some chloroplast-localized proteins are critical for the maintenance of PS II activity in response to salt stress. Rubredoxin (RUB) is a small, non-heme protein, which attaches to the thylakoid membranes and is exposed to the stroma, acting as an electron carrier in a variety of biochemical processes, such as the detoxification of ROS [75]. Mutant analysis of the *RUB* gene in the green alga *Chlamydomonas reinhardtii* suggests that RUB may be necessary for normal PS II activity in a diverse set of organisms that perform oxygenic photosynthesis. Knockout mutants of *RUB* orthologs in the cyanobacterium *S.* sp. PCC 6803 and Arabidopsis also imply that the activity and stability of PS II are specifically affected [76]. In addition, the overexpression of the *PutRUB* gene from alkaligrass in Arabidopsis increased the tolerance to NaCl and $NaHCO_3$ stress. This was probably due to decreasing $H_2O_2$ accumulation for chloroplast redox balance [39]. Similarly, a *RCI* gene encoding a chloroplast membrane protein has also been reported to be involved in the regulation of PS II activity. RCI is a homolog of a plasma membrane protein 3 family and contains two putative transmembrane domains. Arabidopsis plants overexpressing the wheat *RCI* performed better than the wild-type under salinity stress. Transgenic plants showed significantly higher PS II activity in terms of the maximum photochemical efficiency, and higher proline and chlorophyll contents than wild-type plants [40].

Under salt stress, the maintenance of stable $CO_2$ assimilation would be important to salinity tolerance. $NADP^+$-dependent malate dehydrogenase (NADP-MDH) is responsible for the reduction of oxaloacetate (OAA) to malate in chloroplasts, which was suggested to be crucial for $CO_2$ fixation (Figure 2B, Table 1). In C3 plants, NADP-MDH is essential for the balance of reducing equivalents between chloroplasts and cytoplasm via the malate/oxaloacetate shuttle. In C4 plants, NADP-MDH is located exclusively in the mesophyll chloroplasts for C4-photosynthesis. *NADP-MDH* transcripts in the facultative Crassulacean acid metabolism (CAM) plant *M. crystallinum* significantly accumulated in response to salt stress. The salinity-induced expression of *NADP-MDH* suggests that $CO_2$ fixation is enhanced, which implies that the facultative, halophytic *M. crystallinum* shifts the photosynthesis carbon fixation mode from a C3 to a CAM in response to salinity [41]. In addition, the glyceraldehyde 3-phosphate dehydrogenase beta subunit (GAPB) is a key enzyme for the conversion of glycerate-3-phosphate (3-PGA) to glyceraldehyde-3-phosphate interacting with ATP and NADPH. 3-PGA can accept electrons from NADPH, preventing the ROS-induced deceleration of PS II repair (Figure 2B, Table 1). Arabidopsis overexpressing *ThGAPB* exhibited higher recycling rates of ADP and $NADP^+$. This would reduce ROS production, contributing to the maintenance of photosynthetic efficiency under salinity conditions [42].

Additionally, salt stress-induced stomatal closure limits the $CO_2$ concentration in cells, which causes the over-reduction of photosynthetic ETC. Under such a condition, salinity-responsive ribulose-1, 5-bisphosphate carboxylase/oxygenase (Rubisco) operates as an oxygenase, and photorespiration is activated in response to salt stress. Phosphoglycerate and $CO_2$ generated from photorespiration enter the Calvin cycle, and the consumption of NADPH and ATP may contribute to the dissipation of excess light energy or reducing power, for preventing the over-reduction of ETC. Interestingly, transgenic rice overexpressing a chloroplastic glutamine synthetase encoding gene *GS2* increased their photorespiration capacity for improving salt tolerance (Figure 2B, Table 1) [43].

## 4. Osmotic and Ion Homeostasis

To withstand salinity-induced osmotic stress, plants synthesize and accumulate compatible solutes/osmoprotectants for stabilizing proteins, membranes, and even transcriptional and translational machineries in the cells. Major osmoprotectants include betaines, amino acids (e.g., proline), non-reducing sugars (e.g., trehalose and arabitol), and polyols [77]. Betaine is a quaternary ammonium compound found in a wide variety of plants, animals, and microorganisms [78]. Salinity-induced betaine accumulates in the chloroplasts of many halotolerant plants, stabilizing the quaternary structure of enzymes and protein complexes, as well as the highly-ordered structure of membranes in photosynthetic machinery [79]. For example, it can stabilize the Rubisco and PS II oxygen-evolving complex and accelerate the repair of photodamaged PS II under salt or other abiotic stress [80,81]. In plants, betaine is synthesized by several betaine-biosynthetic enzymes, such as choline monooxygenase (CMO) and betaine aldehyde dehydrogenase (BADH) (Figure 2C, Table 1). Genes encoding CMO and BADH have been cloned and targeted to the chloroplast genome of various plant species (Table 1). In transgenic plants, the enzyme activities and betaine levels were increased, leading to an improved photosynthetic performance and enhanced tolerance to salt stress [44–47]. Importantly, trehalose and inositol also act as important osmolytes in chloroplasts, which can significantly enhance the salt tolerance of the transgenic plants. Chloroplast-localized trehalose-6-phosphate phosphatase (TPP) catabolizes trehalose-6-phosphate to generate trehalose, which is the final step of trehalose metabolism (Figure 2C, Table 1). Arabidopsis plants deficient in *AtTPPD* were hypersensitive, whereas plants overexpressing *AtTPPD* were tolerant to high salinity [48]. In addition, the co-expression of *PcINO1* and *McIMT1* in chloroplasts allowed the transgenic tobacco plants to perform better in terms of growth potential and photosynthesis rates with increased levels of *myo*-inositol and methylated inositol under salt stress [49].

Besides osmotic balance, ion homeostasis is also vital for plant cells to cope with salinity. Extra $Na^+$ in chloroplasts destroys the thylakoid membrane structure, causing the inactivation of PS II and PS I and inhibiting the repair of photodamaged PS II, as well as decreasing photosynthesis electron transport, and therefore, the $Na^+$ concentration in chloroplasts must be well-controlled. *AtCHX23* encodes a putative $Na^+(K^+)/H^+$ exchanger, which is localized in the chloroplast envelope and functions by putatively regulating the homeostasis of the cytoplasmic and stromal pH and $Na^+$ concentration through the sequestering of $Na^+$ entering into chloroplasts. Arabidopsis *CHX23* mutants displayed a high sensitivity to NaCl [50]. Moreover, *NHD1* encodes a sodium hydrogen antiporter, which is localized to the chloroplast envelope. Arabidopsis NHD1 functions as a chloroplast sodium exporter, protecting chloroplasts from deleterious $Na^+$ accumulation after salt exposure. *NHD1* T-DNA insertion mutants showed high $Na^+$ levels in chloroplasts, resulting in a markedly impaired photosynthetic performance as revealed by a lower quantum yield of PS II and increased non-photochemical quenching [51]. In addition, several members of the putative $K^+$-efflux antiporters (KEA), such as KEA1, KEA2, and KEA3, function as chloroplast $K^+/H^+$ antiporters for regulating the osmotic pressure, pH, and ion homeostasis of chloroplasts. KEA1 and KEA2 are chloroplast inner envelope membrane proteins, and function to release $K^+$ from the chloroplasts in exchange for $H^+$ influx, whereas KEA3 is a thylakoid membrane protein that uptakes $K^+$ into the thylakoid lumen. Arabidopsis loss-of-function mutants of these three genes have been analyzed, and the higher-order mutants showed increasingly impaired photosynthesis, along with altered chloroplast pH homeostasis [52].

In addition, a chloroplast stroma-localized YL1 protein is an YqeH-type GTPase, involved in the regulation of $Na^+$ delivery in response to salt stress. The expression of Arabidopsis *YL1* was markedly reduced under high salinity. Shoots of the *YL1* mutant accumulated significantly higher levels of $Na^+$ than wild type under salt stress. The expression of *Abscisic acid insensitive 4 (ABI4)* was increased and *high-affinity* $K^+$ *transporter 1 (HKT1)* was suppressed in the mutant shoots. *HKT1*, encoding a $K^+/Na^+$ symporter, is an important regulator that can directly retrieve $Na^+$ from the xylem sap back to the phloem of the shoot and unload it in the root for ion homeostasis [82]. ABI4 has recently been reported to act as a negative regulator that could directly bind to the promoter region and inhibit *HKT1* expression in Arabidopsis [83]. It is speculated that the reduction of *YL1* and the induction of *ABI4* may be an adaptive mechanism to achieve $Na^+$ equilibrium in the entire plant, which needs to be further investigated [53].

### 5. ABA and Kinase Signaling Pathways

The investigation of cellular ABA level fluctuation, ABA perception, and ABA-mediated signaling in response to salinity is pivotal for understanding plant stress tolerance. All except the last two steps of ABA de novo biosynthesis occur in plastids [84]. Most of the genes involved in ABA biosynthesis have been identified, among which, *ABA1*, *ABA4*, and *NCED* encode plastid-localized zeaxanthin epoxidase, neoxanthin synthase, and 9-*cis*-epoxycarotenoid dioxygenase, respectively (Figure 2D, Table 1) [54,55,85,86]. Salt stress has some influence on the cellular ABA content, and it was reported that *AtABA1* and *VuNCED1* were up-regulated by salt stress [54,55].

In addition, *MDA1* encoding a transcription termination factor family protein, is likely required for a proper ABA response in chloroplasts. An Arabidopsis *MDA1* mutant showed reduced sensitivity to ABA, enhanced salt tolerance, and some salt-responsive gene expression [56]. *MDA1* deficit in plants may disrupt chloroplast homeostasis and negatively affect ABA retrograde signaling, which elicits the nuclear-encoded functions required for coping with salt stress [56]. Moreover, a gene encoding stroma-localized cold-responsive (COR) protein, *COR15B*, was up-regulated under salt stress in an ABA-dependent manner. Therefore, *COR15B* was proposed to be a potential member in the ABA signaling regulatory network in the Arabidopsis response to salt stress [57].

Protein kinases are major components of intracellular signal transduction, which mediate various signaling pathways that enable plant cells to rapidly acclimate and prevail in challenging environmental conditions. MsK4 is a plastid-localized novel alfalfa glycogen synthase kinase 3 like kinase, which was reported to be a novel signaling component in the regulation of starch metabolism and the salt stress response. Starch metabolism is highly sensitive to environmental changes and its accurate regulation is crucial for the adaptation of carbon and energy flow in response to stress conditions. Fluxes of carbon into and out of starch are extremely complex and must be highly controlled. Evidence is emerging that reversible protein phosphorylation is crucial for the regulation of starch-associated carbon metabolism, and the phosphorylation of starch metabolic enzymes has been shown to control their activities and protein complex formation. It has been found that MsK4 can bind to some isoforms of soluble and granule-bound starch synthases in vitro, suggesting that the enzymes involved in starch metabolism might be direct targets of the MsK4 action (Figure 2D, Table 1). High-salinity induced the activity of MsK4 kinase in alfalfa and Arabidopsis. Moreover, *MsK4* overexpressing transgenic plants showed an enhanced salt tolerance and significantly increased starch content. This opens new perspectives as to how metabolic carbon flux can be regulated in response to salinity, and provides links between stress signaling and metabolic adaptation [58].

## 6. Chloroplast Gene Expression and Protein Turnover

The salt-responsive gene expression of the chloroplast genome must be very precisely controlled, and the targeting of nuclear-encoded proteins into the chloroplasts is important in regulating chloroplast gene expression. Chloroplast gene expression is mainly regulated at the level of posttranscriptional RNA metabolism, including RNA processing, editing, splicing, decay, and translational control [87]. The functions of most RNA molecules rely on a well-defined three-dimensional structure, and the correct folding of RNA molecules requires the assistance of diverse RNA binding proteins (RBPs) [88]. A variety of nuclear-encoded RBPs are targeted to chloroplasts and play indispensable roles in the plant salt response through the posttranscriptional regulation of RNA metabolism and gene expression in chloroplasts, such as DEAD-box RNA helicases (RHs) and RNA-recognition motifs (RRMs)-containing proteins (Figure 2E, Table 1). Arabidopsis chloroplast-localized RH3 has been demonstrated to function in intron splicing and ribosome biogenesis. *RH3* mutants displayed more sensitivity to salt stress and the splicing of several intron-containing chloroplast genes was inhibited [59]. Besides, Arabidopsis CRP1 and S-RBP11, two of the chloroplast-localized RRM containing proteins, were also shown to be involved in the salt stress response. Seed germination of the *CRP1* mutant was delayed compared with that of the wild-type seeds under salt stress [60]. Similarly, transgenic Arabidopsis overexpressing *S-RBP11* showed an increased salt tolerance, whereas *S-RBP11* mutants were shown to be more sensitive to salt stress [61].

Moreover, protein synthesis, processing, and degradation in chloroplasts are also important for plant salt stress adaption (Figure 2E, Table 1). The expression of pea *TufA*, which encodes a chloroplast translation elongation factor (EF-Tu), was down-regulated in response to salinity [62]. EF-Tu is an essential component for polypeptide elongation during protein synthesis. It is reported that EF-Tu might have a chaperone-like property of refolding the denatured proteins or preventing their aggregation under heat stress [89].

Salt-induced excess excitation energy may cause the photodamage of PS II. The D1 protein is one of the core proteins in the PS II reaction center, which is the main target of oxidative damage. The de novo synthesis, assembly, and rapid degradation of the D1 protein are necessary for the efficient PS II repair in salt-stressed plants [63,90]. In cyanobacterium *S.* sp. PCC 6803, the transcription and translation of the D1 protein encoding gene *psbA* was decreased, and the repair of the photodamaged D1 protein was also inhibited under salt stress [63]. Moreover, a prokaryotic trypsin-type Deg/Htr serine protease (DegP2) was found to be vital for the photodamaged D1 protein degradation under salt stress [91]. DegP2 is peripherally associated with the outer surface of the thylakoid membrane, and involved in the primary cleavage of the photodamaged D1 protein on the stromal DE loop. The expression of Arabidopsis *DegP2* was down-regulated under high salt stress (400 mM NaCl), while the amount of DegP2 protein was increased significantly under stress [64]. This would induce its molecular chaperone and proteolytic activities to enhance the repair of oxygen-evolving PS II in salinity-stressed plants.

## 7. Conclusions

Chloroplasts have evolved fine-tuned pathways for the salt response. In this paper, we reviewed and discussed the salt-responsive genes encoding chloroplast-localized proteins, which represent several crucial pathways in chloroplasts in response to salinity, such as chloroplast ROS scavenging (water–water cycle, stromal AsA-GSH cycle, Trx/Prx pathway, and non-enzymatic scavenging system), photosynthetic thylakoid membrane modulation, $CO_2$ assimilation, the synthesis of osmoprotectant, and ion homeostasis regulation, as well as chloroplast gene expression and protein turnover. These provide important molecular information for better understanding the chloroplast salt response, and allow us to expand our knowledge on the adaptation of this photosynthetic apparatus to salinity stress. However, the photosynthetic machinery and chloroplast metabolic pathways for salt tolerance are too complicated to be interpreted by the genes characterized so far. Although a large number of candidate genes/proteins have been identified in chloroplasts from plants under salt treatments using large-scale genomic, transcriptomic, and proteomic approaches [9–12,92], further investigations of their biological functions in salinity tolerance are needed. More importantly, a deeper analysis of the post-translational modifications and protein-protein interactions of these salt-responsive proteins will facilitate a thorough understanding of the complicated salt-responsive networks in chloroplasts.

**Acknowledgments:** This work is supported by the Fundamental Research Funds for the Central Universities (No. 2572015AA03) to Jinwei Suo. Capacity Construction Project of Local Universities, Shanghai, China (No. 14390502700) and Specially-Appointed Professors (Oriental Scholars) for Shanghai's Universities to Shaojun Dai.

## Abbreviations

| | |
|---|---|
| 3-PGA | Glycerate-3-phosphate |
| ABA | Abscisic acid |
| ABC1K | Activity of bc1 complex-like kinase |
| ABI4 | Abscisic acid insensitive 4 |
| APX/tAPX | Ascorbate peroxidase/Thylakoid ascorbate peroxidase |
| AsA | Ascorbate |
| BADH | Betaine aldehyde dehydrogenase |
| CAM | Crassulacean acid metabolism |
| CAT | Catalase |
| CMO | Choline monooxygenase |
| $CO_2$ | Carbon dioxide |
| COR | Cold-responsive protein |
| Cu/Zn SOD | Copper/zinc superoxide dismutase |
| DegP2 | Prokaryotic trypsin-type Deg/Htr serine protease |
| DGDG | Digalactodiacylglycerol |
| DHA | Dehydroascorbate |
| DHAR | Dehydroascorbate reductase |
| EF-Tu | Translation elongation factor |
| ETC | Electron transport chain |
| GAPB | Glyceraldehyde 3-phosphate dehydrogenase beta subunit |
| GPAT | Glycerol-3-phosphate acyltransferase |
| GPX | Glutathione peroxidase |
| GR | Glutathione reductase |
| GS2 | Glutamine synthetase |
| GSH | Reduced glutathione |
| GSSG | Oxidized glutathione |
| $H_2O_2$ | Hydrogen peroxide |
| HKT1 | High-affinity $K^+$ transporter 1 |
| KEA | $K^+$-efflux antiporter |
| MDHAR | Monodehydroascorbate reductase |
| Met | Methionine |
| MGD | Monogalactosyldiacylglycerol synthase |
| MGDG | Monogalactosyldiacylglycerol |
| MSR | Methionine sulfoxide reductase |
| NADP-MDH | NADP+-dependent malate dehydrogenase |
| NDPK2 | Nucleoside diphosphate kinase 2 |
| NTRC | NADPH thioredoxin reductase |
| $^1O_2$ | Singlet oxygen |
| $O_2$ | Oxygen |

$O_2^-$ Superoxide anion
C Oxaloacetate
OH• Hydroxyl radical
PS II Photosystem II
$\gamma$-TMT $\gamma$-tocopherol methyltransferase
RBP RNA binding protein
RH DEAD-box RNA helicase
ROS Reactive oxygen species
RRM RNA-recognition motif
RUB Rubredoxin
Rubisco Ribulose-1,5-bisphosphate carboxylase/oxygenase
TPP Trehalose-6-phosphate phosphatase
Trx/Prx Thioredoxin/Peroxiredoxin
TSPO 18 kDa translocator protein

## References


1. Munns, R.; Tester, M. Mechanisms of salinity tolerance. *Annu. Rev. Plant Biol.* **2008**, *59*, 651–681. [CrossRef] [PubMed]
2. Tuteja, N. Chapter twenty-four-mechanisms of high salinity tolerance in plants. *Methods Enzymol.* **2007**, *428*, 419–438. [PubMed]
3. Van Wijk, K.J. Proteomics of the chloroplast: Experimentation and prediction. *Trends Plant Sci.* **2000**, *5*, 420–425. [CrossRef]
4. Baginsky, S.; Gruissem, W. Chloroplast proteomics: Potentials and challenges. *J. Exp. Bot.* **2004**, *55*, 1213–1220. [CrossRef] [PubMed]
5. Abdallah, F.; Salamini, F.; Leister, D. A prediction of the size and evolutionary origin of the proteome of chloroplasts of Arabidopsis. *Trends Plant Sci.* **2000**, *5*, 141–142. [CrossRef]
6. Arabidopsis Genome Initiative. Analysis of the genome sequence of the flowering plant *Arabidopsis thaliana*. *Nature* **2000**, *408*, 796–815.
7. Leister, D. Chloroplast research in the genomic age. *Trends Genet.* **2003**, *19*, 47–56. [CrossRef]
8. Peltier, J.B.; Emanuelsson, O.; Kalume, D.E.; Ytterberg, J.; Friso, G.; Rudella, A.; Liberles, D.A.; Söderberg, L.; Roepstorff, P.; von Heijne, G.; et al. Central functions of the lumenal and peripheral thylakoid proteome of Arabidopsis determined by experimentation and genome-wide prediction. *Plant Cell* **2002**, *14*, 211–236. [CrossRef] [PubMed]
9. Wang, L.; Liang, W.; Xing, J.; Tan, F.; Chen, Y.; Huang, L.; Cheng, C.L.; Chen, W. Dynamics of chloroplast proteome in salt-stressed mangrove *Kandelia candel* (L.) Druce. *J. Proteome Res.* **2013**, *12*, 5124–5136. [CrossRef] [PubMed]
10. Fan, P.; Feng, J.; Jiang, P.; Chen, X.; Bao, H.; Nie, L.; Jiang, D.; Lv, S.; Kuang, T.; Li, Y. Coordination of carbon fixation and nitrogen metabolism in *Salicornia europaea* under salinity: Comparative proteomic analysis on chloroplast proteins. *Proteomics* **2011**, *11*, 4346–4367. [CrossRef] [PubMed]
11. Zörb, C.; Herbst, R.; Forreiter, C.; Schubert, S. Short-term effects of salt exposure on the maize chloroplast protein pattern. *Proteomics* **2009**, *9*, 4209–4220. [CrossRef] [PubMed]
12. Do Amaral, M.N.; Arge, L.W.P.; Benitez, L.C.; Danielowski, R.; da Silveira Silveira, S.F.; da Rosa Farias, D.; de Oliveira, A.C.; da Maia, L.C.; Braga, E.J.B. Comparative transcriptomics of rice plants under cold, iron, and salt stresses. *Funct. Integr. Genom.* **2016**, *16*, 567–579. [CrossRef] [PubMed]
13. Edreva, A. Generation and scavenging of reactive oxygen species in chloroplasts: A submolecular approach. *Agr. Ecosyst. Environ.* **2005**, *106*, 119–133. [CrossRef]
14. Miller, G.; Suzuki, N.; Ciftci-Yilmaz, S.; Mittler, R. Reactive oxygen species homeostasis and signalling during drought and salinity stresses. *Plant Cell Environ.* **2010**, *33*, 453–467. [CrossRef] [PubMed]
15. Asada, K. Production and scavenging of reactive oxygen species in chloroplasts and their functions. *Plant Physiol.* **2006**, *141*, 391–396. [CrossRef] [PubMed]
16. Wu, J.; Zhang, J.; Li, X.; Xu, J.; Wang, L. Identification and characterization of a *PutCu/Zn-SOD* gene from *Puccinellia tenuiflora* (Turcz.) Scribn. et Merr. *Plant Growth Regul.* **2016**, *79*, 55–64. [CrossRef]

17. Badawi, G.H.; Yamauchi, Y.; Shimada, E.; Sasaki, R.; Kawano, N.; Tanaka, K.; Tanaka, K. Enhanced tolerance to salt stress and water deficit by overexpressing superoxide dismutase in tobacco (*Nicotiana tabacum*) chloroplasts. *Plant Sci.* **2004**, *166*, 919–928. [CrossRef]
18. Jing, X.; Hou, P.; Lu, Y.; Deng, S.; Li, N.; Zhao, R.; Sun, J.; Wang, Y.; Han, Y.; Lang, T.; et al. Overexpression of copper/zinc superoxide dismutase from mangrove *Kandelia candel* in tobacco enhances salinity tolerance by the reduction of reactive oxygen species in chloroplast. *Front. Plant Sci.* **2015**, *5*, 23. [CrossRef] [PubMed]
19. Tseng, M.J.; Liu, C.W.; Yiu, J.C. Enhanced tolerance to sulfur dioxide and salt stress of transgenic Chinese cabbage plants expressing both superoxide dismutase and catalase in chloroplasts. *Plant Physiol. Biochem.* **2007**, *45*, 822–833. [CrossRef] [PubMed]
20. Luo, X.; Wu, J.; Li, Y.; Nan, Z.; Guo, X.; Wang, Y.; Zhang, A.; Wang, Z.; Xia, G.; Tian, Y. Synergistic effects of *GhSOD1* and *GhCAT1* overexpression in cotton chloroplasts on enhancing tolerance to methyl viologen and salt stresses. *PLoS ONE* **2013**, *8*, e54002. [CrossRef] [PubMed]
21. Badawi, G.H.; Kawano, N.; Yamauchi, Y.; Shimada, E.; Sasaki, R.; Kubo, A.; Tanaka, K. Over-expression of ascorbate peroxidase in tobacco chloroplasts enhances the tolerance to salt stress and water deficit. *Physiol. Plant.* **2004**, *121*, 231–238. [CrossRef] [PubMed]
22. Sheptovitsky, Y.G.; Brudvig, G.W. Isolation and characterization of spinach photosystem II membrane-associated catalase and polyphenol oxidase. *Biochemistry* **1996**, *35*, 16255–16263. [CrossRef] [PubMed]
23. Kavitha, K.; George, S.; Venkataraman, G.; Parida, A. A salt-inducible chloroplastic monodehydroascorbate reductase from halophyte *Avicennia marina* confers salt stress tolerance on transgenic plants. *Biochimie* **2010**, *92*, 1321–1329. [CrossRef] [PubMed]
24. Le Martret, B.; Poage, M.; Shiel, K.; Nugent, G.D.; Dix, P.J. Tobacco chloroplast transformants expressing genes encoding dehydroascorbate reductase, glutathione reductase, and glutathione-*S*-transferase, exhibit altered anti-oxidant metabolism and improved abiotic stress tolerance. *Plant Biotechnol. J.* **2011**, *9*, 661–673. [CrossRef] [PubMed]
25. Wu, T.M.; Lin, W.R.; Kao, Y.T.; Hsu, Y.T.; Yeh, C.H.; Hong, C.Y.; Kao, C.H. Identification and characterization of a novel chloroplast/mitochondria co-localized glutathione reductase 3 involved in salt stress response in rice. *Plant Mol. Biol.* **2013**, *83*, 379–390. [CrossRef] [PubMed]
26. Wu, T.M.; Lin, W.R.; Kao, C.H.; Hong, C.Y. Gene knockout of glutathione reductase 3 results in increased sensitivity to salt stress in rice. *Plant Mol. Boil.* **2015**, *87*, 555–564. [CrossRef] [PubMed]
27. Jing, L.W.; Chen, S.H.; Guo, X.L.; Zhang, H.; Zhao, Y.X. Overexpression of a chloroplast-located peroxiredoxin Q gene, *SsPrxQ*, increases the salt and low-temperature tolerance of Arabidopsis. *J. Integr. Plant Biol.* **2006**, *48*, 1244–1249. [CrossRef]
28. Serrato, A.J.; Pérez-Ruiz, J.M.; Spínola, M.C.; Cejudo, F.J. A novel NADPH thioredoxin reductase, localized in the chloroplast, which deficiency causes hypersensitivity to abiotic stress in *Arabidopsis thaliana*. *J. Biol. Chem.* **2004**, *279*, 43821–43827. [CrossRef] [PubMed]
29. Zhai, C.Z.; Zhao, L.; Yin, L.J.; Chen, M.; Wang, Q.Y.; Li, L.C.; Xu, Z.S.; Ma, Y.Z. Two wheat glutathione peroxidase genes whose products are located in chloroplasts improve salt and $H_2O_2$ tolerances in Arabidopsis. *PLoS ONE* **2013**, *8*, e73989. [CrossRef] [PubMed]
30. Yang, S.; Zhang, Q.; Li, T.; Du, J.; Yang, S.; Yang, C. *AtSIA1*, an ABC1-like kinase, regulates salt response in Arabidopsis. *Biologia* **2012**, *67*, 1107–1111. [CrossRef]
31. Jin, S.; Daniell, H. Expression of $\gamma$-tocopherol methyltransferase in chloroplasts results in massive proliferation of the inner envelope membrane and decreases susceptibility to salt and metal-induced oxidative stresses by reducing reactive oxygen species. *Plant Biotechnol. J.* **2014**, *12*, 1274–1285. [CrossRef] [PubMed]
32. Ye, W.; Hu, S.; Wu, L.; Ge, C.; Cui, Y.; Chen, P.; Wang, X.; Xu, J.; Ren, D.; Dong, G.; et al. *White stripe leaf 12* (*WSL12*), encoding a nucleoside diphosphate kinase 2 (OsNDPK2), regulates chloroplast development and abiotic stress response in rice (*Oryza sativa* L.). *Mol. Breed.* **2016**, *36*, 1–15. [CrossRef] [PubMed]
33. Kim, Y.H.; Lim, S.; Yang, K.S.; Kim, C.Y.; Kwon, S.Y.; Lee, H.S.; Wang, X.; Zhou, Z.; Ma, D.; Yun, D.J.; et al. Expression of Arabidopsis *NDPK2* increases antioxidant enzyme activities and enhances tolerance to multiple environmental stresses in transgenic sweetpotato plants. *Mol. Breed.* **2009**, *24*, 233–244. [CrossRef]

34. Balsemão-Pires, E.; Jaillais, Y.; Olson, B.J.; Andrade, L.R.; Umen, J.G.; Chory, J.; Sachetto-Martins, G. The Arabidopsis translocator protein (AtTSPO) is regulated at multiple levels in response to salt stress and perturbations in tetrapyrrole metabolism. *BMC Plant Biol.* **2011**, *11*, 108. [CrossRef] [PubMed]
35. Guo, X.; Wu, Y.; Wang, Y.; Chen, Y.; Chu, C. OsMSRA4.1 and OsMSRB1.1, two rice plastidial methionine sulfoxide reductases, are involved in abiotic stress responses. *Planta* **2009**, *230*, 227–238. [CrossRef] [PubMed]
36. Zhang, J.T.; Zhu, J.Q.; Zhu, Q.; Liu, H.; Gao, X.S.; Zhang, H.X. Fatty acid desaturase-6 (Fad6) is required for salt tolerance in *Arabidopsis thaliana*. *Biochem. Biophys. Res. Commun.* **2009**, *390*, 469–474. [CrossRef] [PubMed]
37. Sun, Y.L.; Li, F.; Su, N.; Sun, X.L.; Zhao, S.J.; Meng, Q.W. The increase in unsaturation of fatty acids of phosphatidylglycerol in thylakoid membrane enhanced salt tolerance in tomato. *Photosynthetica* **2010**, *48*, 400–408. [CrossRef]
38. Wang, S.; Uddin, M.I.; Tanaka, K.; Yin, L.; Shi, Z.; Qi, Y.; Mano, J.; Matsui, K.; Shimomura, N.; Sakaki, T.; et al. Maintenance of chloroplast structure and function by overexpression of the *OsMGD* gene leads to enhanced salt tolerance in tobacco. *Plant Physiol.* **2014**, *165*, 1144–1155. [CrossRef] [PubMed]
39. Li, Y.; Liu, P.; Takano, T.; Liu, S. A chloroplast-localized rubredoxin family protein gene from *Puccinellia tenuiflora* (*PutRUB*) increases NaCl and $NaHCO_3$ tolerance by decreasing $H_2O_2$ accumulation. *Int. J. Mol. Sci.* **2016**, *17*, 804. [CrossRef] [PubMed]
40. Khurana, N.; Chauhan, H.; Khurana, P. Characterization of a chloroplast localized wheat membrane protein (TaRCI) and its role in heat, drought and salinity stress tolerance in *Arabidopsis thaliana*. *Plant Gene* **2015**, *4*, 45–54. [CrossRef]
41. Cushman, J.C. Molecular cloning and expression of chloroplast NADP-malate dehydrogenase during Crassulacean acid metabolism induction by salt stress. *Photosynth. Res.* **1993**, *35*, 15–27. [CrossRef] [PubMed]
42. Chang, L.; Guo, A.; Jin, X.; Yang, Q.; Wang, D.; Sun, Y.; Huang, Q.; Wang, L.; Peng, C.; Wang, X. The β subunit of glyceraldehyde 3-phosphate dehydrogenase is an important factor for maintaining photosynthesis and plant development under salt stress-based on an integrative analysis of the structural, physiological and proteomic changes in chloroplasts in *Thellungiella halophila*. *Plant Sci.* **2015**, *236*, 223–238. [PubMed]
43. Hoshida, H.; Tanaka, Y.; Hibino, T.; Hayashi, Y.; Tanaka, A.; Takabe, T.; Takabe, T. Enhanced tolerance to salt stress in transgenic rice that overexpresses chloroplast glutamine synthetase. *Plant Mol. Biol.* **2000**, *43*, 103–111. [CrossRef] [PubMed]
44. Bao, Y.; Zhao, R.; Li, F.; Tang, W.; Han, L. Simultaneous expression of *Spinacia oleracea* chloroplast choline monooxygenase (CMO) and betaine aldehyde dehydrogenase (BADH) genes contribute to dwarfism in transgenic *Lolium perenne*. *Plant Mol. Biol. Report.* **2011**, *29*, 379–388. [CrossRef]
45. Zhang, J.; Tan, W.; Yang, X.H.; Zhang, H.X. Plastid-expressed choline monooxygenase gene improves salt and drought tolerance through accumulation of glycine betaine in tobacco. *Plant Cell Rep.* **2008**, *27*, 1113–1124. [CrossRef] [PubMed]
46. Kumar, S.; Dhingra, A.; Daniell, H. Plastid-expressed betaine aldehyde dehydrogenase gene in carrot cultured cells, roots, and leaves confers enhanced salt tolerance. *Plant Physiol.* **2004**, *136*, 2843–2854. [CrossRef] [PubMed]
47. Yang, X.; Liang, Z.; Wen, X.; Lu, C. Genetic engineering of the biosynthesis of glycinebetaine leads to increased tolerance of photosynthesis to salt stress in transgenic tobacco plants. *Plant Mol. Biol.* **2008**, *66*, 73–86. [CrossRef] [PubMed]
48. Krasensky, J.; Broyart, C.; Rabanal, F.A.; Jonak, C. The redox-sensitive chloroplast trehalose-6-phosphate phosphatase *AtTPPD* regulates salt stress tolerance. *Antioxid. Redox Signal.* **2014**, *21*, 1289–1304. [CrossRef] [PubMed]
49. Patra, B.; Ray, S.; Richter, A.; Majumder, A.L. Enhanced salt tolerance of transgenic tobacco plants by co-expression of *PcINO1* and *McIMT1* is accompanied by increased level of *myo*-inositol and methylated inositol. *Protoplasma* **2010**, *245*, 143–152. [CrossRef] [PubMed]
50. Song, C.P.; Guo, Y.; Qiu, Q.; Lambert, G.; Galbraith, D.W.; Jagendorf, A.; Zhu, J.K. A probable $Na^+(K^+)/H^+$ exchanger on the chloroplast envelope functions in pH homeostasis and chloroplast development in *Arabidopsis thaliana*. *Proc. Natl. Acad. Sci. USA* **2004**, *101*, 10211–10216. [CrossRef] [PubMed]
51. Müller, M.; Kunz, H.H.; Schroeder, J.I.; Kemp, G.; Young, H.S.; Neuhaus, H.E. Decreased capacity for sodium export out of Arabidopsis chloroplasts impairs salt tolerance, photosynthesis and plant performance. *Plant J.* **2014**, *78*, 646–658. [CrossRef] [PubMed]

52. Kunz, H.H.; Gierth, M.; Herdean, A.; Satoh-Cruz, M.; Kramer, D.M.; Spetea, C.; Schroeder, J.I. Plastidial transporters KEA1, -2, and-3 are essential for chloroplast osmoregulation, integrity, and pH regulation in Arabidopsis. *Proc. Natl. Acad. Sci. USA* **2014**, *111*, 7480–7485. [CrossRef] [PubMed]
53. Li, P.C.; Huang, J.G.; Yu, S.W.; Li, Y.Y.; Sun, P.; Wu, C.A.; Zheng, C.C. Arabidopsis YL1/BPG2 is involved in seedling shoot response to salt stress through ABI4. *Sci. Rep.* **2016**, *6*, 30163. [CrossRef] [PubMed]
54. Xiong, L.; Lee, H.; Ishitani, M.; Zhu, J.K. Regulation of osmotic stress-responsive gene expression by the *LOS6/ABA1* locus in Arabidopsis. *J. Biol. Chem.* **2002**, *277*, 8588–8596. [CrossRef] [PubMed]
55. Iuchi, S.; Kobayashi, M.; Yamaguchi-Shinozaki, K.; Shinozaki, K. A stress-inducible gene for 9-*cis*-epoxycarotenoid dioxygenase involved in abscisic acid biosynthesis under water stress in drought-tolerant cowpea. *Plant Physiol.* **2000**, *123*, 553–562. [CrossRef] [PubMed]
56. Robles, P.; Micol, J.L.; Quesada, V. Arabidopsis MDA1, a nuclear-encoded protein, functions in chloroplast development and abiotic stress responses. *PLoS ONE* **2012**, *7*, e42924. [CrossRef] [PubMed]
57. Liu, D.; Hou, L.; Li, W.C.; Cheng, J.F.; Fu, Y.Q. *COR15B* expression is affected by chloroplast functionality and its role in response to salt stress in *Arabidopsis thaliana*. *Biol. Plant.* **2014**, *58*, 667–675. [CrossRef]
58. Kempa, S.; Rozhon, W.; Samaj, J.; Erban, A.; Baluška, F.; Becker, T.; Haselmayer, J.; Schleiff, E.; Kopka, J.; Hirt, H.; et al. A plastid-localized glycogen synthase kinase 3 modulates stress tolerance and carbohydrate metabolism. *Plant J.* **2007**, *49*, 1076–1090. [CrossRef] [PubMed]
59. Gu, L.; Xu, T.; Lee, K.; Lee, K.H.; Kang, H. A chloroplast-localized DEAD-box RNA helicase AtRH3 is essential for intron splicing and plays an important role in the growth and stress response in *Arabidopsis thaliana*. *Plant Physiol. Biochem.* **2014**, *82*, 309–318. [CrossRef] [PubMed]
60. Xu, T.; Sy, N.D.; Lee, H.J.; Kwak, K.J.; Gu, L.; Kim, J.I.; Kang, H. Functional characterization of a chloroplast-targeted RNA-binding protein CRP1 in *Arabidopsis thaliana* under abiotic stress conditions. *J. Plant Biol.* **2014**, *57*, 349–356. [CrossRef]
61. Lee, S.Y.; Seok, H.Y.; Tarte, V.N.; Woo, D.H.; Le, D.H.; Lee, E.H.; Moon, Y.H. The Arabidopsis chloroplast protein S-RBP11 is involved in oxidative and salt stress responses. *Plant Cell Rep.* **2014**, *33*, 837–847. [CrossRef] [PubMed]
62. Singh, B.N.; Mishra, R.N.; Agarwal, P.K.; Goswami, M.; Nair, S.; Sopory, S.K.; Reddy, M.K. A pea chloroplast translation elongation factor that is regulated by abiotic factors. *Biochem. Biophys. Res. Commun.* **2004**, *320*, 523–530. [CrossRef] [PubMed]
63. Allakhverdiev, S.I.; Nishiyama, Y.; Miyairi, S.; Yamamoto, H.; Inagaki, N.; Kanesaki, Y.; Murata, N. Salt stress inhibits the repair of photodamaged photosystem II by suppressing the transcription and translation of *psbA* genes in Synechocystis. *Plant Physiol.* **2002**, *130*, 1443–1453. [CrossRef] [PubMed]
64. Haußühl, K.; Andersson, B.; Adamska, I. A chloroplast DegP2 protease performs the primary cleavage of the photodamaged D1 protein in plant photosystem II. *EMBO J.* **2001**, *20*, 713–722. [PubMed]
65. Sevilla, F.; Camejo, D.; Ortiz-Espín, A.; Calderón, A.; Lázaro, J.J.; Jiménez, A. The thioredoxin/peroxiredoxin/sulfiredoxin system: Current overview on its redox function in plants and regulation by reactive oxygen and nitrogen species. *J. Exp. Bot.* **2015**, *66*, 2945–2955. [CrossRef] [PubMed]
66. Gill, S.S.; Tuteja, N. Reactive oxygen species and antioxidant machinery in abiotic stress tolerance in crop plants. *Plant Physiol. Biochem.* **2010**, *48*, 909–930. [CrossRef] [PubMed]
67. Martinis, J.; Glauser, G.; Valimareanu, S.; Kessler, F. A chloroplast ABC1-like kinase regulates vitamin E metabolism in Arabidopsis. *Plant Physiol.* **2013**, *162*, 652–662. [CrossRef] [PubMed]
68. Yang, K.A.; Moon, H.; Kim, G.; Lim, C.J.; Hong, J.C.; Lim, C.O.; Yun, D.J. NDP kinase 2 regulates expression of antioxidant genes in Arabidopsis. *Proc. Jpn. Acad. Ser. B* **2003**, *79*, 86–91. [CrossRef]
69. Upchurch, R.G. Fatty acid unsaturation, mobilization, and regulation in the response of plants to stress. *Biotechnol. Lett.* **2008**, *30*, 967–977. [CrossRef] [PubMed]
70. Loll, B.; Kern, J.; Saenger, W.; Zouni, A.; Biesiadka, J. Lipids in photosystem II: Interactions with protein and cofactors. *Biochim. Biophys. Acta* **2007**, *1767*, 509–519. [CrossRef] [PubMed]
71. Lee, A.G. Membrane lipids: It's only a phase. *Curr. Biol.* **2000**, *10*, R377–R380. [CrossRef]
72. Jordan, P.; Fromme, P.; Witt, H.T.; Klukas, O.; Saenger, W.; Krauß, N. Three-dimensional structure of cyanobacterial photosystem I at 2.5 Å resolution. *Nature* **2001**, *411*, 909–917. [CrossRef] [PubMed]
73. Mizusawa, N.; Wada, H. The role of lipids in photosystem II. *Biochim. Biophys. Acta* **2012**, *1817*, 194–208. [CrossRef] [PubMed]

74. Shimojima, M.; Ohta, H. Critical regulation of galactolipid synthesis controls membrane differentiation and remodeling in distinct plant organs and following environmental changes. *Prog. Lipid Res.* **2011**, *50*, 258–266. [CrossRef] [PubMed]
75. Kurtz, D.M. Microbial detoxification of superoxide: The non-heme iron reductive paradigm for combating oxidative stress. *Acc. Chem. Res.* **2004**, *37*, 902–908. [CrossRef] [PubMed]
76. Calderon, R.H.; García-Cerdán, J.G.; Malnoë, A.; Cook, R.; Russell, J.J.; Gaw, C.; Dent, R.M.; de Vitry, C.; Niyogi, K.K. A conserved rubredoxin is necessary for photosystem II accumulation in diverse oxygenic photoautotrophs. *J. Biol. Chem.* **2013**, *288*, 26688–26696. [CrossRef] [PubMed]
77. Majumder, A.L.; Sengupta, S.; Goswami, L. Osmolyte regulation in abiotic stress. In *Abiotic Stress Adaptation in Plants*; Pareek, A., Sopory, S.K., Bohnert, H.J., Eds.; Springer: Dordrecht, The Netherlands, 2009; pp. 349–370.
78. Rhodes, D.; Hanson, A.D. Quaternary ammonium and tertiary sulfonium compounds in higher plants. *Annu. Rev. Plant Physiol. Plant Mol. Biol.* **1993**, *44*, 357–384. [CrossRef]
79. Papageorgiou, G.C.; Murata, N. The unusually strong stabilizing effects of glycine betaine on the structure and function of the oxygen-evolving photosystem II complex. *Photosynth. Res.* **1995**, *44*, 243–252. [CrossRef] [PubMed]
80. Chen, T.H.; Murata, N. Glycinebetaine: An effective protectant against abiotic stress in plants. *Trends Plant Sci.* **2008**, *13*, 499–505. [CrossRef] [PubMed]
81. Tian, F.; Wang, W.; Liang, C.; Wang, X.; Wang, G.; Wang, W. Overaccumulation of glycine betaine makes the function of the thylakoid membrane better in wheat under salt stress. *Crop J.* **2017**, *5*, 73–82. [CrossRef]
82. Berthomieu, P.; Conéjéro, G.; Nublat, A.; Brackenbury, W.J.; Lambert, C.; Savio, C.; Uozumi, N.; Oiki, S.; Yamada, K.; Cellier, F. Functional analysis of *AtHKT1* in Arabidopsis shows that $Na^+$ recirculation by the phloem is crucial for salt tolerance. *EMBO J.* **2003**, *22*, 2004–2014. [CrossRef] [PubMed]
83. Shkolnik-Inbar, D.; Adler, G.; Bar-Zvi, D. *ABI4* downregulates expression of the sodium transporter *HKT1;1* in Arabidopsis roots and affects salt tolerance. *Plant J.* **2013**, *73*, 993–1005. [CrossRef] [PubMed]
84. Nambara, E.; Marion-Poll, A. Abscisic acid biosynthesis and catabolism. *Annu. Rev. Plant Biol.* **2005**, *56*, 165–185. [CrossRef] [PubMed]
85. Tan, B.C.; Joseph, L.M.; Deng, W.T.; Liu, L.; Li, Q.B.; Cline, K.; McCarty, D.R. Molecular characterization of the Arabidopsis 9-*cis* epoxycarotenoid dioxygenase gene family. *Plant J.* **2003**, *35*, 44–56. [CrossRef] [PubMed]
86. Xu, Z.Y.; Kim, D.H.; Hwang, I. ABA homeostasis and signaling involving multiple subcellular compartments and multiple receptors. *Plant Cell Rep.* **2013**, *32*, 807–813. [CrossRef] [PubMed]
87. Jung, H.J.; Park, S.J.; Kang, H. Regulation of RNA metabolism in plant development and stress responses. *J. Plant Biol.* **2013**, *56*, 123–129. [CrossRef]
88. Rajkowitsch, L.; Chen, D.; Stampfl, S.; Semrad, K.; Waldsich, C.; Mayer, O.; Jantsch, M.F.; Konrat, R.; Bläsi, U.; Schroeder, R. RNA chaperones, RNA annealers and RNA helicases. *RNA Biol.* **2007**, *4*, 118–130. [CrossRef] [PubMed]
89. Caldas, T.D.; El Yaagoubi, A.; Richarme, G. Chaperone properties of bacterial elongation factor EF-Tu. *J. Biol. Chem.* **1998**, *273*, 11478–11482. [CrossRef] [PubMed]
90. Takahashi, S.; Badger, M.R. Photoprotection in plants: A new light on photosystem II damage. *Trends Plant Sci.* **2011**, *16*, 53–60. [CrossRef] [PubMed]
91. Nixon, P.J.; Barker, M.; Boehm, M.; de Vries, R.; Komenda, J. FtsH-mediated repair of the photosystem II complex in response to light stress. *J. Exp. Bot.* **2005**, *56*, 357–363. [CrossRef] [PubMed]
92. Fan, P.; Nie, L.; Jiang, P.; Feng, J.; Lv, S.; Chen, X.; Bao, H.; Guo, J.; Tai, F.; Wang, J.; et al. Transcriptome analysis of *Salicornia europaea* under saline conditions revealed the adaptive primary metabolic pathways as early events to facilitate salt adaptation. *PLoS ONE* **2013**, *8*, e80595. [CrossRef] [PubMed]

2

# The Mitochondrial Pentatricopeptide Repeat Protein PPR18 is Required for the *cis*-Splicing of *nad4* Intron 1 and Essential to Seed Development in Maize

**Rui Liu [1], Shi-Kai Cao [1], Aqib Sayyed [1], Chunhui Xu [1], Feng Sun [1], Xiaomin Wang [2] and Bao-Cai Tan [1,*]**

[1] Key Laboratory of Plant Development and Environment Adaptation Biology, Ministry of Education, School of Life Sciences, Shandong University, Qingdao 266237, China; liuxiaoshuang_6@163.com (R.L.); caoshk5233@163.com (S.-K.C.); aqib.sayyed@yahoo.com (A.S.); chunhuixu@sdu.edu.cn (C.X.); epusun@sdu.edu.cn (F.S.)

[2] Key Laboratory of Cell Activities and Stress Adaptations, Ministry of Education, School of Life Sciences, Lanzhou University, Lanzhou 730000, China; wangxiaomin@lzu.edu.cn

* Correspondence: bctan@sdu.edu.cn

**Abstract:** Pentatricopeptide repeat (PPR) protein comprises a large family, participating in various aspects of organellar RNA metabolism in land plants. There are approximately 600 PPR proteins in maize, but the functions of many PPR proteins remain unknown. In this study, we defined the function of PPR18 in the *cis*-splicing of *nad4* intron 1 in mitochondria and seed development in maize. Loss function of *PPR18* seriously impairs embryo and endosperm development, resulting in the *empty pericarp* (*emp*) phenotype in maize. *PPR18* encodes a mitochondrion-targeted P-type PPR protein with 18 PPR motifs. Transcripts analysis indicated that the splicing of *nad4* intron 1 is impaired in the *ppr18* mutant, resulting in the absence of *nad4* transcript, leading to severely reduced assembly and activity of mitochondrial complex I and dramatically reduced respiration rate. These results demonstrate that PPR18 is required for the *cis*-splicing of *nad4* intron 1 in mitochondria, and critical to complex I assembly and seed development in maize.

**Keywords:** PPR protein; mitochondrial complex I; *nad4*; RNA splicing; seed development; maize

## 1. Introduction

Mitochondria are originated from α-proteobacteria ancestors via endosymbiosis. During evolution, the majority of the bacterial ancestral genes from mitochondrial genome have been lost or transferred to host nucleus [1]. In angiosperms, mitochondrial genomes contain up to 60 genes, which are involved in biogenesis of respiratory complex subunits, ribosomal proteins, ribosomal RNAs (rRNAs), and transfer RNAs (tRNAs) [2,3]. Maize mitochondrial genome contains 58 identified genes that encode 22 proteins of the electron transport chain, 9 ribosomal proteins, a maturase (MatR), a transporter protein (MttB), 3 ribosomal RNAs (5S, 18S, and 26S), and 21 tRNAs [4].

Mature mitochondrial transcripts undergo extensive post-transcriptional processing events, among which the most reported are RNA editing and RNA splicing [5–7]. In flowering plants, RNA editing usually alters cytidine to uridine through a deamination reaction in mitochondria and plastids and RNA splicing is a processing event in which noncoding segments (intron) of precursor RNA are removed and coding sequences are joined. Based on the distinctive structures, introns are divided into two families, group I and group II [4,8]. In flowering plants, most of organellar transcripts contain group II introns with conserved secondary structure, consisting of six domains extending from a central hub [9]. In bacteria, the splicing of a group II intron is self-facilitated by its cognate maturase encoded in the intron domain IV [10], but in plants, nearly all introns lost the maturase gene with only one intron maturase gene (*matK*) remains in the plastid genome and one *matR* gene in the mitochondrial genome [11,12]. Instead, four maturase genes (*nMat1* to *4*) are found in the nuclear genome in *Arabidopsis* [13–15]. Besides, numerous additional nucleus-encoded splicing co-factors have been reported to be involved in the splicing of organellar introns, such as the chloroplast RNA splicing and ribosome maturation (CRM) domain-containing proteins [16,17], RNA helicase [18,19], mitochondrial transcription termination factors (mTERF) [20,21], plant organellar RNA recognition (PORR) domain proteins [22,23], regulator of chromosome condensation (RCC1) domain proteins [24], and the pentatricopeptide repeat (PPR) proteins [5,25,26].

PPR proteins are a large family of RNA binding proteins, with more than 400 members in angiosperms [5,27]. PPR proteins contain multiple 35-amino-acid tandem repeats and each repeat forms a helix-loop-helix structure. Based on domain constitution, PPR proteins are divided into PLS (repeat P–L–S motif)-class proteins and P-class proteins [27]. The PLS-subclass PPR proteins contain characteristic triplets of P, L, and S motifs with additional E, E+, DYW, or other domains at the C-terminus, whereas the P-subclass PPR proteins contain arrays of only P motifs [5]. The PLS-class PPR proteins are implicated in the C-to-U RNA editing that in most cases is to restore the evolutionary conserved amino acids [28]. Functions of the P-subclass PPR proteins are diverse, which includes RNA cleavage, RNA splicing, RNA stabilization and maturation, and translation initiation [5]. Most PPR proteins are localized in mitochondria or chloroplasts. They bind RNA in a sequence specific manner that one PPR motif binds to one nucleotide of the target RNA. The recognition nucleotides were determined by the different combinations of the amino acid residues at position 5th and 35th of each PPR repeat, which is known as the PPR codes [29–31]. In plant mitochondria, most of group II introns are present in genes that code for subunits of mitochondrial complex I. In maize mitochondria, out of 22 identified group II introns, 19 resides in *nad1*, *nad2*, *nad4*, *nad5*, and *nad7* transcripts, while 3 in *rps3*, *cox2*, and *ccmFc* transcripts [32,33]. Accurate splicing of these group II introns is critical to mitochondrial function and biogenesis, which is important for plant growth and development. For instance, EMP11 and DEK2 are involved in the splicing of *nad1* introns, and the loss of function mutation of *Emp11* and *Dek2* affects the assembly of complex I with severely arrested embryo and endosperm development [34,35]. EMP10, EMP12, EMP16, DEK37, and PPR20 are responsible for the splicing of *nad2* introns in maize. These mutations result in a loss of mitochondrial complex I assembly and activity, impairing the mitochondrial function and embryogenesis and endosperm development [25,36–39].

In this study, we characterized a maize seed mutant *ppr18*, which exhibits arrested embryo and endosperm development phenotype. *PPR18* encodes a mitochondrion-targeted P-type PPR protein with 18 PPR motifs. The loss of *PPR18* function leads to the splicing deficiency of *nad4* intron 1, severely reduced assembly and activity of mitochondrial complex I, resulting in the impairment of mitochondrial function and seed development in maize.

## 2. Results

### *2.1. PPR18 Is a Mitochondrion-Localized P-Type PPR Protein*

*PPR18* (GRMZM2G438456) is an intronless gene, encoding an 85 kDa protein with 768 amino acid residues (Figure 1A). Motif prediction analysis by algorithm TPRpred (http://tprpred.tebingen.mpg.de/tprpred) revealed that PPR18 contained 18 tandemly repeated PPR motifs without any other domains, suggesting that PPR18 is a canonical P-type PPR protein (Figure 1A,B). A phylogenetic analysis based on the maize PPR18 and its homologous proteins revealed extensive conservation in the sequences in both monocots and dicots (Figure S1). Most of PPRs are localized in organelles, either chloroplasts or/and mitochondria, except GRP23 and PNM1, which both have nucleus localized signals [5,40,41]. To determine the subcellular localization of PPR18, the 550 amino acid residues of the N-terminal PPR18 were fused to the green fluorescent protein (GFP) in the binary vector pGWB5, then transiently expressed in the tobacco leaves via *Agrobacterium* EHA105 infiltration. Confocal laser-scanning microscopy revealed that the strong green fluorescence signals of PPR18$^{N550}$-GFP are merged with the red signals of MitoTracker (Figure 1C), indicating that PPR18 is localized in mitochondria.

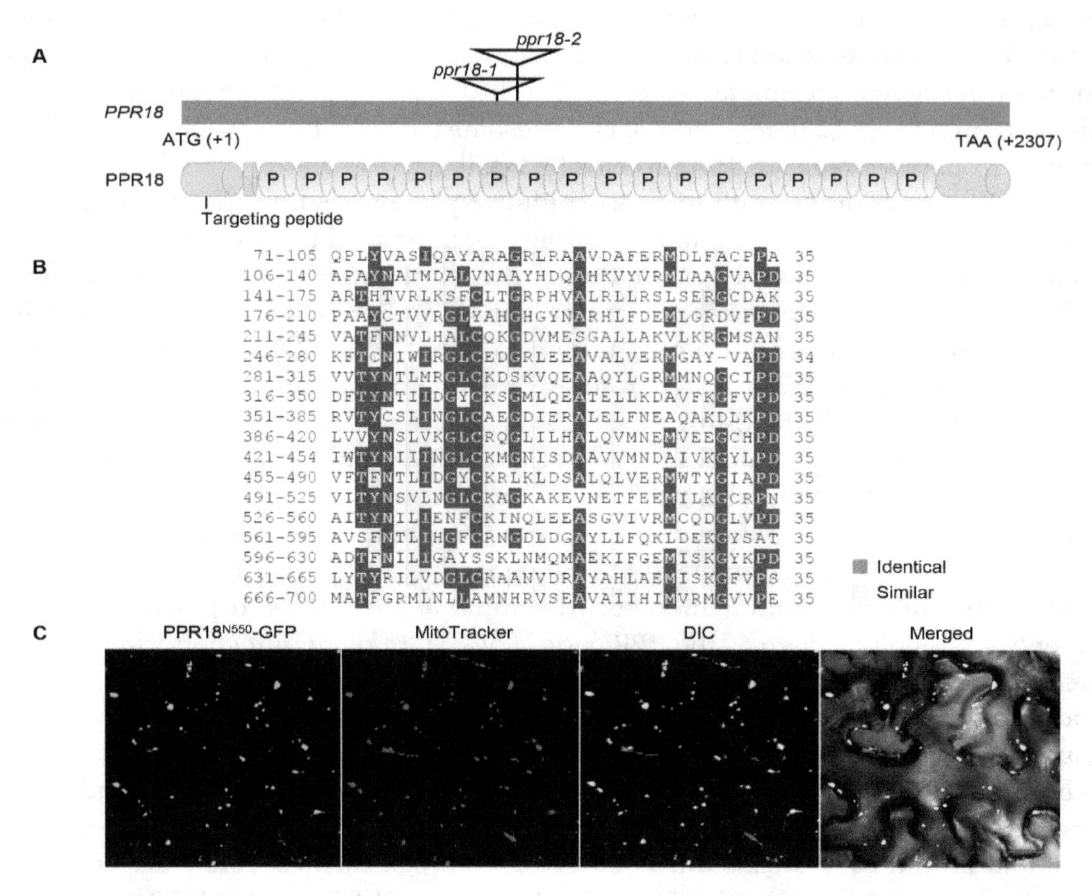


**Figure 1.** PPR18 is a mitochondrion-localized P-type pentatricopeptide repeat (PPR) protein. (**A**) Schematic illustrating the genomic structure and protein structure of *PPR18*. The locations of the *Mu* insertions are marked with triangles in two independent alleles. P, P-type PPR motif. (**B**) Alignment analysis of 18 PPR motifs in PPR18 protein. Identical amino acids are highlighted in dark gray and similar ones in yellow. (**C**) Localization of PPR18$^{N550}$-GFP in tobacco mesophyll cells. Green fluorescence, red fluorescence, and yellow fluorescence (merge) show green fluorescent protein (GFP) fluorescence, MitoTracker stained mitochondria, and localization, respectively. DIC, differential interference contrast. Scale bar, 20 μm.

## 2.2. *Embryo and Endosperm Development Are Arrested in ppr18*

To characterize the function of PPR18, we isolated two independent *Mutator* (*Mu*) insertional mutants containing *Mu* insertions at 880 bp and 939 bp downstream from the start codon of *PPR18* from the UniformMu population in the inbred W22 genetic background, named *ppr18-1* and *ppr18-2*, respectively [42]. The selfed progeny of both *ppr18-1*/+ and *ppr18-2*/+ heterozygotes segregated at a 3:1 ratio of wild-type (WT) and *empty pericarp* (*emp*) kernels, suggesting that both mutants are monogenic nuclear recessive mutations and homozygous lethal (Figure 2A and Figure S2A). Co-segregation analysis of a small isolated population with 72 individuals was performed to test the linkage of *ppr18-1* by genomic PCR using *PPR18*-R1 and *Mu* TIR8 primers [43]. The results showed that all the self-pollinated progenies of *ppr18-1*/+ plants produced *emp* kernels, indicating that the *Mu* insertion is tightly linked to the mutation (Figure S3). Crosses between *ppr18-1*/+ and *ppr18-2*/+ heterozygotes produced heteroallelic progeny *ppr18-1*/*ppr18-2* with approximately 25% *emp* kernels (Figure S2B), confirming that the *ppr18* phenotype results from the disruption of GRMZM2G438456.

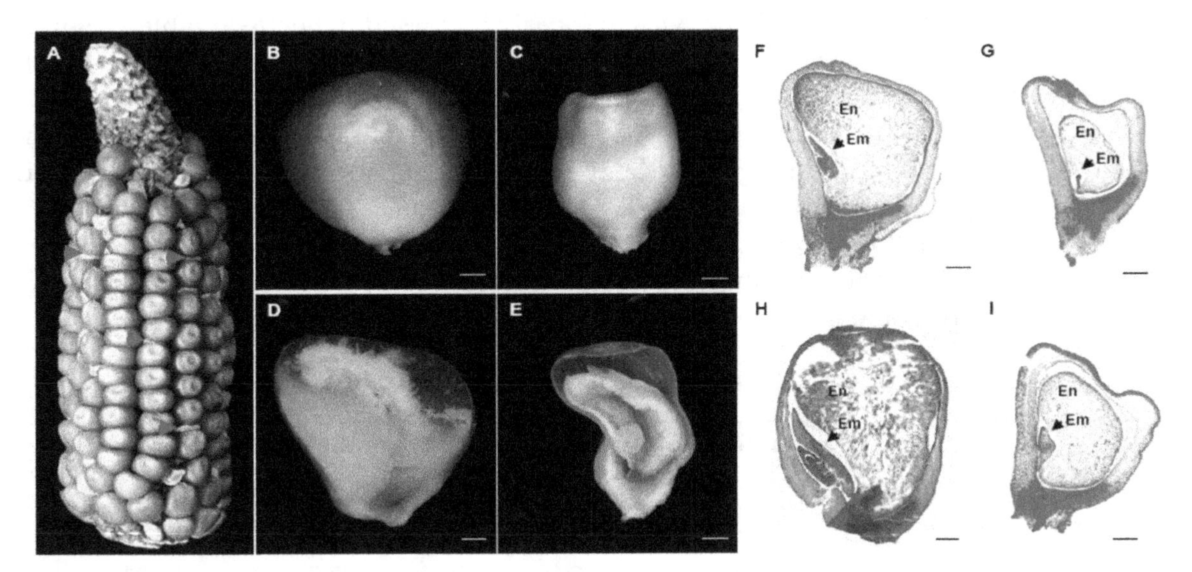


**Figure 2.** Mutant *ppr18-1* kernels abort early in embryogenesis and endosperm development. (**A**) The mature ear segregates 3:1 for wild-type (WT) and *ppr18-1* kernels at 35 days after pollination (DAP). The red arrows indicate the *ppr18-1* kernels. (**B**,**C**) Mature wild-type (**B**) and *ppr18-1* kernels (**C**) from (**A**). (**D**,**E**) Dissection of mature wild-type (**D**) and *ppr18-1* kernels (**E**). (**F**–**I**) Paraffin sections of wild-type (**F**,**H**) and *ppr18-1* kernels (**G**,**I**) at 9 DAP and 14 DAP. Wild-type kernels at 9 DAP (**F**) and 14 DAP (**H**); *ppr18-1* kernels at 9 DAP (**G**) and 14 DAP (**I**). En, endosperm; Em, embryo. Scale bar, 1 mm in (**B**–**I**).

The developing kernels phenotype of WT and *ppr18-1* in the same segregating ear are compared in Figure 2B,C. The *ppr18-1* mutant kernels are remarkably smaller than the wild type, which exhibited pale, half-translucent, and collapsed appearance at 35 days after pollination (DAP; Figure 2B–E). Compared with the WT siblings in the same segregating ear, the embryo and endosperm development are arrested in the *ppr18-1* kernels (Figure 2B–E). To examine the developmental arrest of embryogenesis in *ppr18-1*, we examined the embryo and endosperm development process between *ppr18-1* and WT siblings in a segregating ear by light microscopy. At 9 DAP, the WT embryos reached the coleoptilar stage, whereas the *ppr18-1* embryos remained at the pre-embryo stage (Figure 2F,G). At 14 DAP, the WT embryos reached late embryogenesis stage, while the *ppr18-1* embryos stayed at the transition stage without any discernable differentiation (Figure 2H,I). These results indicate that the loss of *PPR18* severely arrests both embryo and endosperm development.

To assess the impact of the *Mu* insertion on the *PPR18* expression, we analyzed the transcript level of *PPR18* in two *ppr18* alleles by reverse transcription PCR (RT-PCR). Results showed that no transcript of *PPR18* was detected in both alleles (Figure S4A), indicating that both alleles are probably null mutations. In wild type, *PPR18* transcripts can be detected in all vegetative and reproductive tissues by quantitative real-time PCR (qRT-PCR; Figure S4B). Relative high mRNA expression of *PPR18* was in bract and low expression in root, flower and kernel at developmental stages, indicating that *PPR18* is a constitutively expressed gene throughout growth and development in maize, rather than a seed specific gene. As the mutants are embryo lethal, impacts on other tissues and during development cannot be determined.

### 2.3. Loss of PPR18 Affects Mitochondrial Respiratory Activity

Previous reports showed the mutation of mitochondrion-localized PPR proteins with arrested seed development often suffers defects in mitochondrial respiration [44–46]. Thus, we investigated whether the loss of *PPR18* affects mitochondrial respiratory activity in maize by determining the respiratory activity, as shown by three mitochondrial respiratory rates, including total respiratory ($V_t$), cytochrome respiratory capacity ($V_{cyt}$), and alternative respiratory capacity ($V_{alt}$). The ratio of $V_{cyt}/V_t$ was significantly reduced in the *ppr18-1* mutant compared with WT (Table 1), indicating that loss of *PPR18* resulted in a severe reduction of the cytochrome pathway and impaired mitochondrial respiration. Meanwhile, the ratio of $V_{alt}/V_t$ was markedly increased in *ppr18-1* (Table 1), supporting that alternative respiratory pathway was enhanced in the *ppr18-1* mutant. Moreover, we detected the expression of alternative oxidase (AOX) protein by Western blot assay using the specific antibody of AOX. Results showed that the abundance of AOX was increased drastically in the *ppr18-1* mutant compared to WT (Figure 3C), confirming that mutation of *PPR18* enhances the expression of alternative oxidases. The maize genome contains three *AOX* genes, *AOX1*, *AOX2*, and *AOX3*. Both RT-PCR and qRT-PCR assays showed that the expression of *AOX2* and *AOX3* was dramatically increased in the two *ppr18* alleles (Figure 3A,B), indicating that excessive accumulation of AOX is caused by upregulation of *AOX2* and *AOX3* expression in the *ppr18* alleles.

**Table 1.** Alteration of the respiration rate of the wild type (WT) and *ppr18-1* kernels.

| | Respiration Rate (nmol $O_2$ $min^{-1}$ $g^{-1}$ Fresh Weight) | | | | |
|---|---|---|---|---|---|
| | $V_t$ | $V_{alt}$ | $V_{cyt}$ | $V_{alt}/V_t$ (%) | $V_{cyt}/V_t$ (%) |
| WT | 824.96 ± 77.23 | 155.90 ± 14.10 | 734.30 ± 60.97 | 18.90 | 89.01 |
| *ppr18-1* | 175.62 ± 2.85 | 136.61 ± 6.55 | 39.00 ± 4.26 | 77.79 | 22.21 |

Mitochondrial total respiration rate ($V_t$), the alternative pathway ($V_{alt}$), and the capacity of the cytochrome pathway ($V_{cyt}$) were indicated by the oxygen consumption of nmol $O_2$ $min^{-1}$ $g^{-1}$ fresh weight of the maize kernels at 11 DAP using a Clark-type oxygen electrode. Data are mean values ± SEs from three independent biological samples.

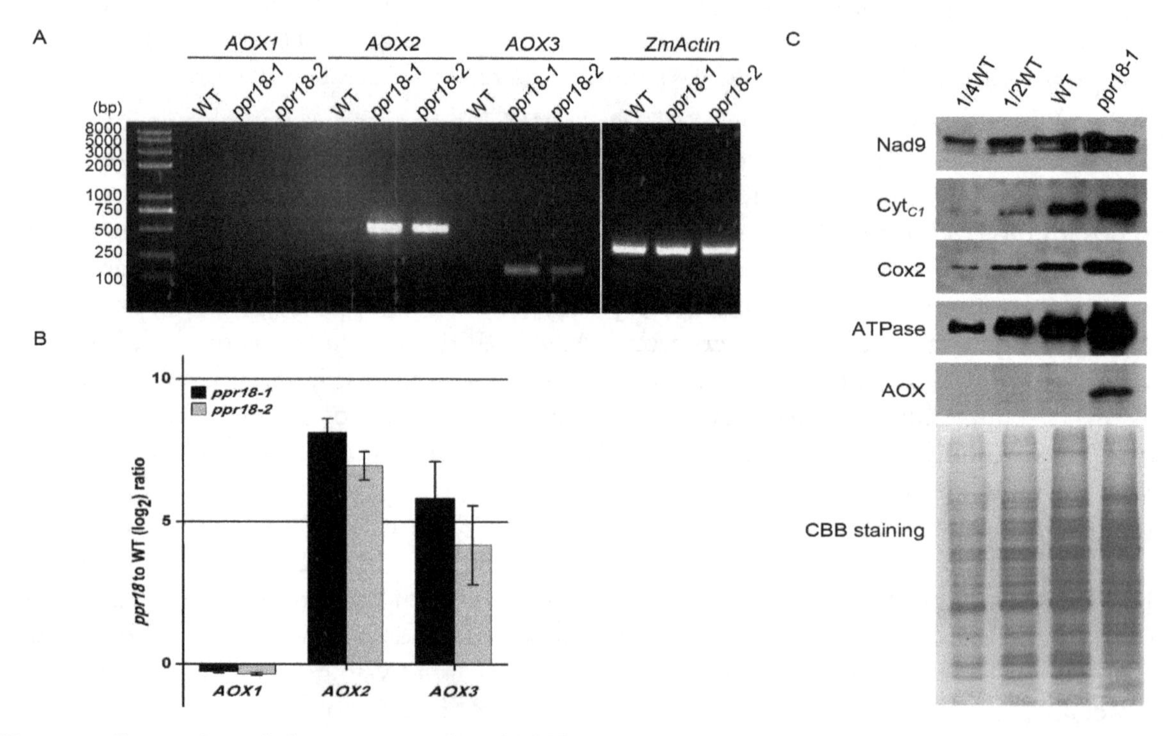


**Figure 3.** Expression of alternative oxidase (*AOX*) genes and analysis of protein expression abundance in the *PPR18* mutants. (**A**) The transcript levels of *AOX* genes in *ppr18-1* and *ppr18-2* kernels at 12 days after pollination (DAP). The expression levels were normalized to *ZmActin* (GRMZM2G126010). (**B**) qRT-PCR analysis of *AOX* gene expression in *ppr18-1* and *ppr18-2* kernels at 12 DAP. (**C**) Western blot analysis with antibodies against Nad9, $Cyt_{C1}$, Cox2, ATPase a subunit, and AOX. Crude mitochondrial extracts from endosperm and embryo of *ppr18-1* and WT immature kernels at 11 DAP. CBB (Coomassie Brilliant Blue) staining demonstrates that equal amounts of mitochondrial proteins were loaded.

*2.4. Loss of PPR18 Affects the Assembly and Activity of Complex I*

The limited cytochrome pathway is closely relevant to the defective transfer electrons from mitochondrial respiratory complexes, complex I to IV [47,48]. To determine the assembly and abundance of mitochondrial respiratory complexes, we performed blue native (BN)-PAGE using crude mitochondria from *ppr18* alleles and WT maize kernels (Figure 4). As indicated by Coomassie Brilliant Blue (CBB) staining, complex III and V was substantially accumulated, whereas complex I and supercomplex I + $III_2$ were depleted in both *ppr18* alleles compared to WT (Figure 4A), indicating that loss of *PPR18* affects the assembly of mitochondrial complex I. Furthermore, we analyzed the NADH dehydrogenase activity of complex I by in-gel NADH activity assay, which showed a consistent result with the CBB staining. As shown in Figure 4B, the dehydrogenase activity of complex I and supercomplex I + $III_2$ were completely deficient in both *ppr18* alleles. Besides, we detected the assembly of complex III, IV, and V by Western blot analysis using the anti-$Cyt_{C1}$, anti-Cox2, and anti-ATP synthase $\alpha$-subunit antibody, respectively. Results showed that complex III, IV, and V were increased in the *ppr18-1* mutant (Figure 4C–E). Collectively, these results imply that PPR18 is important for the assembly and activity of mitochondrial complex I in maize.

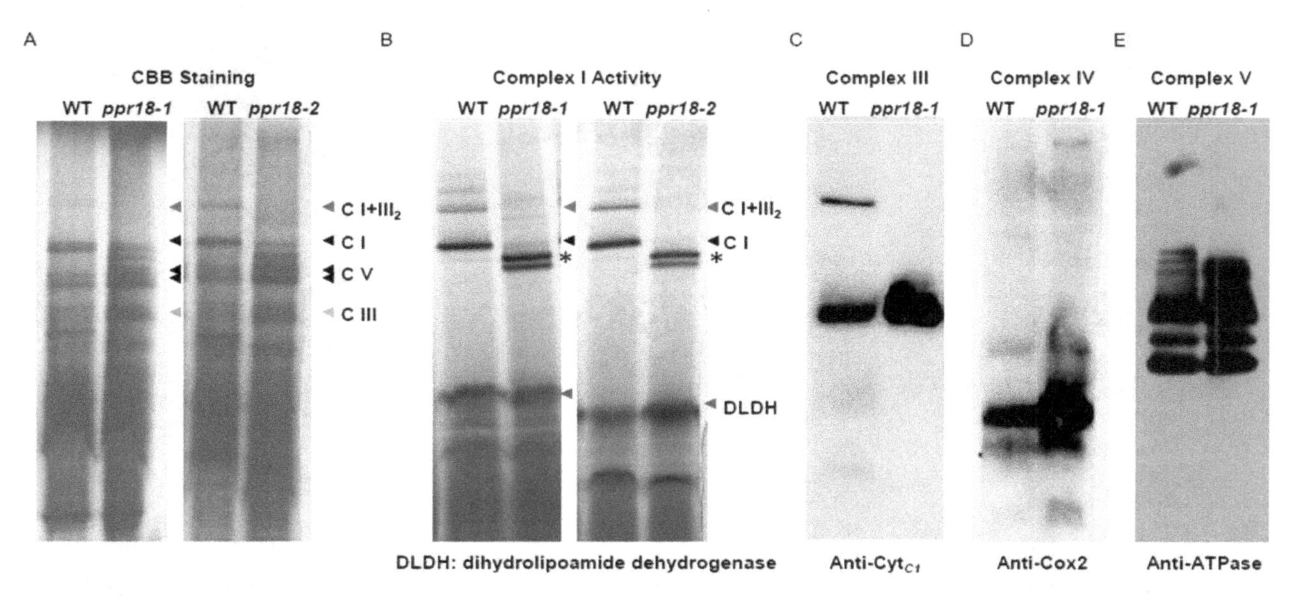


**Figure 4.** BN-PAGE analysis of mitochondrial complexes in *ppr18-1* and *ppr18-2* mutants. (**A**) About 130 mg of mitochondrial protein was loaded to a 3%–12% BN-PAGE. Blue native (BN) gels were stained with Coomassie brilliant blue (CBB). The positions of complex I, III, V, and super complex I + $III_2$ are indicated. (**B**) In-gel NADH dehydrogenase activity of complex I. The activity of dihydrolipoamide dehydrogenase (DLDH) was used as a loading control. Asterisks indicate partially assembled complex I. (**C–E**) Accumulation of respiratory chain complex III (**C**), IV (**D**), and V (**E**) in *ppr18* mutant kernels. The BN gels were probed with antibodies against $Cyt_{C1}$, Cox2, and ATPase ($\alpha$-subunit).

In addition, we determined the abundance of the mitochondrial complex proteins in *ppr18-1* and WT maize kernels by Western blot analysis using antibodies against Nad9 (complex I), $Cyt_{C1}$ (complex III), Cox2 (complex IV), and ATPase (complex V). As shown in Figure 3C, the protein abundance of Nad9, $Cyt_{C1}$, Cox2, and ATPase was increased in *ppr18-1*, speculating that the lack of *PPR18* may enhance the expression of subunit from complex I, III, IV, and V in a feedback mechanism.

### *2.5. PPR18 Is Required for the Splicing of nad4 Intron 1*

Previous study showed that most P-type PPR proteins function on intron splicing, RNA maturation, RNA stabilization or RNA cleavage in organelles [5]. To reveal the molecular function of PPR18, we analyzed the transcript levels of 35 mitochondrion-encoded genes between WT and *ppr18* alleles by RT-PCR and qRT-PCR. The results showed that the expression level of most mitochondrion-encoded genes was indistinguishable between the WT and *ppr18* alleles, only the expression of *nad4* was obviously different between WT and *ppr18* alleles (Figure 5 and Figure S5). The mature *nad4* transcript was not detectable in both *ppr18* alleles. Instead, a band larger than the mature *nad4* transcript was dramatically increased in both *ppr18* mutants (Figure 5). The sequencing results of the larger fragments showed that these fragments contain unspliced *nad4* intron 1. These results indicate that the loss-of-function in PPR18 abolishes the splicing of *nad4* intron 1 in mitochondria.

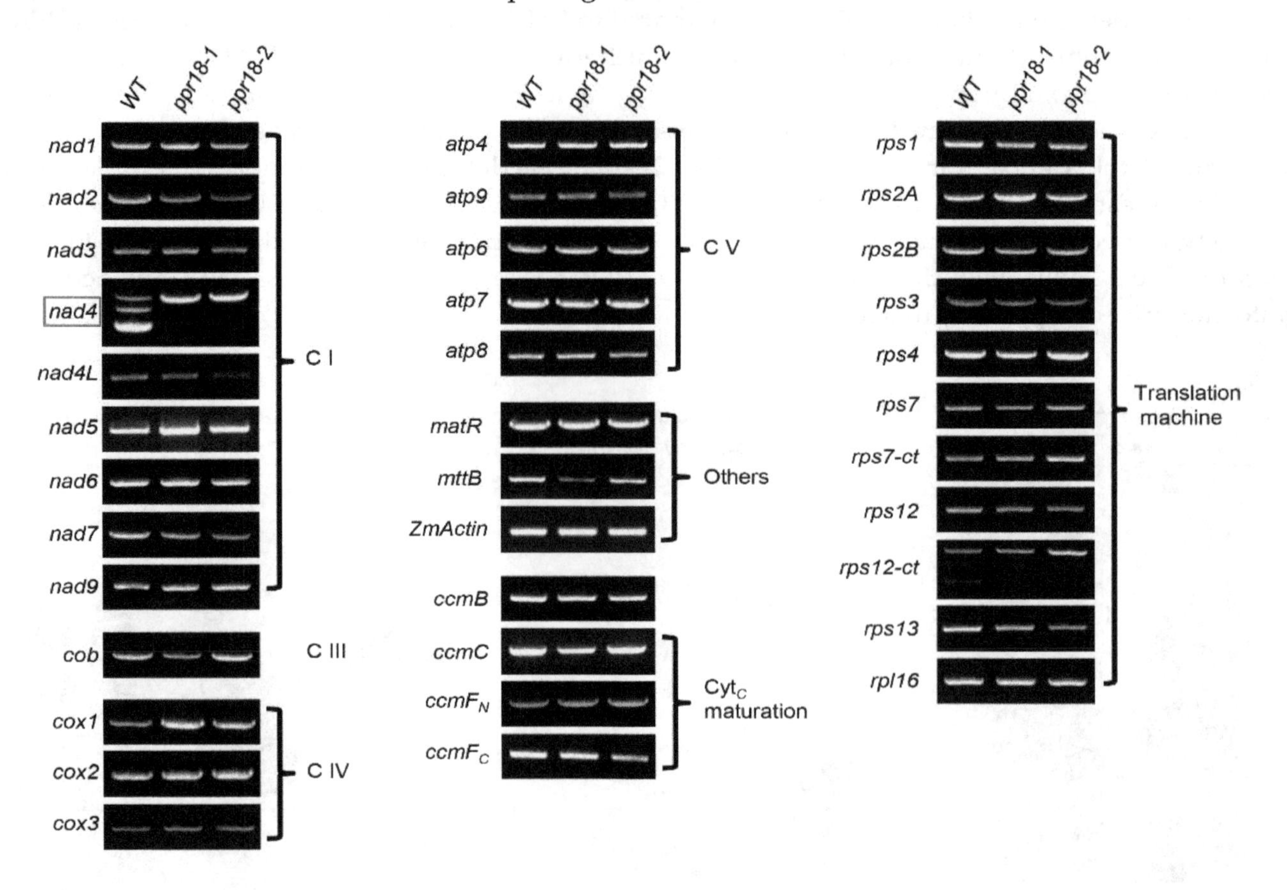


**Figure 5.** The *ppr18* mutants only affect the expression of mitochondrial *nad4* mature transcript. The transcript levels of 35 mitochondrion-encoded genes in *ppr18-1 and ppr18-2* mutant kernels. The RNA was isolated from the same ear segregating for WT and *ppr18* mutants at 13 days after pollination (DAP). The expression levels were normalized to *ZmActin* (GRMZM2G126010). The absence of *nad4* transcripts occurs in both *ppr18* alleles (red box indicated). C I: Complex I, C II: Complex II, C III: Complex III, C IV: Complex IV, C V: Complex V.

The *nad4* precursor RNA contains three *cis*-splicing introns (Figure 6A). To confirm the function of *PPR18* on the *nad4* intron 1 splicing, we amplified fragments containing each of the three introns in the *nad4* transcript by RT-PCR using specific primers. As shown in Figure 6B, only the splicing of *nad4* intron 1 was impaired in both *ppr18* alleles. In addition, we analyzed the splicing efficiency of 22 group II introns in mitochondria by qRT-PCR. Results showed that the splicing efficiency of *nad4* intron 1 was dramatically decreased in the *ppr18* alleles (Figure 7). These data suggest that PPR18 is indeed required for the splicing of mitochondrial *nad4* intron 1. Previous report showed that PPR proteins bind specific RNA via a modular recognition code in which the nucleotide specificity primarily relies on combination at the 5th and 35th amino acid residues of each PPR motif [29–31]. Based on PPR recognition code, potential binding sites of PPR18 in mitochondrial *nad4* intron 1 were predicted (Figure 6C). Results showed that the nucleotides of *nad4* intron 1 are well aligned to the combinatorial codes. We predicted the secondary structure of *nad4* intron 1 and mapped the putative binding site of PPR18 in domain I of *nad4* intron 1 (Figure S6). A phylogenetic analysis based on genomic DNAs in the GenBank, the putative binding site of PPR18 in *nad4* intron 1 appeared to be highly conserved in both monocots and dicots (Figure S7).

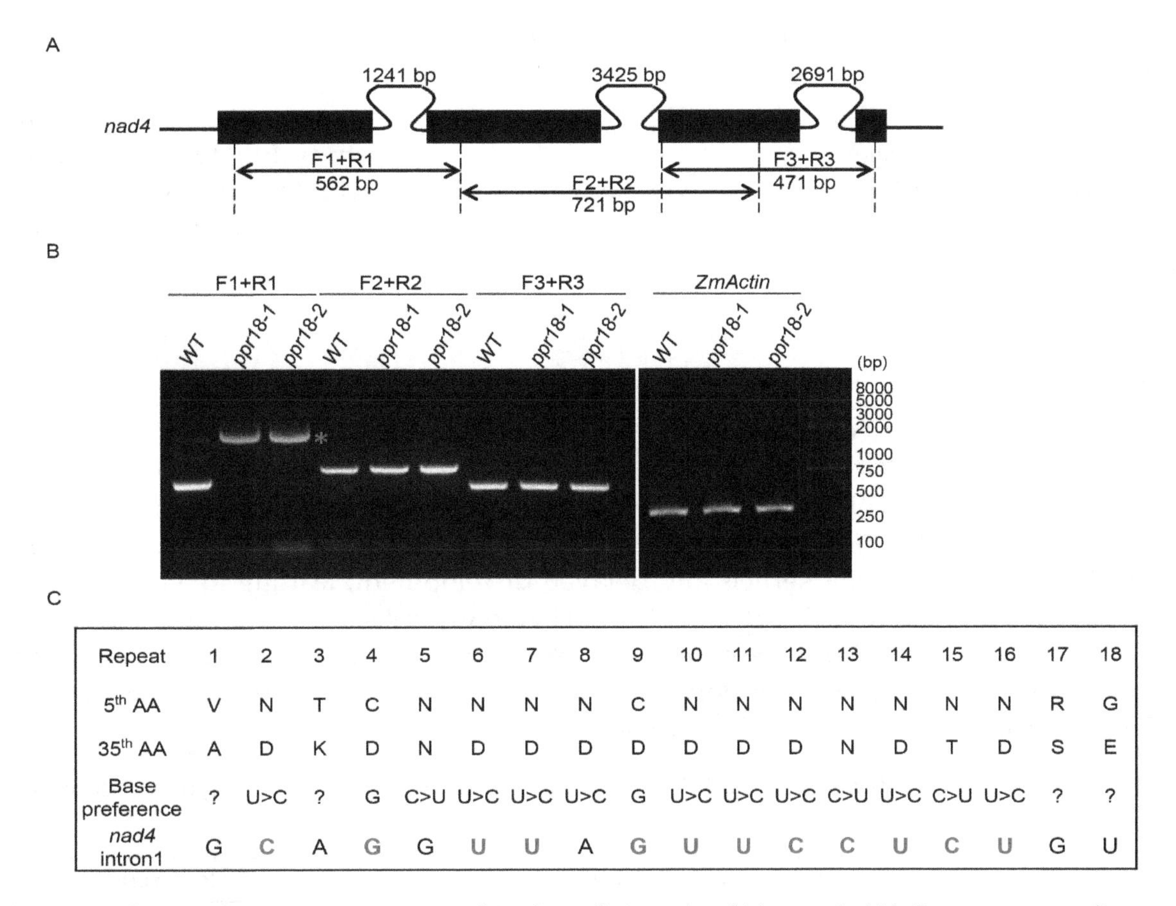


| Repeat | 1 | 2 | 3 | 4 | 5 | 6 | 7 | 8 | 9 | 10 | 11 | 12 | 13 | 14 | 15 | 16 | 17 | 18 |
|---|---|---|---|---|---|---|---|---|---|---|---|---|---|---|---|---|---|---|
| 5th AA | V | N | T | C | N | N | N | N | C | N | N | N | N | N | N | N | R | G |
| 35th AA | A | D | K | D | N | D | D | D | D | D | D | D | N | D | T | D | S | E |
| Base preference | ? | U>C | ? | G | C>U | U>C | U>C | U>C | G | U>C | U>C | U>C | C>U | U>C | C>U | U>C | ? | ? |
| *nad4* intron1 | G | C | A | G | G | U | U | A | G | U | U | C | C | U | C | U | G | U |

**Figure 6.** The *ppr18* mutants are impaired in the splicing of *nad4* intron 1. (**A**) Gene structure diagram of the maize *nad4* gene. Exons are shown as filled black boxes. All the three introns of *nad4* are *cis*-introns. The primers and expected amplification products using are indicated. F: Forward primer, R: Reverse primer. (**B**) RT-PCR analysis of the intron splicing of *nad4* introns using primers as indicated in (**A**). Asterisk indicates the unspliced PCR products of *nad4* intron 1. (**C**) Binding predictions for the PPR18 proteins on the respective targets *nad4* intron 1 referred to Barkan et al., 2012; Yin et al., 2013; and Gully et al., 2015. The amino acid (AA) residues at position 5th and 35th in PPR motifs 1–18 are indicated. Nucleotides matching the amino acid combination are indicated in red. "?" indicates an unidentified nucleotide.

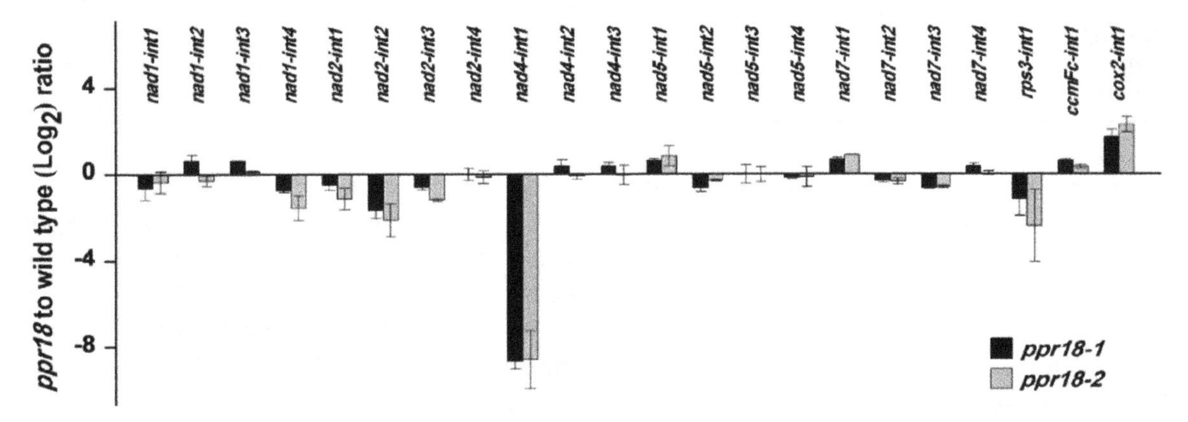


**Figure 7.** Splicing efficiency of mitochondrial introns in *ppr18* mutants. qRT-PCR analysis of the splicing efficiency of all 22 group II introns of maize mitochondrial genes in *ppr18* mutants. The ratio of spliced to unspliced fragments was used to measure splicing efficiency. Data are means (± SE) of three biological replicates.

### 2.6. PPR18 Does Not Show a Direct Interaction with DEK35, EMP8, and EMP602 in Yeast Two-Hybrid Assays

Previous studies reported that three PPR proteins DEK35, EMP8, and EMP602 are involved in the splicing of mitochondrial *nad4* intron 1 in maize [46,49,50]. In this study, PPR18 is also required for the splicing of *nad4* intron 1. To determine whether PPR18 interacts with DEK35, EMP8, and EMP602, we performed a yeast two-hybrid assay. Results showed that PPR18 has no directly physical interaction with these proteins (Figure S8).

## 3. Discussion

### 3.1. A Role of PPR18 on nad4 Intron 1 Splicing and the Assembly of Complex I

The maize mitochondria contain a total of 22 group II introns in 8 genes (*nad1, nad2, nad4, nad5, nad7, ccmFc, cox2,* and *rps3*) [4]. Some PPR proteins have been reported to be responsible for the splicing of group II introns in maize mitochondria (Table 2). Loss-of-function of these PPR proteins usually results in *empty pericarp* (*emp*) or *defective kernel* (*dek*) phenotype in maize. For example, disruption of *Dek2* and *Dek37* causes small kernels and delayed development, leading to a *dek* phenotype of maize [35,38]. The mutation of *Emp16* and *Emp10* severely arrests seed development, resulting in embryo lethality and an *emp* phenotype of maize [25,39]. Similarly, the loss of *PPR18* function severely arrests the embryo and endosperm development (Figure 2), leading to the *emp* phenotype, suggesting that PPR18 is crucial for maize kernel development.

**Table 2.** List of mitochondrion-localized PPR proteins are involved in intron splicing of mitochondrial genes in maize.

| Protein | Target Transcript | Reference |
|---|---|---|
| EMP16 | *nad2*-int4 | (Xiu et al., 2016) |
| EMP10 | *nad2*-int1 | (Cai et al., 2017) |
| DEK35 | *nad4*-int1 | (Chen et al., 2017) |
| DEK2 | *nad1*-int1 | (Qi et al., 2017) |
| EMP11 | *nad1*-int1, 2, 3, 4 | (Ren et al., 2017) |
| DEK37 | *nad2*-int1 | (Dai et al., 2018) |
| EMP8 | *nad1*-int4, *nad2*-int1, *nad4*-int1 | (Sun et al., 2018) |
| EMP12 | *nad2*-int1, 2, 4 | (Sun et al., 2019) |
| DEK41/DEK43 | *nad4*-int3 | (Ren et al., 2019; Zhu et al., 2019) |
| PPR-SMR1 | *nad1*-int1, 2, 3, 4, *nad2*-int1, 2, 3, 4, *nad4*-int1, 2, 3, *nad5*-int1, 3, 4, *nad7*-int2, *rps3*-int1 | (Chen et al., 2019) |
| EMP602 | *nad4*-int1, 3 | (Ren et al., 2019) |
| PPR20 | *nad2*-int3 | (Yang et al., 2019) |
| PPR18 | *nad4*-int1 | This study |

As shown in Table 2, these PPR proteins (EMP8, EMP10, EMP11, EMP12, EMP16, EMP602, DEK2, DEK35, DEK37, DEK41, DEK43, PPR-SMR1, and PPR20) are reported to participate in the splicing of *nad1, nad2, nad4, nad5, nad7,* and *rps3* introns. These *nad* genes encode the core subunits of mitochondrial complex I and lack of these Nad proteins causes the disassembly and reduced activity of complex I in maize [51]. For example, EMP11 is responsible for the intron splicing of *nad1* and the dysfunction of *Emp11* affects the assembly and stability of mitochondrial complex I [34]. EMP12 and PPR20 are essential for the splicing of *nad2* introns. The mutation of EMP12 and PPR20 results in disassembly of mitochondrial complex I and a significant reduction in the dehydrogenase activity [36,37]. In this study, loss of *PPR18* impaired the *cis*-splicing of *nad4* intron 1 and the accumulation of mature *nad4* transcript (Figure 5), leading to severely defective assembly and activity of mitochondrial complex I in the *ppr18* mutants, suggesting the importance of PPR18 in the intron splicing of *nad4* transcript and the assembly and activity of mitochondrial complex I. The splicing of *nad4* intron 1 was dramatically reduced in *dek35*, and completely abolished in *emp8, emp602,* and *ppr18* [46,49,50], causing a deficiency in the mature *nad4* transcript and severely arrested embryo and endosperm development in maize. Together, these results indicate that expression of *nad4* is essential to the mitochondrial function and kernel development in maize. A co-evolution between the mitochondrial *nad4* intron 1 and the nuclear *PPR18* genes are implicated as indicated by the highly conserved sequences between PPR18 and its putative binding site in *nad4* intron 1 in both monocots and dicots (Figures S1 and S7).

In Figure 4, the in-gel NADH activity assay showed that two bands with a smaller size than mature complex I were produced in *ppr18* alleles, which are probably partially assembled complex I. Additionally, the two smaller size complexes had the dehydrogenase activity (Figure 4B), indicating the two partially assembled complex I are stable intermediate of the mitochondrial complex I assembly pathway in maize. Similar cases were also found in the *dek35*, *emp602*, *dek41*, and *dek43* mutants, which are defective in the splicing of *nad4* introns and [49,50,52,53], supporting that Nad4 is assembled into the complex I at a late stage. As reported in *Arabidopsis*, Nad4 is located in the $P_D$ module of the membrane arm of mitochondrial complex I, which is associated with assembled intermediate to form the mature complex I [51], indicating that Nad4 plays a crucial role during the last phase of the complex I assembly process both in monocots and dicots.

Previous reports showed that impairments in the electron transfer chain (ETC) can enhance AOX pathway in mitochondria [54–56]. In the *ppr18* alleles, the ratio of $V_{alt}/V_t$, the abundance of AOX protein, and the expression of *AOX2* and *AOX3* transcripts were notably increased (Figure 3), indicating that AOX pathway is significantly enhanced in *ppr18*. A retrograde signalling pathway is strongly implicated as the AOXs are nucleus-encoded proteins.

### *3.2. Multiple Splicing Factors Participate in the Splicing of nad4 Intron 1*

The defective splicing of *nad4* intron 1 was firstly reported in *Nicotiana sylvestris nms1* mutant [57]. In maize, defects of splicing of *nad4* intron 1 have been found in some *ppr* mutants, e.g., *dek35*, *emp8*, and *emp602* [46,49,50]. Our data show that PPR18 also specifically functions on the splicing of *nad4* intron 1 in maize. These splicing factors share common intron target with PPR18, implying that splicing of a single intron requires multiple splicing factors. Based on the yeast two-hybrid analyses, however, PPR18 does not directly interact with DEK35, EMP8, and EMP602 (Figure S8).

It is possible that these PPR proteins function independently by binding to the specific sequences of *nad4* intron 1 to maintain a splicing-competent structure, and these proteins do not have protein interactions with each other. Proteins could promote group II intron folding to form the native structure [58]. PPR proteins are RNA binding proteins that could guide intron folding by sequence-specific interactions [59]. PPR18 may play a role in folding of *nad4* intron 1 to participate in the intron splicing of *nad4* transcript.

As the intron splicing mechanism in mitochondria is not clear yet, all co-factors may have not been identified. The possibility of a ribonucleoprotein complex similar to the nuclear spliceosome exists for some introns. Thus, it is possible that PPR18 may interact with other splicing co-factors. The tested PPR proteins may interact with some key splicing factors exist in splicing complexes to mediate splicing of *nad4* intron 1. For example, PPR-SMR1 interacts with Zm-mCSF1 to participate in the splicing of several mitochondrial group II introns, speculating that PPR-SMR1 and Zm-mCSF1 might be the core splicing factors to mediate multiply group II introns splicing [60]. Further studies are necessary to unravel the splicing mechanism of plant organellar group II introns.

## 4. Materials and Methods

### *4.1. Plant Materials*

The *ppr18* alleles (UFMu-06715 and UFMu-11033) were obtained from the Maize Genetics Cooperation Stock Center (Urbana, IL, USA). For developmental analyses and population generation, the maize plants were cultivated in the experimental filed of Shandong University in Qingdao. Tobacco

(*Nicotiana tabacum*) was grown in climate chambers at 25 °C/22 °C day/night on a 12 h light/12 h dark regime.

### 4.2. Subcellular Localization

To express PPR18 $^{N550}$-GFP fusion proteins, the amplified PPR18 N-terminal coding sequence (1650 bp) was cloned into binary vector pGWB5 driven by the 35S promoter. *Agrobacterium tumefaciens* strain (EHA105) harboring this construct was infiltrated into tobacco (*Nicotiana tabacum*) leaf epidermis with a syringe, as previously described [61]. After incubation at 24 °C for 24–30 h, the GFP signals were observed and imaged using ZEISS LSM 880 confocal microscope (Carl-Zeiss, Jena, Germany). MitoTracker Red (ThermoFisher Scientific, Waltham, MA, USA) was used as the mitochondrion marker with a working concentration of 100 nM.

### 4.3. Light Microscopy of Cytological Sections

*ppr18-1*/+ heterozygotes were identified by PCR with *Mu* primer TIR8 and gene primer *PPR18*-R1. Immature WT and *ppr18-1* kernels were harvested from the self-pollinated heterozygous segregating ear at 9 and 14 days after pollination (DAP). The fixed material and sections were performed as described previously [62]. Paraffin sections were stained with 1% Johansen's Safranin O and imaged with a stereo microscope (Carl-Zeiss, Jena, Germany).

### 4.4. RNA Extraction, RT-PCR, and qRT-PCR

Total RNA was extracted from fresh kernels by removing the pericarp using TRIzol reagent (ThermoFisher Scientific, Waltham, MA, USA). After DNaseI digestion (NEB) to eliminate DNA contamination, reverse transcription was conducted with random primers according to the manufacturer's instructions (TransGen Company, Beijing, China). Quantitative real time PCR (qRT-PCR) was performed with using LightCycler 96 (Roche, Basel, Switzerland) with three biological replicates. The maize actin gene *ZmActin* (GRMZM2G126010) was used as the reference gene. For functional analysis of *PPR18*, RT-PCR and qRT-PCR were performed with primers as previously described [45,49].

### 4.5. Measurements of Respiration Rate

The respiration rates were determined according to the previous report with some modifications [63]. The respiratory activities were measured in a reaction medium (50 mM phosphate buffer, pH 6.8) at 25 °C in the dark using a Chlorolab II liquid oxygen electrode (Hansatech, King's Lynn, UK, http://hansatech-instruments.com). The alternative pathway capacity ($V_{alt}$) and cytochrome pathway capacity ($V_{cyt}$) are defined as $O_2$ uptake rate in the presence of 2 mM potassium cyanide (KCN; Sigma-Aldrich, St. Louis, MO, USA, catalog number: 207810) and 2 mM salicylhydroxamic acid (SHAM; Sigma-Aldrich, St. Louis, MO, USA, catalog number: S607), respectively. All the respiration rates were indicated by the oxygen consumption of nmol $O_2$ min$^{-1}$ g$^{-1}$ fresh weight of the maize kernels.

### 4.6. Blue Native (BN)-PAGE and Complex I Activity Assay

Crude mitochondrial proteins were extracted from immature maize kernels with the pericarp removed at 11 DAP. Blue native (BN)-PAGE and measurement of NADH dehydrogenase activity were performed as previously described [64]. Of maize mitochondrial proteins 130 μg was separated by 3%–12.5% gradient gel (ThermoFisher Scientific, Waltham, MA, USA). The gel strips were stained by Coomassie Blue R-250 and assayed for complex I activity in nitroblue tetrazolium (NBT)-NADH buffer (25 mg of NBT, 100 μL of NADH (10 mg mL$^{-1}$), and 10 mL of 5 mM Tris–HCl, pH 7.4). The gel strips were transferred to the nitrocellulose membrane and Western blotting with specific antibodies cytochrome $c_1$ (Cyt$_{C1}$, a gift from G. Schatz, University of Basel, Switzerland), *Arabidopsis* Cox2 (Agrisera, Vännäs, Sweden, http://www.agrisera.com), and ATPase (ATPase α-subunit, MBL Beijing Biotech, China) for detection of complex III, VI, and V, respectively [64].

### 4.7. Immunoblot Analysis

Proteins extracted from immature maize kernels were determined by 12.5% SDS-PAGE and transferred to the Polyvinylidene Fluoride (PVDF, GE healthcare, Freiburg im Breisgau, Germany) membrane and Western blotting using antiserum against $Cyt_{C1}$ (1:5000), Cox2 (1:5000), ATPase (1:10,000), AOX (1:10,000) [25], and wheat Nad9 (1:3000) [65], as previously described [44]. The membrane was treated with ECL reagents (Pierce, ThermoFisher Scientific, Waltham, MA, USA). Signals were visualized on X-ray films (Kodak, Tokyo, Japan) and imaged using a Tanon-5200 system (Tanon, Shanghai, China).

### 4.8. Yeast Two-Hybrid Analysis

Yeast two-hybrid analysis was performed according to the manual of Matchmaker™ Gold Yeast Two-Hybrid System (Clontech, Mountain View, CA, USA). The coding sequence (CDS) without signal peptide sequences of PPR18, DEK35, EMP8, and EMP602 were cloned pGADT7 vector and pGBKT7 vector, respectively. The primer sequences are listed in Table S1. Combinations of plasmids were co-transformed into the yeast strain Y2H Gold (Clontech, Mountain View, CA, USA) and placed on SD/–Leu/–Trp (Minimal Media Double Dropouts, DDO) mediums and growth of diploid yeast colonies on SD/–Ade/–His/–Leu/–Trp (Minimal Media Quadruple Dropouts, QDO) mediums for 4 days at 30 °C to reveal protein–protein interactions.

### 4.9. Prediction of PPR18 Binding Site

PPR motif prediction alignment analysis of PPR18 protein was used by algorithm TPRpred (http://tprpred.tebingen.mpg.de/tprpred). The alignment of PPR18 to its *nad4* intron 1 binding site was generated as follows the PPR codes [29–31]. The recognition nucleotides for the PPR18 protein were predicted by the arrangements of the amino acids at position 5th and 35th of each PPR repeat from PPR18. The respective recognition nucleotides were listed and aligned with *nad4* intron 1 to find the best matched nucleotide sites as potential binding sites of PPR18 in mitochondrial *nad4* intron 1.

### 4.10. Phylogenetic Analysis

The full-length amino acid sequences of PPR18, the putative binding site of PPR18 in *nad4* intron 1, and their orthologs in plant species were downloaded from NCBI database (https://blast.ncbi.nlm.nih.gov). The phylogenetic tree was constructed using MEGA7 software by the maximum likelihood method [66].

## 5. Conclusions

In this study, we characterized a maize seed mutant *ppr18*, which exhibits an arrested embryo and endosperm development phenotype. Through a molecular characterization of the *PPR18* gene, we elucidated its function in the *cis*-splicing of *nad4* intron 1 in mitochondria and seed development in maize. The lack of splicing of *nad4* intron 1 results in the absence of *nad4* transcript, leading to severely reduced assembly and activity of mitochondrial complex I. The profiles of complex I assembly, activity, and component accumulation in the *ppr18* mutants shed lights to the assembly process of complex I in maize. Despite PPR proteins have been reported in intron splicing, our study provides additional information on a new PPR protein in intron splicing, complex I assembly, and its essential role in maize seed development.

**Supplementary Materials:** Figure S1: Phylogenetic analysis of PPR18 homologs, Figure S2: Phenotypes of the *ppr18-2* mutant and the ear of *ppr18-1* × *ppr18-2*, Figure S3: The linkage analysis of the *Mu* insertion in *PPR18-1* and the phenotype in the selfed progeny, Figure S4: Expression of *PPR18*, Figure S5: Transcript levels of the 35 mitochondrial genes in *ppr18* alleles in developing kernels, Figure S6: The predicted secondary structure of maize *nad4* intron 1 and the location of the putative binding site, Figure S7: Alignment of the putative binding site of PPR18 in *nad4* intron 1 in different plant species, Figure S8: Interaction assay of PPR18 and the related splicing factors, Table S1: Primers used in this study.

**Author Contributions:** R.L. and B.-C.T. conceived and designed the experiments. R.L., S.-K.C., A.S., F.S., and X.W. performed the experiments. R.L., S.-K.C., and B.-C.T. analyzed data and wrote the manuscript. C.X. contributed to reagents/materials/analysis tools. All authors have read and agreed to the published version of the manuscript.

**Acknowledgments:** We appreciate Tsuyoshi Nakagawa (Shimane University, Japan) for providing the pGWB vectors.

## References

1. Timmis, J.N.; Ayliffe, M.A.; Huang, C.Y.; Martin, W. Endosymbiotic gene transfer: Organelle genomes forge eukaryotic chromosomes. *Nat. Rev. Genet.* **2004**, *5*, 123–135. [CrossRef] [PubMed]
2. Unseld, M.; Marienfeld, J.R.; Brandt, P.; Brennicke, A. The mitochondrial genome of *Arabidopsis thaliana* contains 57 genes in 366,924 nucleotides. *Nat. Genet.* **1997**, *15*, 57–61. [CrossRef] [PubMed]
3. Notsu, Y.; Masood, S.; Nishikawa, T.; Kubo, N.; Akiduki, G.; Nakazono, M.; Hirai, A.; Kadowaki, K. The complete sequence of the rice (*Oryza sativa L.*) mitochondrial genome: Frequent DNA sequence acquisition and loss during the evolution of flowering plants. *Mol. Genet. Genomics* **2002**, *268*, 434–445. [CrossRef] [PubMed]
4. Clifton, S.W.; Minx, P.; Fauron, C.M.; Gibson, M.; Allen, J.O.; Sun, H.; Thompson, M.; Barbazuk, W.B.; Kanuganti, S.; Tayloe, C.; et al. Sequence and comparative analysis of the maize NB mitochondrial genome. *Plant Physiol.* **2004**, *136*, 3486–3503. [CrossRef]
5. Barkan, A.; Small, I. Pentatricopeptide repeat proteins in plants. *Annu. Rev. Plant Biol.* **2014**, *65*, 415–442. [CrossRef]
6. Delannoy, E.; Stanley, W.A.; Bond, C.S.; Small, I.D. Pentatricopeptide repeat (PPR) proteins as sequence-specificity factors in post-transcriptional processes in organelles. *Biochem. Soc. Trans.* **2007**, *35*, 1643–1647. [CrossRef]
7. Kramer, M.C.; Anderson, S.J.; Gregory, B.D. The nucleotides they are a-changin': Function of RNA binding proteins in post-transcriptional messenger RNA editing and modification in *Arabidopsis*. *Curr. Opin. Plant Biol.* **2018**, *45*, 88–95. [CrossRef]
8. Bonen, L. *Cis*- and *trans*-splicing of group II introns in plant mitochondria. *Mitochondrion* **2008**, *8*, 26–34. [CrossRef]
9. Lambowitz, A.M.; Zimmerly, S. Mobile group II introns. *Annu. Rev. Genet.* **2004**, *38*, 1–35. [CrossRef]
10. Cech, T.R. Self-splicing of group I introns. *Annu. Rev. Biochem.* **1990**, *59*, 543–568. [CrossRef]
11. Zoschke, R.; Nakamura, M.; Liere, K.; Sugiura, M.; Borner, T.; Schmitz-Linneweber, C. An organellar maturase associates with multiple group II introns. *Proc. Natl. Acad. Sci. USA* **2010**, *107*, 3245–3250. [CrossRef] [PubMed]
12. Vogel, J.; Börner, T.; Hess, W.R. Comparative analysis of splicing of the complete set of chloroplast group II introns in three higher plant mutants. *Nucleic Acids Res.* **1999**, *27*, 3866–3874. [CrossRef] [PubMed]
13. Cohen, S.; Zmudjak, M.; Colas des Francs-Small, C.; Malik, S.; Shaya, F.; Keren, I.; Belausov, E.; Many, Y.; Brown, G.G.; Small, I.; et al. nMAT4, a maturase factor required for *nad1* pre-mRNA processing and maturation, is essential for holocomplex I biogenesis in *Arabidopsis* mitochondria. *Plant J.* **2014**, *78*, 253–268. [CrossRef] [PubMed]
14. Keren, I.; Tal, L.; des Francs-Small, C.C.; Araujo, W.L.; Shevtsov, S.; Shaya, F.; Fernie, A.R.; Small, I.; Ostersetzer-Biran, O. nMAT1, a nuclear-encoded maturase involved in the *trans*-splicing of *nad1* intron 1, is essential for mitochondrial complex I assembly and function. *Plant J.* **2012**, *71*, 413–426. [CrossRef]
15. Keren, I.; Bezawork-Geleta, A.; Kolton, M.; Maayan, I.; Belausov, E.; Levy, M.; Mett, A.; Gidoni, D.; Shaya, F.; Ostersetzer-Biran, O. AtnMat2, a nuclear-encoded maturase required for splicing of group-II introns in *Arabidopsis* mitochondria. *RNA* **2009**, *15*, 2299–2311. [CrossRef]
16. Till, B.; Schmitz-Linneweber, C.; Williams-Carrier, R.; Barkan, A. CRS1 is a novel group II intron splicing factor that was derived from a domain of ancient origin. *RNA* **2001**, *7*, 1227–1228. [CrossRef]
17. Zmudjak, M.; Colas des Francs-Small, C.; Keren, I.; Shaya, F.; Belausov, E.; Small, I.; Ostersetzer-Biran, O. mCSF1, a nucleus-encoded CRM protein required for the processing of many mitochondrial introns, is involved in the biogenesis of respiratory complexes I and IV in *Arabidopsis*. *New Phytol.* **2013**, *199*, 379–394. [CrossRef]

18. Asakura, Y.; Galarneau, E.; Watkins, K.P.; Barkan, A.; van Wijk, K.J. Chloroplast RH3 DEAD box RNA helicases in maize and *Arabidopsis* function in splicing of specific group II introns and affect chloroplast ribosome biogenesis. *Plant Physiol.* **2012**, *159*, 961–974. [CrossRef]
19. Bobik, K.; McCray, T.N.; Ernest, B.; Fernandez, J.C.; Howell, K.A.; Lane, T.; Staton, M.; Burch-Smith, T.M. The chloroplast RNA helicase ISE2 is required for multiple chloroplast RNA processing steps in *Arabidopsis thaliana*. *Plant J.* **2017**, *91*, 114–131. [CrossRef]
20. Hammani, K.; Barkan, A. An mTERF domain protein functions in group II intron splicing in maize chloroplasts. *Nucleic Acids Res.* **2014**, *42*, 5033–5042. [CrossRef]
21. Hsu, Y.W.; Wang, H.J.; Hsieh, M.H.; Hsieh, H.L.; Jauh, G.Y. *Arabidopsis* mTERF15 is required for mitochondrial *nad2* intron 3 splicing and functional complex I activity. *PLoS ONE* **2014**, *9*, e112360. [CrossRef] [PubMed]
22. Colas des Francs-Small, C.; Kroeger, T.; Zmudjak, M.; Ostersetzer-Biran, O.; Rahimi, N.; Small, I.; Barkan, A. A PORR domain protein required for *rpl2* and *ccmF*$_C$ intron splicing and for the biogenesis of *c*-type cytochromes in *Arabidopsis* mitochondria. *Plant J.* **2012**, *69*, 996–1005. [CrossRef] [PubMed]
23. Kroeger, T.S.; Watkins, K.P.; Friso, G.; van Wijk, K.J.; Barkan, A. A plant-specific RNA-binding domain revealed through analysis of chloroplast group II intron splicing. *Proc. Natl. Acad. Sci. USA* **2009**, *106*, 4537–4542. [CrossRef] [PubMed]
24. Kuhn, K.; Carrie, C.; Giraud, E.; Wang, Y.; Meyer, E.H.; Narsai, R.; des Francs-Small, C.C.; Zhang, B.; Murcha, M.W.; Whelan, J. The RCC1 family protein RUG3 is required for splicing of *nad2* and complex I biogenesis in mitochondria of *Arabidopsis thaliana*. *Plant J.* **2011**, *67*, 1067–1080. [CrossRef] [PubMed]
25. Xiu, Z.; Sun, F.; Shen, Y.; Zhang, X.; Jiang, R.; Bonnard, G.; Zhang, J.; Tan, B.C. EMPTY PERICARP16 is required for mitochondrial *nad2* intron 4 *cis*-splicing, complex I assembly and seed development in maize. *Plant J.* **2016**, *85*, 507–519. [CrossRef]
26. Schmitz-Linneweber, C.; Williams-Carrier, R.E.; Williams-Voelker, P.M.; Kroeger, T.S.; Vichas, A.; Barkan, A. A pentatricopeptide repeat protein facilitates the *trans*-splicing of the maize chloroplast *rps12* pre-mRNA. *Plant Cell* **2006**, *18*, 2650–2663. [CrossRef]
27. Lurin, C.; Andres, C.; Aubourg, S.; Bellaoui, M.; Bitton, F.; Bruyere, C.; Caboche, M.; Debast, C.; Gualberto, J.; Hoffmann, B.; et al. Genome-wide analysis of *Arabidopsis* pentatricopeptide repeat proteins reveals their essential role in organelle biogenesis. *Plant Cell* **2004**, *16*, 2089–2103. [CrossRef]
28. Fujii, S.; Small, I. The evolution of RNA editing and pentatricopeptide repeat genes. *New Phytol.* **2011**, *191*, 37–47. [CrossRef]
29. Barkan, A.; Rojas, M.; Fujii, S.; Yap, A.; Chong, Y.S.; Bond, C.S.; Small, I. A combinatorial amino acid code for RNA recognition by pentatricopeptide repeat proteins. *PLoS Genet.* **2012**, *8*, e1002910. [CrossRef]
30. Gully, B.S.; Cowieson, N.; Stanley, W.A.; Shearston, K.; Small, I.D.; Barkan, A.; Bond, C.S. The solution structure of the pentatricopeptide repeat protein PPR10 upon binding *atpH* RNA. *Nucleic Acids Res.* **2015**, *43*, 1918–1926. [CrossRef]
31. Yin, P.; Li, Q.; Yan, C.; Liu, Y.; Liu, J.; Yu, F.; Wang, Z.; Long, J.; He, J.; Wang, H.W.; et al. Structural basis for the modular recognition of single-stranded RNA by PPR proteins. *Nature* **2013**, *504*, 168–171. [CrossRef] [PubMed]
32. Burger, G.; Gray, M.W.; Lang, B.F. Mitochondrial genomes: Anything goes. *Trends Genet.* **2003**, *19*, 709–716. [CrossRef] [PubMed]
33. Berrisford, J.M.; Sazanov, L.A. Structural basis for the mechanism of respiratory complex I. *J. Biol. Chem.* **2009**, *284*, 29773–29783. [CrossRef] [PubMed]
34. Ren, X.; Pan, Z.; Zhao, H.; Zhao, J.; Cai, M.; Li, J.; Zhang, Z.; Qiu, F. EMPTY PERICARP11 serves as a factor for splicing of mitochondrial *nad1* intron and is required to ensure proper seed development in maize. *J. Exp. Bot.* **2017**, *68*, 4571–4581. [CrossRef]
35. Qi, W.; Yang, Y.; Feng, X.; Zhang, M.; Song, R. Mitochondrial function and maize kernel development requires Dek2, a pentatricopeptide repeat protein involved in *nad1* mRNA splicing. *Genetics* **2017**, *205*, 239–249. [CrossRef]
36. Sun, F.; Xiu, Z.; Jiang, R.; Liu, Y.; Zhang, X.; Yang, Y.Z.; Li, X.; Zhang, X.; Wang, Y.; Tan, B.C. The mitochondrial pentatricopeptide repeat protein EMP12 is involved in the splicing of three *nad2* introns and seed development in maize. *J. Exp. Bot.* **2019**, *70*, 963–972. [CrossRef]

37. Yang, Y.Z.; Ding, S.; Wang, Y.; Wang, H.C.; Liu, X.Y.; Sun, F.; Xu, C.; Liu, B.; Tan, B.C. PPR20 is required for the *cis*-splicing of mitochondrial *nad2* intron 3 and seed development in maize. *Plant Cell Physiol.* **2020**, *61*, 370–380. [CrossRef]
38. Dai, D.; Luan, S.; Chen, X.; Wang, Q.; Feng, Y.; Zhu, C.; Qi, W.; Song, R. Maize *Dek37* encodes a P-type PPR protein that affects *cis*-splicing of mitochondrial *nad2* intron 1 and seed development. *Genetics* **2018**, *208*, 1069–1082. [CrossRef]
39. Cai, M.; Li, S.; Sun, F.; Sun, Q.; Zhao, H.; Ren, X.; Zhao, Y.; Tan, B.C.; Zhang, Z.; Qiu, F. *Emp10* encodes a mitochondrial PPR protein that affects the *cis*-splicing of *nad2* intron 1 and seed development in maize. *Plant J.* **2017**, *91*, 132–144. [CrossRef]
40. Ding, Y.H.; Liu, N.Y.; Tang, Z.S.; Liu, J.; Yang, W.C. *Arabidopsis* GLUTAMINE-RICH PROTEIN23 is essential for early embryogenesis and encodes a novel nuclear PPR motif protein that interacts with RNA polymerase II subunit III. *Plant Cell* **2006**, *18*, 815–830. [CrossRef]
41. Hammani, K.; Gobert, A.; Hleibieh, K.; Choulier, L.; Small, I.; Giegé, P. An *Arabidopsis* dual-Localized pentatricopeptide repeat protein interacts with nuclear proteins involved in gene expression regulation. *Plant Cell* **2011**, *23*, 730–740. [CrossRef] [PubMed]
42. McCarty, D.R.; Settles, A.M.; Suzuki, M.; Tan, B.C.; Latshaw, S.; Porch, T.; Robin, K.; Baier, J.; Avigne, W.; Lai, J.; et al. Steady-state transposon mutagenesis in inbred maize. *Plant J.* **2005**, *44*, 52–61. [CrossRef] [PubMed]
43. Tan, B.C.; Chen, Z.; Shen, Y.; Zhang, Y.; Lai, J.; Sun, S.S. Identification of an active new mutator transposable element in maize. *G3 (Bethesda)* **2011**, *1*, 293–302. [CrossRef] [PubMed]
44. Sun, F.; Wang, X.; Bonnard, G.; Shen, Y.; Xiu, Z.; Li, X.; Gao, D.; Zhang, Z.; Tan, B.C. *Empty pericarp7* encodes a mitochondrial E-subgroup pentatricopeptide repeat protein that is required for $ccmF_N$ editing, mitochondrial function and seed development in maize. *Plant J.* **2015**, *84*, 283–295. [CrossRef] [PubMed]
45. Li, X.J.; Zhang, Y.F.; Hou, M.; Sun, F.; Shen, Y.; Xiu, Z.H.; Wang, X.; Chen, Z.L.; Sun, S.S.; Small, I.; et al. *Small kernel 1* encodes a pentatricopeptide repeat protein required for mitochondrial *nad7* transcript editing and seed development in maize (*Zea mays*) and rice (*Oryza sativa*). *Plant J.* **2014**, *79*, 797–809. [CrossRef] [PubMed]
46. Sun, F.; Zhang, X.; Shen, Y.; Wang, H.; Liu, R.; Wang, X.; Gao, D.; Yang, Y.Z.; Liu, Y.; Tan, B.C. The pentatricopeptide repeat protein EMPTY PERICARP8 is required for the splicing of three mitochondrial introns and seed development in maize. *Plant J.* **2018**, *95*, 919–932. [CrossRef] [PubMed]
47. Acin-Perez, R.; Fernandez-Silva, P.; Peleato, M.L.; Perez-Martos, A.; Enriquez, J.A. Respiratory active mitochondrial supercomplexes. *Mol. Cell* **2008**, *32*, 529–539. [CrossRef]
48. Millar, A.H.; Whelan, J.; Soole, K.L.; Day, D.A. Organization and regulation of mitochondrial respiration in plants. *Annu. Rev. Plant Biol.* **2011**, *62*, 79–104. [CrossRef]
49. Chen, X.; Feng, F.; Qi, W.; Xu, L.; Yao, D.; Wang, Q.; Song, R. *Dek35* encodes a PPR protein that affects *cis*-splicing of mitochondrial *nad4* intron 1 and seed development in maize. *Mol. Plant* **2017**, *10*, 427–441. [CrossRef]
50. Ren, Z.; Fan, K.; Fang, T.; Zhang, J.; Yang, L.; Wang, J.; Wang, G.; Liu, Y. Maize *empty pericarp602* encodes a P-type PPR protein that is essential for seed development. *Plant Cell Physiol.* **2019**, *60*, 1734–1746. [CrossRef]
51. Ligas, J.; Pineau, E.; Bock, R.; Huynen, M.A.; Meyer, E.H. The assembly pathway of complex I in *Arabidopsis thaliana*. *Plant J.* **2019**, *97*, 447–459. [CrossRef] [PubMed]
52. Zhu, C.; Jin, G.; Fang, P.; Zhang, Y.; Feng, X.; Tang, Y.; Qi, W.; Song, R. Maize pentatricopeptide repeat protein DEK41 affects *cis*-splicing of mitochondrial *nad4* intron 3 and is required for normal seed development. *J. Exp. Bot.* **2019**, *70*, 3795–3808. [CrossRef] [PubMed]
53. Ren, R.C.; Wang, L.L.; Zhang, L.; Zhao, Y.J.; Wu, J.W.; Wei, Y.M.; Zhang, X.S.; Zhao, X.Y. DEK43 is a P-type PPR protein responsible for the *cis*-splicing of *nad4* in maize mitochondria. *J. Integr. Plant Biol.* **2020**, *62*, 299–313. [CrossRef] [PubMed]
54. De Longevialle, A.F.; Meyer, E.H.; Andres, C.; Taylor, N.L.; Lurin, C.; Millar, A.H.; Small, I.D. The pentatricopeptide repeat gene OTP43 is required for *trans*-splicing of the mitochondrial *nad1* intron 1 in *Arabidopsis thaliana*. *Plant Cell* **2007**, *19*, 3256–3265. [CrossRef]
55. Karpova, O.V.; Kuzmin, E.V.; Elthon, T.E.; Newton, K.J. Differential expression of alternative oxidase genes in maize mitochondrial mutants. *Plant Cell* **2002**, *14*, 3271–3284. [CrossRef]

56. Toda, T.; Fujii, S.; Noguchi, K.; Kazama, T.; Toriyama, K. Rice *MPR25* encodes a pentatricopeptide repeat protein and is essential for RNA editing of *nad5* transcripts in mitochondria. *Plant J.* **2012**, *72*, 450–460. [CrossRef]
57. Brangeon, J.; Sabar, M.; Gutierres, S.; Combettes, B.; Bove, J.; Gendy, C.; Chétrit, P.; Des Francs-Small, C.C.; Pla, M.; Vedel, F.; et al. Defective splicing of the first *nad4* intron is associated with lack of several complex I subunits in the *Nicotiana sylvestris* NMS1 nuclear mutant. *Plant J.* **2000**, *21*, 269–280. [CrossRef]
58. Barkan, A. Intron splicing in plant organelles. *Mol. Biol. Biotechnol. Plant Organelles* **2004**, *11*, 295–322.
59. Brown, G.G.; Colas des Francs-Small, C.; Ostersetzer-Biran, O. Group II intron splicing factors in plant mitochondria. *Front. Plant Sci.* **2014**, *5*, 35. [CrossRef]
60. Chen, Z.; Wang, H.C.; Shen, J.; Sun, F.; Wang, M.; Xu, C.; Tan, B.C. PPR-SMR1 is required for the splicing of multiple mitochondrial introns, interacts with Zm-mCSF1, and is essential for seed development in maize. *J. Exp. Bot.* **2019**, *70*, 5245–5258. [CrossRef]
61. Shen, Y.; Li, C.; McCarty, D.R.; Meeley, R.; Tan, B.C. *Embryo defective12* encodes the plastid initiation factor 3 and is essential for embryogenesis in maize. *Plant J.* **2013**, *74*, 792–804. [CrossRef] [PubMed]
62. Zhang, Y.F.; Hou, M.M.; Tan, B.C. The requirement of WHIRLY1 for embryogenesis is dependent on genetic background in maize. *PLoS ONE* **2013**, *8*, e67369. [CrossRef] [PubMed]
63. Wang, X.M.; Chang, N.; Bi, Y.R.; Tan, B.C. Measurement of mitochondrial respiration rate in maize (*Zea mays*) leaves. *Bio-Protocol* **2015**, *5*, e1483. [CrossRef]
64. Meyer, E.H.; Tomaz, T.; Carroll, A.J.; Estavillo, G.; Delannoy, E.; Tanz, S.K.; Small, I.D.; Pogson, B.J.; Millar, A.H. Remodeled respiration in *ndufs4* with low phosphorylation efficiency suppresses *Arabidopsis* germination and growth and alters control of metabolism at night. *Plant Physiol.* **2009**, *151*, 603–619. [CrossRef] [PubMed]
65. Lamattina, L.; Gonzalez, D.; Gualberto, J.; Grienenberger, J.M. Higher plant mitochondria encode an homologue of the nuclear-encoded 30-kDa subunit of bovine mitochondrial complex I. *Eur. J. Biochem.* **1993**, *217*, 831–838. [CrossRef] [PubMed]
66. Kumar, S.; Stecher, G.; Tamura, K. MEGA7: Molecular evolutionary genetics analysis version 7.0 for bigger datasets. *Mol. Biol. Evol.* **2016**, *33*, 1870–1874. [CrossRef]

# 3

# Mitochondrial Transcriptome Control and Intercompartment Cross-Talk during Plant Development

**Adnan Khan Niazi [1,2,†], Etienne Delannoy [3,†], Rana Khalid Iqbal [1,†], Daria Mileshina [1], Romain Val [1], Marta Gabryelska [4], Eliza Wyszko [4], Ludivine Soubigou-Taconnat [3], Maciej Szymanski [5], Jan Barciszewski [4,6], Frédérique Weber-Lotfi [1], José Manuel Gualberto [1] and André Dietrich [1,*]**

[1] Institute of Plant Molecular Biology (IBMP), CNRS and University of Strasbourg, 12 rue du Général Zimmer, 67084 Strasbourg, France; adnan1753@yahoo.com (A.K.N.); rkiqbal@etu.unistra.fr (R.K.I.); d.mileshina@gmail.com (D.M.); r.val@arvalis.fr (R.V.); lotfif@unistra.fr (F.W.-L.); jose.gualberto@ibmp-cnrs.unistra.fr (J.M.G.)
[2] Centre of Agricultural Biochemistry and Biotechnology (CABB), University of Agriculture, Faisalabad 38000, Pakistan
[3] Institute of Plant Sciences Paris-Saclay IPS2, CNRS, INRA, Université Paris-Sud, Université Evry, Université Paris-Saclay, Paris Diderot, Sorbonne Paris-Cité, 91405 Orsay, France; etienne.delannoy@inra.fr (E.D.); ludivine.soubigou-taconnat@inra.fr (L.S.-T.)
[4] Institute of Bioorganic Chemistry, Polish Academy of Sciences, Ul. Z. Noskowskiego 12/14, 61-704 Poznan, Poland; marta.gabryelska@gmail.com (M.G.); eliza.wyszko@ibch.poznan.pl (E.W.); Jan.Barciszewski@ibch.poznan.pl (J.B.)
[5] Department of Computational Biology, Institute of Molecular Biology and Biotechnology, A. Mickiewicz University Poznan, Ul. Umultowska 89, 61-614 Poznan, Poland; mszyman@amu.edu.pl
[6] NanoBioMedical Centre of the Adam Mickiewicz University, Umultowska 85, 61614 Poznan, Poland
* Correspondence: diandid.5926@orange.fr
† These authors contributed equally to this project.

**Abstract:** We address here organellar genetic regulation and intercompartment genome coordination. We developed earlier a strategy relying on a tRNA-like shuttle to mediate import of nuclear transgene-encoded custom RNAs into mitochondria in plants. In the present work, we used this strategy to drive *trans*-cleaving hammerhead ribozymes into the organelles, to knock down specific mitochondrial RNAs and analyze the regulatory impact. In a similar approach, the tRNA mimic was used to import into mitochondria in *Arabidopsis thaliana* the *orf77*, an RNA associated with cytoplasmic male sterility in maize and possessing sequence identities with the *atp9* mitochondrial RNA. In both cases, inducible expression of the transgenes allowed to characterise early regulation and signaling responses triggered by these respective manipulations of the organellar transcriptome. The results imply that the mitochondrial transcriptome is tightly controlled by a "buffering" mechanism at the early and intermediate stages of plant development, a control that is released at later stages. On the other hand, high throughput analyses showed that knocking down a specific mitochondrial mRNA triggered a retrograde signaling and an anterograde nuclear transcriptome response involving a series of transcription factor genes and small RNAs. Our results strongly support transcriptome coordination mechanisms within the organelles and between the organelles and the nucleus.

**Keywords:** anterograde regulation; cytoplasmic male sterility (CMS); plant mitochondria; retrograde regulation; ribozyme; RNA trafficking; signaling

---

## 1. Introduction

Mitochondria massively import proteins encoded by the nuclear genome, but still contain their own genetic system, distinct from that of the nucleo-cytosolic compartment. Mammalian mitochondria possess an extremely compact 16.5 kb circular genome with a single non-coding region that carries the promoters for transcription in both orientations [1]. In plants, mitochondrial genomes are large (usually 200–700 kb) and have a low gene density. Nevertheless, due to the occurrence of multiple promoters spread all along, they are still almost entirely transcribed, including the large non-coding regions [2]. The mitochondrial DNA (mtDNA) encodes essential subunits of the different oxidative phosphorylation (OXPHOS) complexes and one can thus presume that the mitochondrial transcriptome needs to be tightly kept under control. Studies on the regulation of mammalian mitochondrial gene expression so far highlighted a major role of post-transcriptional processes, including coordination of mitochondrial and cytosolic translation [3–6]. Translation regulation and efficiency, rather than mtDNA copy number or transcription rate, were proposed to be crucial for the synthesis of the OXPHOS subunits [7]. Methylation of the mtDNA and epigenetic-like modifications of the mitochondrial transcription factor A (TFAM) were suggested to participate in establishing the mitochondrial gene expression profiles [8]. Known nuclear gene expression regulators turned out to be also involved in mtDNA transcription [9]. Conversely, models for mtDNA transcriptional regulation by regular nuclear-encoded mitochondrial factors were proposed [1]. In plant mitochondria as well, the major genetic control points are generally considered to be post-transcriptional, although there is some transcriptional control of gene expression [4,10]. A number of processes driven by nuclear-encoded protein factors, such as 5′- and 3′-end maturation of transcripts, intron splicing and RNA editing are all potential sites of regulation [11].

As a further level of complexity, mitochondrial biogenesis and functioning involve both nuclear gene expression and mitochondrial gene expression, especially to build the protein complexes of the electron transport chain (ETC). This is likely to require a coordinated communication between the organelles and the nucleus [12]. Intercompartment coordination mechanisms include both anterograde (nucleus to organelle) and retrograde (organelle to nucleus) signals [13]. Anterograde mechanisms control organelle gene expression in response to endogenous or exogenous signals that are perceived by the nucleus. Indeed, expression of the mitochondrial genome, from DNA replication and transcription to transcript processing, editing and translation, is entirely carried out by proteins encoded by nuclear-located genes [3,14]. Retrograde regulation refers to signals sent by the organelles to communicate their functional and developmental state to the nucleus, which can then modulate anterograde control and cellular metabolism accordingly [11,15]. Mitochondrial retrograde regulation (MRR) is an important mechanism of communication between mitochondria and the nucleus and is conserved among yeasts, mammals and plants. However, the signaling molecules and the signal transduction mechanisms are diverse [15].

The present work aims to further explore these pathways with a special interest in mitochondrial genetic expression control and signaling during development. Plants offer in this respect a relevant and easily accessible working model. Establishment of the plant mitochondrial transcriptome has been documented in detail along seed germination in *A. thaliana*, but little is known about the spatial and temporal control of the organellar transcripts during post-germination development [16–18]. Our experiments were based on the analysis and manipulation of the organellar transcriptome in *Arabidopsis thaliana* through a novel approach set up in our laboratory and based on the existence of a natural process of transfer RNA (tRNA) import from the cytosol into mitochondria in plant cells (reviewed in [19]). Circumventing the lack of methodologies for mitochondrial transformation, we showed previously that a tRNA mimic can be used in vivo as a shuttle for importing into plant mitochondria cargo RNAs expressed from nuclear transgenes [20]. Taking *trans*-cleaving hammerhead ribozymes as cargo sequences allowed to specifically knock down mitochondrial RNAs and demonstrate that the mt-DNA-encoded MATR protein is a mitochondrial maturase involved in the splicing of multiple organellar introns [20,21]. We now used tRNA mimic-mediated shuttling of cargo RNAs to affect the mitochondrial transcriptome in *A. thaliana* plants at different developmental stages, so as to investigate the resulting regulation processes. Two strategies were applied. First, individual mitochondrial mRNAs encoding subunits of OXPHOS complexes were chosen as targets for specific *trans*-cleaving hammerhead ribozymes driven into the organelles by the tRNA mimic. In a second strategy, the tRNA mimic was used to drive into mitochondria a sequence associated with cytoplasmic male sterility (CMS). The results altogether imply that the mitochondrial transcriptome is tightly controlled by a buffering mechanism at early and intermediate stages of development. On the other hand, knocking down a major mitochondrial mRNA triggered a retrograde signaling and an anterograde response from the nucleus. Whereas it is currently considered that mitochondrial regulation processes occur essentially at the post-transcriptional stage (see above), our results support mRNA coordination mechanisms within the organelles and between the organelles and the nucleus.

## 2. Materials and Methods

### 2.1. Preparation of Gene Constructs for In Vivo Expression

For the catalytic RNA-mediated knockdown strategy, the complete sequences encoding the *trans*-ribozymes, the linker L, the *Turnip yellow mosaic virus* (TYMV) PKTLS import shuttle and the *Hepatitis delta virus* (HDV) antigenomic *cis*-cleaving ribozyme (cHDV) were assembled by polymerase chain reaction (PCR) amplification from the pCK-PSTYPKTLScHDV plasmid [20]. Direct megaprimers contained a 5′ *Xho*I site followed by the desired ribozyme and linker sequences upstream of a sequence corresponding to the 5′ region of the PKTLS. The reverse primer contained a 5′ *Spe*I site followed by a sequence annealing to the 3′ region of the cHDV [22]. The obtained PCR products were cloned into the pGEM-T vector (Promega, Madison, WI, USA) for sequencing, re-excized with the *Xho*I and *Spe*I restriction enzymes and cloned into the *Xho*I and *Spe*I sites of the inducible transcription unit ($O_{LexA}$-46 estradiol-inducible promoter) of the pER8 vector [23], raising a series of pER8-Rzxxxx-L-PKTLS-cHDV plasmids. These were used to transform *A. thaliana* or *Nicotiana tabacum*. Sequence details can be found in Figure 1. As previously [20,21], the linker (L) to provide spacing between the *trans*-ribozyme and PKTLS motifs was selected from a pool of random sequences, based on MFOLD [24] predictions of weak RNA secondary structures unlikely to interfere either with PKTLS formation or *trans*-ribozyme binding to the target sequence.

For the CMS RNA strategy, a PCR product comprising the complete *orf77* sequence with four nucleotides of 5′-UTR and 92 nucleotides of 3′-UTR was amplified using total DNA from CMS-S maize as a template. The forward primer comprised a *Hind*III site and a sequence encoding a 6xHis tag. The reverse primer had a *Bam*HI site. The resulting product was cloned into the *Hind*III and *Bam*HI sites upstream of the sequences encoding the TYMV PKTLS and the HDV *cis*-ribozyme in an available derivative of the pUCAP plasmid [25] carrying these sequences. After sequencing of the recombinant plasmids, the assembled *orf77*-PKTLS-HDV construct was re-excized upon digestion with *Asc*I and *Pac*I and cloned into the *Asc*I and *Pac*I restriction sites of the estradiol-inducible transcription unit of the pER8 vector [23]. The resulting plasmid called pER8-*orf77*-PKTLS-cHDV was used for transformation of *A. thaliana*. For control, plants were transformed with empty pER8 plasmid. Sequence details can be found in Figure 2.

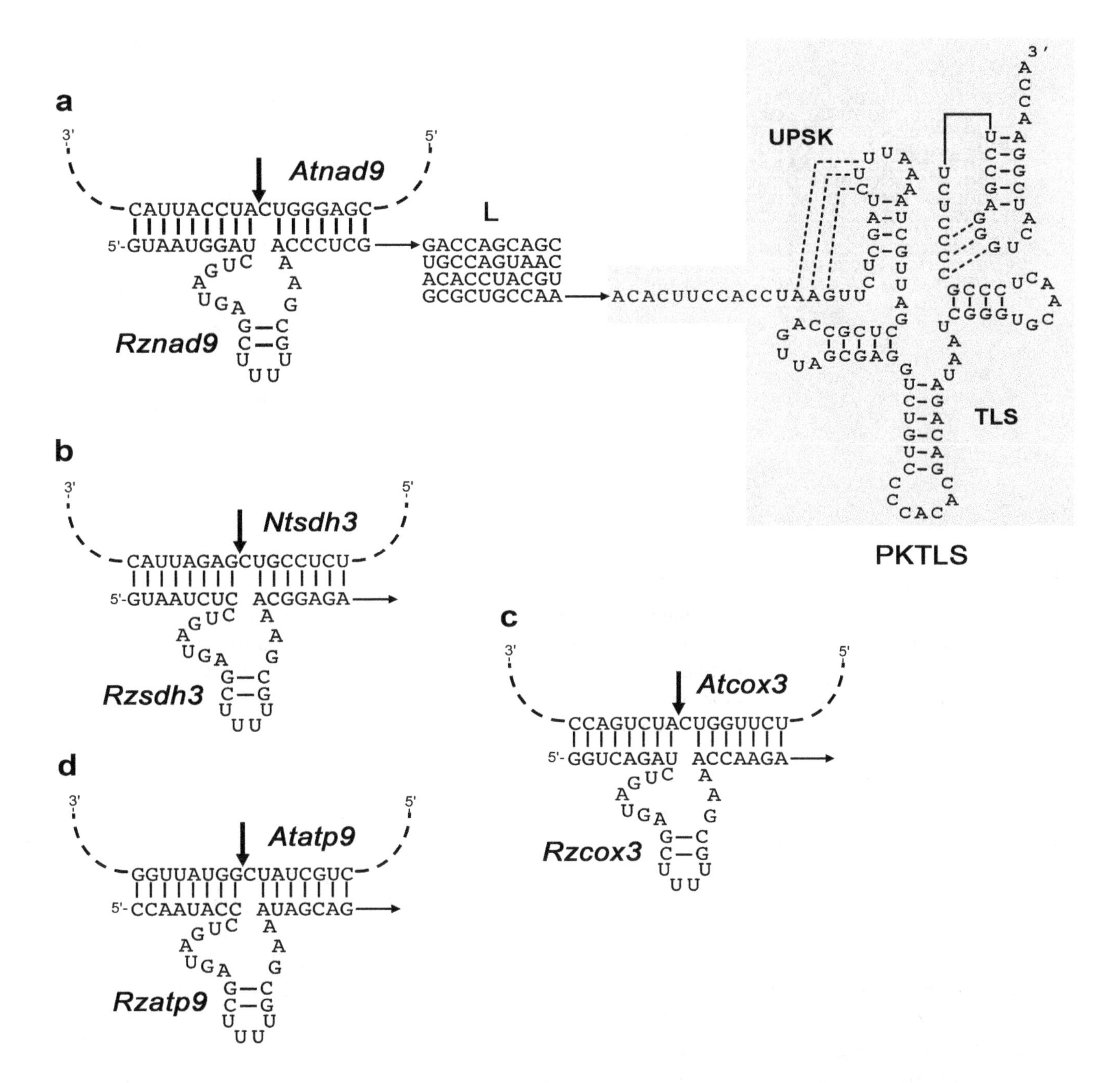


**Figure 1.** Structure of the ribozyme-PKTLS chimeric RNAs targeted to mitochondria in the present study. (**a**) Structure of the *Rznad9*-L-PKTLS RNA. The *trans*-cleaving ribozyme *Rznad9* directed against the *A. thaliana* mitochondrial *nad9* mRNA (*Atnad9*) is attached to the 5′-end of the PKTLS shuttle via a 40 nucleotide linker (L) selected from a pool of random sequences. The *Rznad9* hammerhead sequence is annealed to its target sequence motif in the *Atnad9* mRNA. The *Rzsdh3* (**b**), *Rzcox3* (**c**) and *Rzatp9* (**d**) ribozymes, directed against the *N. tabacum sdh3* (*Ntsdh3*) mRNA, the *A. thaliana cox3* (*Atcox3*) mRNA and the *A. thaliana atp9* (*Atatp9*) mRNA are attached to the same L-PKTLS moiety to generate the *Rzsdh3*-L-PKTLS, *Rzcox3*-L-PKTLS and *Rzatp9*-L-PKTLS, respectively. Ribozyme cleavage sites are indicated in the target sequence motifs by thick arrows. Their precise location is after position 421 in the *A. thaliana nad9* coding sequence (24662-25234 in accession JF729201), after position 216 in the *N. tabacum sdh3* coding sequence (77198-77524 in accession BA000042, complementary strand), after position 685 in the *A. thaliana cox3* coding sequence (328926-329723 in accession JF729201, complementary strand), and after position 99 in the *A. thaliana atp9* coding sequence (269920-270177 in accession JF729201, complementary strand). The *Rzsdh3* and *Rzatp9* ribozymes are as described earlier in Sultan et al. [21] and Val et al. [20], respectively.

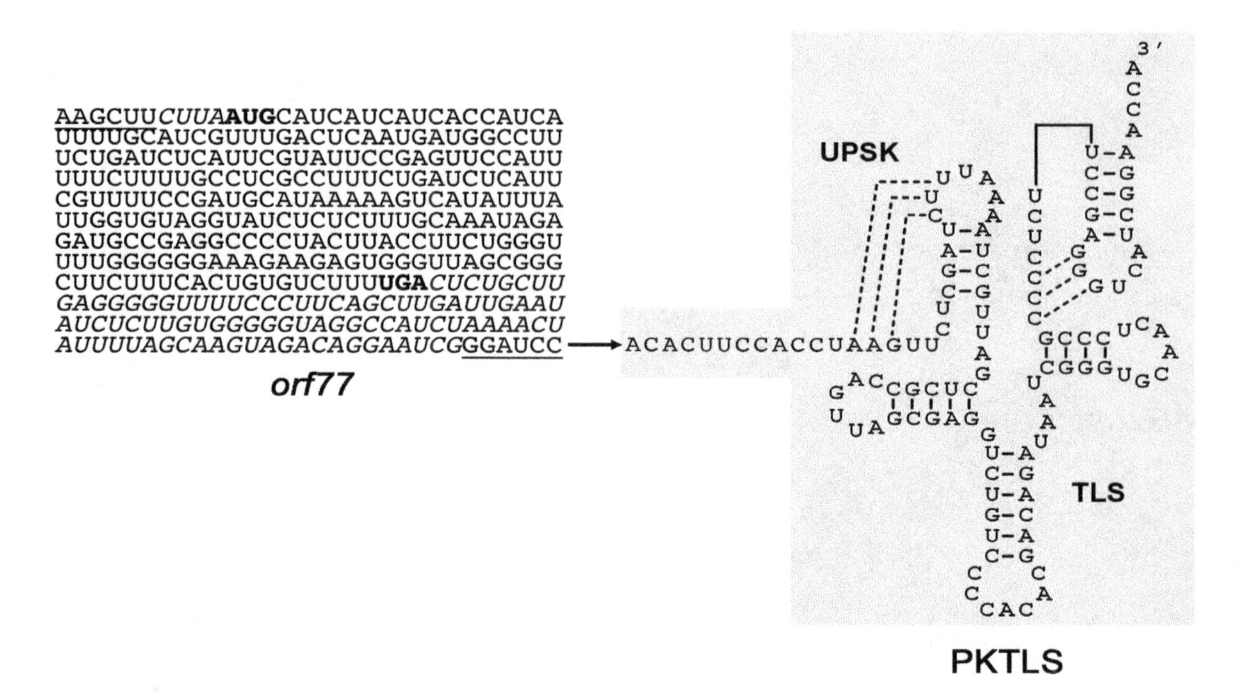


**Figure 2.** Structure of the *orf77*-PKTLS RNA targeted to mitochondria in the present study. The *Zea mays* CMS-S-specific *orf77* coding sequence (initiation codon and termination codon in bold) with its four nucleotide upstream and 92 nucleotide downstream sequences (italics) is directly attached to the 5′-end of the PKTLS shuttle. Short additional sequences deriving from the *Hind*III and *Bam*HI restriction sites introduced for cloning purposes are underlined. The chimeric RNA was expressed from a nuclear transgene and driven into the organelles by the PKTLS moiety.

### *2.2. Nuclear Transformations*

*A. thaliana* plants (ecotype Col-0) were transformed via *Agrobacterium tumefaciens* through floral dip [26]. Seeds from transformed plants were recovered and germinated on hygromycin medium under long day conditions (16 h light/8 h dark) for transformant selection. The presence and integrity of the constructs was confirmed by PCR and homozygous lines of the relevant transformants were selected.

To generate tobacco transformants, leaf discs from *N. tabacum* plants grown in vitro on germination medium (4.3 g/L DUCHEFA micro- and macroelements M0238, 10 g/L agar, 1% *w*/*v* sucrose, pH 5.7) were exposed to the relevant recombinant *A. tumefaciens* suspension and subsequently cultured in vitro on coculture medium (4.3 g/L DUCHEFA micro- and macroelements M0238, 825 mg/L $NH_4NO_3$, 2 mg/mL glycine, 100 mg/L myo-inositol, 0.5 mg/L nicotinic acid, 0.5 mg/L putrescine, 0.1 mg/L thiamine, 3% *w*/*v* sucrose, 0,8% *w*/*v* bacto-agar, 2 mg/mL benzylaminopurine (BAP), 0.05 mg/L naphthaleneacetic acid (NAA), pH 5.8) for 3–4 days at 28 °C (12 h light/12 h dark). The leaf discs were then transferred to selection medium (4.3 g/L DUCHEFA micro- and macroelements M0238, 825 mg/L $NH_4NO_3$, 2 mg/mL glycine, 100 mg/L myo-inositol, 0.5 mg/L nicotinic acid, 0.5 mg/L putrescine, 0.1 mg/L thiamine, 3% *w*/*v* sucrose, 0,8% *w*/*v* bacto-agar, 350 mg/L cefotaxime, 20 mg/L hygromycin, pH 5.8) at 25 °C (12 h light/12 h dark). The medium was changed every two to three weeks. Calli appearing on the edges were transferred to individual pots with selection medium when young leaves were formed, and planted into mineral-rich soil when roots had emerged. After acclimatization in growth chambers (25 °C, 12 h light/12 h dark) the plantlets were grown in the greenhouse. Recovered seeds were further selected through germination on hygromycin medium.

### *2.3. Seedling Growth and Trangene Induction*

*A. thaliana* or *N. tabacum* transgenic seedlings were grown under long day light conditions (16h light/8h dark) or in the dark on solid agar medium. For transgene induction, the agar layer was overlaid with liquid medium containing 10 μM β-estradiol [23] Alternatively, plants grown on solid agar in the light were transferred to culture plates containing liquid medium supplemented with 10 µM ß-estradiol. Only the roots were dipping into the medium. Developmental stages considered are defined in the "Results" section. At all stages and in all growth conditions, samples of 3 to 4 entire plants were harvested each day till day 4 after induction. To avoid variations that would be due to the circadian clock, plants were always induced and collected at the same time of the day. Whole plant samples were immediately frozen in liquid nitrogen. Control plants (see the "Results" section) were grown and treated in the same way to take into account any side effects.

### *2.4. RNA Extraction, Northern Bloting and RT-qPCR Analyses*

Total RNA was extracted from *A. thaliana or N. tabacum* seedlings following standard TRI Reagent protocols (Molecular Research Center). Northern blot analyses were carried out through standard protocols using non-radioactive digoxygenin-labeled probes prepared with the PCR DIG Probe Synthesis kit (Roche, Basel, Switzerland). Prior to reverse transcription, RNA samples were treated up to 3 times with RNase-free DNase 1 (Thermo Scientific, Waltham, MA, USA). Reverse transcription was carried out with SuperScript III (Invitrogen, Carlsbad, CA, USA), RevertAid (Fermentas, Waltham, MA, USA) or GoScript (Promega, Madison, WI, USA) reverse transcriptase according to the corresponding recommended protocols, using 2 pmol of a specific primer or 250 ng of a random hexanucleotide mixture. Reactions were stopped by 15 min incubation at 70 °C and the RT samples served directly for standard or real-time PCR. The absence of residual DNA contamination in the original RNA samples was assessed by the absence of PCR product amplification from cDNAs resulting from reverse transcription assays run without the enzyme.

Real time PCR (qPCR) was performed in 384-well optical plates on an iCycler (BioRad, Hercules, CA, USA) or a LightCycler 480 II (Roche) using 5 µL of PCR master mix containing 480 SYBER Green I fluorescent reporter (Roche) with 2.5 µM forward and reverse specific primers. DNA or cDNA template and water were added to a total volume of 10 µL. Each sample was performed in triplicate. Reactions were run in the following conditions: pre-heating at 95 °C for 10 min, followed by 40 cycles of 15 sec at 95 °C, 30 sec at 60 °C and 15 sec at 72 °C. Melting curves were run to assess the specificity of the products.

The nuclear genes encoding actin 2 (accession AT3G18780; AB158612), glyceraldehyde 3-phosphate dehydrogenase c (GAPDH, accession AT1G13440; AY049259) and RGS1-HXK1 interacting protein 1 (RHIP1, accession AT4G26410; BT002964), as well as the mitochondrial gene encoding mitochondrial ribosomal protein L2 (rpl2, accession BA000042, region 361051-363948) served as reference genes. To design qPCR primers, we used the Universal ProbeLibrary and the ProbeFinder software (Roche) (https://lifescience.roche.com/en_fr/articles/Universal-ProbeLibrary-System-Assay-Design.html#ProbeFinder). For short sequences that were unsuccessful with such an approach, we used the Primer 3 software [27].

RT-qPCR data obtained with RNAs from transformants expressing ribozyme-PKTLS transcripts were analyzed with the Student's *t*-test. Statistical analyses of the mitochondrial gene expression RT-qPCR data obtained with RNAs from the *orf77*-PKTLS expressing transformants were carried out using the GraphPad Prism version 7.01 software (www.graphpad.com/). Correlation analyses, one way ANOVA, multiple comparison tests and Tukey tests were used to calculate the $p$ values. $p$ values $< 0.05$ were considered statistically significant.

### *2.5. Isolation of Mitochondria and Import Assessment*

To confirm that the *orf77*-PKTLS RNA expressed in the transformants was targeted to the organelles, mitochondria were isolated according to established protocols [28] from light-grown plants at day 2 after induction of transgene expression with β-estradiol. Isolated organelles were extensively treated with RNase in buffer containing 100 µg/mL RNase A and 750 U/mL RNase T1. Mitochondrial RNA was extracted following standard TRI Reagent protocols (Molecular Research Center), with 150 µL of Trizol reagent (Invitrogen) added to an amount of mitochondria equivalent to 200 µg of proteins.

Total RNA was prepared in parallel from the same plants. Mitochondrial and total RNAs were probed by RT-qPCR for *orf77*-PKTLS, for specific nuclear transcripts and for selected mitochondrial transcripts, so as to characterize mitochondrial enrichment and confirm the absence of significant contamination.

### 2.6. Microarray Assays

Microarray analyses were carried out on the INRA transcriptomic platform, first at the Unité de Recherche en Génomique Végétale (URGV, Evry, France) and subsequently at the Institute of Plant Sciences Paris-Saclay (IPS2, Orsay, France), using CATMAv6.2 (Roche-NimbleGen technology, Madison, WI, USA) and CATMAv7 (AGILENT technology, Santa Clara, CA, USA) arrays. High density CATMAv6.2 microarray slides contained per chamber 270,000 primers representing all *A. thaliana* genes, i.e., 30,834 probes referring to the TAIRv8 annotation (including 476 probes of mitochondrial and chloroplast genes) (www.arabidopsis.org/), plus 1289 probes corresponding to EUGENE software predictions, 5352 probes corresponding to repeated elements, 658 probes for miRNAs, 342 probes for other non-coding RNAs (rRNAs, tRNAs, snRNAs, soRNAs) and finally 36 control probes. CATMAv7 slides contained per chamber 180,000 primers corresponding to 38,360 probes referring to the TAIRv8 annotation (including primers for mitochondrial and chloroplast genes), plus primers corresponding to EUGENE software specific predictions, primers corresponding to repeats, primers for miRNAs and other ncRNAs (rRNAs, tRNAs, snRNAs, snoRNAs) and the 36 controls. Each of these primers was designed in both orientations (forward strand in triplicate and reverse strand without duplication).

RNA samples from plants collected each day from Day 0 to Day 4 post-induction (i.e., 5 test samples and 5 control samples) were submitted to RT-qPCR analysis to assess *nad9* knockdown and individually used for microarray assays. Two to three independent biological replicates were analyzed and fluorochrome reversal (dye swap) was systematically applied. Total RNA extracted from *A. thaliana* plants following TRI-reagent protocols (Molecular Research Center) and treated twice with RNase-free DNase I (Fermentas) was finally purified using NucleoSpin RNA kits (Macherey-Nagel). Further steps were mostly adapted from Lurin et al. [29]. RNA samples were checked for quality on a bioanalyser (Agilent), quantified with a Quant-iT *RiboGreen* RNA Assay Kit (Thermo Fisher Scientific) and amplified with the Complete Whole Transcriptome Amplification kit (WTA2, Sigma-Aldrich) using random hexamer primers. After purification, 30 pmoles of cDNA per sample and per slide were hybridized overnight at 42 °C in the presence of formamide. Two micron scanning was performed with an InnoScan900 scanner (InnopsysR, Carbonne, France) and raw data were extracted using the MapixR software (InnopsysR, Carbonne, France).

### 2.7. Microarray Data Analyses

Microarray raw data comprised the logarithm of median feature pixel intensity at wavelengths 635 nm (red) and 532 nm (green). To correct the dye bias, global intensity-dependent normalization was run using the LOESS procedure [30]. Differential analysis was based on log-ratio averaging over the duplicate probes and over the technical replicates. The data were submitted to *limma* moderated t-statistics [31,32] using the R software package (www.r-project.org). No evidence that the specific variances would vary between probes was raised by *limma* and consequently the moderated t-statistics was assumed to follow a standard normal distribution. The squeezeVar function of the *limma* library was used to smooth the specific variances by computing empirical Bayes posterior means. To control the false discovery rate, adjusted *p*-values were calculated using the optimized FDR (false discovery rate) approach and the kerfdr library [33,34]. Probes with an adjusted *p*-value $\leq 0.05$ were considered as differentially expressed. Complementary analyses were developed with the MapMan software (www.mapman.gabipd.org). To link the levels of the *nad9* transcript to nuclear transcripts, a sparse PLS (Partial Least Squares) regression approach was performed using the spls (v2.2) R package with the defaukt parameters, except eta = 0.5, K = 1, kappa = 0.5 [35]. The 99% confidence interval of the coefficient of each gene was calculated by bootstrapping (N = 1000). Only genes for which the confidence interval did not include 0 were considered as associated with *nad9* expression. As we have directly manipulated the *nad9* transcript levels, this association can be interpreted as a regulation of these genes by the *nad9* transcript levels. The sparse PLS analyses were performed separately for the etiolated and the light-grown seedlings using the log2 fold-changes. Microarray data from this manuscript were deposited into the CATdb database [36,37] (tools.ips2.u-psud.fr/CATdb/, Projects: RS13-02_Mitomanip and RS14-01_Mitomanip2) and into the Gene Expression Omnibus (GEO) repository [38] at the National Center for Biotechnology Information (NCBI) (accession numbers GSE127756 and GSE93122), according to the "Minimum Information About a Microarray Experiment" standards [39].

## 3. Results

### 3.1. *Trans-Ribozyme-Mediated Modulation of Mitochondrial RNA Steady State Levels Depends on the Plant Developmental Stage*

Four mitochondrial mRNAs, *nad9*, *sdh3*, *cox3* and *atp9*, were initially chosen as targets for specific *trans*-cleaving hammerhead ribozymes. The catalytic RNAs, *Rznad9*, *Rzsdh3*, *Rzcox3* and *Rzatp9* were designed on the same basis as in previous studies [20,21], i.e., with only 2 base pairs in helix II and a UUUU tetraloop. The *Rzatp9* ribozyme was already described before [20]. The ribozyme sequences were attached as 5′-trailors to the previously set up PKTLS mitochondrial RNA shuttle [20] through a weakly structured linker (L). The PKTLS sequence corresponded to the last 120 nucleotides of the *Turnip yellow mosaic virus* (TYMV) genomic RNA and included the tRNA-like structure and an upstream pseudoknot considered to improve the interaction with the aminoacyl-tRNA synthetase [40], which is a prerequisite for mitochondrial import [41,42]. Figure 1 shows the resulting design. The specifically designed *trans*-ribozymes associated with the linker and the PKTLS shuttle were expressed from nuclear transgenes in stably transformed *A. thaliana* plants (for cleavage of the *nad9*, *cox3* and *atp9* mitochondrial targets) and *N. tabacum* plants (for cleavage of the *sdh3* mitochondrial target). To master the expression, the sequences were placed under the control of the commonly used estradiol-inducible promoter system of the pER8 vector [23]. In this system, a chimeric transcription activator (XVE), taking together the DNA-binding domain of the bacterial repressor LexA (X), the acidic transactivating domain of the *Herpes simplex virus* VP16 factor (V) and the regulatory region of the human estrogen receptor (E), is expressed under the G10-90 strong constitutive promoter. The *trans*-activating activity of the chimeric XVE factor is triggered upon estradiol binding. The XVE-estradiol complex activates the target promoter, which consists of eight copies of the LexA operator fused upstream of a minimal 35S promoter. The activated target promoter drives the expression of the transgene, which is thus strictly dependent on estradiol delivery to the cells [23]. The XVE factor is constitutively expressed, with possible side effects. In our hands, estradiol delivery to the plants showed by itself a slight activating effect on mitochondrial genes. Also, our ribozyme-PKTLS constructs include a tRNA-like moiety that is recognized by the cognate aminoacyl-tRNA synthetase and enters tRNA pathways. Untransformed plants not exposed to estradiol were thus not appropriate controls to take into account all possible side effects and we took advantage of the fact that the SDH3 subunit is not encoded by a mitochondrial gene in *A. thaliana*, so that the *Rzsdh3* chimeric ribozyme directed against the *N. tabacum* mitochondrial *sdh3* mRNA has no RNA target in *A. thaliana* mitochondria. Transformant *A. thaliana* expressing the *Rzsdh3*-L-PKTLS chimeric ribozyme upon estradiol treatment was thus used as a control, as in previous studies [21]. Expression of the ribozyme transgenes was induced with estradiol at different developmental stages of light-grown plants and in dark-grown seedlings. Test lines and the *Rzsdh3* "no target" control line were grown, treated and harvested in parallel in strictly identical conditions. Expected knockdown of the corresponding target RNAs in test lines was analyzed every day for up to four days after onset of estradiol induction. For both the test lines and the control line, the level of the target RNA of interest at Day 1, Day 2, Day 3 and Day 4 post-induction was compared to the level at Day 0. At each growth stage, the kinetics obtained with the test lines were subsequently compared day by day to the kinetics of the *Rzsdh3* control line. Initial comparison to

Day 0 and day-by-day normalization against the control at the same growth stage always gave the same profiles. *Rznad9* and *Rzatp9* were the most active in vivo and were exploited at all growth stages. In all our assays, expression of the ribozymes had no visually detectable phenotypic effects within the four-day time frame of the experiments (See for example, Figure S1).

Induced expression of the previously described *Rzatp9*-L-PKTLS RNA in *A. thaliana* seedlings at an early stage of growth (maximum 4 true leaves) was efficient (Figure 3a), but failed to trigger a significant decrease in the steady-state level of the *atp9* target mRNA within four days in a day by day analysis (Figure 3b). A more detailed kinetic analysis along 16 h of initial phase of ribozyme expression led to the same absence of significant decrease of the target RNA (Figure S3). Similarly, expression of the *Rznad9*-L-PKTLS RNA at the early stage of growth was unable to knock down the level of *nad9* mRNA in a 4-day (Figure 3c) or 16 h (Figure S3) analysis. Thus, at an early stage of development, *A. thaliana* seedlings seemed to be able to maintain the steady-state level of mRNAs like *atp9* and *nad9* in the presence of the corresponding *trans*-cleaving ribozymes. This suggests that mitochondrial function is of primary importance at that stage of growth and that a still active transcription might efficiently "buffer" transcriptome changes through replacement of cleaved RNAs.

At an intermediate stage of *A. thaliana* development (up to 10 true leaves), expression (Figure 3d) and mitochondrial import of the *Rznad9*-L-PKTLS RNA resulted in a striking "bounce-back" profile of the *nad9* transcript level. Figure 3e shows that, following initial efficient knockdown of the *nad9* mRNA at Day 1 after onset of *Rznad9*-L-PKTLS expression, the steady-state level of the target RNA went back to normal at Day 2, before dropping again at Day 3 and 4. The profile established by RT-qPCR was confirmed by northern blot (Figure S2). The *atp9* transcript level also showed a marked bounce-back profile upon expression and mitochondrial import of the *Rzatp9*-L-PKTLS RNA at the intermediate stage of *A. thaliana* development (Figure 3f). Further analyses showed a similar bounce-back kinetics for the steady-state level of the *cox3* mitochondrial mRNA (Figure 3h) upon induced expression of the corresponding *Rzcox3*-L-PKTLS ribozyme in light-grown *A. thaliana* plants at the intermediate stage of development (Figure 3g). Thus, the bounce-back profile of the RNA steady-state level was not restricted to a given target, but appeared to be potentially a general response of mitochondrial RNAs to the expression and organellar import of specific ribozymes at intermediate stages of growth. Moreover, the phenomenon was not either restricted to *A. thaliana*, as the mitochondrial *sdh3* mRNA level responded in the same way to the expression of the *Rzsdh3*-L-PKTLS RNA in *N. tabacum* plants at the equivalent developmental stage (Figure 3i,j). These observations altogether imply that the buffering mechanism that might operate at the early stage of development (Figure 3b,c and Figure S3) becomes looser at a later stage, but disturbance of the mitochondrial transcriptome through the knockdown of an individual mitochondrial RNA is still sensed and transiently complemented. Notably, ribozyme-mediated knockdown of the mitochondrial *matR* mRNA encoding an organellar splicing factor was previously shown to be efficient and continuous in *A. thaliana* plants at an intermediate developmental stage [21]. Transcriptomic buffering might thus mainly apply to mRNAs encoding subunits of the OXPHOS complexes.

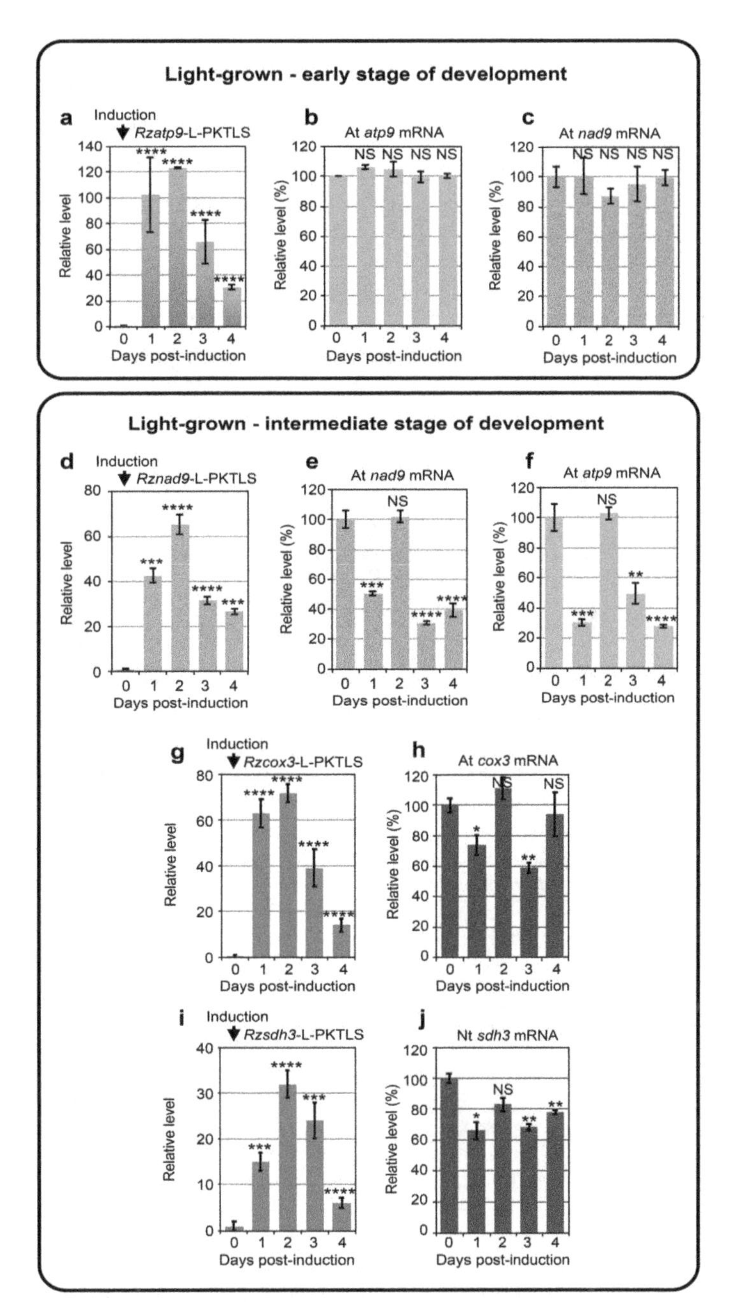


**Figure 3.** Chimeric ribozyme expression and knockdown of steady-state levels of mitochondrial target RNAs in transformed seedlings at different stages of growth. (**a**–**h**) *A. thaliana* control seeds and seeds carrying the *Rzatp9*-L-PKTLS, *Rznad9*-L-PKTLS or *Rzcox3*-L-PKTLS transgene were germinated in the light on solid MS-agar medium. Plants at early stage (upper panel) or intermediate stage (lower panel) of development were transferred at Day 0 to wells in culture plates containing liquid medium supplemented with estradiol for transgene induction. Kinetics of induced expression of the *Rzatp9*-L-PKTLS (**a**), *Rznad9*-L-PKTLS (**d**) or *Rzcox3*-L-PKTLS (**g**) RNA and of the steady-state level of the mitochondrial *atp9* (**b**,**f**), *nad9* (**c**,**e**) or *cox3* (**h**) target RNA were analyzed by RT-qPCR with total RNA from plant samples collected each day from Day 0 to Day 4 post-induction. (**i**,**j**) Transformant *N. tabacum* carrying the *Rzsdh3*-L-PKTLS transgene was germinated in the light on solid MS-agar medium and transferred at intermediate stage of development to liquid medium supplemented with estradiol for transgene induction. Kinetics of induced expression of the *Rzsdh3*-L-PKTLS RNA (**i**) and of the steady-state level of the mitochondrial *sdh3* target RNA (**j**) were analyzed by RT-qPCR with total RNA from transformed plant samples collected each day from Day 0 to Day 4 post-induction. Data from three independent biological replicates were analyzed with the Student's *t*-test; NS = not significant; * = $p \leq 0.05$; ** = $p \leq 0.01$; *** = $p \leq 0.001$; **** = $p \leq 0.0001$.

Continuous knockdown of the *nad9* target RNA (Figure 4b), or of the *atp9* target RNA (Figure 4c) was recorded upon *Rznad9*-L-PKTLS (Figure 4a) or *Rzatp9*-L-PKTLS expression in *A. thaliana* plants at the bolting stage of development (up to 18 true leaves). Mitochondrial transcriptome surveillance and buffering thus seems to be progressively released at later developmental stages. Notably, continuous and strong knockdown of the *atp9* mRNA was also obtained earlier in flowering plants expressing the *Rzatp9*-L-PKTLS RNA [20].

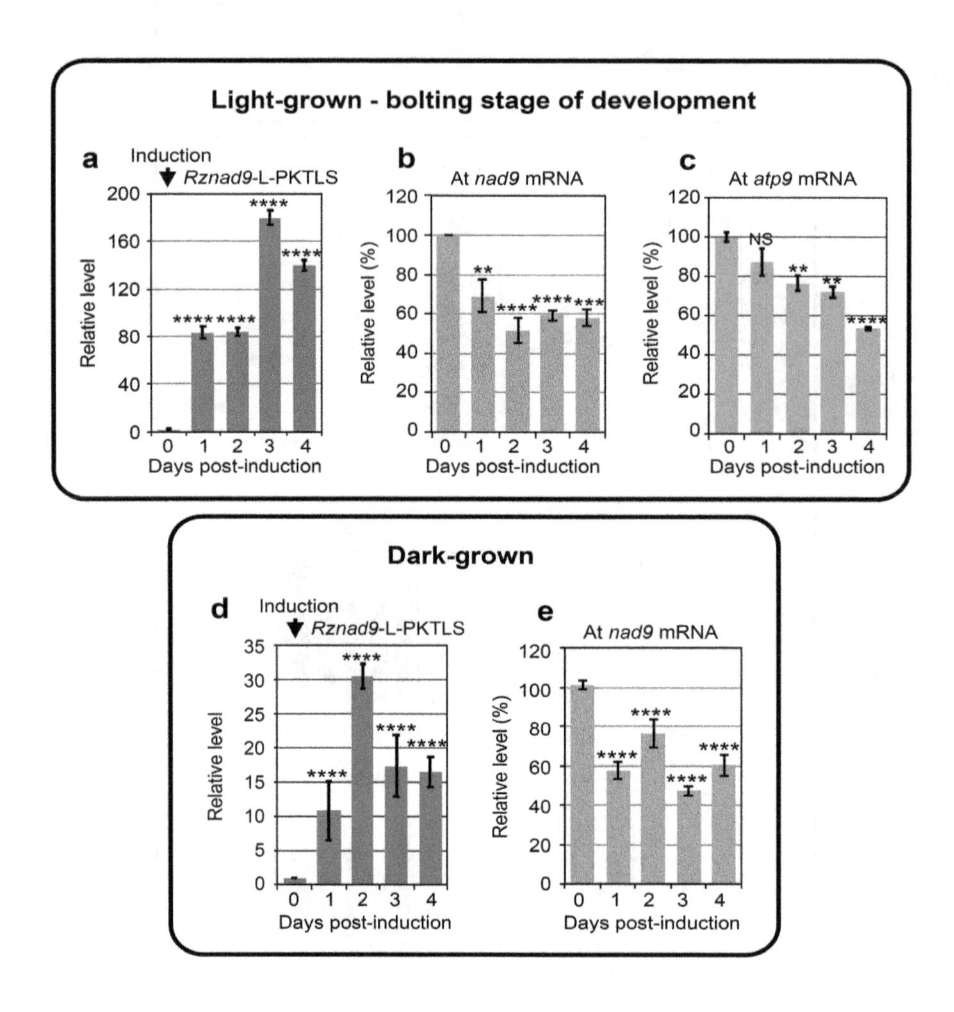


**Figure 4.** Chimeric ribozyme expression and knockdown of steady-state levels of mitochondrial target RNAs in transformed seedlings in different conditions. (**a**–**e**) *A. thaliana* control seedlings and seedlings carrying the *Rznad9*-L-PKTLS, or the *Rzatp9*-L-PKTLS transgene were grown in the light (upper panel) or in the dark (lower panel) on solid MS-agar medium. Plants at bolting stage of development (upper panel) were transferred at Day 0 to wells in culture plates containing liquid medium supplemented with estradiol for transgene induction. Kinetics of induced expression of the *Rznad9*-L-PKTLS RNA (**a**), of the steady-state level of the mitochondrial *nad9* target RNA (**b**) and of the steady-state level of the mitochondrial *atp9* target RNA (**c**) were analyzed by RT-qPCR with total RNA from plant samples collected each day from Day 0 to Day 4 post-induction. Plates with ten-day-old seedlings grown in the dark (lower panel) were overlayed with liquid medium supplemented with estradiol for transgene induction. Kinetics of induced expression of the *Rznad9*-L-PKTLS RNA (**d**) and of the steady-state level of the mitochondrial *nad9* target RNA (**e**) were analyzed by RT-qPCR with total RNA from plant samples collected each day from Day 0 to Day 4 post-induction. Data from three independent biological replicates were analyzed with the Student's *t*-test; NS = not significant; * = $p \leq 0.05$; ** = $p \leq 0.01$; *** = $p \leq 0.001$; **** = $p \leq 0.0001$.

Remarkably, a bounce-back profile of the *nad9* mRNA level (Figure 4e) was also observed when expressing the *Rznad9*-L-PKTLS transcript in 10-day-old *A. thaliana* seedlings grown in the dark (Figure 4d), indicating that the phenomenon was related to the age and importance of the mitochondrial function rather than to the growth conditions *per se*.

### 3.2. CMS RNA-Triggered Modulation of Mitochondrial RNA Steady State Levels Depends on the Plant Developmental Stage

The possibility that physiological conditions might impair ribozyme cleavage activity at some developmental stages cannot be completely ruled out. In particular, even when optimized, ribozymes require a minimal free $Mg^{++}$ concentration. We thus developed a parallel strategy to affect the mitochondrial transcriptome and test its reactivity. The concept was to use the PKTLS shuttle to drive a CMS RNA into the mitochondria of *A. thaliana* plants. Mitochondria of maize lines with the CMS-S cytoplasm express characteristic transcripts that carry a specific open reading frame called *orf77* ([43] and references therein). The *orf77* is a chimera combining sequences of unknown origin with sequences derived from the mitochondrial *atp9* gene (Figure S4). Due to these similarities, it is hypothesized that the *orf77* transcript can impair *atp9* expression at the RNA or protein level. ATP9 is a core component of the $F_0$ moiety of the mitochondrial ATP synthase located at the inner membrane. Ten or more ATP9 subunits assemble to form an inner membrane-embedded, hydrophobic ring that functions as a rotor pumping protons from the intermembrane space to the matrix side. ATP9 is thus a major protein for mitochondrial function.

In the present work, we imported the maize *orf77* CMS-S RNA into mitochondria in *A. thaliana* plants, so as to analyze the resulting impact on the whole mitochondrial transcriptome. Similarly, to the above *trans*-cleaving ribozymes, the *orf77* sequence was attached to the PKTLS shuttle as a 5′-trailor (Figure 2) and placed under the control of the estradiol-inducible promoter system in the pER8 expression vector [23]. The *orf77*-PKTLS RNA was expressed from a nuclear transgene in stably transformed *A. thaliana* plants (Figure S6). Expression was induced with estradiol at different developmental stages of light-grown plants and in dark-grown plants. In this strategy, plants transformed with a pER8 plasmid deprived of transgene served as a control. The resulting impact on the whole mitochondrial transcriptome was analyzed day by day for up to four days after onset of estradiol induction. At all stages, *orf77*-PKTLS expression had no visually detectable phenotypic effects on the plants within this time frame (Figure 5 and Figure S5). In this case, the cargo sequence attached to the PKTLS was longer and we could not rely on a well-defined activity to confirm organellar targeting. Purified, RNase-treated mitochondria were thus prepared from *A. thaliana* plants expressing the transgene and organellar import of the *orf77*-PKTLS transcript was confirmed by RT-qPCR probing, which showed about tenfold enrichment of the *orf77*-PKTLS RNA in the mitochondrial fraction *versus* the actin and GAPDH nuclear transcripts (Figure S7).

**Table 1.** Mitochondrial RNAs whose level increased (in black) or decreased (in purple) significantly in relation with the expression and mitochondrial import of the *orf77*-PKTLS RNA.

| Conditions and Developmental Stage | Day 1 | Day 2 | Day 3 | Day 4 |
|---|---|---|---|---|
| Light *Early stage* | | ↓nad5a*<br>↓atp4**<br>↑rps7*<br>↑AOX1d* | ↓AOX1d*** | ↓AOX1d*** |
| Light *Intermediate stage* | ↑nad5b**<br>↓ccmFN2** | ↑nad5a**<br>↑atp4*<br>↑rpl2*<br>↑rpl5* | ↑cob**<br>↓ccmFN2***<br>↑AOX1d*** | ↑rps7*<br>↑rpl2**<br>↑rpl5*<br>↑AOX1d*** |
| Light *Bolting stage* | ↓nad2b**<br>↑nad5a**<br>↓nad5b**<br>↓nad9**<br>↓cox1** ↓ccmB**<br>↓mttB*<br>↓rpl5*<br>↓AOX1d** | ↓atp4*<br>↓atp9****<br>↓rps4**<br>↓rps7****<br>↓rpl2****<br>↓rpl5**<br>↓rrn18**<br>↓AOX1d* | ↑nad1b*<br>↓nad2b*<br>↓nad4**<br>↑nad5a***<br>↓nad5b*<br>↓nad6*<br>↓nad9**<br>↓cob***<br>↓cox1*<br>↓cox2**<br>↓cox3****<br>↑atp9*<br>↓ccmB***<br>↓ccmFC*<br>↓ccmFN1***<br>↓ccmFN2**<br>↓mttB**<br>↓rrn18***<br>↓rrn26**<br>↓AOX1a*<br>↓AOX1d* | ↓cox1****<br>↓atp9*<br>↓ccmFC***<br>↑mttB*<br>↑rpl5**<br>↓rrn26* |
| Dark | ↑nad2b**<br>↓nad5a**<br>↑nad5b****<br>↑nad9*<br>↓cox2*<br>↓matR*<br>↑rps7**<br>↓rpl5*<br>↓rrn26* | ↑nad4*<br>↑nad5b*<br>↑cox1*<br>↓ccmFN2**<br>↑AOX1d** | ↓nad1b***<br>↓nad2b**<br>↑nad5b**<br>↓ccmFN2**<br>↓mttB**<br>↓rrn26** | ↓ccmFC***<br>↓matR*<br>↓mttB*<br>↓rrn18****<br>↑rrn26** |

Significant variations of nuclear RNAs coding for the mitochondrially targeted enzyme alternative oxidase (AOX1, underlined) are also included. Expression of the *orf77*-PKTLS transgene was induced at Day 0 and sampling was run every day until day 4 post-induction. Analyses were done by RT-qPCR with total RNA. Mitochondrial RNAs showing level variations versus control at Day 0 were not further considered. Data from three independent biological replicates were analyzed with the Graph Pad Prism software; * = $p \leq 0.05$; ** = $p \leq 0.01$; *** = $p \leq 0.001$; **** = $p \leq 0.0001$.

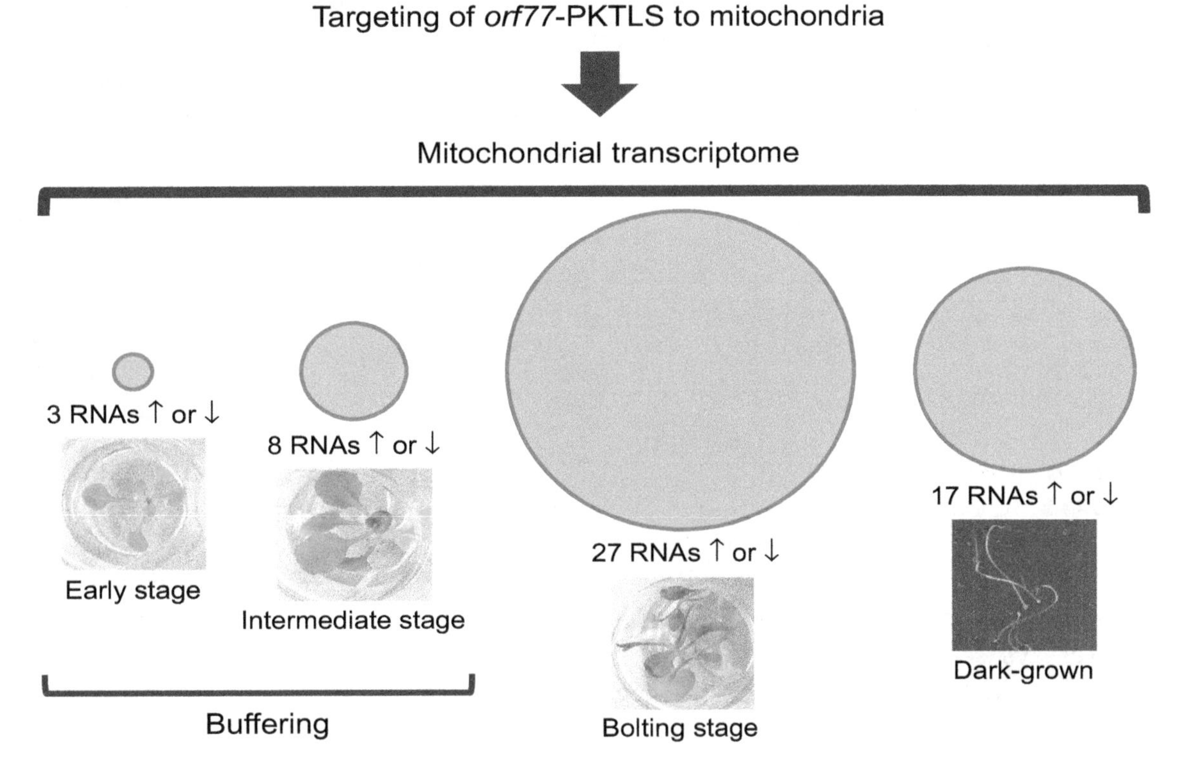


**Figure 5.** Synthetic scheme representing the impact of *orf77*-PKTLS expression on the mitochondrial transcriptome in transformed *A. thaliana* plants at different stages of growth in the light and at 10 days of growth in the dark. Plants were grown to the appropriate stage and the transgene was induced with estradiol as in Figures 3 and 4. Following transgene induction, samples were collected each day from Day 0 to Day 4 post-induction and RNAs were analyzed by RT-qPCR. The total numbers of positively or negatively affected transcripts combining all daily samples for a given growth stage are indicated and represented by proportional circular areas. Detailed day-by-day and gene-by-gene results are given in Table 1. Data from three independent biological replicates were analyzed with the Graph Pad Prism version 7.01 software.

As for ribozymes, the effect of *orf77*-PKTLS expression and mitochondrial import on the levels of essentially all mitochondrial RNAs was characterized in transformed *A. thaliana* plants at different growth stages in the light and in dark-grown seedlings. A synthetic scheme of the results is given in Figure 5, while detailed data are compiled in Table 1. The analyses confirmed the transcriptome buffering process highlighted above in the case of *trans*-ribozyme-mediated RNA knockdown. Induced expression of the *orf77*-PKTLS RNA in transgenic *A. thaliana* seedlings at an early stage of development (maximum 4 true leaves) had almost no effect on the mitochondrial transcriptome, with only 3 RNAs positively or negatively affected and only at Day 2. At an intermediate stage of development, i.e., the stage where ribozyme-mediated target knockdown showed a bounce-back profile (Figure 3e,f,h,j), transcriptome buffering became somehow looser and up to eight mitochondrial RNAs were affected as a whole. Finally, at the bolting stage buffering was essentially released, as up to 27 mitochondrial RNAs were positively or negatively affected in response to *orf77*-PKTLS expression and organellar import. Expressing the *orf77*-PKTLS RNA in dark-grown seedlings affected up to 17 mitochondrial RNAs.

As mentioned, *orf77* contains sequences from the *atp9* gene (Figure S4), so that the *orf77*-PKTLS transcript was expected to interfere with the *atp9* mRNA. Remarkably, such an effect was indeed observed, but only at the bolting stage, i.e., at a stage where the transcriptome buffering was released. The level of *atp9* decreased at Day 2 and Day 4, while increasing at Day 3 (Table 1), further suggesting a dynamic process. Like *atp9*, *mttB* and *rpl5* showed some modulation at the bolting stage, with an initial decrease and subsequent increase at Day 4.

Except in early stage light-grown seedlings, all genetic functions encoded by the mtDNA, i.e., OXPHOS subunits, biogenesis co-factors (especially cytochrome c maturation factors), as well as ribosomal proteins and rRNAs, were represented in the organellar transcriptome response to *orf77*-PKTLS expression and mitochondrial import. Strikingly, the mitochondrial genes affected at the bolting stage were essentially downregulated, while the limited set of reactive genes at the intermediate stage was upregulated, with the exception of the cytochrome c maturation gene *ccmFN2* (Table 1). It is tempting to speculate that upregulation at the intermediate stage might be representative for transcriptome buffering, as proposed above, while general downregulation at the bolting stage would reflect the release of such a compensation process. Due to these differences, *nad5b*, *cob*, *atp4*, *rpl2*, *rpl5* and *rps7* showed an opposite behaviour between the intermediate and bolting stages. For instance, *atp4* was upregulated by about 50% at the intermediate stage and downregulated by 50% at the bolting stage.

The response to *orf77*-PKTLS expression and mitochondrial import in 10-day-old dark-grown seedlings was more balanced between up- and downregulation. As for light-grown intermediate stage samples, this led to an opposite behaviour of a number of transcripts, like for example the different *nad* mRNAs or *cox1*, *versus* the bolting stage of light-grown plants (Table 1). Notably, as intermediate stage light-grown samples, dark-grown seedlings also showed a bounce-back profile in ribozyme-mediated knockdown assays (see above, § *3.1.* and Figure 4e). When affected, *ccmFN2* was always downregulated, whichever the growth conditions or growth stage (Table 1).

### *3.3. Ribozyme-Mediated Knockdown of a Mitochondrial mRNA Triggers a Nuclear Transcriptome Response*

Microarray analyses revealed that continuous ribozyme-directed knockdown of the *nad9* mitochondrial mRNA in light-grown *A. thaliana* plants at the bolting stage of development affected the level of multiple nuclear-encoded RNAs. However, the extent of *nad9* knockdown differed from one kinetics to the other for the samples from biological replicates dedicated to the microarray assays, so that we decided to perform a sparse PLS analysis of the data (Supplementary File 2). The latter highlighted the significant regulation of 208 genes by the ribozyme-directed knockdown of *nad9* (Supplementary File 3). Notably, these included as many as 12 upregulated (Table 2) and 7 downregulated (Table 3) transcription factor genes, as well as 12 upregulated (Table 2) and 8 downregulated (Table 3) small nucleolar RNA (snoRNA) genes. Downregulation of the questioned miR414 microRNA was also noticed (Table 3), as well as up- or downregulation of genes involved in signaling, RNA metabolism or translation (Tables 2 and 3). Successful knockdown of *nad9*, i.e., a major mitochondrial mRNA encoding an OXPHOS subunit, in light-grown mature plants thus appeared to trigger a complex nuclear response driven by retrograde/anterograde regulation processes.

**Table 2.** Selected *A. thaliana* nuclear-encoded RNAs upregulated upon knockdown of the mitochondrial *nad9* mRNA.

| Modulation | Annotation / Function / Organellar Localization | Gene |
|---|---|---|
| Upregulated (opposite to *nad9*) | Basic helix-loop-helix (bHLH) DNA-binding superfamily protein; transcription factor; response to ethylene | AT1G05710 |
| | ABI3-interacting protein 3, AIP3, PFD4, PREFOLDIN 4; protein chaperone; ABI3 is an auxin-inducible transcription factor | AT1G08780 |
| | ATMYB60, Myb domain protein 60, MYB60; transcription factor; response to abscisic acid, jasmonic acid, salicylic acid | AT1G08810 |
| | Agamous-like 87, AGL87, MADS-box family protein; transcription factor | AT1G22590 |
| | GL2, GLABRA 2, HD-ZIP IV family of homeobox-leucine zipper protein with lipid-binding START domain; transcription factor | AT1G79840 |
| | DREB subfamily A-6 of ERF/AP2 transcription factor family; one AP2 domain; ethylene-activated | AT2G22200 |
| | AP2/B3-like transcriptional factor family protein; transcription factor | AT2G33720 |
| | FMA (FAMA), basic helix-loop-helix (bHLH) DNA-binding superfamily protein; transcription factor/ transcriptional activator | AT3G24140 |
| | EDF3, ethylene response DNA-binding factor 3; transcription factor; AP2 domain; ethylene responding | AT3G25730 |
| | IAA30, indole-3-acetic acid inducible protein 30; transcription factor; response to auxin | AT3G62100 |
| | B-BOX domain protein 23, BBX23; transcription factor | AT4G10240 |
| | PUCHI, ethylene response factor (ERF) subfamily B-1 of ERF/AP2 transcription factor family; one AP2 domain; ethylene response | AT5G18560 |
| | ARGOS, Auxin-regulated gene involved in organ size; response to ethylene, auxin; membrane, cytoplasm, mitochondrion | AT3G59900 |
| | EMB3103, Embryo-defective 3103, PDM2, Pigment-Defective Mutant2; pentatricopeptide repeat (PPR) superfamily protein; endonuclease; chloroplast/mitochondrion | AT1G10910 |
| | RNH1C, RNase H family protein, RNase H domain-containing protein; chloroplast | AT1G24090 |
| | RING-finger, DEAD-like helicase, PHD and SNF2 domain-containing protein | AT2G40770 |
| | AT-SR34B, Serine/arginine-rich protein splicing factor 34B, SR34B | AT4G02430 |
| | Mitochondrial nuclease 1, MNU1; putative endonuclease or glycosyl hydrolase; mitochondrial RNA 5'-end processing; chloroplast/mitochondrion | AT5G64710 |
| | EMB2394, Embryo-defective 2394; structural constituent of chloroplast ribosome; response to cytokinin | AT1G05190 |
| | rRNA, cytosolic small ribosomal subunit | AT2G01010 |
| | 5.8S rRNA, cytosolic large ribosomal subunit | AT2G01020 |
| | 40S ribosomal protein S23 (RPS23A), ribosomal protein S12/S23 family protein; cytosolic small ribosomal subunit | AT3G09680 |

**Table 2.** *Cont.*

| Modulation | Annotation / Function / Organellar Localization | Gene |
| --- | --- | --- |
| | Ribosomal protein L15, RPL15; large subunit of the chloroplast ribosome; response to cytokinin | AT3G25920 |
| | EMB3126, Embryo-defective 3126, plastid ribosomal protein L1, PRPL1; ribosomal protein L1p/L10e family; chloroplast large ribosomal subunit | AT3G63490 |
| | 60S ribosomal protein L31, RPL31B; cytosolic large ribosomal subunit | AT4G26230 |
| | U3 ribonucleoprotein, Utp family protein; rRNA processing | AT5G08600 |
| | MA3 domain-containing translation regulatory factor 1, MRF1; colocalizes with cytosolic large ribosomal subunit; isomerase activity | AT5G63190 |
| | snoRNA | AT1G03743; AT1G19373; AT1G19376; AT1G75166; AT2G35387; AT3G27865; AT3G47342; AT3G47347; AT3G58193; AT3G58196; AT4G39366; AT5G44286 |

Data taken from sparse PLS analysis of microarray results obtained with RNAs from light-grown *A. thaliana* plants at the bolting stage. Individual Day 0 to Day 4 RNA samples from three independent biological replicates were analyzed (i.e., 15 test samples and 15 control samples). Organelllar localization is mentioned where annotated. Selection was for genes of potential relevance in regulation mechanisms, including transcription factors, hormone signaling or RNA metabolism.

To cast off the effects of photosynthesis processes and chloroplast pathways, and try to simplify the picture, microarray assays were continued with RNAs from 10-day-old dark-grown *A. thaliana* seedlings expressing the *Rznad9*-L-PKTLS chimeric ribozyme. In this case, the mitochondrial transcriptome was constrained by buffering, so that the response was also expected to be representative for the moderate bounce-back profile of *nad9* knockdown in dark-grown plants (see Figure 4e). Microarray analyses of the RNAs from dark-grown *A. thaliana* seedlings expressing the *Rznad9*-L-PKTLS ribozyme (Supplementary File 4) revealed that the number of differentially expressed genes reached 3046 at Day 1 and then dropped to 1622 and 416 at Day 2 and Day 3, before re-increasing to 959 at Day 4. Sparse PLS sorting of these data (Supplementary File 5) showed a regulation of 49 genes by the ribozyme-directed knockdown of *nad9* (Supplementary File 3), including 3 downregulated transcription factor genes, as well as genes involved in signaling or RNA metabolism (Table 4). In parallel, we mined the original microarray data for modulation of transcripts encoding proteins shown or predicted to localize to mitochondria (Supplementary File 6). Interestingly, the analysis pointed to upregulation of a series of members of the large "small auxin upregulated RNAs" (SAUR) family, as well as to upregulation of transcripts for cytokinin and auxin response factors (Table 5), suggesting a hormone signaling-related process. MapMan analysis of the data also pointed to hormone metabolism pathways and especially to auxin-regulated genes. Conversely, downregulation of a jasmonate-responding factor, of a set of pentatricopeptide repeat (PPR) proteins and of further RNA metabolism factors was noticed (Table 5). Also, to be noted, the knockdown of *nad9* in dark-grown *A. thaliana* seedlings negatively affected the level of a series of mitochondrial mRNAs encoding subunits of OXPHOS complexes or ribosomal proteins, while mRNAs coding for biogenesis co-factors like CCM factors or MATR were little affected (Figure S8).

**Table 3.** Selected *A. thaliana* nuclear-encoded RNAs downregulated upon knockdown of the mitochondrial *nad9* mRNA.

| Modulation | Annotation/Function/Organellar Localization | Gene |
|---|---|---|
| Downregulated (as *nad9*) | MicroRNA414, mir414, short open reading frame 16, SORF16; miRNA; identified as a translated small open reading frame by ribosome profiling | AT1G67195 |
| | B-BOX domain protein 25, BBX25, Salt tolerance homologue, STH; transcription factor; zinc ion binding | AT2G31380 |
| | ATIBH1, IBH1, ILI1 binding BHLH 1 ILI1 binding bHLH 1; transcription factor; brassinosteroid signaling; gibberellic acid signaling | AT2G43060 |
| | Arabidopsis thaliana response regulator 2, ARR5, ATRR2, IBC6, induced by cytokinin 6, response regulator 5, RR5; transcription repressor; cytokinin signaling | AT3G48100 |
| | Basic helix-loop-helix (bHLH) DNA-binding superfamily protein; transcription factor | AT4G01460 |
| | KELP; homodimers or heterodimers with the kiwi protein; transcriptional co-activator | AT4G10920 |
| | WRKY DNA-binding protein 24, ATWRKY24, WRKY24; WRKY transcription factor group II-c | AT5G41570 |
| | BOA, Brother of lux ARRHYTHMO; transcription factor; circadian clock; mRNA cell-to-cell mobile | AT5G59570 |
| | SAUR53, small auxin-upregulated RNA 53, SAUR-like auxin-responsive protein family; <u>mitochondrion</u> | AT1G19840 |
| | SAUR65, small auxin-upregulated RNA 65, SAUR-like auxin-responsive protein family; membrane; <u>mitochondrion</u> | AT1G29460 |
| | ATCLE19, CLAVATA3/ESR-related 19, CLE19, embryo surrounding region 19, ESR19; receptor binding; signal transduction; <u>mitochondrion</u> | AT3G24225 |
| | VQ motif-containing protein 29, VQ29; response to hypoxia | AT4G37710 |
| | RNA-binding (RRM/RBD/RNP motifs) family protein | AT1G33470 |
| | ATRNS1, Ribonuclease 1, RNS1; endoribonuclease | AT2G02990 |
| | Eukaryotic translation initiation factor 2 (eIF-2) family protein; cytosol | AT1G76820 |
| | EMB3113, Embryo-defective 3113, ribosomal protein S5, RPS5, SCA1, SCABRA 1; structural component of the 70S <u>chloroplast</u> ribosome; <u>mitochondrion</u> small ribosomal subunit | AT2G33800 |
| | U3 containing 90S pre-ribosomal complex subunit | AT2G43110 |
| | ATRAB8D, ATRABE1B, RAB GTPase homolog E1B, RABE1B; translation elongation factor; membrane; <u>chloroplast</u> | AT4G20360 |
| | snoRNA | AT2G35382; AT2G43137; AT2G43138; AT2G43139; AT2G43141; AT4G02550; AT4G02555; AT4G13245 |

Data taken from sparse PLS analysis of microarray results obtained with RNAs from light-grown *A. thaliana* plants at the bolting stage. Individual Day 0 to Day 4 RNA samples from three independent biological replicates were analyzed (i.e., 15 test samples and 15 control samples). Organelllar localization is mentioned where annotated. Selection was for genes of potential relevance in regulation mechanisms, including transcription factors, hormone signaling or RNA metabolism.

**Table 4.** Selected *A. thaliana* nuclear-encoded RNAs up- or downregulated upon knockdown of the mitochondrial *nad9* mRNA.

| Modulation | Annotation/Function/Organellar Localization | Gene |
|---|---|---|
| Upregulated (opposite to *nad9*) | Small nuclear RNA U6acat, mRNA splicing | AT5G40395 |
| Downregulated (as *nad9*) | ANAC028, NAC domain-containing protein 28, NAC028; transcription factor | AT1G65910 |
| | Transcription elongation factor Spt5; KOW domain | AT2G34210 |
| | MYR2, homeodomain-like superfamily protein; transcription factor | AT3G04030 |
| | HVA22-like protein F, HVA22F; membrane protein; response to abscisic acid | AT2G42820 |
| | ALY2 RNA-binding (RRM/RBD/RNP motifs) family protein; mRNA transport | AT5G02530 |

Data taken from sparse PLS analysis of microarray results obtained with RNAs from 10-day-old dark-grown *A. thaliana* seedlings. Individual Day 0 to Day 4 RNA samples from two independent biological replicates were analyzed (i.e., 10 test samples and 10 control samples). Selection was for genes of potential relevance in regulation mechanisms, including transcription factors, hormone signaling or RNA metabolism.

**Table 5.** Selected *A. thaliana* nuclear RNAs up- or downregulated upon knockdown of the mitochondrial *nad9* mRNA and coding for proteins predicted to be mitochondrion-targeted.

| Modulation | Annotation/Function/Organellar Localization | Gene |
|---|---|---|
| Upregulated (opposite to *nad9*) | auxin-responsive family protein (SAUR72) | AT3G12830 |
| | auxin-responsive protein, putative (SAUR9) | AT4G36110 |
| | auxin-responsive family protein (SAUR41) | AT1G16510 |
| | auxin-responsive protein-related (SAUR77) | AT1G17345 |
| | auxin-responsive family protein (SAUR53); mitochondrion | AT1G19840 |
| | auxin-responsive family protein (SAUR71) | AT1G56150 |
| | auxin-responsive family protein (SAUR52); mitochondrion | AT1G75590 |
| | auxin-responsive protein-related (SAUR36); mitochondrion/nucleus (nucleus confirmed by GFP targeting experiments) | AT2G45210 |
| | auxin-responsive family protein (SAUR59); mitochondrion | AT3G60690 |
| | auxin-responsive protein, putative (SAUR25); mitochondrion | AT4G13790 |
| | auxin-responsive family protein (SAUR1); chloroplast | AT4G34770 |
| | auxin-responsive protein, putative (SAUR23) | AT5G18060 |
| | MIF1 (MINI ZINC FINGER 1); transcription factor; response to abscisic acid, auxin, brassinosteroid, cytokinin, gibberellin; nucleus (cytosol reported from GFP data) | AT1G74660 |
| | PLS (POLARIS), cytokinin and auxin responses; mitochondrion | AT4G39403 |
| | Tetratricopeptide repeat (TPR)-like superfamily protein; mitochondrion | AT1G28690 |

**Table 5.** *Cont.*

| Modulation | Annotation/Function/Organellar Localization | Gene |
|---|---|---|
| | S-RBP11, SMALL RNA-BINDING PROTEIN 11; salt stress response; chloroplast | AT5G06210 |
| Downregulated (as *nad9*) | ARGAH2, Arginine amidohydrolase 2, response to jasmonate; chloroplast/mitochondrion (confirmed by MS data) | AT4G08870 |
| | Tetratricopeptide repeat (TPR)-like superfamily protein; chloroplast/mitochondrion (chloroplast confirmed by MS data) | AT2G37230 |
| | pentatricopeptide (PPR) repeat-containing protein | AT3G62470 |
| | pentatricopeptide (PPR) repeat-containing protein | AT4G01030 |
| | EMB1417 (embryo-defective 1417), PPR protein; RNA binding; endonuclease activity | AT4G21190 |
| | pentatricopeptide (PPR) repeat-containing protein | AT4G21880 |
| | pentatricopeptide (PPR) repeat-containing protein | AT5G65560 |
| | EMB1586, Embryo-defective 1586, increased size exclusion limit 1, ISE1; DEAD-box RNA helicase; chloroplast/mitochondrion (mitochondrion confirmed by GFP data) | AT1G12770 |
| | AGS1, AHG2-1 suppressor 1, bacterial-type poly(A) polymerase; mRNA polyadenylation; chloroplast/mitochondrion | AT2G17580 |
| | ATTRM2A, TRM2A, tRNA methyltransferase 2A, RNA methyltransferase family protein; cytoplasm | AT3G21300 |
| | ATP-dependent RNA helicase | AT5G39840 |
| | ARFB1A, ATARFB1A (ADP-ribosylation factor B1A); GTP binding; protein transport; Golgi apparatus | AT2G15310 |

Data taken from microarray analysis of RNAs from 10-day-old dark-grown *A. thaliana* seedlings. Individual Day 0 to Day 4 RNA samples from two independent biological replicates were analyzed (*i.e.*, 10 test samples and 10 control samples). All proteins included have a prediction consensus for mitochondrial targeting in the SUBA database [44] (http://suba.plantenergy.uwa.edu.au/). Experimentally reported localization compiled in the SUBA database is also indicated. Selection was for genes of potential relevance in regulation mechanisms, including hormone signaling or RNA metabolism.

## 4. Discussion

All complexes of the mitochondrial OXPHOS chain consist of subunits encoded by the nuclear genome and subunits expressed from mitochondrial genes, hence the idea of a coordinated expression of the two genomes. Nevertheless, according to the available data mtDNA transcription and mitochondrial transcript levels did not seem to be major regulation issues by themselves, so that mitochondrial genetic control is considered to occur mainly at the post-transcriptional level [3,4]. However, over the past decades, there have been a number of reports describing changes in the abundance of specific mitochondrial transcripts during plant growth [45], such as during the development of wheat leaves [46] or seed germination in maize and rice [47–49]. Variations in the steady state levels of a number of mitochondrial mRNAs were also observed with *A. thaliana* nuclear mutants deficient for the editing or splicing of a given organellar RNA [50–53]. The question of a control of mitochondrial gene expression at the transcriptome level, connected with mitochondrial retrograde pathways, can thus still be raised. So-called intergenomic signaling has been characterized in yeast [54]. Unaffected by respiration and metabolic signals, the process requires mtDNA and appears as a direct communication between the mitochondrial and nuclear genetic systems. Similarly, the idea of an involvement of chloroplast transcription in signaling to the nucleus has been put forward [55,56].

The present work follows this line. We disturbed the mitochondrial transcriptome in plants at different stages of development through specific in vivo knockdown of different organellar target mRNAs or introduction of a CMS RNA into mitochondria. The ribozymes and the CMS RNA were expressed from an inducible promoter, so that at each growth stage the plants were in a regular physiological configuration before induction. This enabled to reveal early and specific events, a notable advantage over the use of mutants that would have to cope with the mutation and adapt permanently their pathways. Also, to be noted, transgene expression upon single estradiol induction is transient and lasts only a few days. The question of the existence of transcriptome control within mitochondria and of intergenic coordination could then be addressed by analyzing the mitochondrial and nuclear responses. Remarkably, the observations highlighted RNA control mechanisms in mitochondria with relation to the plant developmental stage. The mitochondrial transcriptome appeared to be tightly constrained in young, actively developing seedlings, as ribozyme-directed target RNA knockdown was not observed at that stage and the *orf77* CMS RNA triggered almost no effect. This suggests in particular that in young seedlings high mitochondrial transcription rates [57] might compensate negative variations of RNA levels and "protect" the transcriptome at a growth stage where efficient mitochondrial activity is essential. A corollary of such a hypothesis would be that changes affecting the organellar transcriptome can be "sensed". Compensation became only transient at an intermediate stage of development, perhaps due to limited RNA synthesis, hence the bounce-back effect. Finally, at the bolting stage, ribozyme-mediated knockdown of target RNAs became efficient. Both *nad9* knockdown and organellar targeting of the *orf77* CMS RNA (presumed competitor of *atp9*) were then associated with a decrease of many mitochondrial RNAs, especially of mRNAs encoding subunits of OXPHOS complexes, suggesting a coordination and cohesion of the mitochondrial transcriptome. According to complementary analyses, the observed decrease of mitochondrial RNAs was not associated with variations in mtDNA copy number or in the number of mitochondria per cell. Notably, directed knockdown of *nad9*, i.e., a mitochondrial mRNA encoding a subunit of an OXPHOS complex, negatively affected the level of many other mitochondrial mRNAs encoding OXPHOS subunits or ribosome subunits. By contrast, in earlier experiments knockdown of the *matR* mitochondrial mRNA, encoding a maturase, affected intron splicing, but had limited effect on the level of intronless mitochondrial RNAs [21], suggesting that the mitochondrial transcriptome might be somehow coordinated to optimize expression of OXPHOS complex subunits.

As revealed by microarray analyses, successful ribozyme-mediated knockdown of *nad9*, i.e., a mitochondrial mRNA encoding a subunit of an OXPHOS complex, positively or negatively affected the level of multiple nuclear-encoded RNAs. This implies that directed alteration of the mitochondrial transcriptome resulted in a retrograde/anterograde response. The effect was fast, starting right away from the first day of ribozyme expression and knockdown of the target RNA, which implies a very fast retrograde signaling and raises the question of the nature of such a signaling. So far described retrograde signals are essentially metabolic signals sent to inform the nucleo-cytosolic compartment about the status of the organelles or resulting from impaired organellar function or biogenesis [54,58–60]. In particular, different signaling pathways respond to OXPHOS chain dysfunction. Our earlier studies showed that the profile of the mitochondrial OXPHOS complexes was still unaffected in plant cells four days after onset of ribozyme-mediated knockdown of the *atp9* mRNA [20]. Also, the level of MATR protein started to decrease significantly only from day 4 after onset of ribozyme-mediated knockdown of the *matR* mRNA [21]. With such a slow turnover of mitochondrial proteins, it seems unlikely that mitochondrial respiratory and metabolic dysfunction would occur fast enough to generate the signals underlying the observed retrograde response to the knockdown of the *nad9* mRNA. The data presented here thus support the idea of a fast communication between the mitochondrial and nuclear genomes through yet to be identified signals. Mitochondrial-nuclear communication through intergenomic signaling described in yeast was suggested to involve the Abf1p transcription factor and seemed indeed to be distinct from regulation activated by respiration dysfunction [54]. On the other hand, it was shown that sigma factors utilized by chloroplast-encoded RNA polymerase to transcribe specific sets of plastid genes are a source of retrograde signals to the nucleus [56]. However, no significant up- or downregulation of the nuclear genes for mitochondrial RNA polymerases ever appeared in our RT-qPCR and microarray analyses in response to *nad9* knockdown. It can still be hypothesized that other regular nuclear-encoded factors involved in plant mitochondrial transcription play a regulatory role.

Whichever the mitochondrial signal, ribozyme-mediated *nad9* knockdown triggered an anterograde response involving a series of nuclear transcription factors, either in light-grown or dark-grown seedlings. A number of these belonged to families highlighted in previous studies as taking part in mitochondrion-nucleus communication during plant growth, development, defense or stress (MYB, WRKY, ARR, AP2/EREBP, IAA, bHLH, ANAC/NAM) [14,61,62]. Many of them are responsive to hormone signaling. Notably, a meta-analysis of available datasets identified the plant hormones abscissic acid, auxin, cytokinin, jasmonate, and salicylic acid as main regulators of mitochondrial functions [61]. Modulation of the miR414 micro RNA (miRNA) level in response to *nad9* knockdown was also in line. miR414 was proposed earlier to be involved in the regulation of the genes participating in auxin response pathways and energy metabolism during plant embryonic development [63] and a number of its target genes were reported to play an important role in hormone signal transduction, especially in the auxin and ethylene pathways [64]. miR164 is a further miRNA highlighted in previous mitochondrion-nucleus communication analyses [61]. Also striking is the up- or downregulation of a number of small nucleolar RNAs (snoRNAs) in response to *nad9* knockdown. Beyond their fundamental role in the modification and processing of ribosomal RNAs, snoRNAs recently appeared to be a source of short regulatory RNAs [65,66].

Mining the data to identify proteins potentially sent to the mitochondria upon *nad9* knockdown in dark-grown seedlings in turn highlighted some interesting features. Firstly, the results suggest an upregulation of putative mitochondrial auxin-responsive proteins encoded by small auxin upregulated RNAs (SAUR), as well as further hormone-responsive factors (Table 5). The family of SAUR auxin-responsive genes has about 60–140 members in most higher plant species [67] and plays a central role in fine-tuning of growth in response to auxin or other hormones. Whether the SAUR proteins pointed out in our analyses are indeed targeted into mitochondria and have a specific functional or regulatory activity in there remains to be investigated in the future. The second remarkable feature highlighted by the analyses was a downregulation of a number of putative mitochondrial RNA metabolism factors, especially pentatricopeptide repeat (PPR) proteins (Table 5). PPR proteins in plants are encoded by a large family of nuclear genes with up to 800 members and most of them are targeted to the organelles [68–70]. Aiding in RNA editing, maturation, stabilisation, intron splicing, transcription and translation of organellar genes, they are major RNA metabolism factors in both mitochondria and chloroplasts. The AT4G01030 PPR gene highlighted in our plant study was annotated as coding for an editing factor, while endonuclease activity was attributed to the protein encoded by the AT4G21190 gene. The function of the remaining ones (Table 5) remains to be investigated and experimental evidence is needed to confirm mitochondrial targeting.

## 5. Conclusions

In conclusion, establishing directed manipulation of the mitochondrial transcriptome brought evidence that RNA levels are controlled and coordinated in plant mitochondria, and that interfering with specific mitochondrial transcripts causes a retrograde response that affects the nuclear transcriptome and triggers an anterograde response. The exact mechanisms underlying the observed transcriptome control and intercompartment cross-talk are still to be deciphered, but they do not seem to rely on signals that would be generated by mitochondrial metabolic dysfunction. While the control of mitochondrial gene expression had so far mostly been documented at the post-transcriptional level, our observations introduce a broader and more integrated view of the genetic regulation in the organelles.

## 6. Patents

Commercial use of the ribozyme/PKTLS strategy to manipulate mitochondrial transcripts is covered by patent Number WO 2010/031918.

**Supplementary Materials:** Supplementary File 1 containing: Figure S1. Phenotype of *A. thaliana* transgenic seedlings expressing the target-deprived *Rzsdh3* ribozyme or the *Rznad9* ribozyme; Figure S2. *Rznad9*-L-PKTLS-mediated knockdown of *nad9* steady-state levels in transformed *A. thaliana* seedlings at intermediate stage of growth; Figure S3. Short time kinetics of ribozyme expression and target RNA steady-state level in *A. thaliana* plants at the early stage of development; Figure S4. Common regions between the *orf77* and *atp9* sequences; Figure S5. Phenotype of *A. thaliana* transgenic seedlings expressing the *orf77*-PKTLS RNA *versus* control seedlings; Figure S6. Kinetics of *orf77*-PKTLS RNA expression at the early stage of development in light-grown transformed *A. thaliana* plants; Figure S7. The *orf77*-PKTLS RNA is recovered in the mitochondrial fraction; Figure S8. Knockdown of *nad9* negatively impacts the level of further mitochondrial mRNAs coding for subunits of oxidative phosphorylation complexes; Supplementary File 2: sparse PLS analysis of microarray data from light-grown seedlings; Supplementary File 3: significant results from sparse PLS analyses of microarray data from light-grown and dark-grown seedlings; Supplementary File 4: microarray analysis of samples from dark-grown seedlings; Supplementary File 5: sparse PLS analysis of microarray data from dark-grown seedlings; Supplementary File 6: mining microarray data from dark-grown seedlings for transcripts encoding proteins shown or predicted to localize to mitochondria

**Author Contributions:** Conceptualization, A.D.; Formal analysis, E.D.; Investigation, A.K.N., R.K.I., D.M., R.V., M.G., E.W. and L.S.-T.; Methodology, M.S.; Project administration, A.D.; Supervision, J.B., F.W.-L., J.M.G. and A.D.; Writing—original draft, A.D.

**Funding:** This research was funded by grants from the French State program "Investments for the Future" (LABEX ANR-11-LABX-0057_MITOCROSS and ANR-10-LABX-0040-SPS), from the French National Research Agency (ANR-06-MRAR-037-02, ANR-09-BLAN-0240-01), from the Polish National Science Centre (UMO-2013/09/B/NZ1/03359), as well as by regular funding from the French National Center for Scientific Research (CNRS-UPR2357) and the University of Strasbourg. R.V. was supported by a fellowship shared by the CNRS and the Région Alsace. Cooperation between the IBMP and the CABB is supported by the PERIDOT Program developed in France by the Ministère de l'Europe et des Affaires Etrangères (MEAE) and the Ministère de l'Enseignement Supérieur, de la Recherche et de l'Innovation (MESRI) and in Pakistan by the Higher Education Commission (HEC).

**Acknowledgments:** We thank Anne Cosset, Anne-Marie Duchêne and Clara Chicois for their help at some stages of this work.

## References

1. Shokolenko, I.N.; Alexeyev, M.F. Mitochondrial transcription in mammalian cells. *Front. Biosci.* **2017**, *22*, 835–853.
2. Holec, S.; Lange, H.; Kühn, K.; Alioua, M.; Börner, T.; Gagliardi, D. Relaxed transcription in *Arabidopsis* mitochondria is counterbalanced by RNA stability control mediated by polyadenylation and polynucleotide phosphorylase. *Mol. Cell. Biol.* **2006**, *26*, 2869–2876. [CrossRef] [PubMed]
3. Rackham, O.; Mercer, T.R.; Filipovska, A. The human mitochondrial transcriptome and the RNA-binding proteins that regulate its expression. *Wiley Interdiscip. Rev. RNA* **2012**, *3*, 675–695. [CrossRef] [PubMed]
4. Small, I.D.; Rackham, O.; Filipovska, A. Organelle transcriptomes: Products of a deconstructed genome. *Curr. Opin. Microbiol.* **2013**, *16*, 652–658. [CrossRef] [PubMed]
5. Pearce, S.F.; Rebelo-Guiomar, P.; D'Souza, A.R.; Powell, C.A.; Van Haute, L.; Minczuk, M. Regulation of mammalian mitochondrial gene expression: Recent advances. *Trends Biochem. Sci.* **2017**, *42*, 625–639. [CrossRef] [PubMed]
6. Rebelo-Guiomar, P.; Powell, C.A.; Van Haute, L.; Minczuk, M. The mammalian mitochondrial epitranscriptome. *Biochim. Biophys. Acta Gene Regul. Mech.* **2019**, *1862*, 429–446. [CrossRef] [PubMed]
7. Pohjoismaki, J.L.; Kruger, M.; Al-Furoukh, N.; Lagerstedt, A.; Karhunen, P.J.; Braun, T. Postnatal cardiomyocyte growth and mitochondrial reorganization cause multiple changes in the proteome of human cardiomyocytes. *Mol. BioSyst.* **2013**, *9*, 1210–1219. [CrossRef] [PubMed]
8. Mposhi, A.; Van der Wijst, M.G.; Faber, K.N.; Rots, M.G. Regulation of mitochondrial gene expression, the epigenetic enigma. *Front. Biosci. (Landmark Ed)* **2017**, *22*, 1099–1113.
9. Barshad, G.; Marom, S.; Cohen, T.; Mishmar, D. Mitochondrial DNA transcription and its regulation: An evolutionary perspective. *Trends Genet.* **2018**, *34*, 682–692. [CrossRef]
10. Robles, P.; Quesada, V. Transcriptional and post-transcriptional regulation of organellar gene expression (OGE) and its roles in plant salt tolerance. *Int. J. Mol. Sci.* **2019**, *20*, 1056. [CrossRef]
11. Woodson, J.D.; Chory, J. Coordination of gene expression between organellar and nuclear genomes. *Nat. Rev. Genet.* **2008**, *9*, 383–395. [CrossRef] [PubMed]
12. Cogliati, S.; Lorenzi, I.; Rigoni, G.; Caicci, F.; Soriano, M.E. Regulation of mitochondrial electron transport chain assembly. *J. Mol. Biol.* **2018**, *430*, 4849–4873. [CrossRef] [PubMed]
13. Weinhouse, C. Mitochondrial-epigenetic crosstalk in environmental toxicology. *Toxicology* **2017**, *391*, 5–17. [CrossRef] [PubMed]

14. Ng, S.; De Clercq, I.; Van Aken, O.; Law, S.R.; Ivanova, A.; Willems, P.; Giraud, E.; Van Breusegem, F.; Whelan, J. Anterograde and retrograde regulation of nuclear genes encoding mitochondrial proteins during growth, development, and stress. *Mol. Plant* **2014**, *7*, 1075–1093. [CrossRef] [PubMed]
15. Liu, Z.; Butow, R.A. Mitochondrial retrograde signaling. *Annu. Rev. Genet.* **2006**, *40*, 159–185. [CrossRef]
16. Dekkers, B.J.; Pearce, S.; van Bolderen-Veldkamp, R.P.; Marshall, A.; Widera, P.; Gilbert, J.; Drost, H.G.; Bassel, G.W.; Müller, K.; King, J.R.; et al. Transcriptional dynamics of two seed compartments with opposing roles in *Arabidopsis* seed germination. *Plant Physiol.* **2013**, *163*, 205–215. [CrossRef]
17. Law, S.R.; Narsai, R.; Taylor, N.L.; Delannoy, E.; Carrie, C.; Giraud, E.; Millar, A.H.; Small, I.; Whelan, J. Nucleotide and RNA metabolism prime translational initiation in the earliest events of mitochondrial biogenesis during *Arabidopsis* germination. *Plant Physiol.* **2012**, *158*, 1610–1627. [CrossRef]
18. Law, S.R.; Narsai, R.; Whelan, J. Mitochondrial biogenesis in plants during seed germination. *Mitochondrion* **2014**, *19*, 214–221. [CrossRef]
19. Salinas, T.; Duchêne, A.M.; Maréchal-Drouard, L. Recent advances in tRNA mitochondrial import. *Trends Biochem. Sci.* **2008**, *33*, 320–329. [CrossRef] [PubMed]
20. Val, R.; Wyszko, E.; Valentin, C.; Szymanski, M.; Cosset, A.; Alioua, M.; Dreher, T.W.; Barciszewski, J.; Dietrich, A. Organelle trafficking of chimeric ribozymes and genetic manipulation of mitochondria. *Nucleic Acids Res.* **2011**, *39*, 9262–9274. [CrossRef] [PubMed]
21. Sultan, L.D.; Mileshina, D.; Grewe, F.; Rolle, K.; Abudraham, S.; Glodowicz, P.; Niazi, A.K.; Keren, I.; Shevtsov, S.; Klipcan, L.; et al. The reverse transcriptase/RNA maturase protein MatR is required for the splicing of various group II introns in *Brassicaceae* mitochondria. *Plant Cell* **2016**, *28*, 2805–2829. [CrossRef] [PubMed]
22. Perrotta, A.T.; Been, M.D. A pseudoknot-like structure required for efficient self-cleavage of hepatitis delta virus RNA. *Nature* **1991**, *350*, 434–436. [CrossRef] [PubMed]
23. Zuo, J.; Niu, Q.W.; Chua, N.H. Technical advance: An estrogen receptor-based transactivator XVE mediates highly inducible gene expression in transgenic plants. *Plant J.* **2000**, *24*, 265–273. [CrossRef] [PubMed]
24. Mathews, D.H.; Sabina, J.; Zuker, M.; Turner, D.H. Expanded sequence dependence of thermodynamic parameters improves prediction of RNA secondary structure. *J. Mol. Biol.* **1999**, *288*, 911–940. [CrossRef] [PubMed]
25. Van Engelen, F.A.; Molthoff, J.W.; Conner, A.J.; Nap, J.P.; Pereira, A.; Stiekema, W.J. pBINPLUS: An improved plant transformation vector based on pBIN19. *Transgenic Res.* **1995**, *4*, 288–290. [CrossRef] [PubMed]
26. Clough, S.J.; Bent, A.F. Floral dip: A simplified method for *Agrobacterium*-mediated transformation of *Arabidopsis thaliana*. *Plant J.* **1998**, *16*, 735–743. [CrossRef] [PubMed]
27. Untergasser, A.; Cutcutache, I.; Koressaar, T.; Ye, J.; Faircloth, B.C.; Remm, M.; Rozen, S.G. Primer3–new capabilities and interfaces. *Nucleic Acids Res.* **2012**, *40*, e115. [CrossRef]
28. Sweetlove, L.J.; Taylor, N.L.; Leaver, C.J. Isolation of intact, functional mitochondria from the model plant *Arabidopsis thaliana*. *Methods Mol. Biol.* **2007**, *372*, 125–136.
29. Lurin, C.; Andrés, C.; Aubourg, S.; Bellaoui, M.; Bitton, F.; Bruyère, C.; Caboche, M.; Debast, C.; Gualberto, J.; Hoffmann, B.; et al. Genome-wide analysis of *Arabidopsis* pentatricopeptide repeat proteins reveals their essential role in organelle biogenesis. *Plant Cell* **2004**, *16*, 2089–2103. [CrossRef]
30. Yang, Y.H.; Dudoit, S.; Luu, P.; Lin, D.M.; Peng, V.; Ngai, J.; Speed, T.P. Normalization for cDNA microarray data: A robust composite method addressing single and multiple slide systematic variation. *Nucleic Acids Res.* **2002**, *30*, e15. [CrossRef]
31. Matthew, E.R.; Phipson, B.; Wu, D.; Hu, Y.; Law, C.W.; Shi, W.; Smyth, G.K. *limma* powers differential expression analyses for RNA-sequencing and microarray studies. *Nucleic Acids Res.* **2015**, *43*, e47.
32. Smyth, G.K. Linear models and empirical bayes methods for assessing differential expression in microarray experiments. *Stat. Appl. Genet. Mol. Biol.* **2004**, *3*, 3. [CrossRef] [PubMed]
33. Guedj, M.; Robin, S.; Celisse, A.; Nuel, G. Kerfdr: A semi-parametric kernel-based approach to local false discovery rate estimation. *BMC Bioinform.* **2009**, *10*, 84. [CrossRef] [PubMed]
34. Storey, J.D.; Tibshirani, R. Statistical significance for genomewide studies. *Proc. Natl. Acad. Sci. USA* **2003**, *100*, 9440–9445. [CrossRef] [PubMed]
35. Chun, H.; Keleş, S. Sparse partial least squares regression for simultaneous dimension reduction and variable selection. *J. R. Stat. Soc. Series B Stat. Methodol.* **2010**, *72*, 3–25. [CrossRef] [PubMed]

36. Gagnot, S.; Tamby, J.P.; Martin-Magniette, M.L.; Bitton, F.; Taconnat, L.; Balzergue, S.; Aubourg, S.; Renou, J.P.; Lecharny, A.; Brunaud, V. CATdb: A public access to *Arabidopsis* transcriptome data from the URGV-CATMA platform. *Nucleic Acids Res.* **2008**, *36*, D986–D990. [CrossRef] [PubMed]
37. Zaag, R.; Tamby, J.P.; Guichard, C.; Tariq, Z.; Rigaill, G.; Delannoy, E.; Renou, J.P.; Balzergue, S.; Mary-Huard, T.; Aubourg, S.; et al. GEM2Net: From gene expression modeling to–Omics networks, a new CATdb module to investigate *Arabidopsis thaliana* genes involved in stress response. *Nucleic Acids Res.* **2015**, *43*, D1010–D1017. [CrossRef]
38. Barrett, T.; Wilhite, S.E.; Ledoux, P.; Evangelista, C.; Kim, I.F.; Tomashevsky, M.; Marshall, K.A.; Phillippy, K.H.; Sherman, P.M.; Holko, M.; et al. NCBI GEO: Archive for functional genomics data sets–update. *Nucleic Acids Res.* **2013**, *41*, D991–D995. [CrossRef]
39. Brazma, A.; Hingamp, P.; Quackenbush, J.; Sherlock, G.; Spellman, P.; Stoeckert, C.; Aach, J.; Ansorge, W.; Ball, C.A.; Causton, H.C.; et al. Minimum information about a microarray experiment (MIAME)–toward standards for microarray data. *Nat. Genet.* **2001**, *29*, 365–371. [CrossRef]
40. Matsuda, D.; Dreher, T.W. The tRNA-like structure of Turnip yellow mosaic virus RNA is a 3′-translational enhancer. *Virology* **2004**, *321*, 36–46. [CrossRef]
41. Delage, L.; Duchêne, A.M.; Zaepfel, M.; Maréchal-Drouard, L. The anticodon and the D-domain sequences are essential determinants for plant cytosolic tRNA(Val) import into mitochondria. *Plant J.* **2003**, *34*, 623–633. [CrossRef] [PubMed]
42. Dietrich, A.; Maréchal-Drouard, L.; Carneiro, V.; Cosset, A.; Small, I. A single base change prevents import of cytosolic tRNA(Ala) into mitochondria in transgenic plants. *Plant J.* **1996**, *10*, 913–918. [CrossRef]
43. Matera, J.T.; Monroe, J.; Smelser, W.; Gabay-Laughnan, S.; Newton, K.J. Unique changes in mitochondrial genomes associated with reversions of S-type cytoplasmic male sterility in maizemar. *PLoS ONE* **2011**, *6*, e23405. [CrossRef] [PubMed]
44. Heazlewood, J.L.; Verboom, R.E.; Tonti-Filippini, J.; Small, I.; Millar, A.H. SUBA: The *Arabidopsis* subcellular database. *Nucleic Acids Res.* **2007**, *35*, D213–D218. [CrossRef]
45. Millar, A.H.; Whelan, J.; Soole, K.L.; Day, D.A. Organization and regulation of mitochondrial respiration in plants. *Annu. Rev. Plant Biol.* **2011**, *62*, 79–104. [CrossRef]
46. Topping, J.F.; Leaver, C.J. Mitochondrial gene expression during wheat leaf development. *Planta* **1990**, *182*, 399–407. [CrossRef]
47. Howell, K.A.; Millar, A.H.; Whelan, J. Ordered assembly of mitochondria during rice germination begins with pro-mitochondrial structures rich in components of the protein import apparatus. *Plant Mol. Biol.* **2006**, *60*, 201–223. [CrossRef] [PubMed]
48. Howell, K.A.; Cheng, K.; Murcha, M.W.; Jenkin, L.E.; Millar, A.H.; Whelan, J. Oxygen initiation of respiration and mitochondrial biogenesis in rice. *J. Biol. Chem.* **2007**, *282*, 15619–15631. [CrossRef] [PubMed]
49. Ishizaki, K.; Larson, T.R.; Schauer, N.; Fernie, A.R.; Graham, I.A.; Leaver, C.J. The critical role of *Arabidopsis* electron-transfer flavoprotein:ubiquinone oxidoreductase during dark-induced starvation. *Plant Cell* **2005**, *17*, 2587–2600. [CrossRef]
50. Sung, T.Y.; Tseng, C.C.; Hsieh, M.H. The SLO1 PPR protein is required for RNA editing at multiple sites with similar upstream sequences in *Arabidopsis* mitochondria. *Plant J.* **2010**, *63*, 499–511. [CrossRef] [PubMed]
51. Hsieh, W.Y.; Liao, J.C.; Chang, C.Y.; Harrison, T.; Boucher, C.; Hsieh, M.H. The SLOW GROWTH3 pentatricopeptide repeat protein is required for the splicing of mitochondrial NADH dehydrogenase subunit 7 intron 2 in *Arabidopsis. Plant Physiol.* **2015**, *168*, 490–501. [CrossRef] [PubMed]
52. Koprivova, A.; Des Francs-Small, C.C.; Calder, G.; Mugford, S.T.; Tanz, S.; Lee, B.R.; Zechmann, B.; Small, I.; Kopriva, S. Identification of a pentatricopeptide repeat protein implicated in splicing of intron 1 of mitochondrial *nad7* transcripts. *J. Biol. Chem.* **2010**, *285*, 32192–32199. [CrossRef] [PubMed]
53. Nakagawa, N.; Sakurai, N. A mutation in At-nMat1a, which encodes a nuclear gene having high similarity to group II intron maturase, causes impaired splicing of mitochondrial NAD4 transcript and altered carbon metabolism in *Arabidopsis thaliana. Plant Cell Physiol.* **2006**, *47*, 772–783. [CrossRef] [PubMed]
54. Woo, D.K.; Phang, T.L.; Trawick, J.D.; Poyton, R.O. Multiple pathways of mitochondrial-nuclear communication in yeast: Intergenomic signaling involves ABF1 and affects a different set of genes than retrograde regulation. *Biochim. Biophys. Acta* **2009**, *1789*, 135–145. [CrossRef] [PubMed]
55. Börner, T. The discovery of plastid-to-nucleus retrograde signaling–A personal perspective. *Protoplasma* **2017**, *254*, 1845–1855. [CrossRef] [PubMed]

56. Woodson, J.D.; Perez-Ruiz, J.M.; Schmit, R.J.; Ecker, J.R.; Chory, J. Sigma factor-mediated plastid retrograde signals control nuclear gene expression. *Plant J.* **2013**, *73*, 1–13. [CrossRef] [PubMed]
57. Holec, S.; Lange, H.; Canaday, J.; Gagliardi, D. Coping with cryptic and defective transcripts in plant mitochondria. *Biochim. Biophys. Acta* **2008**, *1779*, 566–573. [CrossRef]
58. Nam, M.; Akie, T.E.; Sanosaka, M.; Craige, S.M.; Kant, S.; Keaney, J.F., Jr.; Cooper, M.P. Mitochondrial retrograde signaling connects respiratory capacity to thermogenic gene expression. *Sci. Rep.* **2017**, *7*, 2013. [CrossRef]
59. da Cunha, F.M.; Torelli, N.Q.; Kowaltowski, A.J. Mitochondrial retrograde signaling: Triggers, pathways, and outcomes. *Oxid. Med. Cell. Longev.* **2015**, *2015*, 482582. [CrossRef]
60. Jazwinski, S.M. The retrograde response: A conserved compensatory reaction to damage from within and from without. *Prog. Mol. Biol. Transl. Sci.* **2014**, *127*, 133–154.
61. Berkowitz, O.; De Clercq, I.; Van Breusegem, F.; Whelan, J. Interaction between hormonal and mitochondrial signalling during growth, development and in plant defence responses. *Plant Cell Environ.* **2016**, *39*, 1127–1139. [CrossRef] [PubMed]
62. Kühn, K.; Yin, G.; Duncan, O.; Law, S.R.; Kubiszewski-Jakubiak, S.; Kaur, P.; Meyer, E.; Wang, Y.; Colas des Francs-Small, C.; Giraud, E.; et al. Decreasing electron flux through the cytochrome and/or alternative respiratory pathways triggers common and distinct cellular responses dependent on growth conditions. *Plant Physiol.* **2015**, *167*, 228–250. [CrossRef]
63. Zhang, F.; Zhao, J.; Xu, S.; Fang, W.; Chen, F.; Teng, N. MicroRNA and putative target discoveries in *Chrysanthemum* polyploidy breeding. *Int. J. Genomics* **2017**, *2017*, 6790478. [CrossRef] [PubMed]
64. Ma, C.; Yang, J.; Cheng, Q.; Mao, A.; Zhang, J.; Wang, S.; Weng, Y.; Wen, C. Comparative analysis of miRNA and mRNA abundance in determinate cucumber by high-throughput sequencing. *PLoS ONE* **2018**, *13*, e0190691.
65. Scott, M.S.; Ono, M. From snoRNA to miRNA: Dual function regulatory non-coding RNAs. *Biochimie* **2011**, *93*, 1987–1992. [CrossRef] [PubMed]
66. Stepanov, G.A.; Filippova, J.A.; Komissarov, A.B.; Kuligina, E.V.; Richter, V.A.; Semenov, D.V. Regulatory role of small nucleolar RNAs in human diseases. *Biomed. Res. Int.* **2015**, *2015*, 206849. [CrossRef] [PubMed]
67. Stortenbeker, N.; Bemer, M. The SAUR gene family: The plant's toolbox for adaptation of growth and development. *J. Exp. Bot.* **2019**, *70*, 17–27. [CrossRef] [PubMed]
68. Chen, L.; Li, Y.X.; Li, C.; Shi, Y.; Song, Y.; Zhang, D.; Li, Y.; Wang, T. Genome-wide analysis of the pentatricopeptide repeat gene family in different maize genomes and its important role in kernel development. *BMC Plant Biol.* **2018**, *18*, 366. [CrossRef] [PubMed]
69. Gorchs Rovira, A.; Smith, A.G. PPR proteins–Orchestrators of organelle RNA metabolism. *Physiol. Plant* **2019**. [CrossRef]
70. Xing, H.; Fu, X.; Yang, C.; Tang, X.; Guo, L.; Li, C.; Xu, C.; Luo, K. Genome-wide investigation of pentatricopeptide repeat gene family in poplar and their expression analysis in response to biotic and abiotic stresses. *Sci. Rep.* **2018**, *8*, 2817. [CrossRef]

# 4

# Gene Losses and Variations in Chloroplast Genome of Parasitic Plant *Macrosolen* and Phylogenetic Relationships within Santalales

**Liping Nie [1,2], Yingxian Cui [1,2], Liwei Wu [1,2], Jianguo Zhou [1,2], Zhichao Xu [1,2], Yonghua Li [3,*], Xiwen Li [4], Yu Wang [1,2] and Hui Yao [1,2,*]**

1 Key Lab of Chinese Medicine Resources Conservation, State Administration of Traditional Chinese Medicine of the People's Republic of China, Institute of Medicinal Plant Development, Chinese Academy of Medical Sciences and Peking Union Medical College, Beijing 100193, China; nielpforever@sina.com (L.N.); yxcui2017@163.com (Y.C.); 15895996332@163.com (L.W.); jgzhou1316@163.com (J.Z.); xuzhichao830@126.com (Z.X.); ywang@implad.ac.cn (Y.W.)
2 Engineering Research Center of Chinese Medicine Resources, Ministry of Education, Beijing 100193, China
3 College of Pharmacy, Guangxi University of Traditional Chinese Medicine, Nanning 530200, China
4 Institute of Chinese Materia Medica, China Academy of Chinese Medical Sciences, Beijing 100700, China; xwli@icmm.ac.cn
* Correspondence: liyonghua185@126.com (Y.L.); scauyaoh@sina.com (H.Y.)

**Abstract:** *Macrosolen* plants are parasitic shrubs, several of which are important medicinal plants, that are used as folk medicine in some provinces of China. However, reports on *Macrosolen* are limited. In this study, the complete chloroplast genome sequences of *Macrosolen cochinchinensis*, *Macrosolen tricolor* and *Macrosolen bibracteolatus* are reported. The chloroplast genomes were sequenced by Illumina HiSeq X. The length of the chloroplast genomes ranged from 129,570 bp (*M. cochinchinensis*) to 126,621 bp (*M. tricolor*), with a total of 113 genes, including 35 tRNA, eight rRNA, 68 protein-coding genes, and two pseudogenes (*ycf1* and *rpl2*). The simple sequence repeats are mainly comprised of A/T mononucleotide repeats. Comparative genome analyses of the three species detected the most divergent regions in the non-coding spacers. Phylogenetic analyses using maximum parsimony and maximum likelihood strongly supported the idea that Loranthaceae and Viscaceae are monophyletic clades. The data obtained in this study are beneficial for further investigations of *Macrosolen* in respect to evolution and molecular identification.

**Keywords:** *Macrosolen*; *Macrosolen cochinchinensis*; *Macrosolen tricolor*; *Macrosolen bibracteolatus*; Santalales; gene loss; chloroplast genome; phylogenetic relationship

---

## 1. Introduction

The traits of trophic specialization in all parasitic plants are described as "parasitic reduction syndrome". At the genetic level, parasitic reduction syndrome includes the functional and physical reduction of heterotrophs' plastid genomes, where rampant gene loss and an acceleration of molecular evolutionary rates occur [1,2]. Considering the partial or complete absence of their photosynthetic capacity, parasitic plants have to absorb organic nutrients, inorganic nutrients, and water from their hosts [3]. Most parasitic plants are included in the order Santalales and the families Orobanchaceae and Orchidaceae [2]. The first complete chloroplast genome of a parasitic plant was obtained from *Epifagus virginiana*, and all of its photosynthesis and energy producing genes have been lost [4]. Petersen et al. reported the complete plastome sequences of one species of *Osyris* and three species of *Viscum*. These researchers found that these four species have experienced rearrangements, and a number of protein-coding genes and two tRNA genes have been pseudogenised or completely lost [5]. The complete chloroplast genome of *Schoepfia jasminodora* has been reported; *S. jasminodora* represents the early stages of chloroplast genome degradation along with its transition to heterotrophy in related taxa [6]. Li et al. determined the complete chloroplast genome sequences of *Taxillus chinensis* and *Taxillus sutchuenensis*. The results showed that all *ndh* genes, three ribosomal protein genes, seven tRNA genes, four *ycf* genes, and the *infA* gene of these two species have been lost [7]. Previous studies have reported that *Rafflesia lagascae* only contains small fragments of plastid sequences at low coverage depth, and they cannot recover any substantial portions of the chloroplast genome [8]. In the parasitic family Orobanchaceae, the complete chloroplast genomes of some species, including *Cistanche deserticola* [9], *Aureolaria virginica, Lindenbergia philippensis* [10], and *Lathraea squamaria* [11], have been reported. These chloroplast genomes have shown physical and functional gene loss or pseudogenization. The *Balanophora* plastid genomes of *Balanophora laxiflora* and *Balanophora reflexa* [12], at 15.5 kb in size with only 19 genes, are the most reduced plastomes reported thus far, except for the 11.3 and 15.2 kb genomes of two holoparasitic species of *Pilostyles* [13] and the 12.8 kb genome of the myco-heterotroph *Sciaphila thaidanica* [14]. *Rhopalocnemis phalloides* [15], which belongs to the family Balanophoraceae, has also shown highly plastid genome reduction with 18.6 kb in length. In addition, gene loss has also been found in myco-heterotrophs [16], where carbon is obtained from fungi, thus forming mycorrhizal symbiosis with their roots. Photosynthesis-related genes are lost first, followed by housekeeping genes, which eventually results in a highly reduced genome [17].

The chloroplast is an important organelle in plant cells, and it primarily carries out photosynthesis and carbon fixation. The chloroplast genome is independent of nuclear genes, and the chloroplast possesses its own independent transcription and transport system [18,19]. A typical chloroplast genome of most angiosperms consists of four parts, namely a pair of inverted repeats (IRa and IRb), a large single-copy (LSC) region and a small single-copy (SSC) region [20]. The chloroplast genome sequences are highly conserved in gene order and content [21], and they are thus ideal research models for the study of molecular markers [22,23], species identification [24–26], and species evolution [27].

*Macrosolen* plants are parasitic shrubs that belong to the family Loranthaceae. There are approximately 40 species of *Macrosolen*, and most of them are distributed in Southern and Southeastern Asia, whereas five species of *Macrosolen* are dispersed in China [28]. *Macrosolen cochinchinensis, Macrosolen tricolor,* and *Macrosolen bibracteolatus* have been used as folk medicines in China for a long time. *M. cochinchinensis* is used to clear heat and fire, remove blood stasis, and relieve pain. *M. tricolor* is used to dissipate heat and relieve coughing. *M. bibracteolatus* is used to invigorate the liver and kidney, expel wind, remove dampness, and strengthen tendons and bones [29–31]. These species exhibit different medicinal effects. However, they have similar morphologies when they are not in fluorescence (Figure 1), resulting in an extreme difficulty in their identification on the basis of morphological features. The limited reports on *Macrosolen* hinder the related research and development. In this study, we determined the complete chloroplast genome sequences of *M. cochinchinensis*, *M. tricolor* and *M. bibracteolatus*. To reveal the phylogenetic positions of the three species and the evolution of *Macrosolen* within Santalales, we conducted phylogenetic trees using the maximum parsimony (MP) and maximum likelihood (ML) methods on the basis of common protein-coding genes from 16 species. Our results can provide important genetic resources for the study of *Macrosolen*.

**Figure 1.** Plant materials of three *Macrosolen* species. ① *Macrosolen cochinchinensis*; ② *Macrosolen tricolor*; and ③ *Macrosolen bibracteolatus*.

## 2. Results

### 2.1. Complete Chloroplast Genomes of Three Macrosolen Species

The length of the three studied chloroplast genomes ranged from 129,570 bp (*M. cochinchinensis*) to 126,621 bp (*M. tricolor*) with a typical quadripartite structure consisting of a pair of IRs (24,703–25,445 bp) separated by the LSC (70,692–73,052 bp) and the SSC (5320–5724 bp) regions (Figure 2). The three chloroplast genomes were found to highly conserved in GC content, gene content and gene order (Table 1 and Table S1). All three species comprised 113 genes, including 68 protein-coding genes, 35 tRNAs, eight rRNAs and two pseudogenes (*rps12* and *ycf2*). A total of 17 genes were found to be repeated genes, and 79 were found to be unique genes in the chloroplast genomes. Three genes (*clpP, ycf3 and rps12*) contained two introns, whereas 10 genes (*atpF, rpoC1, rpl2, rpl16, petB, petD, trnA-UGC, trnI-GAU, trnK-UUU* and *trnL-UAA*) had only one intron (Table 2 and Table S2).

**Table 1.** Length of chloroplast genome of three *Macrosolen* species and their base composition.

| Species | *M. cochinchinensis* | *M. tricolor* | *M. bibracteolatus* |
|---|---|---|---|
| Accession No. | MH161424 | MH161425 | MH161423 |
| Genome size (bp) | 129,570 | 126,621 | 127,169 |
| LSC length (bp) | 73,052 | 71,895 | 70,692 |
| SSC length (bp) | 5724 | 5320 | 5587 |
| IRs length (bp) | 25,397 | 24,703 | 25,445 |
| GC content (%) | 37.3 | 37.7 | 37.9 |
| Number of genes | 113 | 113 | 113 |
| Number of protein-coding genes | 68 | 68 | 68 |
| Number of tRNAs | 35 | 35 | 35 |
| Number of rRNAs | 8 | 8 | 8 |
| Number of pseudogenes | 2 | 2 | 2 |

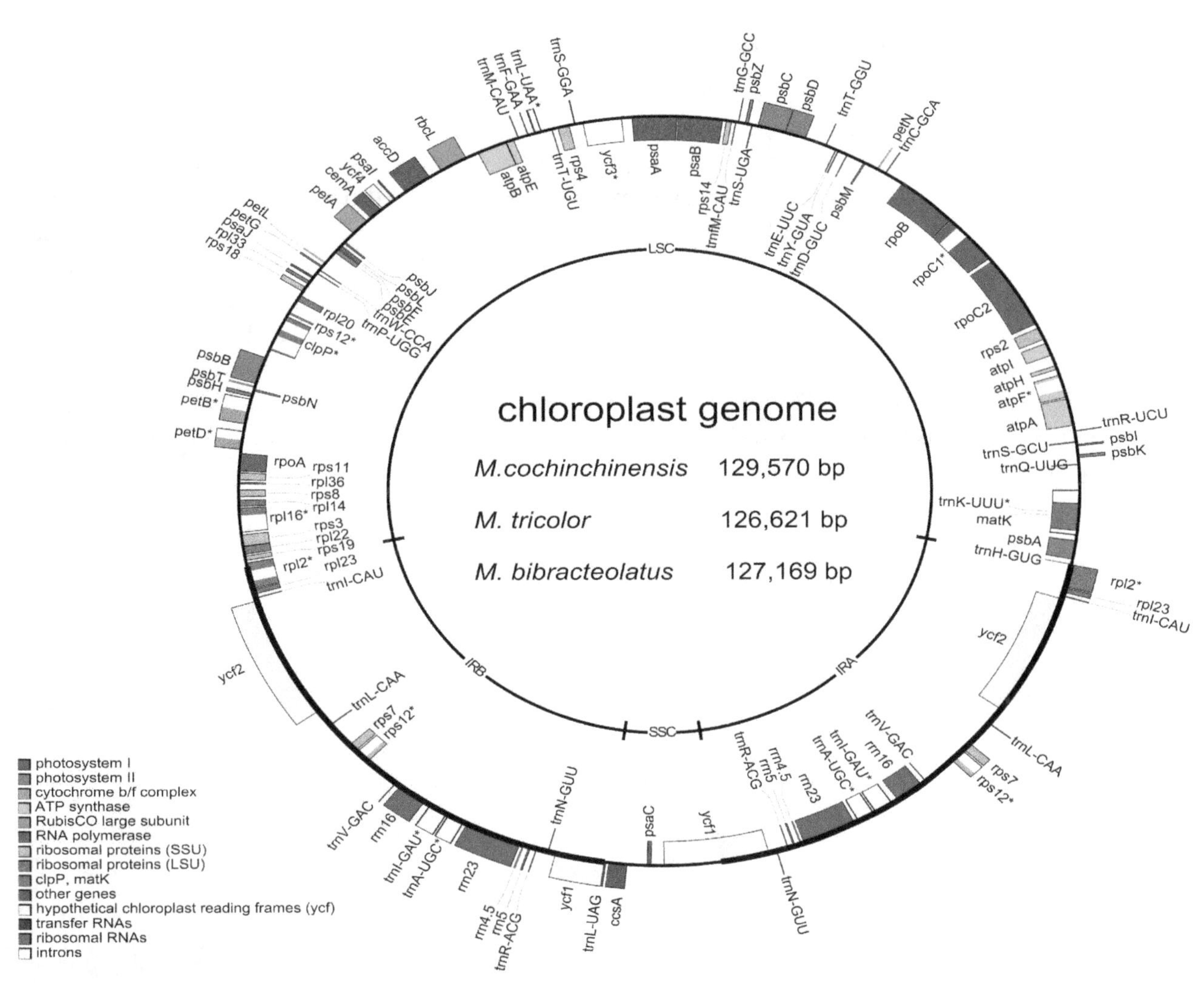


**Figure 2.** Gene map of the complete chloroplast genome of three *Macrosolen* species. Genes outside the large ring circle are transcribed in a counter-clockwise direction, and genes inside the circle are transcribed clockwise. The same color represents the same category of genes. Deep grey in the inner circle represents GC content, and lighter grey represents A/T content.

**Table 2.** Gene list of chloroplast genome of three *Macrosolen* species.

| No. | Group of Genes | Gene Names | Number |
|---|---|---|---|
| 1 | Photosystem I | *psaA, psaB, psaC, psaI, psaJ* | 5 |
| 2 | Photosystem II | *psbA, psbB, psbC, psbD, psbE, psbF, psbH, psbI, psbJ, psbK, psbL, psbM, psbN, psbT, psbZ* | 15 |
| 3 | Cytochrome b/f complex | *petA, petB *, petD *, petG, petL, petN* | 6 |
| 4 | ATP synthase | *atpA, atpB, atpE, atpF *, atpH, atpI* | 6 |
| 5 | NADH dehydrogenase | - | 0 |
| 6 | RubisCO large subuni | *rbcL* | 1 |
| 7 | RNA polymerase | *rpoA, rpoB, rpoC1 *, rpoC2* | 4 |
| 8 | Ribosomal proteins (SSU) | *rps2, rps3, rps4, rps7 (×2), rps8, rps11, rps12 ** (×2), rps14, rps18, rps19* | 12(2) |
| 9 | Ribosomal proteins (LSU) | *rpl2 * (×2), rpl14, rpl16 *, rpl20, rpl22, rpl23 (×2), rpl33, rpl36* | 10(2) |
| 10 | Proteins of unknown function | *ycf1(×2), ycf2(×2), ycf3 **, ycf4* | 6(2) |
| 11 | Transfer RNAs | *35 tRNAs* (4 contain an intron, 7 in the IRs) | 35(7) |
| 12 | Ribosomal RNAs | *rrn4.5 (×2), rrn5(×2), rrn16 (×2), rrn23 (×2)* | 8(4) |
| 13 | Other genes | *accD, clpP **, matK, ccsA, cemA* | 5 |

* One or two asterisks following genes indicate one or two contained introns, respectively. (×2) indicates that the number of the repeat unit is two. The numbers in parenthesis at the line of 'Number' indicate the total number of repeated genes.

### 2.2. Codon Usage Analyses and RNA Editing Sites

Relative synonymous codon usage (RSCU) is the ratio between the use and expected frequencies for a particular codon and a measure of nonuniform synonymous codon usage in coding sequences [32]. On the basis of the sequences of protein-coding genes, the codon usage frequency was estimated for the chloroplast genome of the three *Macrosolen* species (Figure 3). All the protein-coding genes were found to consist of 21,581, 21,598 and 21,520 codons in the chloroplast genomes of *M. cochinchinensis*, *M. tricolor* and *M. bibracteolatus*, respectively (Table S3). Figure 3 shows that the RSCU value increased with the increase in the quantity of codons which coded for a specific amino acid. Most of the amino acid codons show preferences except for methionine and tryptophan. Potential RNA editing sites were also predicted for 29 genes in the chloroplast genomes of the three species. A total of 39 RNA editing sites were identified (Table S4). The amino acid conversion from serine (S) to leucine (L) occurred most frequently, whereas that from proline (P) to serine (S) and from threonine (T) to methionine (M) occurred the least.

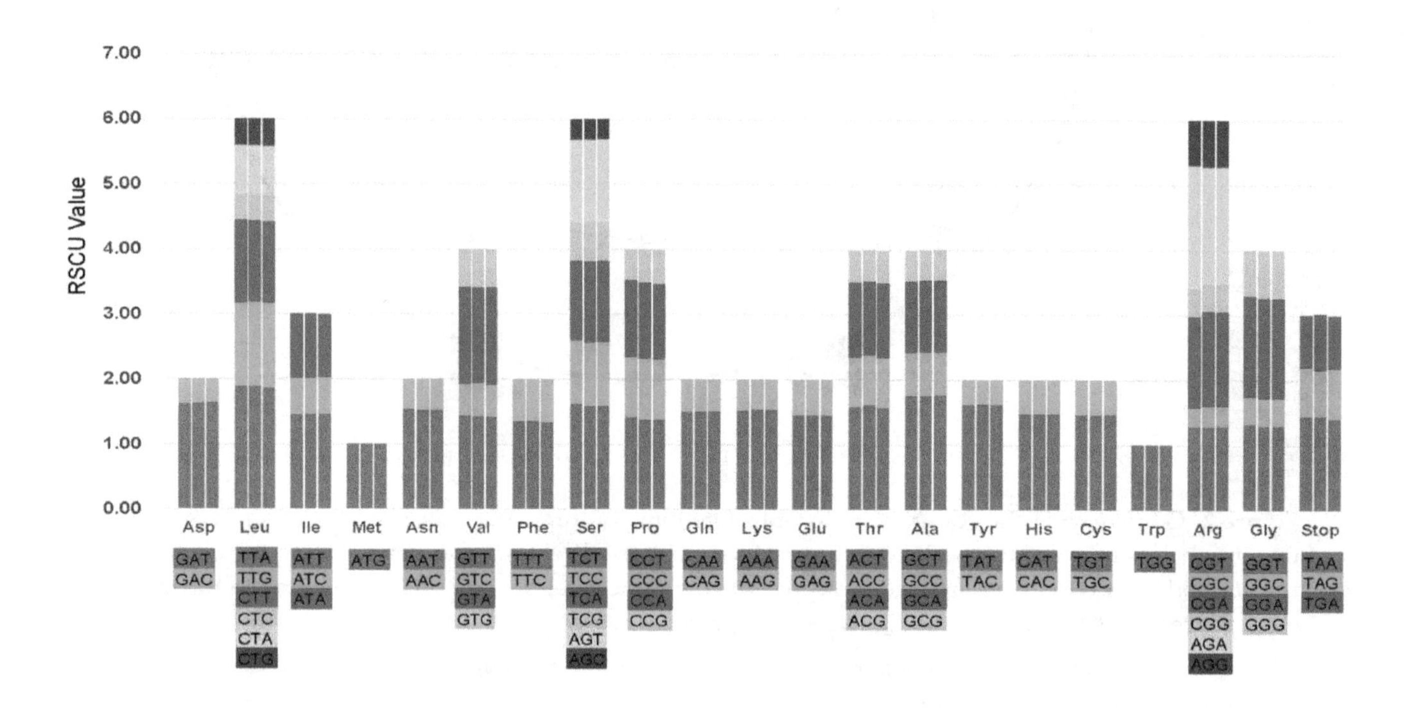


**Figure 3.** Codon content of 20 amino acids and stop codons in all of the protein-coding genes of the chloroplast genomes of three *Macrosolen* species.

### 2.3. IR Constriction and Expansion

Figure 4 shows the comparison of the boundaries of the LSC/IR/SSC regions of three *Macrosolen* species. The LSC/IR/SSC boundaries and gene contents in the chloroplast genomes of the three species were found to be highly conserved, featuring the same sequence structure and differences in length. In the three species, the *rpl2* gene, which is a normal functional gene, crossed the LSC/IRa boundary, but the *rpl2* pseudogene with a length of 1268 bp formed in the IRb region. The SSC/IRb boundaries of *M. cochinchinensis*, *M. tricolor* and *M. bibracteolatus* were found to be located in the complete *ycf1* gene, and their *ycf1* pseudogenes with lengths of 2457, 2455 and 2448 bp, respectively, were found to be produced in IRa.

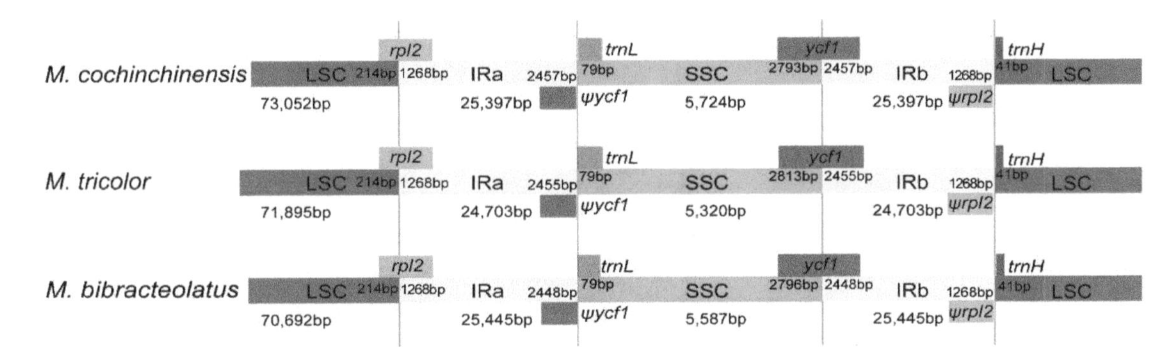


**Figure 4.** Comparison of the borders of the large single-copy (LSC), small single-copy (SSC), and inverted repeats (IR) regions among the chloroplast genomes of three *Macrosolen* species. The number above the gene features means the distance between the ends of genes and the borders sites. These features are not to scale. Ψ: pseudogenes.

### *2.4. Simple Sequence Repeats (SSRs) and Repeat Structure Analyses*

A simple sequence repeat (SSR), which is also known as microsatellite DNA, is a tandem repeat sequence consisting of one to six nucleotide repeat units [22]. SSRs are widely used as molecular markers in species identification, population genetics, and phylogenetic investigations due to their high polymorphism level [33,34]. A total of 238, 226 and 217 SSRs were identified in the chloroplast genomes of *M. cochinchinensis*, *M. tricolor* and *M. bibracteolatus*, respectively (Table 3). Amongst all SSRs, the numbers of mononucleotide repeats were the highest, with values detected at 169, 166 and 162 times in *M. cochinchinensis*, *M. tricolor* and *M. bibracteolatus*, respectively. Amongst these mononucleotide repeats, A/T was found to be the most frequent SSR. In accordance with the number of repeats, mononucleotide and dinucleotide SSRs exhibited a certain base preference that mainly contained A/T units. Long repeat sequences should be >30 bp, and these repeats are mainly distributed in the gene spacer and intron sequences. The result shows that *M. cochinchinensis* presented the highest number, comprising six forward, seven palindromic, four reverse and one complement repeats (Figure 5). Two types of *M. tricolor*, comprising six forward and nine palindromic repeats, were present. *M. bibracteolatus* presented seven forward, six palindromic and two reverse repeats.

**Table 3.** Types and amounts of simple sequence repeats (SSRs) in the chloroplast genomes of three *Macrosolen* species.

| SSR Type | Repeat Unit | Amount | | | Ratio (%) | | |
|---|---|---|---|---|---|---|---|
| | | ① | ② | ③ | ① | ② | ③ |
| mono | A/T | 161 | 159 | 153 | 95.3 | 95.8 | 94.4 |
| | C/G | 8 | 7 | 9 | 4.7 | 4.2 | 5.6 |
| di | AC/GT | 5 | 4 | 4 | 9.6 | 8.5 | 9.3 |
| | AG/CT | 13 | 14 | 13 | 25 | 29.8 | 30.2 |
| | AT/TA | 34 | 29 | 26 | 64.4 | 61.7 | 60.5 |
| tri | AAT/ATT | 4 | 4 | 0 | 66.7 | 66.7 | 0 |
| | ATC/ATG | 2 | 2 | 2 | 33.3 | 33.3 | 100 |
| tetra | AAAG/CTTT | 3 | 3 | 3 | 33.3 | 42.9 | 30 |
| | AATC/ATTG | 1 | 1 | 0 | 11.1 | 14.3 | 0 |
| | ACAG/CTGT | 1 | 1 | 1 | 11.1 | 14.3 | 10 |
| | AAAT/ATTT | 3 | 1 | 3 | 33.3 | 14.3 | 30 |
| | AATG/ATTC | 1 | 0 | 1 | 11.1 | 0 | 10 |
| | AGAT/ATCT | 0 | 1 | 1 | 0 | 14.3 | 10 |
| | ACAT/ATGT | 0 | 0 | 1 | 0 | 0 | 10 |
| penta | AATAT/ATATT | 1 | 0 | 0 | 100 | 0 | 0 |
| hexa | ATATCC/ATATGG | 1 | 0 | 0 | 100 | 0 | 0 |

① *M. cochinchinensis*; ② *M. tricolor*; and ③ *M. bibracteolatus*.

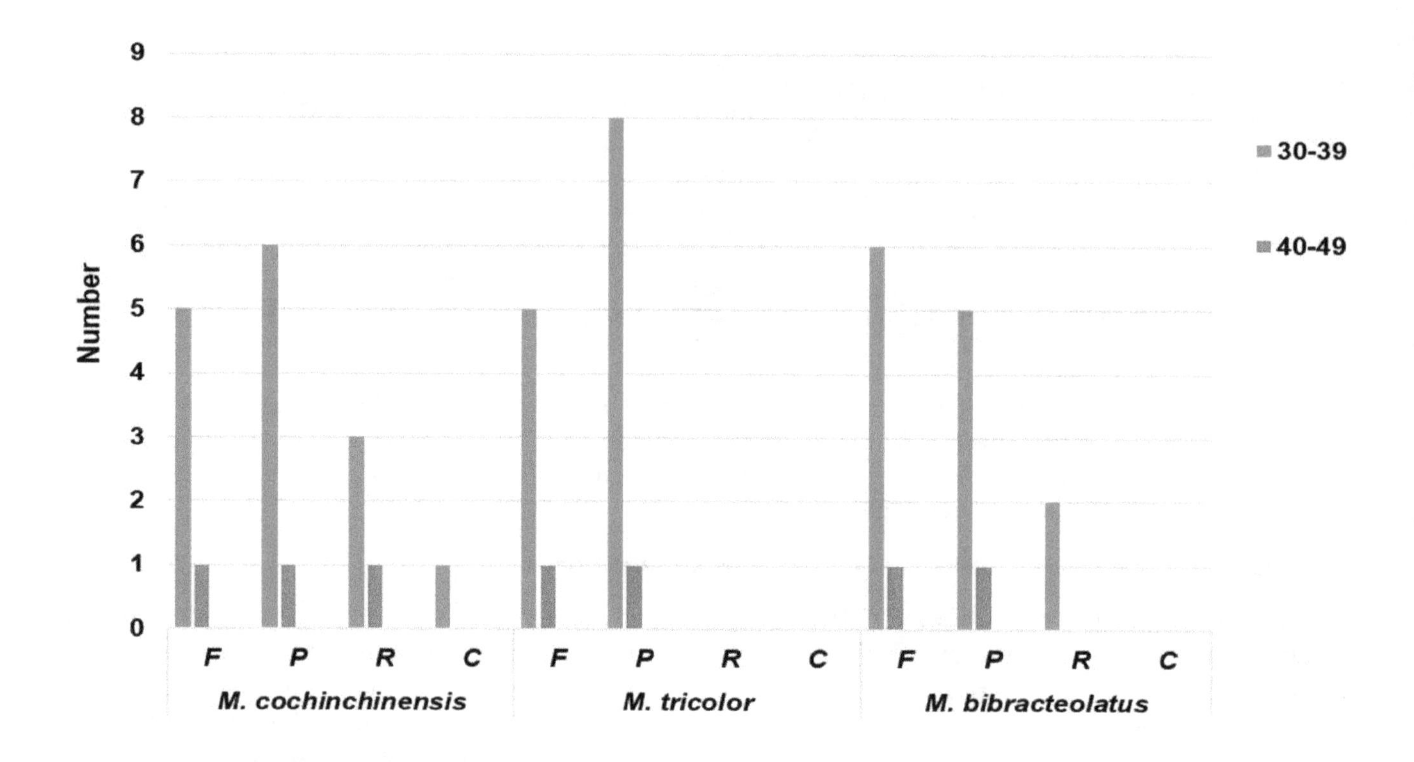


**Figure 5.** Repeat sequences in the chloroplast genomes of three *Macrosolen* species. F, P, R, and C indicate the repeat types F (forward), P (palindrome), R (reverse) and C (complement), respectively. Repeats with different lengths are indicated in different colors.

*2.5. Comparative Genomic Analyses*

The complete chloroplast of the three chloroplast genomes were compared with that of *M. cochinchinensis* as a reference using the mVISTA program. As shown in Figure 6, the *ycf1* and *ccsA* genes were found to be the most mutant genes. Except for these genes, the other genes were found to be highly conserved, and most of them showed similarities of >90%. The variations in the coding regions were smaller than those in the noncoding regions. Amongst the three chloroplast genomes, the most divergent regions were found to be localized in the intergenic spacers such as *trnF-trnM*. The rRNA genes of the three species were highly conservative, and almost no variations were observed. The K values (sequence divergence between species) were calculated, and the sliding windows of the K values were constructed by the DnaSP [35] (Figure 7). Figure 7 shows that the sequence divergence between *M. tricolor* and *M. cochinchinensis* was much higher than the other two K values. *M. bibracteolatus* and *M. tricolor* showed a small divergence ($K < 0.05$). The LSC and SSC regions were more divergent than IRs. Two mutational hotspots were found with high K values, and they were located at the LSC and SSC regions. Combined with genes location and the mVISTA result, the two hotspots were found to be *trnF-trnM* and *ycf1*.

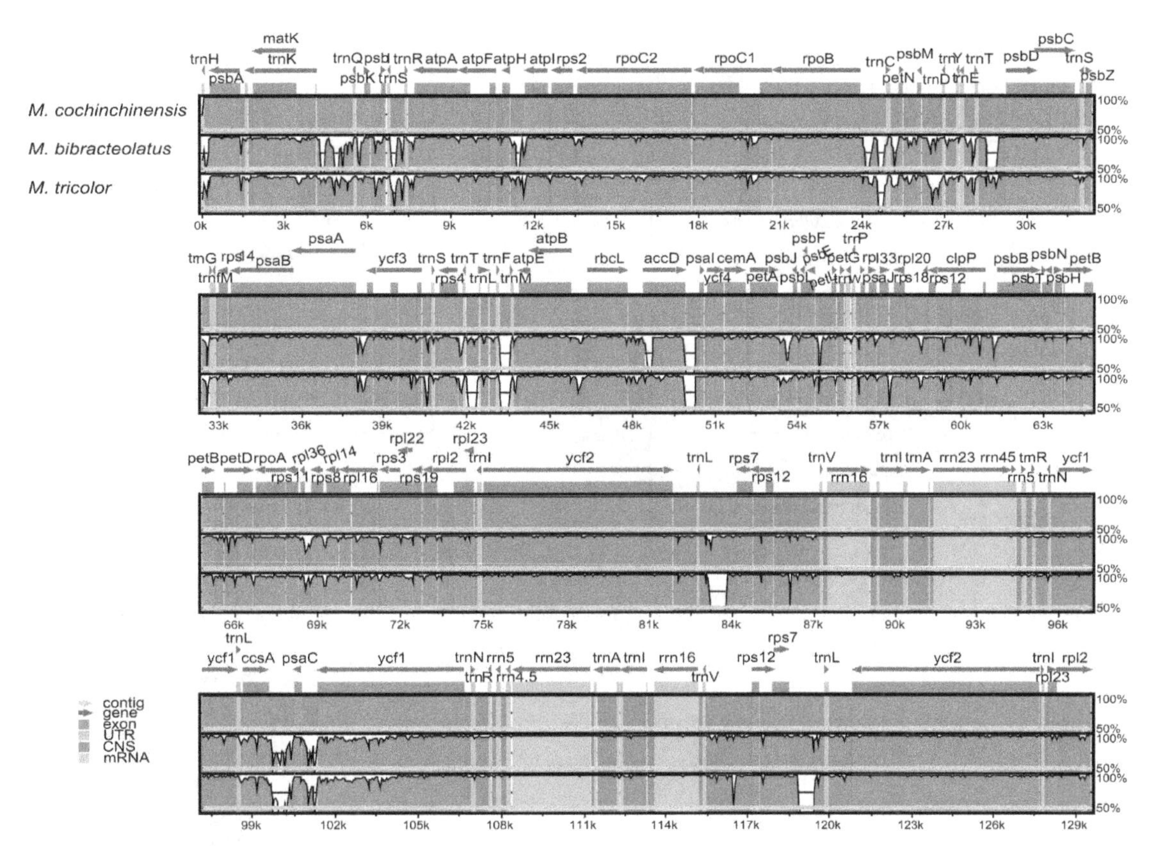

**Figure 6.** Sequence identity plot comparing the three chloroplast genomes with *M. cochinchinensis* as a reference by using mVISTA. Grey arrows and thick black lines above the alignment indicate genes with their orientation and the position of their IRs, respectively. A cut-off of 70% identity was used for the plots, and the Y-scale represents the percent identity ranging from 50% to 100%.

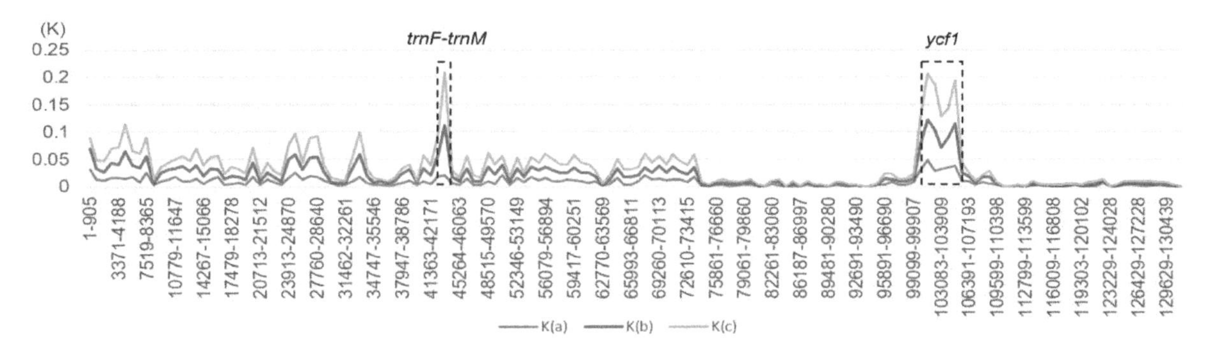

**Figure 7.** Sliding window analyses of the three whole chloroplast genomes. X-axis: position of a window. Y-axis: sequence divergence between species of each window. K(a): K values between *M. bibracteolatus* and *M. tricolor*; K(b): K values between *M. bibracteolatus* and *M. cochinchinensis*; K(c): K values between *M. tricolor* and *M. cochinchinensis*.

## *2.6. Phylogenetic Analyses*

To analyze the phylogenetic relationships of *Macrosolen* in Santalales, we constructed phylogenetic trees using 58 common protein-coding genes of 16 species and *matK* genes of 15 species by the MP and ML methods with a bootstrap of 1000 repetitions. The MP and ML trees were the same whether they were constructed by either common protein-coding genes or *matK* genes (Figure 8). All nodes in all the phylogenetic trees received a >50% bootstrap value. All four phylogenetic trees showed that the three *Macrosolen* species are sister taxa with respect to *S. jasminodora* (Olacaceae). *M. cochinchinensis*, *M. tricolor* and *M. bibracteolatus* were gathered into one branch with a well-supported bootstrap value (100%). The three species within the genus *Viscum* grouped with *Osyris alba* (Santalaceae) and all

Santalales species were clustered within a lineage distinct from the outgroup. As shown in Figure 8, the trees constructed by common protein-coding genes also received a higher bootstrap value than the trees constructed by the *matK* genes.

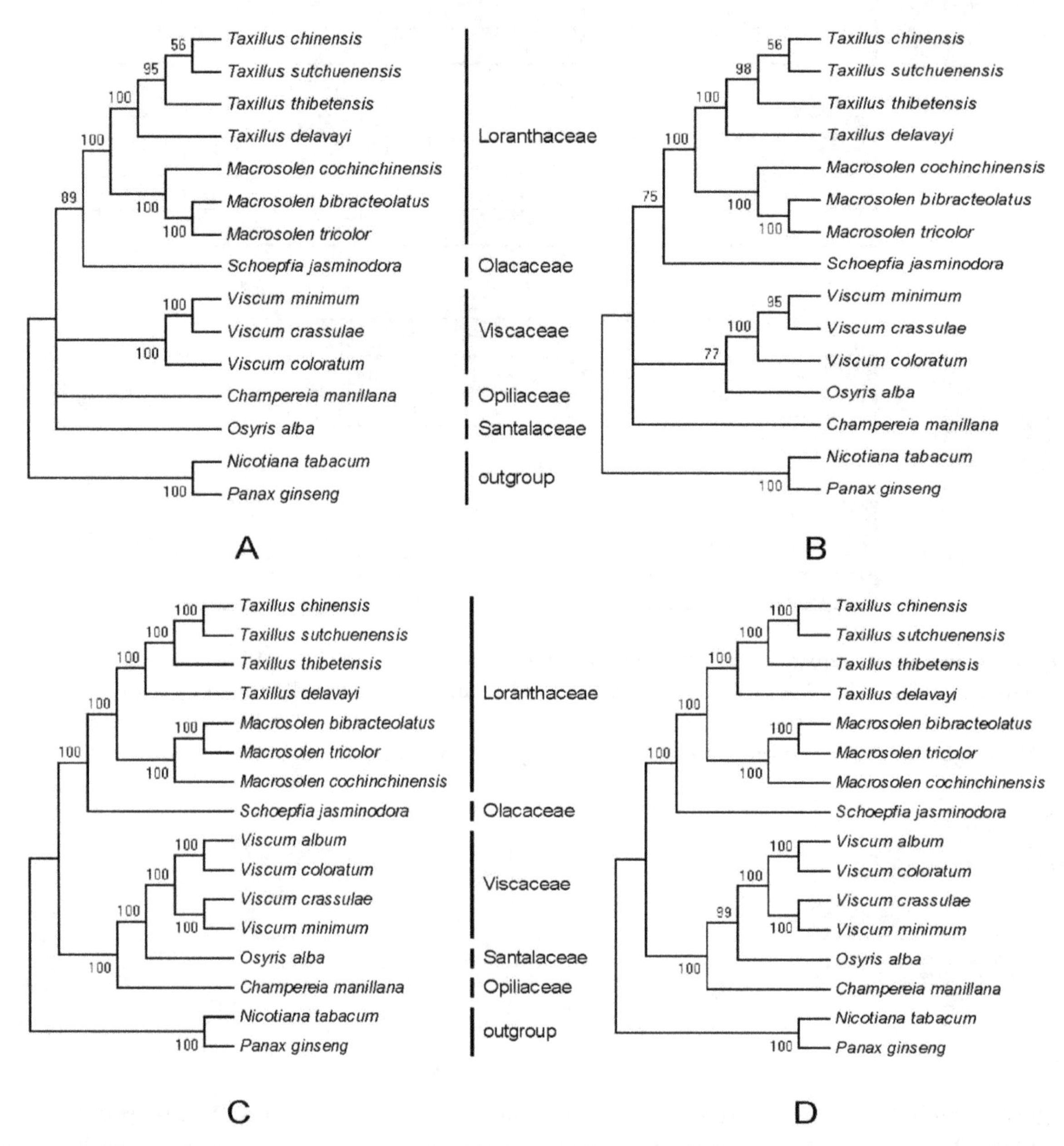


**Figure 8.** Phylogenetic trees constructed with the *matK* genes of 15 species by using the maximum parsimony (MP) (**A**) and maximum likelihood (ML) (**B**) methods. Phylogenetic trees constructed with 58 common protein-coding genes of 16 species using the MP (**C**) and ML (**D**) methods. Numbers at nodes are bootstrap values.

### 3. Discussion

Numerous variations occur in the chloroplast genomes of parasitic plants. However, only a small number of plants within Santalales have been studied. In this study, the complete chloroplast genomes of *M. cochinchinensis*, *M. tricolor* and *M. bibracteolatus* from Santalales were assembled, annotated and analyzed. Compared with the chloroplast genomes of the model plant *Nicotiana tabacum*, all the *ndh* genes of the chloroplast genomes were lost amongst the three species, and the *infA* gene, which codes for a translation initiation factor, was also missing in these species. These cases were similar to those of *T. chinensis* and *T. sutchuenensis* [7]. The *rpl16* and *ycf15* genes were lost in the three species, but they were still present in *T. chinensis* as pseudogenes (Figure 9). However, compared with the results reported by Shin et al. [36], different gene contents of the chloroplast genome were observed in *M. cochinchinensis*. These studies have shown that *M. cochinchinensis* contains the exon 1 fragment of the *ndhB* gene and a fragment of the *infA* gene, whereas the *rpl36* gene is completely lost. However, the *rpl36* gene is still present in the chloroplast genome according to our results. *M. cochinchinensis* has lost the *infA* gene and all *ndh* genes. The number of tRNA genes also differed between the two studies. We annotated 35 tRNA genes, but previous studies only obtained 30 tRNA genes. The evolution of the chloroplast genome in parasitic plants, particularly nonphotosynthetic holoparasites, can lead to significantly reconfigured plastomes [21]. The losses of *ndh* genes are associated with nutritional status or extensive rearrangements of chloroplast structures [37], and they have occurred in the reported chloroplast genomes of parasitic plants [7]. Our study also showed that *ndh* genes were lost in the transformation from autotrophy to heterotrophy [38].

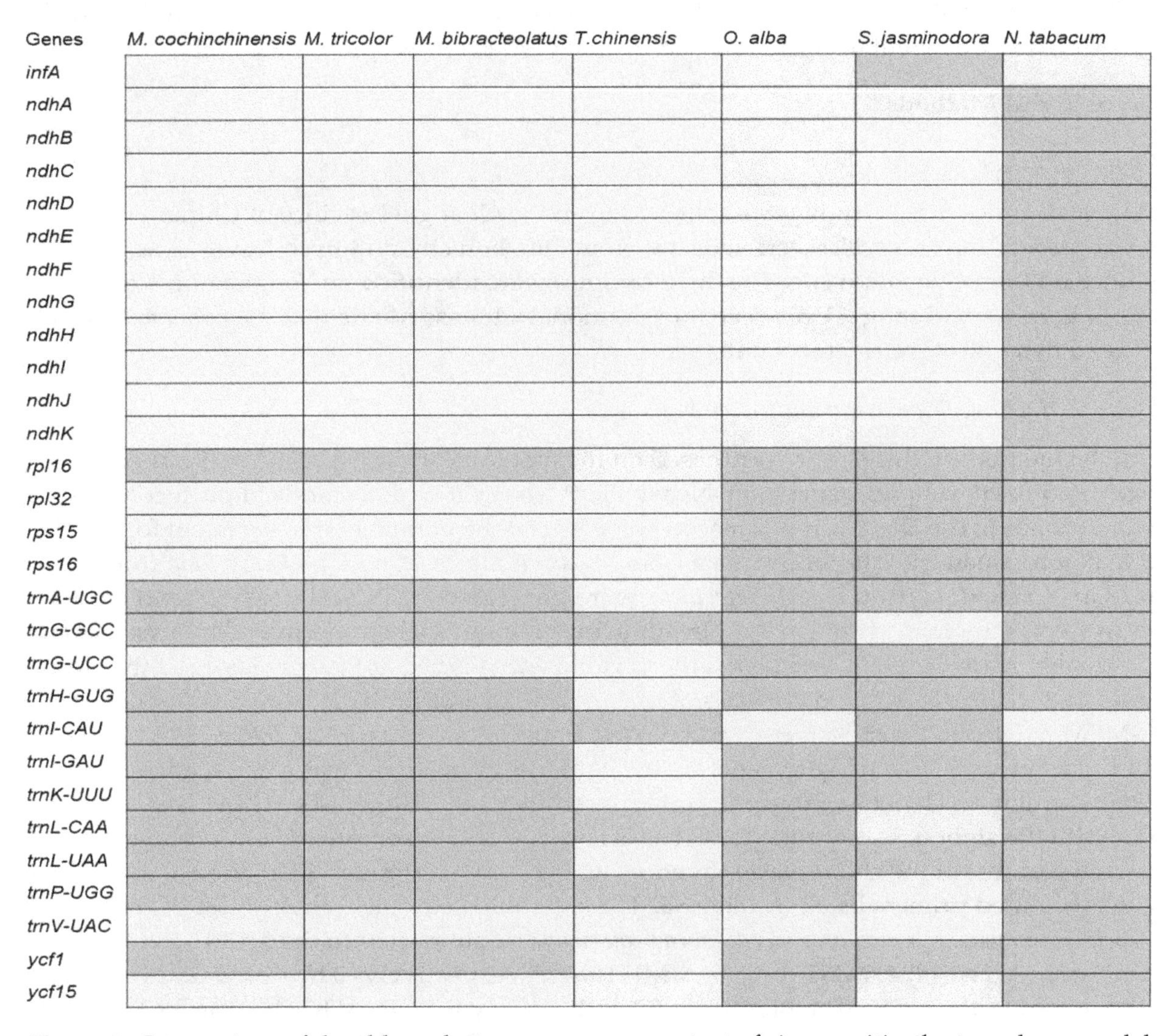


**Figure 9.** Comparison of the chloroplast genome gene content of six parasitic plants and one model plant (*Nicotiana tabacum*). The common existing genes in the complete chloroplast genome of the seven species are not listed. Red boxes indicate each gene present, and green boxes indicate that each gene is considered as a pseudogene. The yellow boxes indicate an absent gene.

The Santalales order consists of a small number of autotrophic species and a large number of parasitic species which are root or aerial (stem) parasites [39]. According to the Engler system, Santalales consists of seven families. We downloaded five families belonging to Santalales, which were available in the National Center for Biotechnology Information (NCBI) at that time, and two species as outgroups to analyze the phylogenetic relationships of *Macrosolen* in Santalales. The present study showed that Loranthaceae is closely related to Olacaceae, whereas Viscaceae is closely related to Santalaceae and Opiliaceae. These results are similar to those of previous studies [13,14]. All the phylogenetic results strongly support that Loranthaceae and Viscaceae diverged independently from each other.

As folk medicine in China, *M. cochinchinensis*, *M. tricolor* and *M. bibracteolatus* have been used to treat diseases for a long time, and their dried stems and branches with leaves are used as medicinal parts. However, *Macrosolen* species are similar in appearance, especially when they are processed into medicinal slices, thereby causing difficulty in their identification. The identification of parasitic medicinal materials has rarely been reported. Though phytochemical approaches have played an important role in species identification [26], they are inadequate because they are limited to the environment and harvest period. Molecular characterization has shown an improved specificity for plants [23,26]. In our study, mutational hotspots such as the *ycf1* gene, the *ccsA* gene and the *trnF-trnM* intergenic region are potential sites for identification of *Macrosolen* species.

## 4. Materials and Methods

### 4.1. Plant Materials

All the samples in this study were collected from the Guangxi Province of China. Fresh leaves of *M. cochinchinensis* and *M. tricolor* were collected from Qinzhou city, and fresh leaves of *M. bibracteolatus* were collected from Chongzuo city. The three samples were identified by Yonghua Li, who is from the College of Pharmacy, Guangxi University of Traditional Chinese Medicine. The collected fresh leaves were stored in a −80 °C refrigerator until use.

### 4.2. DNA Extraction, Sequencing and Assembly

All the methods in this article were based on the methods of Zhou et al. [40]. Total genomic DNA was extracted from samples using the DNeasy Plant Mini Kit with a standard protocol (Qiagen Co., Hilden, Germany). The DNA was sequenced according to the manufacturer's manual for the Illumina Hiseq X. Approximately 6.2 Gb of raw data from *M. cochinchinensis*, 6.5 Gb of raw data from *M. tricolor*, and 6.3 Gb of raw data from *M. bibracteolatus* were generated with 150 bp paired-end read lengths. The software Trimmomatic (version 0.39, Institute for Biology, Aachen, German) [41] was used to filter the low-quality reads of the raw data, and the Q value was defined as Sanger. Then, all the clean reads were mapped to the database on the basis of their coverage and similarity. Burrows–Wheeler Aligner (BWA-MEM, Wellcome Trust Sanger Institute, Wellcome Genome Campus, Cambridge, UK) was used in chloroplast genome assembly to generate the bam files. The depth was calculated using Samtools (Medical Population Genetics Program, Broad Institute, Cambridge, MA, USA) and plotted using Rscript (with the smoothScatter function). The accuracy of the assembly of the four boundaries (SSC, LSC and IR regions) of the chloroplast sequences was confirmed through PCR and Sanger sequencing using the validated primers listed in Table S5. The assembled complete chloroplast genome sequence of *M. cochinchinensis*, *M. tricolor* and *M. bibracteolatus* were submitted to the NCBI, and the accession numbers were MH161424, MH161425 and MH161423, respectively. The raw data of three species were submitted to the NCBI. The Bioproject ID of this study is PRJNA587349. The SRA accession ID of *M. tricolor* is SRR10442639, that of *M. bibracteolatus* is SRR10442640, and that of *M. cochinchinensis* is SRR10442641.

### 4.3. Genome Comparison and Phylogenetic Analyses

The whole-genome alignment for the chloroplast genomes of three *Macrosolen* species were performed and plotted using the mVISTA program (http://genome.lbl.gov/vista/mvista/submit.shtml) [42]. Gene content comparison was analyzed by CPGAVAS2 (Institute of Medicinal Plant Development, Chinese Academy of Medical Sciences and Peking Union Medical College, Beijing,

China) [43] and identified by manual correction. To determine the phylogenetic positions of three *Macrosolen* species within Santalales, we analyzed the chloroplast genomes of 16 species, encompassing 11 other taxa within this lineage, *Viscum album* (KT003925), *V. coloratu* (NC_035414), *V. crassula* (KT070881), *V. minimum* (KJ512176), *Osyris alba* (KT070882), *Schoepfia jasminodora* (KX775962), *Champereia manillana* (NC_034931), *T. chinensis* (KY996492), *T. sutchuenensis* (KY996493), *T. delavayi* (MH161426), and *T. thibetensis* (MH161427). The chloroplast genomes of *Panax ginseng* (AY582139) and *N. tabacum* (NC_001879) were used as outgroups.

### 4.4. Other Analyses

On the basis of the study of Zhou et al. [40], we analyzed the complete chloroplast genome of three *Macrosolen* species, including genome structure analyses (genome length, gene content and GC content), codon usage analyses, RNA editing site prediction, and repeat sequences analyses. The distribution of codon usage was investigated using the CodonW software (University of Texas, Houston, TX, USA) with the RSCU ratio [32]. Potential RNA editing sites were predicted using the Predictive RNA Editor for Plants (PREP-Cp, Center for Plant Science Innovation, University of Nebraska-Lincoln, Lincoln, NE, USA) suite online program [44] with a cutoff value of 0.8. Simple sequence repeats were detected using the MISA software (Pgrc.ipk-gatersleben.de/misa/) [45]. Repeat sequences were identified by REPuter (University of Bielefeld, Bielefeld, Germany) [46].

**Author Contributions:** Conceptualization, H.Y. and Y.L.; methodology, L.N., L.W. and J.Z.; formal analyses, L.N., Y.C. and J.Z.; resources, Y.L. and Y.W.; data curation, L.N., Y.C., Z.X. and X.L.; writing—original draft preparation, L.N. and H.Y.; writing—review and editing, L.N. and H.Y.; funding acquisition, H.Y. and Y.L.

## Abbreviations

| | |
|---|---|
| LSC | Large single copy |
| SSC | Small single copy |
| IR | Inverted repeat |
| MP | Maximum parsimony |
| ML | Maximum likelihood |
| RSCU | Relative synonymous codon usage |
| SSR | Simple Sequence Repeats |
| NCBI | National Center for Biotechnology Information |

## References

1. Wicke, S.; Naumann, J. Molecular evolution of plastid genomes in parasitic flowering plants. *Adv. Bot. Res.* **2018**, *85*, 315–347.
2. Wang, L.; Dong, W.; Zhou, S. Structural mutations and reorganizations in chloroplast genomes of flowering plants. *Acta Bot. Boreali Occident. Sin.* **2012**, *32*, 1282–1288.
3. Huang, X.; Guan, K.; Li, A. Biological traits and their ecological significances of parasitic plants: A review. *Chin. J. Ecol.* **2011**, *30*, 1838–1844.
4. Wolfe, K.H.; Morden, C.W.; Palmer, J.D. Function and evolution of a minimal plastid genome from a nonphotosynthetic parasitic plant. *Proc. Natl. Acad. Sci. USA* **1992**, *89*, 10648–10652. [CrossRef]
5. Petersen, G.; Cuenca, A.; Seberg, O. Plastome evolution in hemiparasitic mistletoes. *Genome Biol. Evol.* **2015**, *7*, 2520–2532. [CrossRef]
6. Su, H.; Hu, J. The complete chloroplast genome of hemiparasitic flowering plant *Schoepfia jasminodora*. *Mitochondrial DNA Part B* **2016**, *1*, 767–769. [CrossRef]
7. Li, Y.; Zhou, J.G.; Chen, X.L.; Cui, Y.X.; Xu, Z.C.; Li, Y.H.; Song, J.Y.; Duan, B.Z.; Yao, H. Gene losses and partial deletion of small single-copy regions of the chloroplast genomes of two hemiparasitic *Taxillus* species. *Sci. Rep.* **2017**, *7*, 12834. [CrossRef]
8. Molina, J.; Hazzouri, K.M.; Nickrent, D.; Geisler, M.; Meyer, R.S.; Pentony, M.M.; Flowers, J.M.; Pelser, P.; Barcelona, J.; Inovejas, S.A.; et al. Possible loss of the chloroplast genome in the parasitic flowering plant *Rafflesia lagascae* (Rafflesiaceae). *Mol. Biol. Evol.* **2014**, *31*, 793–803. [CrossRef]

9. Li, X.; Zhang, T.C.; Qiao, Q.; Ren, Z.; Zhao, J.; Yonezawa, T.; Hasegawa, M.; Crabbe, M.J.; Li, J.; Zhong, Y. Complete chloroplast genome sequence of holoparasite *Cistanche deserticola* (Orobanchaceae) reveals gene loss and horizontal gene transfer from its host *Haloxylon ammodendron* (Chenopodiaceae). *PLoS ONE* **2013**, *8*, e58747. [CrossRef]
10. Frailey, D.C.; Chaluvadi, S.R.; Vaughn, J.N.; Coatney, C.G.; Bennetzen, J.L. Gene loss and genome rearrangement in the plastids of five Hemiparasites in the family Orobanchaceae. *BMC Plant Biol.* **2018**, *18*, 30. [CrossRef]
11. Samigullin, T.H.; Logacheva, M.D.; Penin, A.A.; Vallejo-Roman, C.M. Complete plastid genome of the recent holoparasite *Lathraea squamaria* reveals earliest stages of plastome reduction in orobanchaceae. *PLoS ONE* **2016**, *11*, e0150718. [CrossRef] [PubMed]
12. Su, H.J.; Barkman, T.J.; Hao, W.; Jones, S.S.; Naumann, J.; Skippington, E.; Wafula, E.K.; Hu, J.M.; Palmer, J.D.; de Pamphilis, C.W. Novel genetic code and record-setting AT-richness in the highly reduced plastid genome of the holoparasitic plant *Balanophora*. *Proc. Natl. Acad. Sci. USA* **2019**, *116*, 934–943. [CrossRef] [PubMed]
13. Bellot, S.; Renner, S.S. The plastomes of two species in the endoparasite genus pilostyles (Apodanthaceae) each retain just five or six possibly functional genes. *Genome Biol. Evol.* **2016**, *8*, 189–201. [CrossRef] [PubMed]
14. Petersen, G.; Zervas, A.; Pedersen, H.; Seberg, O. Contracted genes and dwarfed plastome in mycoheterotrophic *Sciaphila thaidanica* (Triuridaceae, Pandanales). *Genome Biol. Evol.* **2018**, *10*, 976–981. [CrossRef] [PubMed]
15. Schelkunov, M.I.; Nuraliev, M.S.; Logacheva, M.D. *Rhopalocnemis phalloides* has one of the most reduced and mutated plastid genomes known. *Peer J.* **2019**, *7*, e7500. [CrossRef]
16. Graham, S.W.; Lam, V.K.; Merckx, V.S. Plastomes on the edge: The evolutionary breakdown of mycoheterotroph plastid genomes. *New Phytol.* **2017**, *214*, 48. [CrossRef]
17. Lallemand, F.; Logacheva, M.; Le Clainche, I.; Berard, A.; Zheleznaia, E.; May, M.; Jakalski, M.; Delannoy, E.; Le Paslier, M.C.; Selosse, M.A. Thirteen new plastid genomes from mixotrophic and autotrophic species provide insights into heterotrophy evolution in Neottieae orchids. *Genome Biol. Evol.* **2019**, *11*, 2457–2467. [CrossRef]
18. Clegg, M.T.; Gaut, B.S.; Learn, G.H., Jr.; Morton, B.R. Rates and patterns of chloroplast DNA evolution. *Proc. Natl. Acad. Sci. USA* **1994**, *91*, 6795–6801. [CrossRef]
19. Dyall, S.D.; Brown, M.T.; Johnson, P.J. Ancient invasions: From endosymbionts to organelles. *Science* **2004**, *304*, 253–257. [CrossRef]
20. Palmer, J.D. Comparative organization of chloroplast genomes. *Annu. Rev. Genet.* **1985**, *19*, 325–354. [CrossRef]
21. Wicke, S.; Schneeweiss, G.M.; dePamphilis, C.W.; Müller, K.F.; Quandt, D. The evolution of the plastid chromosome in land plants: Gene content, gene order, gene function. *Plant Mol. Biol.* **2011**, *76*, 273–297. [CrossRef] [PubMed]
22. Powell, W.; Morgante, M.; Andre, C.; Hanafey, M.; Vogel, J.; Tingey, S.; Rafalski, A. The comparison of RFLP, RAPD, AFLP and SSR (microsatellite) markers for germplasm analysis. *Mol. Breed.* **1996**, *2*, 225–238. [CrossRef]
23. Niu, Z.; Zhu, S.; Pan, J.; Li, L.; Sun, J.; Ding, X. Comparative analysis of *Dendrobium* plastomes and utility of plastomic mutational hotspots. *Sci. Rep.* **2017**, *7*, 2073. [CrossRef]
24. Zhou, Y.; Nie, J.; Xiao, L.; Hu, Z.; Wang, B. Comparative chloroplast genome analysis of rhubarb botanical origins and the development of specific identification markers. *Molecules* **2018**, *23*, 2811. [CrossRef]
25. Chen, X.; Zhou, J.; Cui, Y.; Wang, Y.; Duan, B.; Yao, H. Identification of *Ligularia* herbs using the complete chloroplast genome as a super-barcode. *Front. Pharmacol.* **2018**, *9*, 695. [CrossRef]
26. Niu, Z.; Pan, J.; Xue, Q.; Zhu, S.; Liu, W.; Ding, X. Plastome-wide comparison reveals new SNV resources for the authentication of *Dendrobium huoshanense* and its corresponding medicinal slice (Huoshan Fengdou). *Acta Pharm. Sin. B* **2018**, *8*, 466–477. [CrossRef]
27. Wu, M.; Li, Q.; Hu, Z.; Li, X.; Chen, S. The complete *Amomum kravanh* chloroplast genome sequence and phylogenetic analysis of the commelinids. *Molecules* **2017**, *22*, 1875. [CrossRef]
28. Flora Reipublicae Popularis Sinicae (FRPS). Available online: http://www.iplant.cn/ (accessed on 28 October 2019).
29. Gong, Z.; Wang, Z.; Xu, L.; Xu, G.; Wu, J. Studies on medicinal plants of Loranthaceae in China. *Chin. Wild Plant Resour.* **1996**, *1*, 11–15.
30. Zhao, Q.; Xu, Q.; Zhang, H. Pharmacognostic identification on crude drug of *Macrosolen cochinchinensis*. *Chin. J. Ethnomed. Ethnopharm.* **1998**, *34*, 1–3.

31. Li, Y.; Lu, D.; Zhao, M.; Zhu, K. Research on the developments and applications for medicinal plants of Loranthaceae in Guangxi. *Guangxi Med. J.* **2006**, *28*, 1695–1698.
32. Sharp, P.M.; Li, W.H. The codon Adaptation Index–a measure of directional synonymous codon usage bias, and its potential applications. *Nucleic Acids Res.* **1987**, *15*, 1281–1295. [CrossRef] [PubMed]
33. Powell, W.; Morgante, M.; Mcdevitt, R.; Vendramin, G.G.; Rafalski, J.A. Polymorphic simple sequence repeat regions in chloroplast genomes: Applications to the population genetics of pines. *Proc. Natl. Acad. Sci. USA* **1995**, *92*, 7759–7763. [CrossRef] [PubMed]
34. Akkaya, S.M. Length polymorphism of simple sequence repeat DNA in soybean. *Genetics* **1992**, *132*, 1131–1139. [PubMed]
35. Rozas, J.; Ferrer-Mata, A.; Sanchez-DelBarrio, J.C.; Guirao-Rico, S.; Librado, P.; Ramos-Onsins, S.E.; Sanchez-Gracia, A. DnaSP 6: DNA Sequence Polymorphism Analysis of Large Data Sets. *Mol. Biol. Evol.* **2017**, *34*, 3299–3302. [CrossRef] [PubMed]
36. Shin, H.W.; Lee, N.S. Understanding plastome evolution in Hemiparasitic Santalales: Complete chloroplast genomes of three species, *Dendrotrophe varians*, *Helixanthera parasitica*, and *Macrosolen cochinchinensis*. *PLoS ONE* **2018**, *13*, e0200293. [CrossRef]
37. Kim, H.T.; Kim, J.S.; Moore, M.J.; Neubig, K.M.; Williams, N.H.; Whitten, W.M.; Kim, J.H. Seven new complete plastome sequences reveal rampant independent loss of the *ndh* gene family across orchids and associated instability of the inverted repeat/small single-copy region boundaries. *PLoS ONE* **2015**, *10*, e0142215. [CrossRef]
38. Mcneal, J.R.; Kuehl, J.V.; Boore, J.L.; Pamphilis, C.W.D. Complete plastid genome sequences suggest strong selection for retention of photosynthetic genes in the parasitic plant genus *Cuscuta*. *BMC Plant Biol.* **2007**, *7*, 57. [CrossRef]
39. Der, J.P.; Nickrent, D.L. A molecular phylogeny of Santalaceae (Santalales). *Syst. Bot.* **2008**, *33*, 107–116. [CrossRef]
40. Zhou, J.; Chen, X.; Cui, Y.; Sun, W.; Li, Y.; Wang, Y.; Song, J.; Yao, H. Molecular structure and phylogenetic analyses of complete chloroplast genomes of two *Aristolochia* medicinal species. *Int. J. Mol. Sci.* **2017**, *18*, 1839. [CrossRef]
41. Bolger, A.M.; Lohse, M.; Usadel, B. Trimmomatic: A flexible trimmer for Illumina sequence data. *Bioinformatics* **2014**, *30*, 2114–2120. [CrossRef]
42. Frazer, K.A.; Lior, P.; Alexander, P.; Rubin, E.M.; Inna, D. VISTA: Computational tools for comparative genomics. *Nucleic Acids Res.* **2004**, *32*, W273. [CrossRef] [PubMed]
43. CPGAVAS2. Available online: http://47.96.249.172:16018/analyzer/extractSeq (accessed on 30 July 2019).
44. Mower, J.P. The PREP suite: Predictive RNA editors for plant mitochondrial genes, chloroplast genes and user-defined alignments. *Nucleic Acids Res.* **2009**, *37*, 253–259. [CrossRef] [PubMed]
45. Misa-Microsatellite Identification Tool. Available online: Pgrc.ipk-gatersleben.de/misa/ (accessed on 2 June 2017).
46. Kurtz, S.; Choudhuri, J.V.; Ohlebusch, E.; Schleiermacher, C.; Stoye, J.; Giegerich, R. REPuter: The manifold applications of repeat analysis on a genomic scale. *Nucleic Acids Res.* **2001**, *29*, 4633–4642. [CrossRef] [PubMed]

5

# Characterization and Analysis of the Mitochondrial Genome of Common Bean (*Phaseolus vulgaris*) by Comparative Genomic Approaches

**Changwei Bi [1], Na Lu [1], Yiqing Xu [2], Chunpeng He [1,*] and Zuhong Lu [1,*]**

[1] State Key Laboratory of Bioelectronics, School of Biological Science and Medical Engineering, Southeast University, Nanjing 210096, Jiangsu, China; bichwei@seu.edu.cn (C.B.); nlu@seu.edu.cn (N.L.)
[2] School of Information Science and Technology, Nanjing Forestry University, Nanjing 210037, Jiangsu, China; yiqingxu@njfu.edu.cn
* Correspondence: cphe@seu.edu.cn (C.H.); zhlu@seu.edu.cn (Z.L.)

**Abstract:** The common bean (*Phaseolus vulgaris*) is a major source of protein and essential nutrients for humans. To explore the genetic diversity and phylogenetic relationships of *P. vulgaris*, its complete mitochondrial genome (mitogenome) was sequenced and assembled. The mitogenome is 395,516 bp in length, including 31 unique protein-coding genes (PCGs), 15 transfer RNA (tRNA) genes, and 3 ribosomal RNA (rRNA) genes. Among the 31 PCGs, four genes (*mttB*, *nad1*, *nad4L*, and *rps10*) use ACG as initiation codons, which are altered to standard initiation codons by RNA editing. In addition, the termination codon CGA in the $ccmF_C$ gene is converted to UGA. Selective pressure analysis indicates that the *ccmB*, $ccmF_C$, *rps1*, *rps10*, and *rps14* genes were under evolutionary positive selection. The proportions of five amino acids (Phe, Leu, Pro, Arg, and Ser) in the whole amino acid profile of the proteins in each mitogenome can be used to distinguish angiosperms from gymnosperms. Phylogenetic analyses show that *P. vulgaris* is evolutionarily closer to the Glycininae than other leguminous plants. The results of the present study not only provide an important opportunity to conduct further genomic breeding studies in the common bean, they also provide valuable information for future evolutionary and molecular studies of leguminous plants.

**Keywords:** common bean; *Phaseolus vulgaris*; mitochondrial genome; comparative genomics; phylogeny

## 1. Introduction

Mitochondria (mt) are semi-autonomous organelles that are part of almost all eukaryotic cells (cells with clearly defined nuclei). Their primary function is to produce a steady supply of adenosine triphosphate (ATP). Mitochondria are thus termed the 'powerhouses' or 'energy factories' of cells. Chloroplasts (cp) and mitochondria most likely originated from formerly free-living bacteria through endosymbiotic acquisition, which can explain the presence of their own genomes [1,2]. With rapid developments in sequencing and genome assembly methods, an increasing number of complete organelle genomes have been assembled in the last decade. Thus far, over 4900 complete chloroplast and plastid genomes have been assembled but only 321 plant mitogenomes have been assembled and deposited in GenBank Organelle Genome Resources (as of 14 May 2020; https://www.ncbi.nlm.nih.gov/genome/browse/), suggesting that their assembly is complex and difficult.

Mitochondria are specific to each plant and have complex genome structures [3–5], variable genome sizes [6,7], numerous repetitive sequences [8,9], multiple RNA editing modifications [10,11], and frequent gene gains or losses during evolution [9,12,13]. In seed plant mitogenomes, the genome sizes are highly variable, ranging from an exceptionally small genome of 66 kb in the parasitic plant *Viscum scurruloideum* [14] to the largest multi-chromosomal genome of 11.3 Mb in *Silene conica* [15]. Even if two species are evolutionarily close, their genome sizes may vary considerably. The mitogenome sizes of plants in the subfamily Papilionoideae range from 271 kb in *Medicago truncatula* [16] to 588 kb in *Vicia faba* [17], while the mitogenomes of most papilionoid legumes are approximately 400 kb in length [18]. This wide variation in mitogenome size can be attributed to the proliferation of repetitive sequences and the acquisition of foreign DNA from other organisms during evolution [19,20].

Previous studies have documented that the mitogenomes of seed plants are enriched with repetitive sequences, including simple sequence repeats (SSRs), tandem repeats, and dispersed repeats. The SSRs in plant mitogenomes are commonly used as molecular markers for studying genetic diversity and identifying species [21]. The tandem repeats occur in a broad range of plant mitogenomes, which can also serve as molecular markers for unravelling population processes in plants [22]. Large dispersed repeats are the main causes of genome rearrangements, which may generate multipartite structures [13,23–25].

Although the mitogenome sizes of seed plants are variable, the functional genes of NADH dehydrogenase, ubiquinol cytochrome c reductase, ATP synthase, and cytochrome c biogenesis are quite conservative, except for succinate dehydrogenase genes and ribosomal proteins. Many primordial mt genes have been lost during evolution, which has been found to be closely related to their specific functions. For example, *sdh3* and *sdh4* were lost in all gramineous mitogenomes, the *rps11* gene was lost in the differentiation of gymnosperms and angiosperms [26], and the *cox2* gene was lost in the differentiation of the Phaseoleae and Glycininae [18]. Strikingly, nearly all of the universally present NADH dehydrogenase genes were lost from the mitogenome of *Viscum scurruloideum*, with the loss closely associated with its parasitic lifestyle [14].

The Fabaceae, commonly known as legumes, is an economically and ecologically important family of flowering plants ranging from small annual herbs to giant trees, most of which are herbaceous perennials. This family is the third-largest angiosperm family after the Asteraceae and Orchidaceae [27,28], consisting of about 770 genera and more than 20,000 species. A recent study by the Legume Phylogeny Working Group (LPWG) reclassified the three widely-accepted Fabaceae subfamilies (Caesalpinioideae, Minosoideae, and Papilionoideae) into six new subfamilies (Cercidoideae, Detarioideae, Duparquetioideae, Dialioideae, Caesalpinioideae, and Papilionoideae) based on a taxonomically-comprehensive phylogeny [28]. However, due to the complexity of plant mitogenomes, only 27 mitogenomes of Fabaceae species have been assembled and deposited in the NCBI Nucleotide database (14 May 2020), including 19 species in the Papilionoideae, six species in the Caesalpinioideae, one species of *Cercis canadensis* in the Cercidoideae, and one species of *Tamarindus indica* in the Detarioideae.

In this study, we assembled the complete mitogenome of the common bean *Phaseolus vulgaris*, an herbaceous annual plant grown worldwide for its edible dry seeds or unripe fruit. The common bean is one of the most important grain legumes for human consumption and plays an important role in sustainable agriculture due to its ability to fix atmospheric nitrogen [29]. We analyzed its gene content, repetitive sequences, RNA editing sites, selective pressure, and phylogenetic position, then made comparisons with other plant mitogenomes. The complete mitogenome of *P. vulgaris* will provide important information for the investigation of mitogenomic evolution among the Fabaceae family and aid the functional study of fabaceous mitogenomes. Mitochondrial biogenesis is very important in plant breeding and knowledge of the complete mitogenome provides an opportunity to conduct further important genomic breeding studies in the common bean.

## 2. Materials and Methods

### *2.1. Plant Materials, DNA Extraction, and Sequencing*

An inbred landrace of *P. vulgaris* accession G19833 derived from the Andean pool (Race Peru) was selected for sequencing. Seeds were obtained from the germplasm bank of the Embrapa Arroz e Feijão, Brazil [30]. Root tips obtained from germinated seeds were pre-treated with 2 mM 8-hydroxyquinoline for 18 h at 10 °C, fixed in ethanol-acetic acid (3:1 *v/v*), and stored in fixative at −20 °C for up to several weeks. Total genomic DNA was extracted from root tips using DNAeasy Plant Mini Kits (Qiagen). To construct the shotgun library, DNA was fragmented by nebulization. The raw reads were sequenced with a combination of Roche/454 GS FLX sequencing reads, Illumina HiSeq-2500 sequencing short reads (primarily to correct 454 sequencing errors) and PacBio RS II sequencing long reads (primarily to validate the assembly of the master conformation). The raw reads of *P. vulgaris* used in this study were available in the NCBI Sequence Read Archive (SRA) under accessions SRR069592, SRR5628227, and SRR2912756.

### 2.2. Mitogenome Assembly and Annotation

An efficient procedure for plant mitochondrial genome assembly using whole-genome data from the 454 GS FLX sequencing platform has been applied in many plants, such as *Boea hygrometrica* [31], *Daucus carota* [32], *Gossypium raimondii* [26], and *Salix suchowensis* [33]. Briefly, as shown in Figure S1, we first assembled all the Roche/454 GS FLX sequencing reads using Newbler (version 3.0) [34] with the following parameters: -cpu 20, -het, -sio, -m, -urt, -large, and -s 100. Then, we used custom Perl scripts to construct a draft assembly graph from the file "454AllContigGraph.txt" generated from Newbler. As shown in Figure 1, we obtained six contigs to construct the completed draft mitochondrial graph for assembling the *P. vulgaris* mitogenome. Among the six selected contigs, two (Contig15 and Contig40) were assembled into the mitogenome twice, while the others were assembled only once. To assemble the master conformation (MC), we mapped the PacBio sequencing reads to the mt contigs that spanned repetitive contigs using BLASTN to obtain a major contig relationship map for the repeat regions [35,36].

Specifically, for each repeat pair (Contig15 and Contig40), we built four reference sequences according to Dong et al. [37], each with 200 bp up- and down-stream of the two template sequences (original sequences). Then, we searched the PacBio long reads against the database built up from the reference sequences and extracted the matching reads with a blast identity above 80%, an e-value cut-off of $1e^{-100}$, and a hit length of over 3000 bp. Next, we mapped the best-matched reads to the four reference sequences in MacVector v17.0.7. As shown in Figure 1, we obtained one master genome and two isomeric genomes (ISO) based on the number of PacBio reads that were mapped to both end contigs of the repetitive contigs (Table S1). We then mapped Illumina sequencing reads to the draft MC mitogenome with BWA [38] and SAMtools [39] softwares to correct the homopolymer length errors (especially in A/T enriched regions) from 454 GS FLX Titanium [26]. Finally, the complete mitogenome sequence of *P. vulgaris* was obtained.

The mitogenome was annotated using the public MITOFY analysis web server (http://dogma.ccbb.utexas.edu/mitofy/) [8]. The putative genes were manually checked and adjusted by comparing them with other legume mitogenomes in MacVector v.17.07. All transfer RNA genes were confirmed by using tRNAscan-SE with default settings [40]. The start and stop codons of PCGs were manually adjusted to fit open reading frames. The relative synonymous codon usage (RSCU) values and amino acid composition of PCGs were calculated by MEGA X [41]. The OrganellarGenomeDRAW (OGDRAW) program was used to visualize the circular map of the *P. vulgaris* mitogenome [42].

### 2.3. Selective Pressure Analysis

To reflect the selective pressure of PCGs, we calculated the nonsynonymous ($K_a$) and synonymous ($K_s$) substitution rates of each PCG between *P. vulgaris* and other higher plants. *Arabidopsis thaliana* (*A. thaliana*; Brassicaceae) is a popular model organism in plant biology and genetics. *Citrullus lanatus* (*C. lanatus*; Cucurbitaceae) and *Vitis vinifera* (*V. vinifera*; Vitaceae) are highly cultivated fruits worldwide and belong to the Rosids clade, like leguminous plants. Therefore, we selected the mitogenomes of *A. thaliana*, *V. vinifera*, and *C. lanatus* as references to infer the direction and magnitude of natural selection acting on PCGs during the evolution of *P. vulgaris*. The orthologous gene pairs from *P. vulgaris*, *A. thaliana*, *V. vinifera*, and *C. lanatus* were aligned and formatted by ParaAT2.0 with default parameters [43]. The $K_a$, $K_s$, and $K_a/K_s$ values were calculated using KaKs_Calculator v.2.0 based on the YN method, and Fisher's exact test was performed to justify the validity of the $K_a$ and $K_s$ values [44,45].

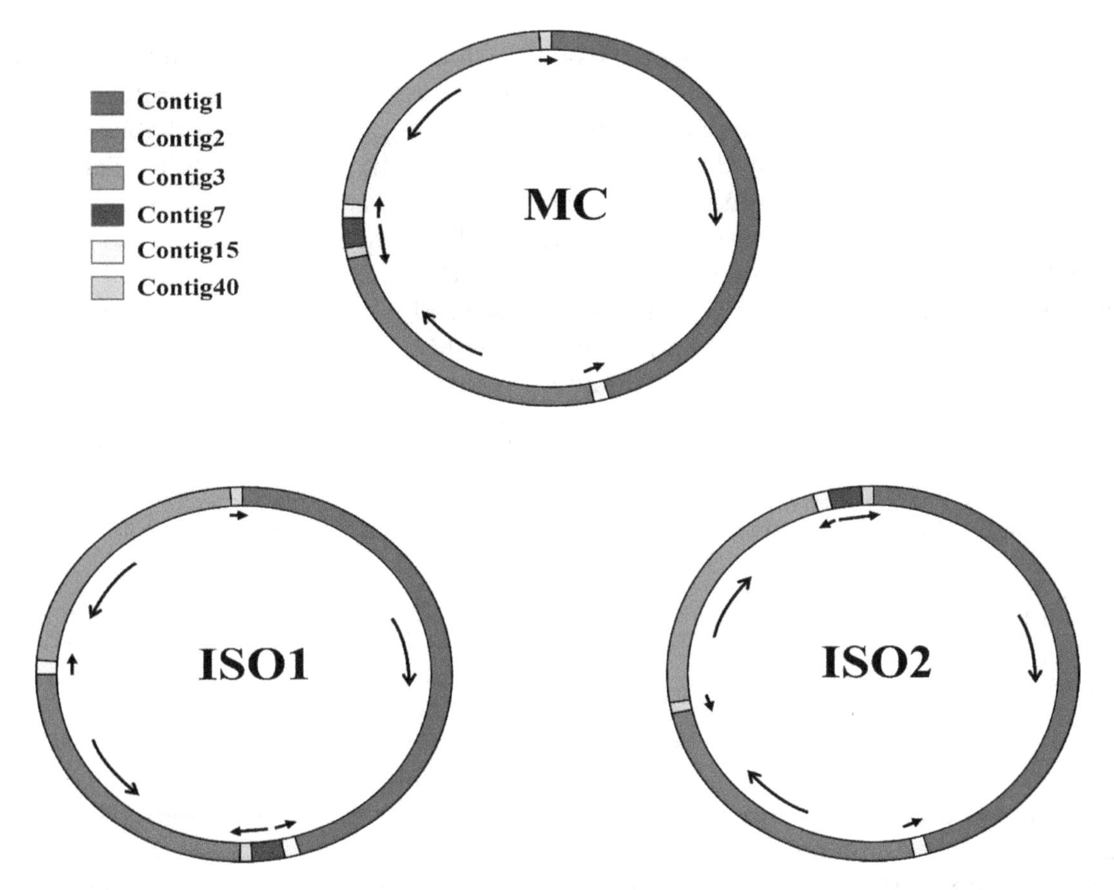


**Figure 1.** The master genome and two isomeric genomes observed from *P. vulgaris* mitogenome mediated by two pairs of large repeats (Contig15 and Contig40). The mt contigs were generated and selected from Newbler assembly software. MC and ISO mean the master and isomeric conformations, respectively. Arrows denote the sequence orientation of assembled contigs.

### 2.4. Prediction of RNA Editing Sites

The online PREP-Mt (predictive RNA editors for plants) suite of servers (http://prep.unl.edu/) was used to predict the possible RNA editing sites in the PCGs of *P. vulgaris* and the other four leguminous mitogenomes (*G. max*, *L. japonicus*, *V. radiata*, and *M. pinnata*). In order to predict more true RNA editing sites, the cut-off for prediction score was set as C = 0.2, which has been proven to be a slight optimum [46]. A low cut-off value will predict more true edit sites but will also increase the probability of misidentifying an unedited site as an edited one.

### 2.5. Identification of Repeat Sequences in P. vulgaris Mitogenome

Three kinds of repeats—SSRs, tandem repeats and dispersed repeats—were detected in the *P. vulgaris* mitogenome. The SSRs were detected using the web-based microsatellite identification tool MISA-web (https://webblast.ipk-gatersleben.de/misa/) [47] according to the methods of previous studies [9,26] with a motif size of one to six nucleotides and thresholds of eight, four, four, three, three, and three, respectively. Tandem repeats were identified using the online tool Tandem Repeats Finder 4.09 with default settings (http://tandem.bu.edu/trf/trf.html) [48]. AB-BLAST 3.0 was utilized to identify and locate dispersed repeats with the following parameters: M = 1, N = −3, Q = 3, R = 3, kap, span, $B = 1 \times 10^9$ and W = 7, which have been proven as effective in many studies [8,9,49]. The BLAST hits with e-values $< 1$ and identities $> 80\%$ were considered as disperse repeats. The number of dispersed repeats was calculated for seven size intervals (30–49, 50–69, 70–99, 100–149, 150–199, 200–999 and ≥1000 bp). Partly or wholly overlapping repeats were considered as a single repeat unit.

### 2.6. Phylogenetic Analyses

In order to accurately infer the phylogenetic relationships of *P. vulgaris* within the Fabaceae family, maximum likelihood (ML) analysis was performed based on the conserved mitochondrial PCGs (amino acid sequences) of 23 higher plants. The NCBI accession numbers and abbreviations of all these observed mitogenomes are listed in Table S2. Apart from the 19 representative Fabaceae species, taxon sampling also included two species of Solanales (*C. annuum* and *N. tabacum*) and two species of Malpighiales (*P. tremula* and *S. suchowensis*) as outgroups. The single-copy orthologous PCGs common among the 23 analysed species were selected with local Perl scripts. All conserved mitochondrial PCGs were extracted from each mitogenome. The conserved gene sequences from the mitogenome were concatenated into a single dataset and aligned using Muscle software with default settings [50]. Poorly-aligned sequences were deleted or manually adjusted for each alignment. Prior to constructing the phylogenetic tree, we applied MEGA X to determine the most appropriate amino acid substitution model [41]. Based on the model selection results, the ML tree based on a JTT + F model with a gamma distribution was constructed using MEGA X. The bootstrap index value (%) in which the associated taxa clustered together was shown next to the branches and was calculated from 1000 replications.

## 3. Results and Discussion

### 3.1. Genomic Features of the P. vulgaris Mitogenome

The complete genomic sequence of the *P. vulgaris* mitogenome was submitted to the NCBI Genome Database (https://www.ncbi.nlm.nih.gov/genome/browse/) under accession number NC_045135.1. The mitogenome was assembled into a typical circular molecule 395,516 bp in length (Figure 2), similar to the mitochondria of some papilionoid legumes such as *G. max* (402,558 bp), *G. soja* (402,545 bp), *L. japonicus* (380,861 bp), *V. angularis* (404,466 bp), and *V. radiata* var. *radiata* (401,262 bp; Table S2). In fact, the mitogenome sizes vary considerably among the papilionoid legumes, ranging from 271,618 bp in *Medicago truncatula* to 588,000 bp in *Vicia faba*. Mitogenome sizes can vary greatly in different cultivars of the same species. For example, the mitogenome size of *G. max* Aiganhuang (N21249) is 402,558 bp, whereas that of *G. max* cultivar Zhonghuang 13 is 513,779 bp [51].

The nucleotide composition of the whole mitogenome is A: 27.37%, C: 22.40%, G: 22.71%, and T: 27.52% (Table 1). The overall GC content is 45.11%, which is consistent with other leguminous plants (*G. max*: 45.03%, *V. faba*: 45.04%, and *V. radiata* var. *radiata*: 45.11%). Strikingly, the GC content of the PCGs is very small compared to those of other regions. As shown in Table 2, a total of 49 unique genes were detected in the *P. vulgaris* mitogenome, comprising 31 PCGs, 15 tRNA genes and 3 rRNA genes. However, none of the genes encodes subunits of Complex II (succinate dehydrogenase), which has also been lost in some other leguminous plants. Additionally, two tRNA genes located in repeat sequences were found to contain two or three copies (*trnC-GCA* and *trnfM-CAU*). The total lengths of the PCGs and *cis*-spliced introns comprise 7.26% and 8.24% of the whole mitogenome, while tRNA and rRNA genes only comprise 0.34% and 1.33% of the mitogenome, respectively. Most PCGs have no introns; however, eight genes (Table 2; *nad1*, *nad2*, *nad4*, *nad5*, *nad7*, *ccmF*$_C$, *rps3*, and *rps10*) were found to contain one or more introns. Three genes (*nad1*, *nad2*, and *nad5*) required trans-splicing to assembly fully-translatable mRNA (Figure 2).

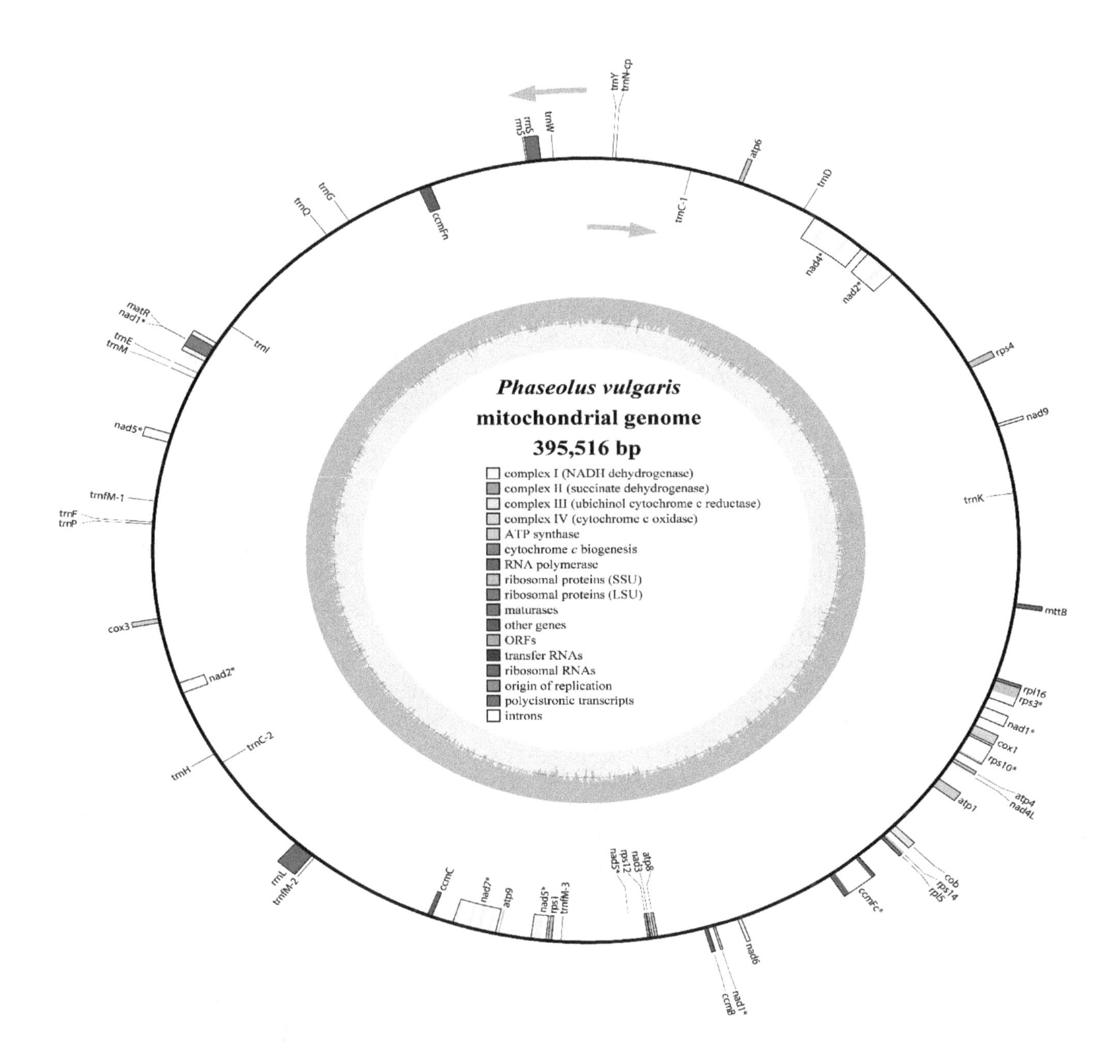


**Figure 2.** Circular map of the *P. vulgaris* mitogenome. Genes shown on the outside of the circle are transcribed clockwise, whereas genes on the inside are transcribed counterclockwise. GC content is represented on the inner circle by the dark gray plot. The asterisks besides genes denote intron-containing genes.

**Table 1.** Genomic features of *P. vulgaris* mitogenome.

| Feature | A % | C % | G % | T % | GC % | Size (bp) | Proportion in Genome (%) |
|---|---|---|---|---|---|---|---|
| Whole genome | 27.37 | 22.40 | 22.71 | 27.52 | 45.11 | 395,516 | 100 |
| Protein-coding genes [a] | 26.34 | 21.29 | 21.62 | 30.75 | 42.91 | 28,725 | 7.26 |
| *cis*-spliced introns [a] | 24.49 | 25.56 | 24.79 | 25.16 | 50.34 | 32,584 | 8.24 |
| tRNA genes [a] | 23.07 | 24.34 | 27.49 | 25.09 | 51.84 | 1335 | 0.34 |
| rRNA genes [a] | 26.14 | 22.62 | 29.04 | 22.20 | 51.66 | 5252 | 1.33 |
| Non-coding regions | 27.78 | 22.18 | 22.48 | 27.56 | 44.65 | 327,620 | 82.83 |

[a] Protein-coding genes, *cis*-spliced introns, tRNAs, and rRNAs belong to coding regions.

**Table 2.** Gene content of *P. vulgaris* mitogenome.

| Group of Genes | Gene Name |
|---|---|
| Complex I (NADH dehydrogenase) | *nad1* *, *nad2* *, *nad3*, *nad4* *, *nad4L*, *nad5* *, *nad6*, *nad7* *, *nad9* |
| Complex II (succinate dehydrogenase) | - |
| Complex III (ubiquinol cytochrome c reductase) | *cob* |
| Complex IV (cytochrome c oxidase) | *cox1*, *cox3* |
| Complex V (ATP synthase) | *atp1*, *atp4*, *atp6*, *atp8*, *atp9* |
| Cytochrome *c* biogenesis | *ccmB*, *ccmC*, *ccmF$_C$* *, *ccmF$_N$* |
| Ribosomal proteins (SSU) | *rps1*, *rps3* *, *rps4*, *rps10* *, *rps12*, *rps14* |
| Ribosomal proteins (LSU) | *rpl5*, *rpl16* |
| Maturases | *matR* |
| Transport membrane protein | *mttB* |
| Ribosomal RNAs | *rrn5*, *rrnS*, *rrnL* |
| Transfer RNAs | *trnC-GCA* (2 copies), *trnD-GUC*, *trnE-UUC*, *trnF-GAA*, *trnG-GCC*, *trnfM-CAU* (3 copies), *trnH-GUG*, *trnI-CAU*, *trnK-UUU*, *trnM-CAU*, *trnN-GUU*, *trnP-UGG*, *trnQ-UUG*, *trnW-CCA*, *trnY-GUA* |

* The asterisks besides genes denotes intron-containing genes.

### 3.2. Codon Usage Analysis of PCGs

In the *P. vulgaris* mitogenome, most of the PCGs use ATG as the start codon, while *mttB* and *nad1* start with ACG (C to U RNA editing on the second site is presumed) as the start codon (Table 3). Four types of stop codons were found in the PCGs: (1) TAA (15 genes; *atp4, atp8, atp9, cox1, nad1, nad2, nad3, nad4L, nad5, nad6, nad9, rpl5, rpl16, rps1,* and *rps4*), (2) TGA (10 genes; *atp1, ccmB, ccmC, ccmF$_N$, cox3, matR, mttB, nad4, rps10,* and *rps12*), (3) TAG (5 genes; *atp6, cob, nad7, rps3,* and *rps14*), and (4) CGA (*ccmF$_C$*; C to U RNA editing on the first site is presumed). As shown in Figure 3, the codon usage analysis revealed that leucine (Leu) and serine (Ser) are the most frequently-used amino acid residues, while cysteine (Cys) and tryptophan (Trp) are the least-used amino acid residues in the plant mitochondrial proteins. By comparison of the composition of *P. vulgaris* with other angiosperms plants, we found that the distribution of amino acid residues across the mitochondrial proteins are very similar in angiosperms (Figure 3). In addition, most of the amino acid residues were found to be very conserved between angiosperms (*P. vulgaris, G. max, L. japonicus, V. radiata, V. faba, A. thaliana, C. lanatus,* and *T. aestivum*) and gymnosperms (*Ginkgo biloba* and *Cycas taitungensis*), except for five of them (Phe, Leu, Pro, Arg, and Ser).

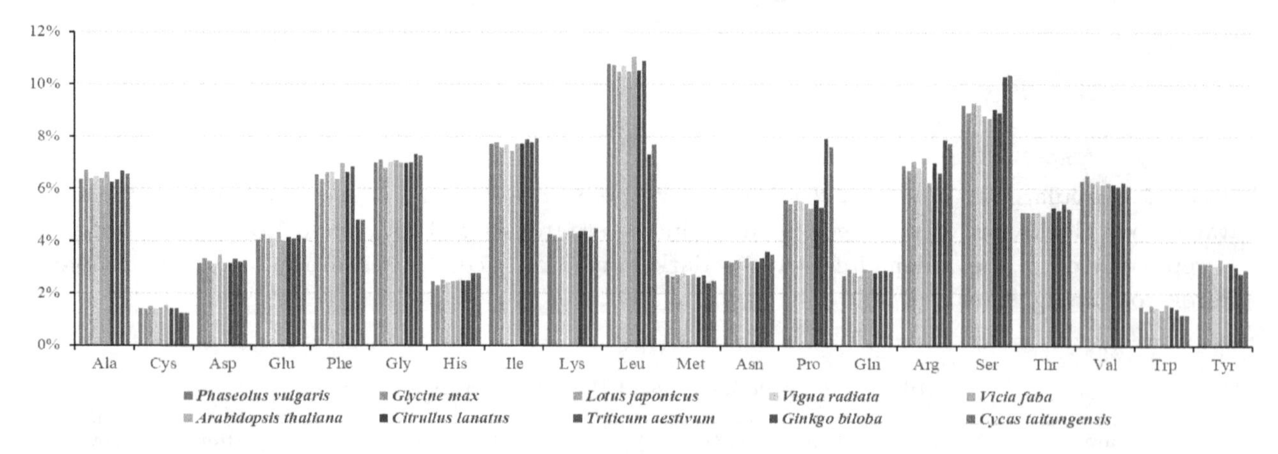


**Figure 3.** Codon usage pattern of *P. vulgaris* mitogenome compared with nine other higher plants. The proportion of each amino acid residues to the whole mitochondrial proteins is shown on the Y-axis. *Ginkgo biloba* and *Cycas taitungensis* are gymnosperms, while others are angiosperms.

The relative synonymous codon usage (RSCU) analysis for the *P. vulgaris* mitogenome is shown in Figure 4, which indicates that all codons are present in the PCGs. Excluding the termination codons, the 31 PCGs in the *P. vulgaris* mitogenome consist of 9545 codons in total. Additionally, the codon usage showed that the RSCU values of the NNT and NNA codons are higher than 1.0 except for Ile (AUA) and Leu (CUA; Figure 4), suggesting a strong As or Ts bias in the third codon position of *P. vulgaris* mitochondrial PCGs, which is a very common phenomenon observed in all studied mitogenomes (Table S3). The codon usage pattern of *P. vulgaris* mitogenome is highly consistent with two other papilionoid legumes. The distributions of some codons encoding Pro (CCU, CCA, and CCG) differ between dicotyledons (*P. vulgaris, G. max, V. angularis, C. lanatus*, and *A. thaliana*) and monocotyledons (*T. aestivum*), and some codons (UCG, AGU, AGC, CCU, CCG, ACG, CGG, and AGA) are distributed differently between angiosperms and gymnosperms.

**Table 3.** Gene profile and organization of PCGs in *P. vulgaris* mitogenome.

| Gene Name | Length | Start Codon | Stop Codon | Direction |
|---|---|---|---|---|
| *atp1* | 1527 | ATG | TGA | F |
| *atp4* | 588 | ATG | TAA | F |
| *atp6* | 726 | ATG | TAG | F |
| *atp8* | 483 | ATG | TAA | R |
| *atp9* | 225 | ATG | TAA | R |
| *ccmB* | 621 | ATG | TGA | F |
| *ccmC* | 741 | ATG | TGA | R |
| $ccmF_C$ | 1329 | ATG | CGA | F |
| $ccmF_N$ | 1740 | ATG | TGA | R |
| *cob* | 1176 | ATG | TAG | F |
| *cox1* | 1584 | ATG | TAA | F |
| *cox3* | 798 | ATG | TGA | F |
| *matR* | 2010 | ATG | TGA | F |
| *mttB* | 723 | ACG | TGA | F |
| *nad1* | 978 | ACG | TAA | F |
| *nad2* | 1467 | ATG | TAA | R |
| *nad3* | 357 | ATG | TAA | R |
| *nad4* | 1488 | ATG | TGA | R |
| *nad4L* | 303 | ACG | TAA | F |
| *nad5* | 2019 | ATG | TAA | F/R |
| *nad6* | 618 | ATG | TAA | F |
| *nad7* | 1185 | ATG | TAG | R |
| *nad9* | 573 | ATG | TAA | F |
| *rpl5* | 558 | ATG | TAA | F |
| *rpl16* | 516 | ATG | TAA | F |
| *rps1* | 618 | ATG | TAA | R |
| *rps3* | 1689 | ATG | TAG | F |
| *rps4* | 1041 | ATG | TAA | F |
| *rps10* | 363 | ACG | TGA | F |
| *rps12* | 378 | ATG | TGA | R |
| *rps14* | 303 | ATG | TAG | F |

### *3.3. Selective Pressure Analysis*

In genetics, the $K_a/K_s$ ratio is useful for inferring the direction and magnitude of natural selection acting on homologous PCGs across diverged species. The ratio is a more powerful test of the neutral model of evolution than many others available in population genetics as it requires fewer assumptions [52]. A $K_a/K_s$ ratio <1 implies purifying or stabilizing selection (acting against change), while a ratio of >1 implies positive or Darwinian selection (driving change) and a ratio of exactly 1 indicates neutral selection. Importantly, the $K_a/K_s$ ratio is unlikely to be significantly above 1 without at least some of the mutations being advantageous.

In this study, the $K_a/K_s$ ratio was determined for all 31 PCGs following comparison of the *P. vulgaris* mitogenome with those of *C. lanatus, V. vinifera* and *A. thaliana* (Figure 5). Nearly all of the $K_a/K_s$ ratios were <1.0, suggesting that most of the PCGs were under stabilizing selection during evolution. Combining the information in Figure 5 and Table 1, the $K_a/K_s$ ratios of all Complex I–V genes were <1, indicating that these genes were highly conserved in the evolutionary process of higher plants.

The large number of mitochondrial genes under stabilizing selection ($K_a/K_s < 1$) may play important roles in stabilizing the normal functioning of mitochondria [53,54].

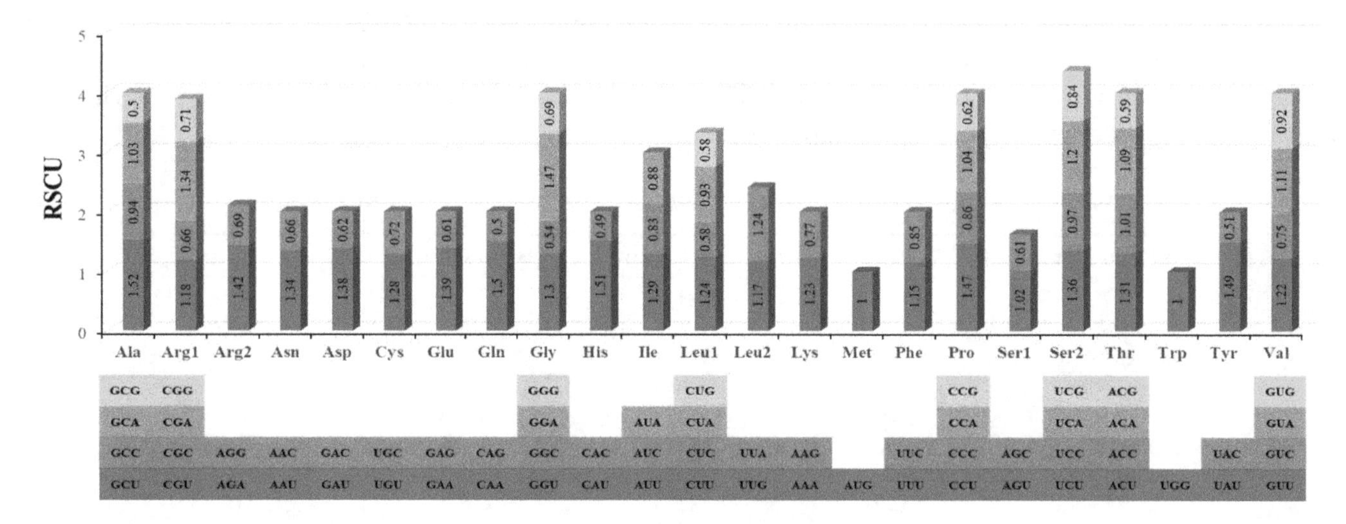


**Figure 4.** Relative synonymous codon usage (RSCU) of *P. vulgaris* mitogenome. Codon families are on the X-axis. RSCU values are the number of times of a particular codon, relative to the number of times that the codon would be observed for a uniform synonymous codon usage.

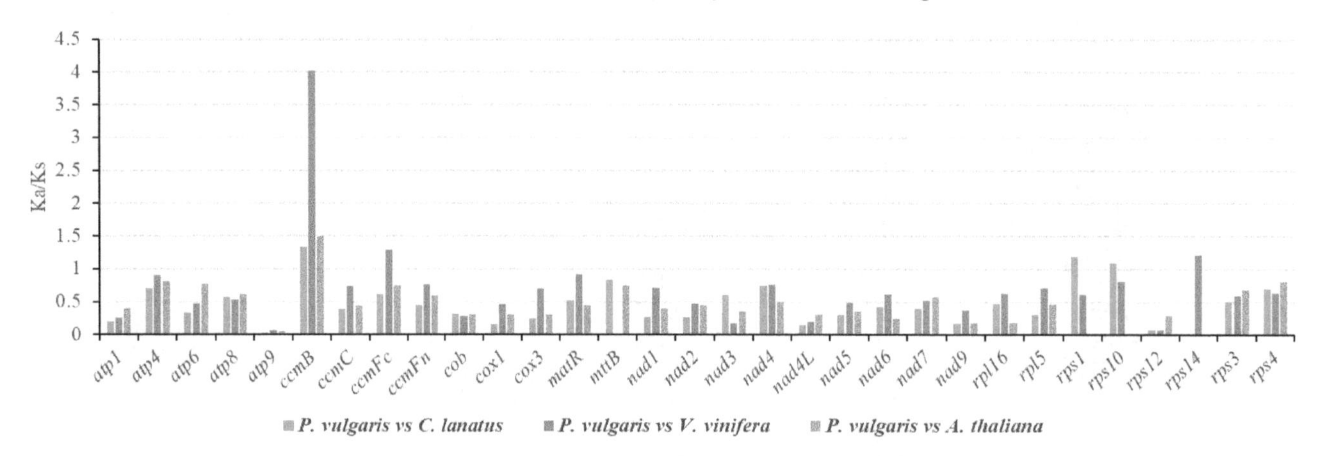


**Figure 5.** $K_a/K_s$ ratios for 31 protein coding genes of *P. vulgaris*, *C. lanatus*, *V. vinifera*, and *A. thaliana*. The blue, orange, and gray boxes indicate $K_a/K_s$ ratios of *P. vulgaris* vs. *C. lanatus*, *P. vulgaris* vs. *V. vinifera*, and *P. vulgaris* vs. *A. thaliana*.

As shown in Figure 5, the $K_a/K_s$ ratios of *ccmB* were >1 between *P. vulgaris* and all of the three selected species, indicating that *ccmB* may have suffered from positive selection since divergence from their last common ancestor. Particularly, the $K_a/K_s$ ratio of *ccmB* between *P. vulgaris* and *V. vinifera* was significantly >1 (4.01), suggesting that some advantage occurred during evolution. Additionally, the $K_a/K_s$ ratios of *ccmF*$_C$, *rps1*, *rps10*, and *rps14* were also >1, indicating that these genes were under positive selection after divergence of the last common ancestor. Since *CcmB* and *ccmF*$_C$ genes encode for some important components of the *c*-type cytochrome maturation pathway in mitochondria, we speculate that the adaptive evolution of *P. vulgaris* is closely related to the roles of *c*-type cytochromes in respiratory and photosynthetic electron transport [55–57]. Additionally, *rps1*, *rps10*, and *rps14* genes encode small mitoribosomal subunit proteins, which have been reported to play crucial roles in various biological processes in eukaryotic organisms, such as embryogenesis, leaf morphogenesis, and the formation of reproductive tissues [58–60]. The high $K_a/K_s$ ratios of *rps* genes observed here may be very important for the evolution of *P. vulgaris*. $K_a/K_s$ ratios >1 have also been reported for some other mitochondrial genes, including *atp8*, *ccmF*$_N$, *matR*, and *mttB* [26,33,61,62], indicating that mitochondrial genes in different plant species may be subjected to diverse selection pressures during evolution. Most importantly, the $K_a/K_s$ ratio of the orthologous gene-pairs is an average over all sites and, even under positive selection, it can be <1 because some sites might be under positive selection while others are under purifying selection [53,61,63].

### 3.4. Prediction of RNA Editing Sites in PCGs

Many previous studies have documented that RNA editing is one of the necessary steps for gene expression in the mitochondrial and chloroplast genomes of higher plants [64–67]. RNA editing is a post-transcriptional modification that converts specific cytidines (C) to uridines (U) and uridines to uridines in the transcripts of nearly all mitochondrial PCGs. Based on the web-based PREP-mt program, we predicted a total of 486 RNA editing sites in 31 PCGs and 100% C-to-U RNA editing. Among the 486 RNA editing sites, 34.57% (168 sites) were predicted at the first base position of the codon and 65.43% (318 sites) were found in the second position, while none were found in the third position. The lack of predicted RNA editing sites in the silent position is probably due to the limitation of the PREP-Mt predictive methodology rather than there being no RNA editing in this position. Since most of the RNA editing sites in third codon positions did not change the amino acid encoded by the codon, the tie-breaking rules used by PREP-Mt could not select the edited state [68]. Therefore, RNA editing in the silent editing position needs to be further identified by experimental methods.

The occurrence of RNA editing can cause alteration of initiation and termination codons in PCGs, and the frequency of their generation is much higher than that of their removal. As shown in Table 3, *mttB*, *nad1*, *nad4L*, and *rps10* genes use ACG as their initiation codons, which may be altered to the normal AUG by RNA editing modification. Additionally, the *ccmF*$_C$ gene uses CGA as its termination codon, which may be altered to UGA by RNA editing modification. As shown in Figure 6, the number of RNA editing sites in different genes varies greatly, and the Complex I (NADH dehydrogenase) and Cytochrome *c* biogenesis genes (*ccmB*, *ccmC*, *ccmF*$_C$, and *ccmF*$_N$) encode the most predicted RNA editing sites. Based on a comparison of the predicted RNA editing sites in five leguminous plants, the *nad4* gene encodes the most RNA editing sites, while *atp1* encodes the fewest (Figure 6).

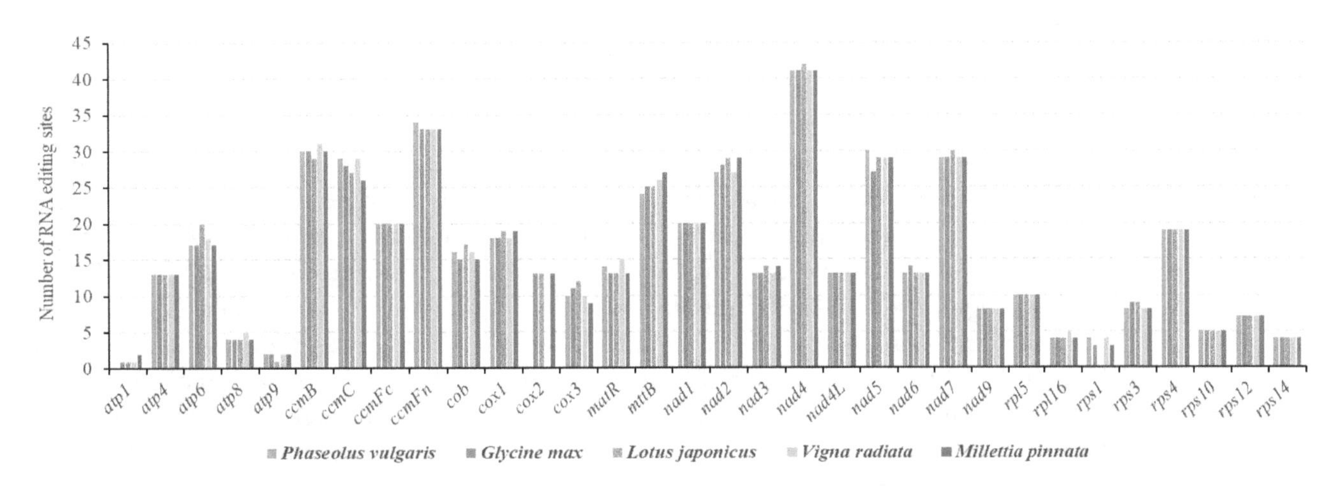


**Figure 6.** Predicted RNA editing sites of the *P. vulgaris* mitogenome compared with four other leguminous plants. RNA-editing sites are predicted on PREP-Mt sites (http://prep.unl.edu/) with the cut-off value of 0.2.

Previous studies have shown that the frequency and type of RNA editing in each organelle is highly lineage-specific [26,65,69,70]. As shown in Figure 6, the number of predicted RNA editing sites in different papilionoid legume mitogenomes is very conserved, ranging from 486 sites in *P. vulgaris* to 503 sites in *Lotus japonicus*, suggesting that they share extremely conserved PCGs. In angiospermous mitogenomes, nearly all of the RNA editing sites are C to U, and the number of RNA editing sites is concentrated between 400 to 500. For example, 463 and 444 RNA-editing sites were found in the

*C. lanatus* and *C. pepo* mitogenomes, of which 394 are shared [8]; 441 and 427 RNA-editing sites were found in *A. thaliana* and *B. napus* mitogenomes, of which 358 are shared [10]. In the gymnosperm *Cycas taitungensis*, 1084 RNA editing sites were found in its mitogenome [71]. The clearly descending number of RNA editing sites is in accordance with gene losses from gymnosperms to angiosperms. In contrast to angiosperms and gymnosperms, both types of C-to-U and U-to-C conversions are found in the mitochondrial transcripts of ferns and hornworts [69,72].

*3.5. Analysis of Repeat Sequences in the P. vulgaris Mitogenome*

The vast majority of the variance in genome size of plant mitogenomes can be explained by differences in the sizes of repeat sequences, which are composed of SSRs, tandem repeats and dispersed repeats. Plant mitogenomes, particularly those of angiosperms, were already well known for its sizeable fractions of repetitive sequences even before any complete mitogenomes were available. SSRs are DNA tracts of tandem-repeated motifs of one to six bases that are useful molecular markers in studying genetic diversity and species identification [21]. In this study, a total of 314 perfect SSRs were identified in the *P. vulgaris* mitogenome, including 139 mono-, 140 di-, 5 tri-, 22 tetra-, 3 penta-, and 5 hexa-nucleotide repeats (Table 4). The mononucleotide repeats of A/T (129 repeats) were found to be more prevalent than other repeat types. The dinucleotides repeats, TA/AT, are the second most numerous (50 repeats), while tri-, tetra-, penta-, and hexa-nucleotide repeats are fewer in number and only observed in intronic or intergenic regions. As shown in Table 5, seven tandem repeats with lengths ranging from 13 bp to 57 bp were also detected in the *P. vulgaris* mitogenome. Among these seven tandem repeats, only one is localized in a coding region (*rrnL*), while the others are all found in intergenic spacers.

Besides SSRs and tandem repeats, 143 dispersed repeats with lengths > 30 bp (total length: 35,000 bp; 8.85% of the genome) were also identified in the *P. vulgaris* mitogenome (Figure 7; Table S4). Most of the repeats (77 repeats, 53.85%) are 30 bp to 59 bp long, and 25 repeats are longer than 100 bp, with only two longer than 1 kb (R1: 4866 bp; R2: 3529 bp). Previous studies have documented the importance of large repeats (>1 kb) in genomic structural changes, and pairwise direct and inverted large repeats may produce two small subgenomic conformations or isomeric conformations, respectively. As shown in Figure 1, the largest repeat was assembled as Contig15 and the second largest was assembled as Contig40, both of which were inverted repeats. By aligning the PacBio long reads to both ends of the two large repeats, we constructed the master circle and two isomeric molecules (Figure 1). Repeats are commonly found in plant mitogenomes but are poorly conserved across species, even within the same family. As shown in Figure 7 and Table S4, the total number of repeats ranges from 59 in *V. angularis* to 215 in *M. pinnata*, and the total length of repeats ranges from 9224 bp (2.28% of the whole genome) in *V. angularis* to 411,265 bp (69.94% of the whole genome) in *V. faba*. Mitogenome enlargement in *V. faba* is mainly caused by the expansion of repeated sequences. Thirteen large (>1 kb) repeats covered 398.8 kb or 68% of the whole mitogenome size [17]. However, when all but single copies of the large repeat sequences were excluded, the *V. faba* mitogenome size is 388.6 kb, which is similar to other Papilionoideae mitogenomes [18]. The extremely complex repeat patterns should be responsible for the various genome sizes of the plant mitogenome. However, genome size is by no means only determined by the size of repeats. The mitochondrial genome of *Vitis vinifera* has only 7% repeats despite a genome size of nearly 773 kb [73], while the moderately-sized (404.5 kb) *V. angularis* genome has fewer and smaller repeats than those found in the much smaller genomes of *Brassica napus* (222 kb) and *Silene latifolia* (253 kb; Table S2) [10,74].

**Table 4.** Frequency of identified SSR motifs in *P. vulgaris* mitogenome.

| Motif Type | Number of Repeats | | | | | | | | | | | Total | Proportion (%) |
|---|---|---|---|---|---|---|---|---|---|---|---|---|---|
| | 3 | 4 | 5 | 6 | 7 | 8 | 9 | 10 | 11 | 12 | 13 | | |
| Monomer | - | - | - | - | - | 91 | 32 | 10 | 4 | 1 | 1 | 139 | 44.27 |
| Dimer | - | 120 | 16 | 4 | 0 | 0 | 0 | 0 | 0 | 0 | 0 | 140 | 44.59 |
| Trimer | - | 5 | 0 | 0 | 0 | 0 | 0 | 0 | 0 | 0 | 0 | 5 | 1.59 |
| Tetramer | 21 | 1 | 0 | 0 | 0 | 0 | 0 | 0 | 0 | 0 | 0 | 22 | 7.01 |
| Pentamer | 3 | 0 | 0 | 0 | 0 | 0 | 0 | 0 | 0 | 0 | 0 | 3 | 0.96 |
| Hexamer | 5 | 0 | 0 | 0 | 0 | 0 | 0 | 0 | 0 | 0 | 0 | 5 | 1.59 |
| Total | 29 | 126 | 16 | 4 | 0 | 91 | 32 | 10 | 4 | 1 | 1 | 314 | 100 |

**Table 5.** Distribution of tandem repeats in *P. vulgaris* mitogenome.

| No | Size (bp) | Start | End | Repeat (bp) × Copy Number | Location |
|---|---|---|---|---|---|
| 1 | 57 | 75,208 | 75,322 | (TTGGATCAAAACGATGTTCAACAACCTTTGCCGCGTCTGTTTCTTGGAGGAAAATAG) × 2 | IGS (*trnD*, *atp6*) |
| 2 | 27 | 92,189 | 92,243 | (AGAGCAGGTCGGTCTAGGTAGTTGAAA) × 2 | IGS (*trnC*, *trnN*) |
| 3 | 38 | 108,846 | 108,922 | (AAAAATATACATAACATATCCCAAACTCTATAGAGATA) × 2 | IGS (*rrn5*, *ccmF*$_N$) |
| 4 | 13 | 231,727 | 231,753 | (TCTTAAGTAAAGT) × 2 | IGS (*nad2*-exon1, *trnH*) |
| 5 | 18 | 252,332 | 252,368 | (CATAGTCGCGAGCTGTTT) × 2 | rrnL |
| 6 | 15 | 315,763 | 315,793 | (GTATAGTATAGTAGG) × 2 | IGS (*nad1*-exon1, *nad6*) |
| 7 | 33 | 356,560 | 356,626 | (CCTTGCCCCCTGCAGAGCCTCAAGCCCCTGAGC) × 2 | IGS (*atp1*, *nad4L*) |

IGS: intergenic pacers.

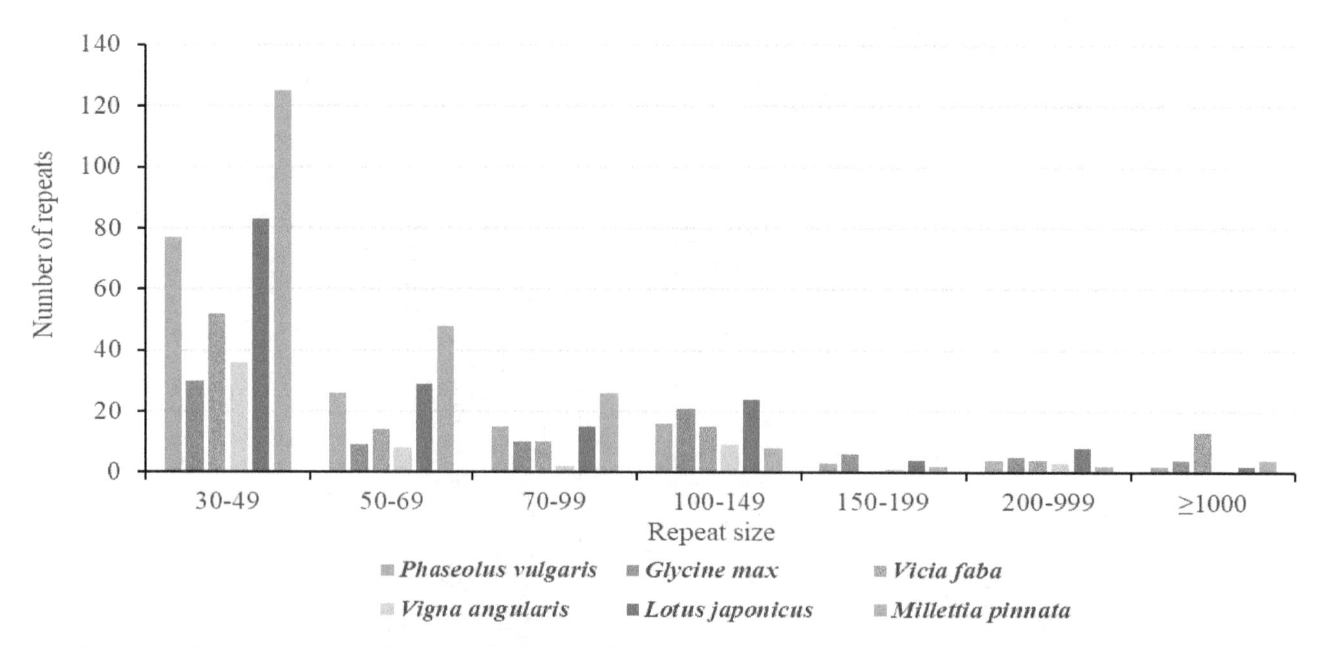


**Figure 7.** Frequency distribution of dispersed repeat in the *P. vulgaris* mitogenome compared with five other leguminous plants. The number of dispersed repeats in *Phaseolus vulgaris, Glycine max Vicia faba, Vigna faba, Vigna angularis, Lotus japonicus,* and *Millettia pinnata* mitogenomes are shown by blue, orange, gray, yellow, blue, and green, respectively.

### *3.6. Phylogenetic Analyses and Multiple Losses of PCGs during Evolution*

With rapid developments in sequencing technology and assembly methods, an increasing number of complete plant mitogenomes has been assembled, providing an important opportunity for phylogenetic analyses using mitogenomes. In this study, to determine the phylogenetic position of *P. vulgaris*, we downloaded 23 plant mitogenomes from the GenBank database (https://www.ncbi.nlm.nih.gov/genome/browse/), including 19 species of Fabales, two species of Solanales, and two species of Malpighiales. A set of 26 conserved single-copy orthologous genes (*atp1, atp4, atp6, atp8, atp9, ccmB, ccmC,* $ccmF_C$, $ccmF_N$, *cob, cox1, cox3, matR, nad1, nad2, nad3, nad4, nad4L, nad5, nad6, nad7, nad9, rps3, rps4,* and *rps12*) present in all of the 23 analyzed mitogenomes was used to construct the phylogenetic tree, and species from the Solanales and Malpighiales were designated as the outgroup. As shown in Figure 8, the bootstrap values of each node are all over 70% supported and 15 nodes are supported 100%. The ML phylogenetic tree strongly supports that *P. vulgaris* is evolutionarily close to the clade formed by two Vigna species. The tree also strongly supports the separation of Fabales from the clade composed of Solanales and Malpighiales (100% bootstrap value), as well as the separation of Papilionoideae from the clade composed of Cercidoideae, Detarioideae, and Caesalpinioideae (100%). The bootstrap value for the separation of Detarioideae and Caesalpinioideae is 80%, and the value for the separation of Cercidoideae from the clade composed of Detarioideae and Caesalpinioideae is 70%.

As described by Richardson et al. [75], the mitochondrial genomes of higher plants vary significantly in genome size, gene content and order. Losses of PCGs occurred frequently during the evolution of higher plants. The phylogenetic tree provides a backdrop for the further analysis of gene loss during evolution, and the gene contents of all observed species are summarized in Figure 9. Most of the PCGs were conserved in different plant mitogenomes, especially for the genes in the groups of Complex I, Complex III, Complex V, cytochrome *c* biogenesis, maturases, and transport membrane protein [13]. The conservation of these genes suggests that they play crucial roles in the function of mitochondria. However, the ribosomal proteins and succinate dehydrogenase genes were highly variable. As shown in Figure 9, the *cox2* gene was only lost in the subfamily Phaseolinae (*V. angularis, V. radiata,* and *P. vulgaris*) but retained in other leguminous plants, suggesting that this gene was lost after separation from the subfamily Glycininae. The *rpl2* gene was lost in most leguminous plants but regained in *A. ligulate*, *L. trichandra*, *H. brasuletto*, and *L. coriaria*, suggesting that this gene was lost before the emergence of Fabales but could be regained in some leguminous plants. Similar phenomena were found in many ribosomal proteins (*rpl10*, *rpl16*, *rps7*, *rps10*, and *rps19*). Additionally, *rpl6* and *rps8* genes were lost from liverworts (*M. polymorpha*) during evolution [76], the *rps11* gene was lost from gymnosperms (*G. biloba*) and liverworts during the divergence of the angiosperms and gymnosperms [77], and the *rpl10* gene was lost in monocots and gymnosperms but regained in dicots [33,78]. The enhanced loss of ribosomal proteins in plant mitogenomes indicates that these genes were encoded partly by mitochondrial native genes and partly by nuclear genes, due to the gene transfer between mitochondria and nucleus [79–81].

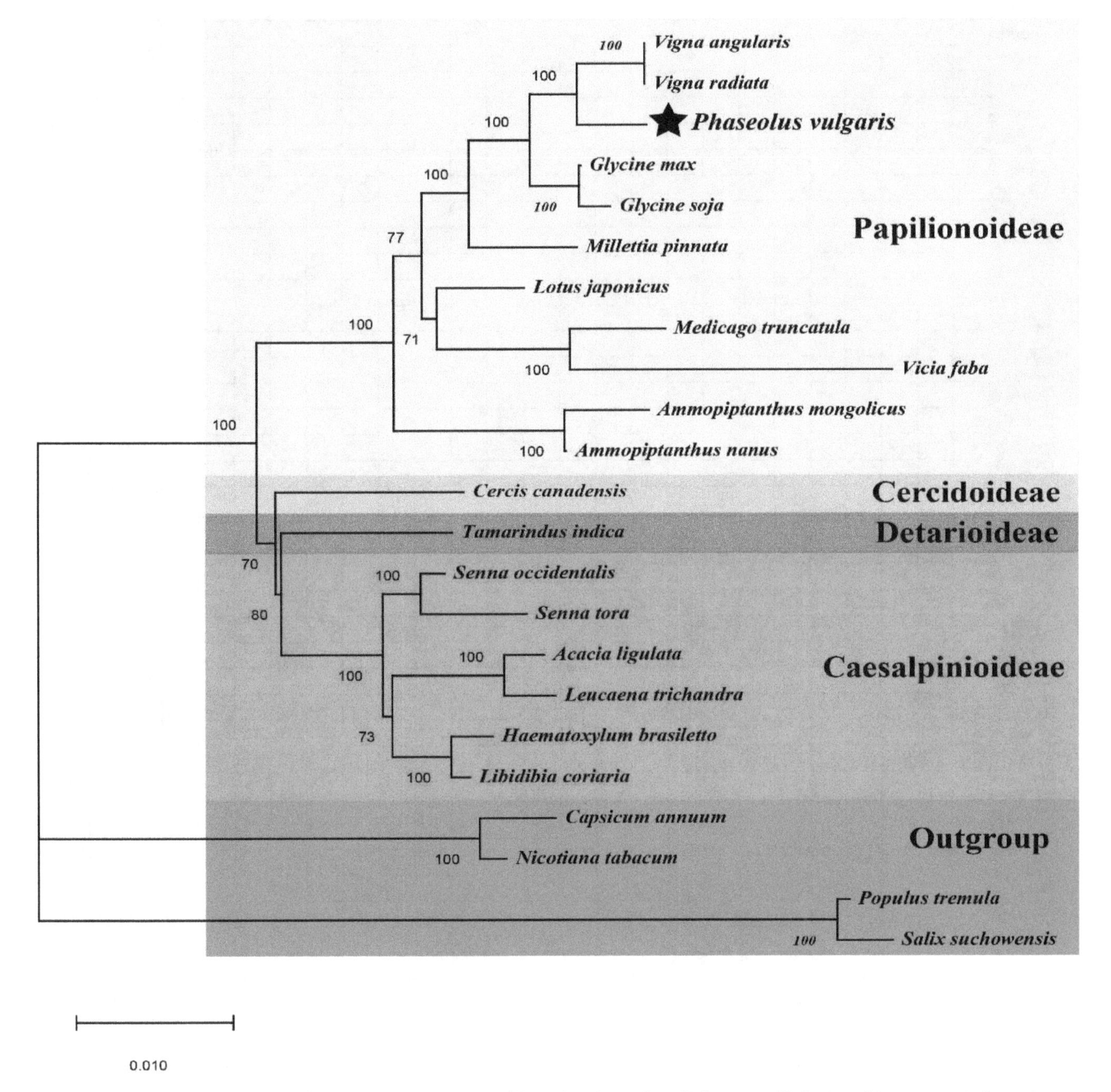


**Figure 8.** Maximum likelihood phylogenies of *P. vulgaris* within Fabaceae. Relationships were inferred employing 26 conserved PCGs of 23 plant mitogenomes. Numbers on each node are bootstrap support values. NCBI accession numbers are listed in Table S2. Scale indicates number of nucleotide substitutions per site.

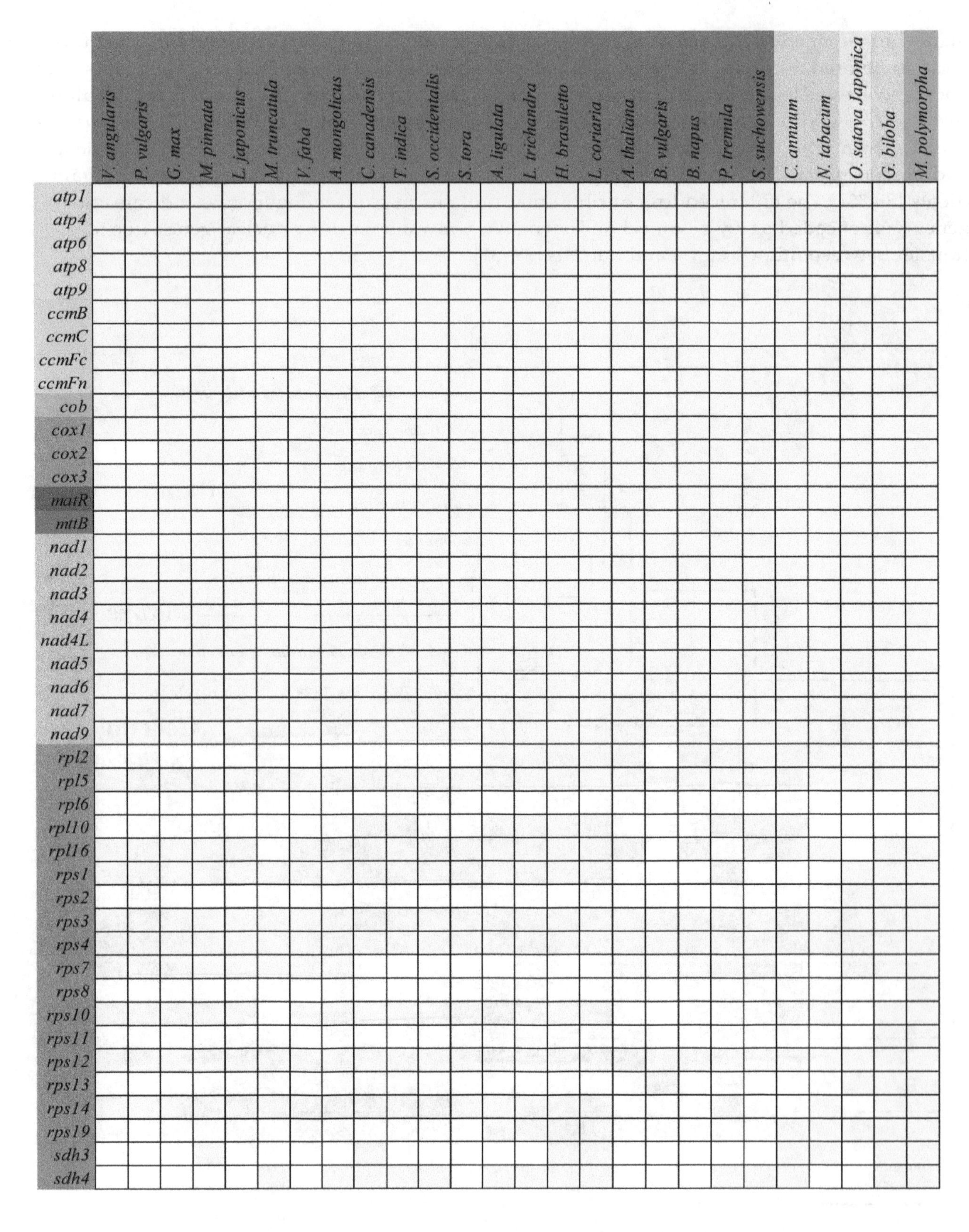


**Figure 9.** Distribution of PCGs in plant mitogenomes. White boxes indicate that the gene is not present in the mitogenome. The colors of genes indicate their corresponding categories. The colors of species represent the classes of rosids (orange), asterids (pink), monocotyledons (gold), gymnosperms (light blue), and liverworts (green).

## 4. Conclusions

In this study, we first assembled and characterized the complete mitogenome of *P. vulgaris*. By aligning the PacBio sequencing reads to the draft mitogenome, one master circle and two isomeric molecules were assembled based on two large repeats. Selective-pressure analysis of PCGs indicates that *ccmB*, *ccmF$_C$*, *rps1*, *rps10*, and *rps14* genes with $K_a/K_s$ ratios > 1 might play important roles during evolution, whereas all Complex I–V genes with $K_a/K_s$ ratios < 1 were highly conserved in the evolutionary process of higher plants. The C-to-U conversions may generate initiation, termination, or internal codons with completely unpredictable functions. The prediction of RNA editing sites in *P. vulgaris* mt PCGs will provide important clues for the investigation of gene functions with novel codons. The comparison of genomic features in all sequenced leguminous plants should contribute to a comprehensive understanding of the evolutionary process of legumes. The sequencing of the *P. vulgaris* mitogenome not only provides an important opportunity to conduct further genomic breeding studies in the common bean, it also provides valuable information for future evolutionary and molecular studies of leguminous plants.

**Supplementary Materials:** Figure S1: The flowchart of *P. vulgaris* mitogenome assembly and annotation, Table S1: PacBio reads of spanning repetitive contigs, Table S2: The abbreviations and genome sizes of studied mitogenomes, Table S3: The relative synonymous codon usage (RSCU) of mitogenome genome across seven higher plants, Table S4: Comparison of dispersed repeats within six mitogenomes of Papilionaceae subfamily.

**Author Contributions:** Conceptualization, C.H.; Formal analysis, C.B. and N.L.; Funding acquisition, Z.L.; Methodology, C.B.; Project administration, Z.L.; C.B., N.L., and Y.X.; Writing—original draft, C.B.; Writing—review and editing, C.H. and Z.L. All authors have read and agreed to the published version of the manuscript.

## References

1. Greiner, S.; Bock, R. Tuning a menage a trois: Co-evolution and co-adaptation of nuclear and organellar genomes in plants. *Bioessays* **2013**, *35*, 354–365. [CrossRef]
2. Timmis, J.N.; Ayliffe, M.A.; Huang, C.Y.; Martin, W. Endosymbiotic gene transfer: Organelle genomes forge eukaryotic chromosomes. *Nat. Rev. Genet.* **2004**, *5*, 123–135. [CrossRef]
3. Hsu, C.L.; Mullin, B.C. Physical characterization of mitochondrial DNA from cotton. *Plant Mol. Biol.* **1989**, *13*, 467–468. [CrossRef]
4. Kubo, T.; Mikami, T. Organization and variation of angiosperm mitochondrial genome. *Physiol. Plant.* **2007**, *129*, 6–13. [CrossRef]
5. Palmer, J.D.; Herbon, L.A. Plant mitochondrial DNA evolves rapidly in structure, but slowly in sequence. *J. Mol. Evol.* **1988**, *28*, 87–97. [CrossRef]
6. Gray, M.W.; Burger, G.; Lang, B.F. Mitochondrial Evolution. *Science* **1999**, *283*, 1476–1481. [CrossRef]
7. Lang, B.F.; Gray, M.W.; Burger, G. Mitochondrial Genome Evolution and the Origin of Eukaryotes. *Annu. Rev. Genet.* **1999**, *33*, 351–397. [CrossRef]
8. Alverson, A.J.; Wei, X.X.; Rice, D.W.; Stern, D.B.; Barry, K.; Palmer, J.D. Insights into the Evolution of Mitochondrial Genome Size from Complete Sequences of *Citrullus lanatus* and *Cucurbita pepo* (Cucurbitaceae). *Mol. Biol. Evol.* **2010**, *27*, 1436–1448. [CrossRef]
9. Alverson, A.J.; Zhuo, S.; Rice, D.W.; Sloan, D.B.; Palmer, J.D. The mitochondrial genome of the legume *Vigna radiata* and the analysis of recombination across short mitochondrial repeats. *PLoS ONE* **2011**, *6*, e16404. [CrossRef]
10. Handa, H. The complete nucleotide sequence and RNA editing content of the mitochondrial genome of rapeseed (*Brassica napus* L.): Comparative analysis of the mitochondrial genomes of rapeseed and *Arabidopsis thaliana*. *Nucleic Acids Res.* **2003**, *31*, 5907–5916. [CrossRef]
11. Mulligan, R.M.; Chang, K.L.; Chou, C.C. Computational analysis of RNA editing sites in plant mitochondrial genomes reveals similar information content and a sporadic distribution of editing sites. *Mol. Biol. Evol.* **2007**, *24*, 1971–1981. [CrossRef]
12. Andre, C.; Levy, A.; Walbot, V. Small repeated sequences and the structure of plant mitochondrial genomes. *Trends Genet.* **1992**, *8*, 128–132. [CrossRef]
13. Chang, S.; Wang, Y.; Lu, J.; Gai, J.; Li, J.; Chu, P.; Guan, R.; Zhao, T. Correction: The Mitochondrial Genome of Soybean Reveals Complex Genome Structures and Gene Evolution at Intercellular and Phylogenetic Levels. *PLoS ONE* **2013**, *8*, e56502. [CrossRef]

14. Skippington, E.; Barkman, T.J.; Rice, D.W.; Palmer, J.D. Miniaturized mitogenome of the parasitic plant *Viscum scurruloideum* is extremely divergent and dynamic and has lost all nad genes. *Proc. Natl. Acad. Sci. USA* **2015**, *112*, E3515–E3524. [CrossRef] [PubMed]
15. Sloan, D.B.; Alverson, A.J.; Chuckalovcak, J.P.; Wu, M.; McCauley, D.E.; Palmer, J.D.; Taylor, D.R. Rapid evolution of enormous, multichromosomal genomes in flowering plant mitochondria with exceptionally high mutation rates. *PLoS Biol.* **2012**, *10*, e1001241. [CrossRef] [PubMed]
16. Bi, C.; Wang, X.; Xu, Y.; Wei, S.; Shi, Y.; Dai, X.; Yin, T.; Ye, N. The complete mitochondrial genome of *Medicago truncatula*. *Mitochondrial DNA Part B* **2016**, *1*, 122–123. [CrossRef]
17. Negruk, V. Mitochondrial genome sequence of the legume *Vicia faba*. *Front. Plant Sci.* **2013**, *4*, 128. [CrossRef]
18. Choi, I.-S.; Schwarz, E.N.; Ruhlman, T.A.; Khiyami, M.A.; Sabir, J.S.; Hajarah, N.H.; Sabir, M.J.; Rabah, S.O.; Jansen, R.K. Fluctuations in Fabaceae mitochondrial genome size and content are both ancient and recent. *BMC Plant Biol.* **2019**, *19*, 448. [CrossRef]
19. Wynn, E.L.; Christensen, A.C. Repeats of Unusual Size in Plant Mitochondrial Genomes: Identification, Incidence and Evolution. *G3 Genes Genomes Genet.* **2019**, *9*, 549–559. [CrossRef]
20. Bergthorsson, U.; Adams, K.L.; Thomason, B.; Palmer, J.D. Widespread horizontal transfer of mitochondrial genes in flowering plants. *Nature* **2003**, *424*, 197–201. [CrossRef]
21. Ma, Q.; Li, S.; Bi, C.; Hao, Z.; Sun, C.; Ye, N. Complete chloroplast genome sequence of a major economic species, *Ziziphus jujuba* (Rhamnaceae). *Curr. Genet.* **2017**, *63*, 117–129. [CrossRef]
22. Sperisen, C.; Büchler, U.; Gugerli, F.; Mátyás, G.; Geburek, T.; Vendramin, G. Tandem repeats in plant mitochondrial genomes: Application to the analysis of population differentiation in the conifer Norway spruce. *Mol. Ecol.* **2001**, *10*, 257–263. [CrossRef]
23. Alverson, A.J.; Rice, D.W.; Dickinson, S.; Barry, K.; Palmer, J.D. Origins and Recombination of the Bacterial-Sized Multichromosomal Mitochondrial Genome of Cucumber. *Plant Cell* **2011**, *23*, 2499–2513. [CrossRef]
24. Backert, S.; Nielsen, B.L.; Börner, T. The mystery of the rings: Structure and replication of mitochondrial genomes from higher plants. *Trends Plant Sci.* **1997**, *2*, 477–483. [CrossRef]
25. Ogihara, Y.; Yamazaki, Y.; Murai, K.; Kanno, A.; Terachi, T.; Shiina, T.; Miyashita, N.; Nasuda, S.; Nakamura, C.; Mori, N. Structural dynamics of cereal mitochondrial genomes as revealed by complete nucleotide sequencing of the wheat mitochondrial genome. *Nucleic Acids Res.* **2005**, *33*, 6235–6250. [CrossRef]
26. Bi, C.; Paterson, A.H.; Wang, X.; Xu, Y.; Wu, D.; Qu, Y.; Jiang, A.; Ye, Q.; Ye, N. Analysis of the complete mitochondrial genome sequence of the diploid cotton *Gossypium raimondii* by comparative genomics approaches. *BioMed. Res. Int.* **2016**, *2016*, 5040598. [CrossRef]
27. Christenhusz, M.J.; Byng, J.W. The number of known plants species in the world and its annual increase. *Phytotaxa* **2016**, *261*, 201–217. [CrossRef]
28. Azani, N.; Babineau, M.; Bailey, C.D.; Banks, H.; Barbosa, A.R.; Pinto, R.B.; Boatwright, J.S.; Borges, L.M.; Brown, G.K.; Bruneau, A.; et al. A new subfamily classification of the Leguminosae based on a taxonomically comprehensive phylogeny: The Legume Phylogeny Working Group (LPWG). *Taxon* **2017**, *66*, 44–77. [CrossRef]
29. Schmutz, J.; McClean, P.E.; Mamidi, S.; Wu, G.A.; Cannon, S.B.; Grimwood, J.; Jenkins, J.; Shu, S.; Song, Q.; Chavarro, C.; et al. A reference genome for common bean and genome-wide analysis of dual domestications. *Nat. Genet.* **2014**, *46*, 707. [CrossRef]
30. Fonsêca, A.; Ferreira, J.; dos Santos, T.R.B.; Mosiolek, M.; Bellucci, E.; Kami, J.; Gepts, P.; Geffroy, V.; Schweizer, D.; dos Santos, K.G. Cytogenetic map of common bean (*Phaseolus vulgaris* L.). *Chrom. Res.* **2010**, *18*, 487–502. [CrossRef]
31. Zhang, T.; Fang, Y.; Wang, X.; Deng, X.; Zhang, X.; Hu, S.; Yu, J. The Complete Chloroplast and Mitochondrial Genome Sequences of *Boea hygrometrica*: Insights into the Evolution of Plant Organellar Genomes. *PLoS ONE* **2012**, *7*, e30531. [CrossRef]
32. Iorizzo, M.; Senalik, D.; Szklarczyk, M.; Grzebelus, D.; Spooner, D.; Simon, P. *De novo* assembly of the carrot mitochondrial genome using next generation sequencing of whole genomic DNA provides first evidence of DNA transfer into an angiosperm plastid genome. *BMC Plant Biol.* **2012**, *12*, 61. [CrossRef]
33. Ye, N.; Wang, X.; Li, J.; Bi, C.; Xu, Y.; Wu, D.; Ye, Q. Assembly and comparative analysis of complete mitochondrial genome sequence of an economic plant *Salix suchowensis*. *PeerJ* **2017**, *5*, e3148. [CrossRef]

34. Nederbragt, A.J. On the middle ground between open source and commercial software-the case of the Newbler program. *Genome Biol.* **2014**, *15*, 113. [CrossRef]
35. Camacho, C.; Coulouris, G.; Avagyan, V.; Ma, N.; Papadopoulos, J.; Bealer, K.; Madden, T.L. BLAST+: Architecture and applications. *BMC Bioinform.* **2009**, *10*, 421. [CrossRef]
36. Zhang, T.; Zhang, X.; Hu, S.; Yu, J. An efficient procedure for plant organellar genome assembly, based on whole genome data from the 454 GS FLX sequencing platform. *Plant Methods* **2011**, *7*, 38. [CrossRef]
37. Dong, S.; Zhao, C.; Chen, F.; Liu, Y.; Zhang, S.; Wu, H.; Zhang, L.; Liu, Y. The complete mitochondrial genome of the early flowering plant *Nymphaea colorata* is highly repetitive with low recombination. *BMC Genom.* **2018**, *19*, 614. [CrossRef]
38. Li, H.; Durbin, R. Fast and accurate short read alignment with Burrows–Wheeler transform. *Bioinformatics* **2009**, *25*, 1754–1760. [CrossRef]
39. Li, H.; Handsaker, B.; Wysoker, A.; Fennell, T.; Ruan, J.; Homer, N.; Marth, G.; Abecasis, G.; Durbin, R. The Sequence Alignment/Map format and SAMtools. *Bioinformatics* **2009**, *25*, 2078–2079. [CrossRef]
40. Schattner, P.; Brooks, A.N.; Lowe, T.M. The tRNAscan-SE, snoscan and snoGPS web servers for the detection of tRNAs and snoRNAs. *Nucleic Acids Res.* **2005**, *33*, W686–W689. [CrossRef]
41. Kumar, S.; Stecher, G.; Li, M.; Knyaz, C.; Tamura, K. MEGA X: Molecular Evolutionary Genetics Analysis across Computing Platforms. *Mol. Biol. Evol.* **2018**, *35*, 1547–1549. [CrossRef]
42. Greiner, S.; Lehwark, P.; Bock, R. OrganellarGenomeDRAW (OGDRAW) version 1.3. 1: Expanded toolkit for the graphical visualization of organellar genomes. *Nucleic Acids Res.* **2019**, *47*, W59–W64. [CrossRef]
43. Zhang, Z.; Xiao, J.; Wu, J.; Zhang, H.; Liu, G.; Wang, X.; Dai, L. ParaAT: A parallel tool for constructing multiple protein-coding DNA alignments. *Biochem. Biophys. Res. Commun.* **2012**, *419*, 779–781. [CrossRef]
44. Yang, Z.; Nielsen, R. Estimating Synonymous and Nonsynonymous Substitution Rates Under Realistic Evolutionary Models. *Mol. Biol. Evol.* **2000**, *17*, 32–43. [CrossRef]
45. Wang, D.; Zhang, Y.; Zhang, Z.; Zhu, J.; Yu, J. KaKs_Calculator 2.0: A Toolkit Incorporating Gamma-Series Methods and Sliding Window Strategies. *Genom. Proteom. Bioinform.* **2010**, *8*, 77–80. [CrossRef]
46. Mower, J.P. The PREP suite: Predictive RNA editors for plant mitochondrial genes, chloroplast genes and user-defined alignments. *Nucleic Acids Res.* **2009**, *37*, W253–W259. [CrossRef]
47. Beier, S.; Thiel, T.; Münch, T.; Scholz, U.; Mascher, M. MISA-web: A web server for microsatellite prediction. *Bioinformatics* **2017**, *33*, 2583–2585. [CrossRef]
48. Benson, G. Tandem repeats finder: A program to analyze DNA sequences. *Nucleic Acids Res.* **1999**, *27*, 573–580. [CrossRef]
49. Liu, G.; Cao, D.; Li, S.; Su, A.; Geng, J.; Grover, C.E.; Hu, S.; Hua, J. The Complete Mitochondrial Genome of *Gossypium hirsutum* and Evolutionary Analysis of Higher Plant Mitochondrial Genomes. *PLoS ONE* **2013**, *8*, e69476. [CrossRef]
50. Edgar, R.C. MUSCLE: Multiple sequence alignment with high accuracy and high throughput. *Nucleic Acids Res.* **2004**, *32*, 1792–1797. [CrossRef]
51. Shen, Y.; Du, H.; Liu, Y.; Ni, L.; Wang, Z.; Liang, C.; Tian, Z. Update soybean Zhonghuang 13 genome to a golden reference. *Sci. China Life Sci.* **2019**, *62*, 1257–1260. [CrossRef] [PubMed]
52. Li, W.-H.; Wu, C.-I.; Luo, C.-C. A new method for estimating synonymous and nonsynonymous rates of nucleotide substitution considering the relative likelihood of nucleotide and codon changes. *Mol. Biol. Evol.* **1985**, *2*, 150–174. [PubMed]
53. Betrán, E.; Bai, Y.; Motiwale, M. Fast Protein Evolution and Germ Line Expression of a Drosophila Parental Gene and Its Young Retroposed Paralog. *Mol. Biol. Evol.* **2006**, *23*, 2191–2202. [CrossRef] [PubMed]
54. Arbiza, L.; Dopazo, J.; Dopazo, H. Positive Selection, Relaxation, and Acceleration in the Evolution of the Human and Chimp Genome. *PLoS Comp. Biol.* **2006**, *2*, e38. [CrossRef] [PubMed]
55. Meyer, E.H.; Giegé, P.; Gelhaye, E.; Rayapuram, N.; Ahuja, U.; Thöny-Meyer, L.; Grienenberger, J.-M.; Bonnard, G. AtCCMH, an essential component of the *c*-type cytochrome maturation pathway in *Arabidopsis* mitochondria, interacts with apocytochrome *c*. *Proc. Natl. Acad. Sci. USA* **2005**, *102*, 16113–16118. [CrossRef] [PubMed]
56. Faivre-Nitschke, S.E.; Nazoa, P.; Gualberto, J.M.; Grienenberger, J.M.; Bonnard, G. Wheat mitochondria *ccmB* encodes the membrane domain of a putative ABC transporter involved in cytochrome *c* biogenesis. *Biochim. Biophys. Acta* **2001**, *1519*, 199–208. [CrossRef]
57. Sanders, C.; Turkarslan, S.; Lee, D.-W.; Daldal, F. Cytochrome *c* biogenesis: The Ccm system. *Trends Microbiol.* **2010**, *18*, 266–274. [CrossRef]

58. Robles, P.; Quesada, V. Emerging Roles of Mitochondrial Ribosomal Proteins in Plant Development. *Int. J. Mol. Sci.* **2017**, *18*, 2595. [CrossRef]
59. Mauro, V.P.; Edelman, G.M. The Ribosome Filter Redux. *Cell Cycle* **2007**, *6*, 2246–2251. [CrossRef]
60. Schippers, J.H.M.; Mueller-Roeber, B. Ribosomal composition and control of leaf development. *Plant Sci.* **2010**, *179*, 307–315. [CrossRef]
61. Cui, P.; Liu, H.; Lin, Q.; Ding, F.; Zhuo, G.; Hu, S.; Liu, D.; Yang, W.; Zhan, K.; Zhang, A.; et al. A complete mitochondrial genome of wheat (*Triticum aestivum* cv. Chinese Yumai), and fast evolving mitochondrial genes in higher plants. *J. Genet.* **2009**, *88*, 299–307. [CrossRef] [PubMed]
62. Feng, L.; Li, N.; Yang, W.; Li, Y.; Wang, C.-M.; Tong, S.-W.; He, J.-X. Analyses of mitochondrial genomes of the genus *Ammopiptanthus* provide new insights into the evolution of legume plants. *Plant Syst. Evol.* **2019**, *305*, 385–399. [CrossRef]
63. Wernegreen, J.J.; Riley, M.A. Comparison of the evolutionary dynamics of symbiotic and housekeeping loci: A case for the genetic coherence of rhizobial lineages. *Mol. Biol. Evol.* **1999**, *16*, 98–113. [CrossRef] [PubMed]
64. Bock, R.; Khan, M.S. Taming plastids for a green future. *Trends Biotechnol.* **2004**, *22*, 311–318. [CrossRef]
65. Chen, H.; Deng, L.; Jiang, Y.; Lu, P.; Yu, J. RNA Editing Sites Exist in Protein-coding Genes in the Chloroplast Genome of *Cycas taitungensis*. *J. Integr. Plant Biol.* **2011**, *53*, 961–970. [CrossRef]
66. Raman, G.; Park, S. Analysis of the Complete Chloroplast Genome of a Medicinal Plant, *Dianthus superbus* var. longicalyncinus, from a Comparative Genomics Perspective. *PLoS ONE* **2015**, *10*, e0141329. [CrossRef]
67. Wakasugi, T.; Hirose, T.; Horihata, M.; Tsudzuki, T.; Kössel, H.; Sugiura, M. Creation of a novel protein-coding region at the RNA level in black pine chloroplasts: The pattern of RNA editing in the gymnosperm chloroplast is different from that in angiosperms. *Proc. Natl. Acad. Sci. USA* **1996**, *93*, 8766–8770. [CrossRef]
68. Mower, J.P. PREP-Mt: Predictive RNA editor for plant mitochondrial genes. *BMC Bioinform.* **2005**, *6*, 96. [CrossRef] [PubMed]
69. Malek, O.; Lättig, K.; Hiesel, R.; Brennicke, A.; Knoop, V. RNA editing in bryophytes and a molecular phylogeny of land plants. *EMBO J.* **1996**, *15*, 1403–1411. [CrossRef]
70. Steinhauser, S.; Beckert, S.; Capesius, I.; Malek, O.; Knoop, V. Plant Mitochondrial RNA Editing. *J. Mol. Evol.* **1999**, *48*, 303–312. [CrossRef]
71. Shu-Miaw, C.; Arthur, C.C.S.; Wang, D.; Yu-Wei, W.; Shu-Mei, L.; The-Yuan, C. The Mitochondrial Genome of the Gymnosperm *Cycas taitungensis* Contains a Novel Family of Short Interspersed Elements, Bpu Sequences, and Abundant RNA Editing Sites. *Mol. Biol. Evol.* **2008**, *25*, 603–615.
72. Hiesel, R.; Combettes, B.; Brennicke, A. Evidence for RNA editing in mitochondria of all major groups of land plants except the Bryophyta. *Proc. Natl. Acad. Sci. USA* **1994**, *91*, 629–633. [CrossRef] [PubMed]
73. Goremykin, V.V.; Salamini, F.; Velasco, R.; Viola, R. Mitochondrial DNA of *Vitis vinifera* and the Issue of Rampant Horizontal Gene Transfer. *Mol. Biol. Evol.* **2009**, *26*, 99–110. [CrossRef] [PubMed]
74. Sloan, D.B.; Alverson, A.J.; Štorchová, H.; Palmer, J.D.; Taylor, D.R. Extensive loss of translational genes in the structurally dynamic mitochondrial genome of the angiosperm *Silene latifolia*. *BMC Evol. Biol.* **2010**, *10*, 274. [CrossRef] [PubMed]
75. Richardson, A.O.; Rice, D.W.; Young, G.J.; Alverson, A.J.; Palmer, J.D. The "fossilized" mitochondrial genome of *Liriodendron tulipifera*: Ancestral gene content and order, ancestral editing sites, and extraordinarily low mutation rate. *BMC Biol.* **2013**, *11*, 29. [CrossRef] [PubMed]
76. Bowman, J.L.; Kohchi, T.; Yamato, K.T.; Jenkins, J.; Shu, S. Insights into Land Plant Evolution Garnered from the *Marchantia polymorpha* Genome. *Cell* **2017**, *171*, 287–304. [CrossRef]
77. Guo, W.; Felix, G.; Fan, W.; Young, G.J.; Volker, K.; Palmer, J.D.; Mower, J.P. *Ginkgo* and *Welwitschia* Mitogenomes Reveal Extreme Contrasts in Gymnosperm Mitochondrial Evolution. *Mol. Biol. Evol.* **2016**, *33*, 1448–1460. [CrossRef]
78. Notsu, Y.; Masood, S.; Nishikawa, T.; Kubo, N.; Akiduki, G.; Nakazono, M.; Hirai, A.; Kadowaki, K. The complete sequence of the rice (*Oryza sativa* L.) mitochondrial genome: Frequent DNA sequence acquisition and loss during the evolution of flowering plants. *Mol. Genet. Genom.* **2002**, *268*, 434–445. [CrossRef]
79. Clifton, S.W.; Minx, P.; Fauron, C.M.-R.; Gibson, M.; Allen, J.O.; Sun, H.; Thompson, M.; Barbazuk, W.B.; Kanuganti, S.; Tayloe, C. Sequence and Comparative Analysis of the Maize NB Mitochondrial Genome. *Plant Physiol.* **2004**, *136*, 3486–3503. [CrossRef]
80. Unseld, M.; Marienfeld, J.R.; Brandt, P.; Brennicke, A. The mitochondrial genome of *Arabidopsis thaliana* contains 57 genes in 366,924 nucleotides. *Nat. Genet.* **1997** *15*, 57–61. [CrossRef]

# 6

# The Role of Chloroplast Gene Expression in Plant Responses to Environmental Stress

**Yi Zhang [1], Aihong Zhang [1], Xiuming Li [2] and Congming Lu [1,*]**

[1] State Key Laboratory of Crop Biology, College of Life Sciences, Shandong Agricultural University, Taian 271018, China; zhangyi@sdau.edu.cn (Y.Z.); ahzhang@sdau.edu.cn (A.Z.)

[2] State Key Laboratory of Crop Biology, College of Horticulture Science and Engineering, Shandong Agricultural University, Taian 271018, China; lixiuming@sdau.edu.cn

* Correspondence: cmlu@sdau.edu.cn

**Abstract:** Chloroplasts are plant organelles that carry out photosynthesis, produce various metabolites, and sense changes in the external environment. Given their endosymbiotic origin, chloroplasts have retained independent genomes and gene-expression machinery. Most genes from the prokaryotic ancestors of chloroplasts were transferred into the nucleus over the course of evolution. However, the importance of chloroplast gene expression in environmental stress responses have recently become more apparent. Here, we discuss the emerging roles of the distinct chloroplast gene expression processes in plant responses to environmental stresses. For example, the transcription and translation of *psbA* play an important role in high-light stress responses. A better understanding of the connection between chloroplast gene expression and environmental stress responses is crucial for breeding stress-tolerant crops better able to cope with the rapidly changing environment.

**Keywords:** chloroplast gene expression; photosynthesis; environmental stress response; transcription; RNA metabolism; translation

## 1. Introduction

Plant often face environmental conditions that are unfavorable for growth and development. These adverse environmental conditions include abiotic and biotic stresses, such as drought, heat, cold, salt, and pathogen infection [1–6]. Environmental stresses pose a great threat to agriculture by limiting crop yields and productivity. The adverse effects of environmental stresses are getting worse due to the increasing worldwide population and climate change. To deal with these environmental stresses, plants rely on their ability to sense and cope with these stresses by regulating the expression of stress-responsive genes in the nucleus, cytoplasm, and organelles.

The chloroplast, a unique plant organelle, is the site of photosynthesis, intracellular signaling, and the production of various compounds important in metabolism, such as amino acids, hormones, nucleotides, vitamins, lipids, and secondary metabolites [7–9]. Chloroplasts also serve as sensors of the external environment. Under stress conditions, chloroplasts send messages to the nucleus through plastid-to-nucleus retrograde signaling, thus optimizing nuclear gene expression based on physiological requirements [7,8]. To date, several possible retrograde signaling pathways have been proposed, including pathways involving intermediates in tetrapyrrole biogenesis [10], the redox state of plastids [11,12], reactive oxygen species [13,14], secondary metabolites in chloroplasts [15,16], and chloroplast gene expression [17–19].

Chloroplasts are semi-autonomous organelles that have retained their own genomes. However, during evolution, most chloroplast genes were lost or transferred to the nucleus: On average, the chloroplast genomes of land plants have retained only 120 genes [20,21]. Nonetheless, these relatively few genes play fundamental roles in chloroplast activities such as energy production and gene expression [22].

Gene expression in chloroplasts is a highly complicated process, far more complex than in their prokaryotic ancestors. This is because chloroplasts have retained a hybrid gene-expression system that combines features of the prokaryotic gene-expression apparatus with eukaryotic innovations (e.g., RNA editing and RNA splicing), and its nascent polycistronic transcripts must undergo many post-transcriptional processing steps [22–25].

The proper expression of chloroplast genes is crucial for chloroplast development and photosynthesis. During the past decade, much effort has been invested in exploring the molecular mechanisms regulating chloroplast gene expression using genetic approaches. Many nucleus-encoded proteins involved in regulating chloroplast gene expression have been identified. However, studies of mutants of these proteins have shown that these mutants are also sensitive to various environmental stresses [8,26–29]. These findings suggest that there is a link between chloroplast gene expression and environmental stress responses, but less attention has been paid to this issue. In this review, we discuss the emerging roles of chloroplast gene expression in plant responses to environmental stresses.

## 2. The Characteristics of Chloroplast Gene Expression

The chloroplast gene-expression system is evolutionarily derived from photosynthetic bacteria that were endocytosed by ancestral eukaryotic plant cells more than 1.5 billion years ago [30]. During evolution, chloroplasts have retained core components of the gene-expression apparatus from their prokaryotic progenitors. In addition, they obtained many eukaryotic properties, such as RNA editing, the prevalence of introns, and complex processing patterns from polycistronic RNA precursors [31]. Here, we briefly describe the processes of chloroplast gene expression in plants (Figure 1).

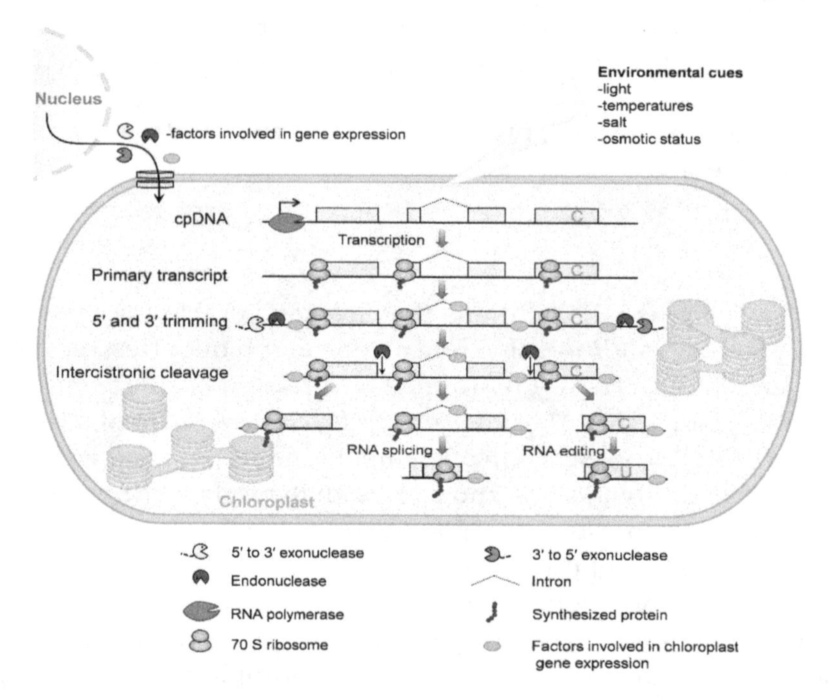


**Figure 1.** Overview of chloroplast gene expression. In plants, most chloroplast genes are organized as operons and are controlled by single promoters (bent arrow). These genes are transcribed by two distinct types of RNA polymerase: Nucleus-encoded RNA polymerase (NEP) and plastid-encoded RNA polymerase (PEP). The resulting primary transcripts require several processing steps to form mature mRNA, including 5′ and 3′ trimming, intercistronic cleavage, RNA splicing, and RNA editing. In order for these events to take place, numerous nucleus-encoded proteins are translated in the cytosol and imported into the chloroplast, where they control and/or regulate chloroplast gene expression. Chloroplast gene translation is conducted by bacterial-type 70S ribosomes, which occurs cotranscriptionally. Since the mRNA turnover rate within chloroplasts is slow, most ribosomes function in posttranscriptional steps. Moreover, chloroplast gene expression is involved in responses to environmental cues.

### 2.1. Transcription

In plants, chloroplast gene transcription is conducted by two distinct types of RNA polymerases: Nucleus-encoded RNA polymerase (NEP) and plastid-encoded RNA polymerase (PEP) [32,33]. In mature chloroplasts, PEP represents the major transcriptional machinery, which transcribes >80% of all primary chloroplast transcripts, while NEP transcribes chloroplast housekeeping genes [34]. NEP is a phage-type RNA polymerase with a single subunit. In Arabidopsis (*Arabidopsis thaliana*), NEP is encoded by two nuclear genes, *rpoTp* and *rpoTmp* [35]. PEP is a bacteria-type RNA polymerase composed of four core enzyme subunits ($\alpha$, $\beta$, $\beta'$, and $\beta''$) and a promoter-recognizing subunit ($\sigma$ factor). The core enzyme subunits of PEP are encoded by a set of genes located in the plastid genome: *rpoA*, *rpoB*, *rpoC1*, and *rpoC2* [33]. By contrast, during evolution, genes for $\sigma$ factors, which provide the necessary promoter specificity to PEP, were transferred to the nuclear genome, perhaps allowing the nucleus to regulate chloroplast gene transcription in response to environmental and developmental cues [36]. PEP and a set of polymerase-associated proteins (PAPs) form a huge protein complex required for transcription. All PAPs are encoded by genes in the nucleus, and most of them are the components of plastid transcriptionally active chromosome (pTAC) [37]. These PAPs are predicted to be involved in DNA and RNA metabolism (PAP1/pTAC3, PAP2/pTAC2, PAP3/pTAC10, PAP5/pTAC12, PAP7/pTAC14, and PAP12/pTAC7), redox regulation from photosynthesis (PAP6/FLN1, PAP10/TrxZ, and PAP12/pTAC7), and protecting the PEP complex from reactive oxygen species (PAP4/FSD3 and PAP9/FSD2) [38]. The transcriptional regulation of chloroplast genes is essential for the proper functioning of chloroplasts and for overall plant growth under both normal and adverse conditions.

### 2.2. RNA Metabolism

Most chloroplast genes in plants are organized as operons. These polycistronic primary RNAs require extensive processing, including 5′ and 3′ trimming, intercistronic cleavage, RNA splicing, and RNA editing [39]. Evidence suggests that 5′ and 3′ trimming and intercistronic cleavage are important for moderating RNA stability and translation within chloroplasts [40–44]. In plants, approximately 20 chloroplast genes (encoding proteins or structural RNAs) are interrupted by introns. RNA splicing removes the intron sequences of genes from primary transcripts to enable the production of mature mRNA with the correct genetic information [45,46]. RNA analyses have shown that RNA editing (mainly in the form of C-to-U base conversions) is highly widespread within the chloroplasts of land plants. During this process, numerous C-to-U conversions alter the coding sequences of chloroplast mRNAs, regulate RNA secondary structures that influences the splicing and/or stability of RNAs, or generate translational start sites (AUG) [47,48]. All of these RNA metabolic events depend on many nucleus-encoded proteins, most of which likely arose during coevolution between the host and endosymbiont. For detailed information on chloroplast RNA metabolism, we direct the reader to recent reviews in this area [39,43,45,48,49].

### 2.3. Translation

Chloroplasts possess a bacterial-type 70S ribosome as well as a full set of transfer RNAs (tRNAs) and ribosomal RNAs (rRNAs), which conduct protein translation [50]. The 70S ribosome comprises two multi-component subunits: The large (50S) and small (30S) subunits. Both subunits contain rRNAs and various plastid- and nucleus-encoded proteins [51–55].

In general, the chloroplast ribosome has a bacterial-type structure, but with some distinctive features. Chloroplast ribosomes contain the complete set of bacterial-type rRNAs (23S, 16S, and 5S rRNA) with functions analogous to those in bacteria. For example, 23S rRNA exhibits peptidyl transferase activity, whereas 16S rRNA functions as the decoding center and serves as a scaffold for other proteins during ribosome assembly [56,57]. However, the chloroplast contains an additional 4.5S rRNA not found in bacteria that is homologous to the 3′ end of prokaryotic 23S rRNA, suggesting that it was derived from fragmentation of this prokaryotic rRNA [58]. Additionally, two post-transcriptional cleavage sites within the 23S rRNA precursor generate mature 23S rRNA fragments. All of these fragments are assembled into the mature 70S ribosome and combined via intermolecular base pairing [52]. During evolution, obvious changes also occurred in the protein composition of the chloroplast ribosome. The homologs of bacterial proteins Rpl25 and Rpl30 were completely lost in chloroplasts [59]. Several new components of the plastid (chloroplast) ribosome, known as plastid-specific ribosomal proteins (PSRPs), have also been identified [59,60]. PSRP5 and PSRP6 in the 50S subunit and PSRP2 and PSRP3 in the 30S subunit are believed to be intrinsic components of the chloroplast ribosome [52].

## 3. Chloroplast Gene Expression and Environmental Stress

To date, genetic analyses have revealed many nucleus-encoded proteins that regulate not only chloroplast gene expression but also responses to environmental stresses. Functional analyses of these nucleus-encoded proteins have indicated that chloroplast gene expression is involved in plant responses to environmental stresses (Table 1).

### *3.1. Transcription and Environmental Stress Responses*

The transcriptional regulation of chloroplast gene expression is crucial not only for photosynthesis but also for plant development. Recent studies have revealed that the transcriptional control of chloroplast gene expression also plays important roles in plant responses to environmental changes. The chloroplast gene *psbA* encodes the D1 reaction center protein of photosystem II (PSII) [61–63]. Due to the nature of PSII photochemistry, D1 protein is continuously subjected to photodamage, which decreases photosynthetic activity (an effect known as photoinhibition). These damaged D1 proteins are replaced by *de novo* synthesized D1 proteins following the partial disassembly of the PSII complex [64,65]. Hence, the capacity to repair photodamaged PSII strongly depends on the ability to generate new D1 protein. Chloroplasts can adjust the transcriptional efficiency of *psbA* during photoinhibition under adverse environmental conditions such as high light and temperature [66–68]. During chloroplast evolution, several nucleus-encoded proteins have developed the ability to regulate *psbA* transcription in order to repair photodamaged PSII under adverse environmental conditions. Tomato (*Solanum lycopersicum*) WHIRLY1 (SlWHY1) was recently found to upregulate *psbA* transcription under chilling conditions. Under these conditions, the chloroplast-localized SlWHY1 promotes the transcription of *psbA* by directly binding to the upstream region of its promoter (the sequence "GTTACCCT"), resulting in increased D1 abundance to relieve photoinhibition [69,70]. Overexpression of *SlWHY1* leads to increased *de novo* synthesis of D1 protein and increased resistance to photoinhibition under chilling conditions [69]. These findings suggest that *psbA* transcription is an important target for regulating PSII activity to adjust plant resistance to environmental stresses.

**Table 1.** Chloroplast gene expression and stress response mutants.

| Gene Symbol Accession No. | Species | Mutant | Mutant Stress Phenotype | Molecular Function | Reference(s) |
|---|---|---|---|---|---|
| | | | **Transcription:** | | |
| *SlWHY1* (*Solyc05g007100*) | *Solanum lycopersicu* | *slwhy1* | Hypersensitivity to chilling | Promotes transcription of tomato *psbA* under chilling conditions | [69] |
| *SIG5* (*AT5G24120*) | *Arabidopsis thaliana* | *sig5* | Hypersensitivity to salt stress and high light | Specifically controls *psbD* transcription in response to circadian rhythms, environmental stresses, and light signals | [71] |
| *mTERF5* (*AT4G14650*) | *Arabidopsis thaliana* | *mterf5* | Decreased sensitivity to salt, ABA, and osmotic stress; altered sugar responses | Serves as a transcriptional pausing factor; specifically regulates the transcription of chloroplast *psbEFLJ* | [72,73] |

**Table 1.** *Cont.*

| Gene Symbol Accession No. | Species | Mutant | Mutant Stress Phenotype | Molecular Function | Reference(s) |
|---|---|---|---|---|---|
| | | | **RNA metabolism:** | | |
| *ORRM1* (*AT3G20930*) | *Arabidopsis thaliana* | *orrm1* | Hypersensitivity to low temperature | Controls RNA editing of 62% (21 of 34) of chloroplast transcripts | [74,75] |
| *DUA1* (*J043016D20*) | *Oryza sativa* | *dua1* | Hypersensitivity to low temperature | Required for RNA editing of the *rps8*-182 site | [76] |
| *WSL* (*Os01g0559500*) | *Oryza sativa* | *wsl* | Hypersensitivity to ABA, salinity, and sugar with increased $H_2O_2$ levels | Required for splicing of chloroplast *rpl2* | [77] |
| *RH3* (*AT5G26742*) | *Arabidopsis thaliana* | *rh3-4* | Hypersensitivity to salt stress and low temperature; reduced ABA content | Involved in splicing of chloroplast *trnA*, *trnI*, *rpl2*, *rps12* intron 1, and *rps12* intron 2 | [78–80] |
| *CP29A* (*AT3G53460*) | *Arabidopsis thaliana* | *cp29a* | Hypersensitivity to low temperature | Required for maintaining the stability of various chloroplast transcripts | [81] |
| *CP31A* (*AT4G24770*) | *Arabidopsis thaliana* | *cp31a* | Hypersensitivity to low temperature | Required for maintaining the stability of various chloroplast transcripts | [81] |
| | | | **Translation:** | | |
| *RPS1* (*AT5G30510*) | *Arabidopsis thaliana* | *rps1* | Heat-sensitive phenotype; perturbed HSF-mediated heat stress response | Component of chloroplast ribosome small subunit; involved in activating cellular heat stress responses | [82] |
| *RPS17* (*GRMZM2G038013*) | *Zea mays* | *hcf60-m1* | Cold-induced bleaching and seedling-lethal phenotype | Component of chloroplast ribosome small subunit | [83] |
| *Rps15* (*GeneID:800489*) | *Nicotiana tabacum* | *Δrps15* | Hypersensitivity to low temperature | Component of chloroplast ribosome large subunit | [84] |
| *Rpl33* (*GeneID:800444*) | *Nicotiana tabacum* | *Δrpl33* | Hypersensitivity to low temperature | Component of chloroplast ribosome large subunit | [85] |

The chloroplast gene *PsbD* encodes the reaction center protein D2 of PSII [86]. The expression of *psbD* is controlled by four PEP promoters. One of these is the blue-light-responsive promoter *psbD* BLRP [87–89]. The structure of *psbD* BLRP is distinct from that of common PEP promoters, which are characterized by conserved −35 and −10 elements. The *psbD* BLRP contains three *cis*-elements, including the AAG box, PGT box, and −10 element, but lacks the conserved −35 element [32,89,90]. This promoter has been well characterized. *psbD* BLRP transcription is specifically regulated by chloroplast-localized sigma factor 5 (SIG5) [71,91]. *psbD* BLRP transcription is also induced by environmental stresses, such as high salinity, low temperature, and osmotic stress [71]. In addition, *psbD* BLRP transcription is modulated in response to the relative proportions of red and far red light in a process mediated by signals from phytochromes [92]. Thus, *psbD* BLRP transcription is modulated during plant responses to environmental stress and sensing of light signals. Indeed, high *psbD* BLRP activity favors the synthesis of D2, thus relieving high-light-induced damage to PSII [93]. On the other hand, *psbD* BLRP transcription mediated by SIG5 shows obvious circadian oscillation, revealing how chloroplast gene expression is involved in the circadian oscillator [94].

*psbD* BLRP transcription may be also involved in biotic stress responses. Pathogens deliver various effectors into plant host cells when pathogens attack plants. These effectors assist pathogen proliferation and suppress plant defense responses [95–101]. Two *Pseudomonas* effectors, HopR1 and HopBB1, has been suggested to be involved in *psbD* transcription by targeting PTF1 (PLASTID TRANSCRIPTION FACTOR 1), a transcription factor for *psbD* BLRP transcription [102,103]. Moreover, the loss of PTF1 leads to more resistant to *Pseudomonas syringae* pv. *tomato* strain DC3000 in Arabidopsis [104]. Thus, *psbD* BLRP transcription may play a role in biotic stress responses.

The *psbEFLJ* operon contains four chloroplast genes: *psbE*, *psbF*, *psbL*, and *psbJ*. These genes encode the α and β subunits of cytochrome b559, PsbL, and PsbJ, respectively, which are crucial for the proper functioning of PSII [105,106]. The transcriptional regulation of *psbEFLJ* was recently investigated. *psbEFLJ* transcription is positively regulated by the nucleus-encoded protein mTERF5 (mitochondrial Transcription Termination Factor 5), which acts as a pausing factor [72,107]. mTERF5 causes transcriptional pausing on *psbEFLJ* by binding to the nucleotides +30 to +51 from the transcription start site and recruits additional pTAC6 into the PEP complex at the pausing region to form an enhanced PEP complex, thus positively regulating *psbEFLJ* transcription. In addition, *mterf5* mutants are less sensitive to NaCl and abscisic acid (ABA) than wild-type plants, indicating that mTERF5 functions as a negative regulator of salt tolerance, perhaps via ABA signaling [73]. These findings point to functional links between *psbEFLJ* transcription and salt tolerance as well as ABA signaling.

### *3.2. RNA Metabolism and Environmental Stress Responses*

RNA metabolism in chloroplasts is remarkably complex, involving a series of steps such as 5′ and 3′ trimming, RNA editing, splicing, and intergenic cleavage [31]. Analyses of mutants with defective RNA editing suggested that RNA editing, splicing, and stability help regulate environmental stress responses in plants [28,29].

An overall deficiency in chloroplast RNA editing (C-to-U base conversion) in Arabidopsis could be caused by the mutation of *ORRM1* (*Organelle RRM Protein 1*), encoding an essential plastid RNA editing factor. *orrm1* mutants exhibited greatly reduced RNA editing efficiency compared to wild-type Arabidopsis at 62% (21 of 34) of the chloroplast editing sites. Among these, the editing efficiency at 12 sites decreased by at least 90%, whereas that of the nine other sites decreased by 10% to 90% in *orrm1* vs. wild-type plants [74]. The reduced RNA editing deficiency at multiple sites in *orrm1* plants did not result in distinctive phenotypes at normal temperatures (22 °C), but the mutants were sensitive to chilling, displaying yellow emerging leaves under chilling conditions (4 °C) [75]. These findings suggest that chloroplast RNA editing confers low-temperature tolerance in Arabidopsis. However, the RNA editing site that confers this improved low-temperature tolerance is unknown.

The *indica* (*Oryza sativa* ssp. *indica*) rice cultivar *Dular*, referred to as *dua1*, is planted in tropical regions of Southeast Asia, including India and the Philippines. *dua1* plants are less tolerant of low temperatures than Nipponbare (*O. sativa* ssp. *japonica*) plants, which are grown in northern areas of Asia, as *dua1* plants display pale leaves under low-temperature conditions (19 °C). A recent study revealed that that the low-temperature sensitivity of *dua1* is caused by defective RNA editing of the plastid ribosome gene *rps8*, which is located 182 nt downstream of the translational start site (*rps8*-182). The edited *rps8* transcripts generate RPS8 protein with altered amino acid hydrophobicity, suggesting that RNA editing at *rps8*-182 improves low-temperature tolerance in rice by moderating the stability of RPS8 protein under low-temperature conditions [76]. Chloroplast genomes have very slow rates of sequence evolution, averaging ~5-fold slower than nuclear genomes [108,109], suggesting that chloroplast RNA editing evolved to improve low-temperature tolerance by increasing protein stability.

*ndhB* encodes the B subunit of the chloroplast NADH dehydrogenase-like complex that is required for cyclic electron flow around photosystem I [110,111]. The defective RNA editing of *ndhB*-2, *ndhB*-3, *ndhB*-4, and *ndhB*-6 sites enhances the disease resistance against fungal pathogens in Arabidopsis [112]. This finding suggests that chloroplast RNA editing is interlinked with plant immunity.

*rpl2* encodes a component of the 50S subunit in the chloroplast ribosome. This gene contains only a group II intron. In rice, the splicing of this intron is specifically regulated by WHITE STRIPE LEAF (WSL), a pentatricopeptide repeat (PPR) protein. Compared to the wild type, *wsl* mutants exhibit a decreased germination rate and reduced shoot and root growth upon treatment with ABA but not with α-naphthaleneacetic acid (NAA, an auxin), gibberellic acid (GA), epi-brassinosteroid (BL), or 6-benzylaminopurine (6-BA, a cytokinin). This finding suggests that the ABA signaling process is specifically affected in *wsl*. These mutants also display decreased germination rates when grown on medium supplemented with sugar and NaCl [77]. Sugar and salinity responses are closely connected with ABA signaling, and several ABA-related genes (e.g., *ABI3* and *WRKY24*) are induced by ABA treatment in *wsl* mutants, suggesting that *rpl2* splicing plays an important role in plant responses to ABA.

In Arabidopsis, the splicing of chloroplast *trnA*, *trnI*, *rpl2*, *rps12* intron 1, and *rps12* intron 2 is regulated by DEAD-BOX RNA HELICASE 3 (RH3) [78]. Null mutants of *RH3* are embryo lethal, whereas the weak allele *rh3-4* displays retarded plant growth and pale-green leaves, along with considerable decreases in the splicing efficiency of *trnA*, *trnI*, *rpl2*, *rps12* intron 1, and *rps12* intron 2. Moreover, the endogenous ABA contents of 1-week-old *rh3-4* seedlings are ~50% lower than those of wild-type plants, suggesting that RH3 plays a role in ABA biosynthesis. The mutation of *RH3* results in the reduced expression of nucleus-encoded gene *ABA1* and *NCDE4*, encoding two crucial enzymes of the ABA biosynthetic pathway, perhaps explaining the decreased ABA contents of *rh3-4* seedlings. Consistent with their decreased ABA contents, *rh3-4* mutants exhibit more severely inhibited plant growth and greening than the wild type under abiotic stress conditions including salinity, cold, and dehydration stress [78–80]. These findings suggest that chloroplast RNA splicing of these genes is required for environmental stress responses in plants, especially responses related to ABA signaling. Yet how chloroplast RNA splicing regulates environmental stress responses is currently unknown. A defect in chloroplast RNA splicing would be likely to result in defective photosynthetic performance, thus leading to enhanced sensitivity to environmental stresses. Alternatively, chloroplast RNA splicing might trigger plastid-to-nucleus retrograde signaling to regulate plant stress responses.

Chloroplast RNA stability is also crucial for the proper expression of chloroplast genes. Increasing evidence indicates that chloroplast RNA stability is involved in plant responses to environmental stresses. Chloroplast ribonucleoproteins CP31A and CP29A are RNA chaperone proteins that associate with large sets of chloroplast transcripts [81,113]. Arabidopsis mutants with deletions of *CP31A* and *CP29A* do not have unusual phenotypes under normal conditions but show bleaching of newly emerging leaves at the bases of the youngest leaves under cold stress (8 °C). Kupsch et al. demonstrated that CP31A and CP29A are required for the accumulation of transcripts of many chloroplast genes under cold stress (8 °C), such as *psaA*, *psbD*, *psbF*, *psbB*, *petB*, *ndhF*, and *rbcL*. This cold-sensitive phenotype could be explained by a decreased stability of chloroplast transcripts in the *cp31a* and *cp29a* mutants [81,113]. DEAD-box RNA helicase 22 (RH22) is another chloroplast RNA chaperone. In cabbage (*Brassica rapa*), *RH22* expression was significantly upregulated by drought, heat, salt, and cold stress but markedly downregulated by UV stress. The overexpression of cabbage *RH22* enhanced the stability of chloroplast transcripts and improved growth and survival in Arabidopsis under drought and salt stress [114]. Moreover, Arabidopsis plants overexpressing cabbage *RH22* displayed better growth and more green leaves upon ABA treatment than the wild type, along with decreased expression of *ABI3*, *ABI4*, and *ABI5*, suggesting that chloroplast RNA stability plays a part in ABA signaling pathways [114]. Chloroplast RNA stability might have a positive role in plant responses to environmental stress by enhancing the translation of chloroplast genes.

### 3.3. Translation and Environmental Stress Responses

Translation is the final step in chloroplast gene expression. Chloroplast gene translation regulates protein accumulation to optimize photosynthetic performance and to attenuate photooxidative damage. Thus, the regulation of chloroplast gene translation represents a unique component of plant responses to internal and external stimuli.

Most plants growing in direct sunlight routinely encounter high-light stress; the resulting high photon flux exceeds the photosynthetic capacity, thereby damaging the chloroplast. To explore the regulation of chloroplast gene translation during the rapid adaptation of plants to high light, a systematic ribosome profiling study was performed to detect changes in chloroplast gene translation efficiency in tobacco seedlings following transfer from moderate light to high light. The ribosome occupancy on *psbA* transcripts (encoding PSII reaction center protein D1) increased in response to high-light treatment [115]. Given that D1 protein is the main site prone to photodamage by high light, the upregulated *psbA* translation should substantially facilitate the repair of PSII under high-light stress. However, the molecular mechanisms underlying the translational activation of *psbA* under these conditions remain to be further explored.

Studies on the functions of chloroplast ribosome proteins have revealed that maintaining sufficiently high chloroplast gene translation efficiency is important for proper chloroplast development at low temperature. Maize (*Zea mays*) mutants with a loss of ribosomal protein RPS17 were pale green when grown at moderate temperature (27 °C) but appeared albino under cool conditions (17 °C) [83]. Tobacco mutants with a loss of the ribosomal protein Rpl33 showed no visible phenotypes at any stage of development under standard conditions, with similar development, growth rates, and onset of flowering to wild-type plants. However, the *Rpl33* knockout mutants were sensitive to cold stress, although not to heat or to low or high light levels. When *Rpl33* knockout mutants were transferred to cold-stress conditions (4 °C), they exhibited strong photooxidative damage symptoms and recovered much more slowly from low-temperature stress than wild-type plants [85]. As with Rpl33, the loss of the ribosomal protein Rps15 in tobacco resulted in a growth phenotype almost identical growth to that of wild-type plants, although young plants grew slightly more slowly and the onset of flowering was slightly delayed. However, the *Rps15* knockout mutants were cold sensitive, with more severe pigment loss and worse photosynthetic performance than wild-type plants [84]. Together, these findings suggest that the maintenance of plastid translational capacity is important in enabling plant tolerance to chilling stress.

In Arabidopsis, the expression of the chloroplast ribosome protein gene *RPS1* was considerably induced by heat stress (2 h at 38 °C). *RPS1* knockdown mutants (*rps1*) displayed retarded growth and slightly pale-green leaves. When *rps1* seedlings were exposed to transient high-temperature conditions (3 h at 45 °C), they were much more heat sensitive than wild-type seedlings, as almost no mutants survived after a 7-d recovery, whereas more than 90% of wild-type seedlings did. However, there were no significant differences between *rps1* and wild-type plants under osmotic and salinity stress. These results suggest that decreased *RPS1* expression alters cellular heat stress responses by disrupting chloroplast gene translation rather than through general physiological defects. RPS1 is required to activate the expression of *HsfA2* (*HEAT STRESS TRANSCRIPTION FACTOR A-2*), a highly heat-shock-inducible gene encoding a transcription factor that is crucial for triggering cellular responses to heat stress. The constitutive expression of *HsfA2* was sufficient to rescue the heat-sensitive phenotype of *rps1* mutants, suggesting that the defective expression of *HsfA2* is responsible for the heat-sensitive phenotype of *rps1* mutants. Like the *rps1* mutant phenotype, treatment with lincomycin, an inhibitor of chloroplast gene translation, also led to an obvious reduction in the expression of *HsfA2* in response to heat stress [82]. These findings reveal a plastid-to-nucleus retrograde signaling pathway that regulates chloroplast gene translational capacity to transcriptionally activate cellular heat stress responses, especially the HsfA2-dependent heat tolerance pathway.

## 4. Conclusions and Future Perspectives

Photosynthesis, one of the most important physiological processes in plants, is highly sensitive to environmental stresses. These stresses often inhibit photosynthesis considerably [116–124]. Many studies have reported that chloroplasts can act as sensors of the external environment [7,19]. Thus, in addition to hosting photosynthesis, chloroplasts play important roles in plant responses to various environmental stresses. Likewise, the proper expression of chloroplast genes is crucial for chloroplast development, photosynthesis, and plant development. However, as summarized in this review, many studies indicate that chloroplast gene expression is also important for plant stress responses.

Further elucidating the roles of chloroplast gene expression in plant responses to various environmental stresses would lay the foundation for genetically improving plant tolerance to the environment. However, the underlying molecular mechanisms remains largely unknown. Most studies have focused on the roles of nucleus-encoded proteins in regulating chloroplast gene expression and plant responses to environmental stress, while the direct connection between chloroplast gene expression and environmental stress responses has been largely ignored. Future studies should therefore address two major issues: How environmental stress triggers chloroplast gene expression, and what roles chloroplast gene expression plays in plant responses to environmental stress. Several specific issues need to be investigated. For example, D1 protein synthesis is important for PSII repair under high-light conditions [71,115,125–128]. Thus, enhanced *psbA* transcription and translation would help increase the tolerance of PSII to high light; however, it is still unclear to what extent high light triggers the transcription and translation of *psbA*. In addition, RNA editing, particularly C-to-U base conversion, is a widespread phenomenon in chloroplasts across nearly all plant species [48]. As discussed above, it is clear that RNA editing is important for plant tolerance of low temperature [74–76], yet the exact editing sites, and how RNA editing at these sites improves plant tolerance to low temperatures, remain to be explored. Finally, translation and splicing of several chloroplast genes trigger plastid-to-nucleus retrograde signaling and ABA signaling [73,77,79,80,114], but it remains unclear how the signals generated from these processes are transferred out of chloroplasts and integrated into these signaling pathways.

New technology is needed to better investigate the molecular mechanism of chloroplast gene expression in response to the environment. Chloroplast transformation is an extremely time-consuming and difficult process that has only been achieved in a few plant species, such as lettuce, poplar, and Arabidopsis. This make it extremely difficult to control chloroplast gene transcription using traditional genetic engineering approaches. To further explore the connection between chloroplast gene transcription and environmental stress responses, novel tools must be designed that are similar to CRISPR/Cas9 and RNA interference to knock out and knock down chloroplast genes. It is also critical to design an artificial RNA editing system to carry out RNA editing of specific sites in chloroplasts. This system would be useful for investigating the specific roles of different RNA editing sites in plant responses to environmental stresses and even creating new RNA editing sites that could improve stress tolerance in plants.

**Author Contributions:** Y.Z. and C.L. conceived and planned this review paper designed the project. Y.Z., C.L., A.Z., and X.L. prepared and drafted the manuscript. Y.Z. and C.L. revised the manuscript. All the authors read and approved the final manuscript. All authors have read and agreed to the published version of the manuscript.

**Acknowledgments:** This work was supported by the National Natural Science Foundation of China (reference numbers 31730102 and 31670238). The authors are grateful to the authors of the excellent papers discussed. The authors also apologize to the authors of the other excellent papers in this area that could not be discussed because of lack of space.

## Abbreviations

| | |
|---|---|
| ABA | Abscisic acid |
| BL | Epi-brassinosteroid |
| CP29A | 29 kD chloroplast protein |
| CP31A | 31 kD chloroplast protein |
| HsfA2 | HEAT STRESS TRANSCRIPTION FACTOR A-2 |
| mTERF5 | Mitochondrial transcription termination factor 5 |
| NAA | α-Naphthaleneacetic acid |
| NEP | Nucleus-encoded RNA polymerase |
| ORRM1 | Organelle RRM Protein 1 |
| PAP | Polymerase-associated protein |
| PEP | Plastid-encoded RNA polymerase |

| | |
|---|---|
| PPR | Pentatricopeptide repeat |
| PSII | Photosystem II |
| PSRP | Plastid-specific ribosomal protein |
| PTF1 | PLASTID TRANSCRIPTION FACTOR 1 |
| RH | DEAD-BOX RNA HELICASE |
| Rpl33 | Ribosomal protein L33 |
| Rpl2 | Ribosomal protein L2 |
| RPS1 | Ribosomal protein S1 |
| Rps8 | Ribosomal protein S8 |
| Rps15 | Ribosomal protein S15 |
| RPS17 | Ribosomal protein S17 |
| rRNA | Ribosome RNA |
| SIG5 | Sigma factor 5 |
| SlWHY1 | Tomato WHIRLY1 |
| tRNA | Transfer RNA |
| WSL | WHITE STRIPE LEAF |
| 6-BA | 6-Benzylaminopurine |

## References


1. Zhu, J.K. Abiotic stress signaling and responses in plants. *Cell* **2016**, *167*, 313–324. [CrossRef]
2. Wang, P.; Yang, C.; Chen, H.; Luo, L.; Leng, Q.; Li, S.; Han, Z.; Li, X.; Song, C.; Zhang, X.; et al. Exploring transcription factors reveals crucial members and regulatory networks involved in different abiotic stresses in *Brassica napus* L. *BMC Plant Biol.* **2018**, *18*, 202. [CrossRef]
3. Xu, Y.; Yu, Z.; Zhang, D.; Huang, J.; Wu, C.; Yang, G.; Yan, K.; Zhang, S.; Zheng, C. CYSTM, a novel non-secreted cysteine-rich peptide family, involved in environmental stresses in *Arabidopsis thaliana*. *Plant Cell Physiol.* **2018**, *59*, 423–438. [CrossRef]
4. Suzuki, N.; Rivero, R.M.; Shulaev, V.; Blumwald, E.; Mittler, R. Abiotic and biotic stress combinations. *New Phytol.* **2014**, *203*, 32–43. [CrossRef]
5. Wang, S.; Li, Q.P.; Wang, J.; Yan, Y.; Zhang, G.L.; Yan, Y.; Zhang, H.; Wu, J.; Chen, F.; Wang, X.; et al. YR36/WKS1-mediated phosphorylation of PsbO, an extrinsic member of photosystem II, inhibits photosynthesis and confers stripe rust resistance in wheat. *Mol. Plant* **2019**, *12*, 1639–1650. [CrossRef]
6. Li, N.; Lin, B.; Wang, H.; Li, X.; Yang, F.; Ding, X.; Yan, J.; Chu, Z. Natural variation in ZmFBL41 confers banded leaf and sheath blight resistance in maize. *Nat. Genet.* **2019**, *51*, 1540–1548. [CrossRef] [PubMed]
7. Chan, K.X.; Phua, S.Y.; Crisp, P.; McQuinn, R.; Pogson, B.J. Learning the languages of the chloroplast: Retrograde signaling and beyond. *Annu. Rev. Plant Biol.* **2016**, *67*, 25–53. [CrossRef] [PubMed]
8. Leister, D.; Wang, L.; Kleine, T. Organellar gene expression and acclimation of plants to environmental stress. *Front. Plant Sci.* **2017**, *8*, 387. [CrossRef] [PubMed]
9. Huang, Y.W.; Zhou, Z.Q.; Yang, H.X.; Wei, C.X.; Wan, Y.Y.; Wang, X.J.; Bai, J.G. Glucose application protects chloroplast ultrastructure in heat-stressed cucumber leaves through modifying antioxidant enzyme activity. *Biol. Plant.* **2015**, *59*, 131–138. [CrossRef]
10. Larkin, R.M. Tetrapyrrole signaling in plants. *Front. Plant Sci.* **2016**, *7*, 1586. [CrossRef]
11. Piippo, M.; Allahverdiyeva, Y.; Paakkarinen, V.; Suoranta, U.M.; Battchikova, N.; Aro, E.M. Chloroplast-mediated regulation of nuclear genes in *Arabidopsis thaliana* in the absence of light stress. *Physiol. Genom.* **2006**, *25*, 142–152. [CrossRef] [PubMed]
12. Pesaresi, P.; Schneider, A.; Kleine, T.; Leister, D. Interorganellar communication. *Curr. Opin. Plant Biol.* **2007**, *10*, 600–606. [CrossRef] [PubMed]
13. Pogson, B.J.; Woo, N.S.; Forster, B.; Small, I.D. Plastid signalling to the nucleus and beyond. *Trends Plant Sci.* **2008**, *13*, 602–609. [CrossRef] [PubMed]
14. Woodson, J.D.; Chory, J. Coordination of gene expression between organellar and nuclear genomes. *Nat. Rev. Genet.* **2008**, *9*, 383–395. [CrossRef]
15. Estavillo, G.M.; Crisp, P.A.; Pornsiriwong, W.; Wirtz, M.; Collinge, D.; Carrie, C.; Giraud, E.; Whelan, J.; David, P.; Javot, H.; et al. Evidence for a SAL1-PAP chloroplast retrograde pathway that functions in drought and high light signaling in *Arabidopsis*. *Plant Cell* **2011**, *23*, 3992–4012. [CrossRef]

16. Xiao, Y.; Savchenko, T.; Baidoo, E.E.; Chehab, W.E.; Hayden, D.M.; Tolstikov, V.; Corwin, J.A.; Kliebenstein, D.J.; Keasling, J.D.; Dehesh, K. Retrograde signaling by the plastidial metabolite MEcPP regulates expression of nuclear stress-response genes. *Cell* **2012**, *149*, 1525–1535. [CrossRef]
17. Jung, H.S.; Chory, J. Signaling between chloroplasts and the nucleus: Can a systems biology approach bring clarity to a complex and highly regulated pathway? *Plant Physiol.* **2010**, *152*, 453–459. [CrossRef]
18. Leister, D.; Wang, X.; Haberer, G.; Mayer, K.F.; Kleine, T. Intracompartmental and intercompartmental transcriptional networks coordinate the expression of genes for organellar functions. *Plant Physiol.* **2011**, *157*, 386–404. [CrossRef]
19. Woodson, J.D. Chloroplast quality control—Balancing energy production and stress. *New Phytol.* **2016**, *212*, 36–41. [CrossRef]
20. Daniell, H.; Lin, C.S.; Yu, M.; Chang, W.J. Chloroplast genomes: Diversity, evolution, and applications in genetic engineering. *Genome Biol.* **2016**, *17*, 134. [CrossRef]
21. Huo, Y.; Gao, L.; Liu, B.; Yang, Y.; Kong, S.; Sun, Y.; Yang, Y.; Wu, X. Complete chloroplast genome sequences of four *Allium* species: Comparative and phylogenetic analyses. *Sci. Rep.* **2019**, *9*, 12250. [CrossRef] [PubMed]
22. Yagi, Y.; Shiina, T. Recent advances in the study of chloroplast gene expression and its evolution. *Front. Plant Sci.* **2014**, *5*, 61. [CrossRef] [PubMed]
23. Leister, D.; Kleine, T. Towards a comprehensive catalog of chloroplast proteins and their interactions. *Cell Res.* **2008**, *18*, 1081–1083. [CrossRef] [PubMed]
24. Huang, M.S.; Friso, G.; Nishimura, K.; Qu, X.; Olinares, P.D.B.; Majeran, W.; Sun, Q.; van Wijk, K.J. Construction of plastid reference proteomes for maize and *Arabidopsis* and evaluation of their orthologous relationships; the concept of orthoproteomics. *J. Proteome Res.* **2013**, *12*, 491–504. [CrossRef]
25. Ren, R.C.; Wang, L.L.; Zhang, L.; Zhao, Y.J.; Wu, J.W.; Wei, Y.M.; Zhang, X.S.; Zhao, X.Y. DEK43 is a P-type pentatricopeptide repeat (PPR) protein responsible for the *Cis*-splicing of *nad4* in maize mitochondria. *J. Integr. Plant Biol.* **2020**, *62*, 299–313. [CrossRef]
26. Chi, W.; Sun, X.; Zhang, L. Intracellular signaling from plastid to nucleus. *Annu. Rev. Plant Biol.* **2013**, *64*, 559–582. [CrossRef]
27. Tadini, L.; Pesaresi, P.; Kleine, T.; Rossi, F.; Guljamow, A.; Sommer, F.; Muhlhaus, T.; Schroda, M.; Masiero, S.; Pribil, M.; et al. GUN1 controls accumulation of the plastid ribosomal protein S1 at the protein level and interacts with proteins involved in plastid protein homeostasis. *Plant Physiol.* **2016**, *170*, 1817–1830. [CrossRef]
28. Nawaz, G.; Kang, H. Chloroplast- or Mitochondria-targeted DEAD-Box RNA helicases play essential roles in organellar RNA metabolism and abiotic stress responses. *Front. Plant Sci.* **2017**, *8*, 871. [CrossRef]
29. Robles, P.; Quesada, V. Transcriptional and post-transcriptional regulation of organellar gene expression (OGE) and its roles in plant salt tolerance. *Int. J. Mol. Sci.* **2019**, *20*, 1056. [CrossRef]
30. Shih, P.M.; Matzke, N.J. Primary endosymbiosis events date to the later Proterozoic with cross-calibrated phylogenetic dating of duplicated ATPase proteins. *Proc. Natl. Acad. Sci. USA* **2013**, *110*, 12355–12360. [CrossRef]
31. Stern, D.B.; Goldschmidt-Clermont, M.; Hanson, M.R. Chloroplast RNA metabolism. *Annu. Rev. Plant Biol.* **2010**, *61*, 125–155. [CrossRef] [PubMed]
32. Liere, K.; Weihe, A.; Borner, T. The transcription machineries of plant mitochondria and chloroplasts: Composition, function, and regulation. *J. Plant Physiol.* **2011**, *168*, 1345–1360. [CrossRef] [PubMed]
33. Borner, T.; Aleynikova, A.Y.; Zubo, Y.O.; Kusnetsov, V.V. Chloroplast RNA polymerases: Role in chloroplast biogenesis. *Biochim. Biophys. Acta* **2015**, *1847*, 761–769. [CrossRef] [PubMed]
34. Zhelyazkova, P.; Sharma, C.M.; Forstner, K.U.; Liere, K.; Vogel, J.; Borner, T. The primary transcriptome of barley chloroplasts: Numerous noncoding RNAs and the dominating role of the plastid-encoded RNA polymerase. *Plant Cell* **2012**, *24*, 123–136. [CrossRef] [PubMed]
35. Hedtke, B.; Borner, T.; Weihe, A. One RNA polymerase serving two genomes. *EMBO Rep.* **2000**, *1*, 435–440. [CrossRef]
36. Chi, W.; He, B.; Mao, J.; Jiang, J.; Zhang, L. Plastid sigma factors: Their individual functions and regulation in transcription. *Biochim. Biophys. Acta* **2015**, *1847*, 770–778. [CrossRef]
37. Steiner, S.; Schroter, Y.; Pfalz, J.; Pfannschmidt, T. Identification of essential subunits in the plastid-encoded RNA polymerase complex reveals building blocks for proper plastid development. *Plant Physiol.* **2011**, *157*, 1043–1055. [CrossRef]

38. Pfannschmidt, T.; Blanvillain, R.; Merendino, L.; Courtois, F.; Chevalier, F.; Liebers, M.; Grubler, B.; Hommel, E.; Lerbs-Mache, S. Plastid RNA polymerases: Orchestration of enzymes with different evolutionary origins controls chloroplast biogenesis during the plant life cycle. *J. Exp. Bot.* **2015**, *66*, 6957–6973. [CrossRef]
39. Del Campo, E.M. Post-transcriptional control of chloroplast gene expression. *Gene Regul. Syst. Biol.* **2009**, *3*, 31–47. [CrossRef]
40. Pfalz, J.; Bayraktar, O.A.; Prikryl, J.; Barkan, A. Site-specific binding of a PPR protein defines and stabilizes 5′ and 3′ mRNA termini in chloroplasts. *EMBO J.* **2009**, *28*, 2042–2052. [CrossRef]
41. Prikryl, J.; Rojas, M.; Schuster, G.; Barkan, A. Mechanism of RNA stabilization and translational activation by a pentatricopeptide repeat protein. *Proc. Natl. Acad. Sci. USA* **2011**, *108*, 415–420. [CrossRef] [PubMed]
42. Zhou, W.; Lu, Q.; Li, Q.; Wang, L.; Ding, S.; Zhang, A.; Wen, X.; Zhang, L.; Lu, C. PPR-SMR protein SOT1 has RNA endonuclease activity. *Proc. Natl. Acad. Sci. USA* **2017**, *114*, E1554–E1563. [CrossRef] [PubMed]
43. Zhang, Y.; Lu, C. The enigmatic roles of PPR-SMR proteins in plants. *Adv. Sci. (Weinh)* **2019**, *6*, 1900361. [CrossRef] [PubMed]
44. Zhang, L.; Zhou, W.; Che, L.; Rochaix, J.D.; Lu, C.; Li, W.; Peng, L. PPR protein BFA2 is essential for the accumulation of the *atpH/F* transcript in chloroplasts. *Front. Plant Sci.* **2019**, *10*, 446. [CrossRef]
45. De Longevialle, A.F.; Small, I.D.; Lurin, C. Nuclearly encoded splicing factors implicated in RNA splicing in higher plant organelles. *Mol. Plant* **2010**, *3*, 691–705. [CrossRef]
46. Wang, X.; Yang, Z.; Zhang, Y.; Zhou, W.; Zhang, A.; Lu, C. Pentatricopeptide repeat protein PHOTOSYSTEM I BIOGENESIS FACTOR2 is required for splicing of *ycf3*. *J. Integr. Plant Biol.* **2020**. [CrossRef]
47. Yan, J.; Zhang, Q.; Yin, P. RNA editing machinery in plant organelles. *Sci. China Life Sci.* **2018**, *61*, 162–169. [CrossRef]
48. Small, I.D.; Schallenberg-Rudinger, M.; Takenaka, M.; Mireau, H.; Ostersetzer-Biran, O. Plant organellar RNA editing: What 30 years of research has revealed. *Plant J.* **2020**, *101*, 1040–1056. [CrossRef]
49. Barkan, A.; Small, I. Pentatricopeptide repeat proteins in plants. *Annu. Rev. Plant Biol.* **2014**, *65*, 415–442. [CrossRef]
50. Tiller, N.; Bock, R. The translational apparatus of plastids and its role in plant development. *Mol. Plant* **2014**, *7*, 1105–1120. [CrossRef]
51. Sharma, M.R.; Wilson, D.N.; Datta, P.P.; Barat, C.; Schluenzen, F.; Fucini, P.; Agrawal, R.K. Cryo-EM study of the spinach chloroplast ribosome reveals the structural and functional roles of plastid-specific ribosomal proteins. *Proc. Natl. Acad. Sci. USA* **2007**, *104*, 19315–19320. [CrossRef] [PubMed]
52. Bieri, P.; Leibundgut, M.; Saurer, M.; Boehringer, D.; Ban, N. The complete structure of the chloroplast 70S ribosome in complex with translation factor pY. *EMBO J.* **2017**, *36*, 475–486. [CrossRef] [PubMed]
53. Graf, M.; Arenz, S.; Huter, P.; Donhofer, A.; Novacek, J.; Wilson, D.N. Cryo-EM structure of the spinach chloroplast ribosome reveals the location of plastid-specific ribosomal proteins and extensions. *Nucleic Acids Res.* **2017**, *45*, 2887–2896. [CrossRef] [PubMed]
54. Perez Boerema, A.; Aibara, S.; Paul, B.; Tobiasson, V.; Kimanius, D.; Forsberg, B.O.; Wallden, K.; Lindahl, E.; Amunts, A. Structure of the chloroplast ribosome with chl-RRF and hibernation-promoting factor. *Nat. Plants* **2018**, *4*, 212–217. [CrossRef]
55. Zoschke, R.; Bock, R. Chloroplast translation: Structural and functional organization, operational control, and regulation. *Plant Cell* **2018**, *30*, 745–770. [CrossRef]
56. Shajani, Z.; Sykes, M.T.; Williamson, J.R. Assembly of bacterial ribosomes. *Annu. Rev. Biochem.* **2011**, *80*, 501–526. [CrossRef]
57. Maier, U.G.; Zauner, S.; Woehle, C.; Bolte, K.; Hempel, F.; Allen, J.F.; Martin, W.F. Massively convergent evolution for ribosomal protein gene content in plastid and mitochondrial genomes. *Genome. Biol. Evol.* **2013**, *5*, 2318–2329. [CrossRef]
58. Edwards, K.; Kossel, H. The rRNA operon from *Zea mays* chloroplasts: Nucleotide sequence of 23S rDNA and its homology with *E. coli* 23S rDNA. *Nucleic Acids Res.* **1981**, *9*, 2853–2869. [CrossRef]
59. Yamaguchi, K.; Subramanian, A.R. The plastid ribosomal proteins. Identification of all the proteins in the 50 S subunit of an organelle ribosome (chloroplast). *J. Biol. Chem.* **2000**, *275*, 28466–28482. [CrossRef]
60. Yamaguchi, K.; von Knoblauch, K.; Subramanian, A.R. The plastid ribosomal proteins. Identification of all the proteins in the 30 S subunit of an organelle ribosome (chloroplast). *J. Biol. Chem.* **2000**, *275*, 28455–28465. [CrossRef]

61. Mulo, P.; Sicora, C.; Aro, E.M. Cyanobacterial *psbA* gene family: Optimization of oxygenic photosynthesis. *Cell. Mol. Life Sci.* **2009**, *66*, 3697–3710. [CrossRef] [PubMed]
62. Mulo, P.; Sakurai, I.; Aro, E.M. Strategies for *psbA* gene expression in cyanobacteria, green algae and higher plants: From transcription to PSII repair. *Biochim. Biophys. Acta.* **2012**, *1817*, 247–257. [CrossRef] [PubMed]
63. Wegener, K.M.; Nagarajan, A.; Pakrasi, H.B. An atypical *psbA* gene encodes a sentinel D1 protein to form a physiologically relevant inactive photosystem II complex in cyanobacteria. *J. Biol. Chem.* **2015**, *290*, 3764–3774. [CrossRef] [PubMed]
64. Takahashi, S.; Badger, M.R. Photoprotection in plants: A new light on photosystem II damage. *Trends Plant Sci.* **2011**, *16*, 53–60. [CrossRef]
65. Lu, Y.; Wang, H.R.; Li, H.; Cui, H.R.; Feng, Y.G.; Wang, X.Y. A chloroplast membrane protein LTO1/AtVKOR involving in redox regulation and ROS homeostasis. *Plant Cell Rep.* **2013**, *32*, 1427–1440. [CrossRef]
66. Pfannschmidt, T.; Nilsson, A.; Tullberg, A.; Link, G.; Allen, J.F. Direct transcriptional control of the chloroplast genes *psbA* and *psaAB* adjusts photosynthesis to light energy distribution in plants. *IUBMB Life* **1999**, *48*, 271–276. [CrossRef]
67. Pfannschmidt, T. Chloroplast redox signals: How photosynthesis controls its own genes. *Trends Plant Sci.* **2003**, *8*, 33–41. [CrossRef]
68. Danilova, M.N.; Kudryakova, N.V.; Andreeva, A.A.; Doroshenko, A.S.; Pojidaeva, E.S.; Kusnetsov, V.V. Differential impact of heat stress on the expression of chloroplast-encoded genes. *Plant Physiol. Biochem.* **2018**, *129*, 90–100. [CrossRef]
69. Zhuang, K.; Kong, F.; Zhang, S.; Meng, C.; Yang, M.; Liu, Z.; Wang, Y.; Ma, N.; Meng, Q. Whirly1 enhances tolerance to chilling stress in tomato via protection of photosystem II and regulation of starch degradation. *New Phytol.* **2019**, *221*, 1998–2012. [CrossRef]
70. Zhuang, K.; Wang, J.; Jiao, B.; Chen, C.; Zhang, J.; Ma, N.; Meng, Q. SlWHIRLY1 maintains leaf photosynthetic capacity in tomato by regulating the expression of *SlRbcS1* under chilling stress. *J. Exp. Bot.* **2020**, *71*, 3653–3663. [CrossRef]
71. Nagashima, A.; Hanaoka, M.; Shikanai, T.; Fujiwara, M.; Kanamaru, K.; Takahashi, H.; Tanaka, K. The multiple-stress responsive plastid sigma factor, SIG5, directs activation of the *psbD* blue light-responsive promoter (BLRP) in *Arabidopsis thaliana*. *Plant Cell Physiol.* **2004**, *45*, 357–368. [CrossRef] [PubMed]
72. Ding, S.; Zhang, Y.; Hu, Z.; Huang, X.; Zhang, B.; Lu, Q.; Wen, X.; Wang, Y.; Lu, C. mTERF5 acts as a transcriptional pausing factor to positively regulate transcription of chloroplast *psbEFLJ*. *Mol. Plant* **2019**, *12*, 1259–1277. [CrossRef] [PubMed]
73. Robles, P.; Micol, J.L.; Quesada, V. Arabidopsis MDA1, a nuclear-encoded protein, functions in chloroplast development and abiotic stress responses. *PLoS ONE* **2012**, *7*, e42924. [CrossRef] [PubMed]
74. Sun, T.; Germain, A.; Giloteaux, L.; Hammani, K.; Barkan, A.; Hanson, M.R.; Bentolila, S. An RNA recognition motif-containing protein is required for plastid RNA editing in *Arabidopsis* and maize. *Proc. Natl. Acad. Sci. USA* **2013**, *110*, E1169–E1178. [CrossRef] [PubMed]
75. Wang, S.; Bai, G.; Wang, S.; Yang, L.; Yang, F.; Wang, Y.; Zhu, J.K.; Hua, J. Chloroplast RNA-binding protein RBD1 promotes chilling tolerance through 23S rRNA processing in *Arabidopsis*. *PLoS Genet.* **2016**, *12*, e1006027. [CrossRef]
76. Cui, X.A.; Wang, Y.W.; Wu, J.X.; Han, X.; Gu, X.F.; Lu, T.G.; Zhang, Z.G. The RNA editing factor DUA1 is crucial to chloroplast development at low temperature in rice. *New Phytol.* **2019**, *221*, 834–849. [CrossRef]
77. Tan, J.; Tan, Z.; Wu, F.; Sheng, P.; Heng, Y.; Wang, X.; Ren, Y.; Wang, J.; Guo, X.; Zhang, X.; et al. A novel chloroplast-localized pentatricopeptide repeat protein involved in splicing affects chloroplast development and abiotic stress response in rice. *Mol. Plant* **2014**, *7*, 1329–1349. [CrossRef]
78. Asakura, Y.; Galarneau, E.; Watkins, K.P.; Barkan, A.; van Wijk, K.J. Chloroplast RH3 DEAD box RNA helicases in maize and *Arabidopsis* function in splicing of specific group II introns and affect chloroplast ribosome biogenesis. *Plant Physiol.* **2012**, *159*, 961–974. [CrossRef]
79. Lee, K.H.; Park, J.; Williams, D.S.; Xiong, Y.; Hwang, I.; Kang, B.H. Defective chloroplast development inhibits maintenance of normal levels of abscisic acid in a mutant of the *Arabidopsis* RH3 DEAD-box protein during early post-germination growth. *Plant J.* **2013**, *73*, 720–732. [CrossRef]
80. Gu, L.; Xu, T.; Lee, K.; Lee, K.H.; Kang, H. A chloroplast-localized DEAD-box RNA helicase AtRH3 is essential for intron splicing and plays an important role in the growth and stress response in *Arabidopsis thaliana*. *Plant Physiol. Biochem.* **2014**, *82*, 309–318. [CrossRef]

81. Kupsch, C.; Ruwe, H.; Gusewski, S.; Tillich, M.; Small, I.; Schmitz-Linneweber, C. *Arabidopsis* chloroplast RNA binding proteins CP31A and CP29A associate with large transcript pools and confer cold stress tolerance by influencing multiple chloroplast RNA processing steps. *Plant Cell* **2012**, *24*, 4266–4280. [CrossRef] [PubMed]
82. Yu, H.D.; Yang, X.F.; Chen, S.T.; Wang, Y.T.; Li, J.K.; Shen, Q.; Liu, X.L.; Guo, F.Q. Downregulation of chloroplast RPS1 negatively modulates nuclear heat-responsive expression of *HsfA2* and its target genes in *Arabidopsis*. *PLoS Genet.* **2012**, *8*, e1002669. [CrossRef] [PubMed]
83. Schultes, N.P.; Sawers, R.J.; Brutnell, T.P.; Krueger, R.W. Maize *high chlorophyll fluorescent 60* mutation is caused by an *Ac* disruption of the gene encoding the chloroplast ribosomal small subunit protein 17. *Plant J.* **2000**, *21*, 317–327. [CrossRef] [PubMed]
84. Fleischmann, T.T.; Scharff, L.B.; Alkatib, S.; Hasdorf, S.; Schottler, M.A.; Bock, R. Nonessential plastid-encoded ribosomal proteins in tobacco: A developmental role for plastid translation and implications for reductive genome evolution. *Plant Cell* **2011**, *23*, 3137–3155. [CrossRef]
85. Rogalski, M.; Schottler, M.A.; Thiele, W.; Schulze, W.X.; Bock, R. Rpl33, a nonessential plastid-encoded ribosomal protein in tobacco, is required under cold stress conditions. *Plant Cell* **2008**, *20*, 2221–2237. [CrossRef]
86. Yao, W.B.; Meng, B.Y.; Tanaka, M.; Sugiura, M. An additional promoter within the protein-coding region of the *psbD-psbC* gene cluster in tobacco chloroplast DNA. *Nucleic Acids Res.* **1989**, *17*, 9583–9591. [CrossRef]
87. Christopher, D.A.; Mullet, J.E. A novel light-regulated promoter is conserved in cereal and dicot chloroplasts. *Plant Cell* **1992**, *4*, 785–798.
88. Hoffer, P.H.; Christopher, D.A. Structure and blue-light-responsive transcription of a chloroplast *psbD* promoter from *Arabidopsis thaliana*. *Plant Physiol.* **1997**, *115*, 213–222. [CrossRef]
89. Kim, M.; Thum, K.E.; Morishige, D.T.; Mullet, J.E. Detailed architecture of the barley chloroplast *psbD-psbC* blue light-responsive promoter. *J. Biol. Chem.* **1999**, *274*, 4684–4692. [CrossRef]
90. Baba, K.; Nakano, T.; Yamagishi, K.; Yoshida, S. Involvement of a nuclear-encoded basic helix-loop-helix protein in transcription of the light-responsive promoter of *psbD*. *Plant Physiol.* **2001**, *125*, 595–603. [CrossRef]
91. Tsunoyama, Y.; Ishizaki, Y.; Morikawa, K.; Kobori, M.; Nakahira, Y.; Takeba, G.; Toyoshima, Y.; Shiina, T. Blue light-induced transcription of plastid-encoded *psbD* gene is mediated by a nuclear-encoded transcription initiation factor, AtSig5. *Proc. Natl. Acad. Sci. USA* **2004**, *101*, 3304–3309. [CrossRef] [PubMed]
92. Belbin, F.E.; Noordally, Z.B.; Wetherill, S.J.; Atkins, K.A.; Franklin, K.A.; Dodd, A.N. Integration of light and circadian signals that regulate chloroplast transcription by a nuclear-encoded sigma factor. *New Phytol.* **2017**, *213*, 727–738. [CrossRef]
93. Tsunoyama, Y.; Morikawa, K.; Shiina, T.; Toyoshima, Y. Blue light specific and differential expression of a plastid sigma factor, Sig5 in *Arabidopsis thaliana*. *FEBS Lett.* **2002**, *516*, 225–228. [CrossRef]
94. Noordally, Z.B.; Ishii, K.; Atkins, K.A.; Wetherill, S.J.; Kusakina, J.; Walton, E.J.; Kato, M.; Azuma, M.; Tanaka, K.; Hanaoka, M.; et al. Circadian control of chloroplast transcription by a nuclear-encoded timing signal. *Science* **2013**, *339*, 1316–1319. [CrossRef] [PubMed]
95. Zhou, J.; Wu, S.; Chen, X.; Liu, C.; Sheen, J.; Shan, L.; He, P. The *Pseudomonas syringae* effector HopF2 suppresses Arabidopsis immunity by targeting BAK1. *Plant J.* **2014**, *77*, 235–245. [CrossRef] [PubMed]
96. Cui, H.; Tsuda, K.; Parker, J.E. Effector-triggered immunity: From pathogen perception to robust defense. *Annu. Rev. Plant Biol.* **2015**, *66*, 487–511. [CrossRef]
97. Lo Presti, L.; Lanver, D.; Schweizer, G.; Tanaka, S.; Liang, L.; Tollot, M.; Zuccaro, A.; Reissmann, S.; Kahmann, R. Fungal effectors and plant susceptibility. *Annu. Rev. Plant Biol.* **2015**, *66*, 513–545. [CrossRef]
98. Xin, X.F.; Nomura, K.; Ding, X.; Chen, X.; Wang, K.; Aung, K.; Uribe, F.; Rosa, B.; Yao, J.; Chen, J.; et al. *Pseudomonas syringae* effector avirulence protein E localizes to the host plasma membrane and down-regulates the expression of the *NONRACE-SPECIFIC DISEASE RESISTANCE1/HARPIN-INDUCED1-LIKE13* gene required for antibacterial immunity in *Arabidopsis*. *Plant Physiol.* **2015**, *169*, 793–802. [CrossRef]
99. Wang, J.; Gao, C.G.; Li, L.; Cao, W.L.; Dong, R.; Ding, X.H.; Zhu, C.X.; Chu, Z.H. Transgenic RXLR effector *PITG_15718.2* suppresses immunity and reduces vegetative growth in potato. *Int. J. Mol. Sci.* **2019**, *20*, 3031. [CrossRef]
100. Li, P.; Zhang, L.; Mo, X.; Ji, H.; Bian, H.; Hu, Y.; Majid, T.; Long, J.; Pang, H.; Tao, Y.; et al. Rice aquaporin PIP1;3 and harpin Hpa1 of bacterial blight pathogen cooperate in a type III effector translocation. *J. Exp. Bot.* **2019**, *70*, 3057–3073. [CrossRef]

101. Yang, B.; Wang, Q.; Jing, M.; Guo, B.; Wu, J.; Wang, H.; Wang, Y.; Lin, L.; Wang, Y.; Ye, W.; et al. Distinct regions of the *Phytophthora* essential effector Avh238 determine its function in cell death activation and plant immunity suppression. *New Phytol.* **2017**, *214*, 361–375. [CrossRef] [PubMed]
102. De Torres Zabala, M.; Littlejohn, G.; Jayaraman, S.; Studholme, D.; Bailey, T.; Lawson, T.; Tillich, M.; Licht, D.; Bolter, B.; Delfino, L.; et al. Chloroplasts play a central role in plant defence and are targeted by pathogen effectors. *Nat. Plants* **2015**, *1*, 15074. [CrossRef] [PubMed]
103. Mukhtar, M.S.; Carvunis, A.R.; Dreze, M.; Epple, P.; Steinbrenner, J.; Moore, J.; Tasan, M.; Galli, M.; Hao, T.; Nishimura, M.T.; et al. Independently evolved virulence effectors converge onto hubs in a plant immune system network. *Science* **2011**, *333*, 596–601. [CrossRef] [PubMed]
104. Wessling, R.; Epple, P.; Altmann, S.; He, Y.; Yang, L.; Henz, S.R.; McDonald, N.; Wiley, K.; Bader, K.C.; Glasser, C.; et al. Convergent targeting of a common host protein-network by pathogen effectors from three kingdoms of life. *Cell Host Microbe* **2014**, *16*, 364–375. [CrossRef]
105. Haley, J.; Bogorad, L. Alternative promoters are used for genes within maize chloroplast polycistronic transcription units. *Plant Cell* **1990**, *2*, 323–333.
106. Swiatek, M.; Regel, R.E.; Meurer, J.; Wanner, G.; Pakrasi, H.B.; Ohad, I.; Herrmann, R.G. Effects of selective inactivation of individual genes for low-molecular-mass subunits on the assembly of photosystem II, as revealed by chloroplast transformation: The *psbEFLJ* operon in *Nicotiana tabacum*. *Mol. Genet. Genom.* **2003**, *268*, 699–710. [CrossRef]
107. Meteignier, L.V.; Ghandour, R.; Meierhoff, K.; Zimmerman, A.; Chicher, J.; Baumberger, N.; Alioua, A.; Meurer, J.; Zoschke, R.; Hammani, K. The Arabidopsis mTERF-repeat MDA1 protein plays a dual function in transcription and stabilization of specific chloroplast transcripts within the *psbE* and *ndhH* operons. *New Phytol.* **2020**, *227*, 1376–1391. [CrossRef]
108. Drouin, G.; Daoud, H.; Xia, J. Relative rates of synonymous substitutions in the mitochondrial, chloroplast and nuclear genomes of seed plants. *Mol. Phylogenet. Evol.* **2008**, *49*, 827–831. [CrossRef]
109. Wu, Z.; Waneka, G.; Broz, A.K.; King, C.R.; Sloan, D.B. MSH1 is required for maintenance of the low mutation rates in plant mitochondrial and plastid genomes. *Proc. Natl. Acad. Sci. USA* **2020**, *117*, 16448–16455. [CrossRef]
110. Joët, T.; Cournac, L.; Horvath, E.M.; Medgyesy, P.; Peltier, G. Increased sensitivity of photosynthesis to antimycin A induced by inactivation of the chloroplast *ndhB* gene. Evidence for a participation of the NADH-dehydrogenase complex to cyclic electron flow around photosystem I. *Plant Physiol.* **2001**, *125*, 1919–1929. [CrossRef]
111. Hashimoto, M.; Endo, T.; Peltier, G.; Tasaka, M.; Shikanai, T. A nucleus-encoded factor, CRR2, is essential for the expression of chloroplast *ndhB* in Arabidopsis. *Plant J.* **2003**, *36*, 541–549. [CrossRef] [PubMed]
112. Garcia-Andrade, J.; Ramirez, V.; Lopez, A.; Vera, P. Mediated plastid RNA editing in plant immunity. *PLoS Pathog.* **2013**, *9*, e1003713. [CrossRef] [PubMed]
113. Nakamura, T.; Ohta, M.; Sugiura, M.; Sugita, M. Chloroplast ribonucleoproteins function as a stabilizing factor of ribosome-free mRNAs in the stroma. *J. Biol. Chem.* **2001**, *276*, 147–152. [CrossRef] [PubMed]
114. Nawaz, G.; Lee, K.; Park, S.J.; Kim, Y.-O.; Kang, H. A chloroplast-targeted cabbage DEAD-box RNA helicase BrRH22 confers abiotic stress tolerance to transgenic *Arabidopsis* plants by affecting translation of chloroplast transcripts. *Plant Physiol. Biochem.* **2018**, *127*, 336–342. [CrossRef] [PubMed]
115. Schuster, M.; Gao, Y.; Schottler, M.A.; Bock, R.; Zoschke, R. Limited responsiveness of chloroplast gene expression during acclimation to high light in tobacco. *Plant Physiol.* **2020**, *182*, 424–435. [CrossRef] [PubMed]
116. Yamori, W. Photosynthetic response to fluctuating environments and photoprotective strategies under abiotic stress. *J. Plant Res.* **2016**, *129*, 379–395. [CrossRef]
117. Demmig-Adams, B.; Stewart, J.J.; Adams, W.W., 3rd. Environmental regulation of intrinsic photosynthetic capacity: An integrated view. *Curr. Opin. Plant Biol.* **2017**, *37*, 34–41. [CrossRef]
118. Fan, X.; Zhang, Z.; Gao, H.; Yang, C.; Liu, M.; Li, Y.; Li, P. Photoinhibition-like damage to the photosynthetic apparatus in plant leaves induced by submergence treatment in the dark. *PLoS ONE* **2014**, *9*, e89067. [CrossRef]
119. Wang, G.; Kong, F.; Zhang, S.; Meng, X.; Wang, Y.; Meng, Q. A tomato chloroplast-targeted DnaJ protein protects Rubisco activity under heat stress. *J. Exp. Bot.* **2015**, *66*, 3027–3040. [CrossRef]
120. Tian, F.; Gong, J.; Zhang, J.; Feng, Y.; Wang, G.; Guo, Q.; Wang, W. Overexpression of monoubiquitin improves photosynthesis in transgenic tobacco plants following high temperature stress. *Plant Sci.* **2014**, *226*, 92–100. [CrossRef]

121. Hou, F.; Jin, L.Q.; Zhang, Z.S.; Gao, H.Y. Systemic signalling in photosynthetic induction of *Rumex* K-1 (*Rumex patientia* x *Rumex tianschaious*) leaves. *Plant Cell Environ.* **2015**, *38*, 685–692. [CrossRef] [PubMed]
122. Sun, Y.; Gao, Y.; Wang, H.; Yang, X.; Zhai, H.; Du, Y. Stimulation of cyclic electron flow around PSI as a response to the combined stress of high light and high temperature in grape leaves. *Funct. Plant Biol.* **2018**, *45*, 1038–1045. [CrossRef] [PubMed]
123. Chovancek, E.; Zivcak, M.; Botyanszka, L.; Hauptvogel, P.; Yang, X.; Misheva, S.; Hussain, S.; Brestic, M. Transient heat waves may affect the photosynthetic capacity of susceptible wheat genotypes due to insufficient photosystem I photoprotection. *Plants (Basel)* **2019**, *8*, 282. [CrossRef] [PubMed]
124. Cheng, D.D.; Zhang, Z.S.; Sun, X.B.; Zhao, M.; Sun, G.Y.; Chow, W.S. Photoinhibition and photoinhibition-like damage to the photosynthetic apparatus in tobacco leaves induced by *pseudomonas syringae* pv. Tabaci under light and dark conditions. *BMC Plant Biol.* **2016**, *16*, 29.
125. Chen, J.H.; Chen, S.T.; He, N.Y.; Wang, Q.L.; Zhao, Y.; Gao, W.; Guo, F.Q. Nuclear-encoded synthesis of the D1 subunit of photosystem II increases photosynthetic efficiency and crop yield. *Nat. Plants* **2020**, *6*, 570–580. [CrossRef] [PubMed]
126. Kong, F.; Deng, Y.; Zhou, B.; Wang, G.; Wang, Y.; Meng, Q. A chloroplast-targeted DnaJ protein contributes to maintenance of photosystem II under chilling stress. *J. Exp. Bot.* **2014**, *65*, 143–158. [CrossRef]
127. Kong, F.; Deng, Y.; Wang, G.; Wang, J.; Liang, X.; Meng, Q. LeCDJ1, a chloroplast DnaJ protein, facilitates heat tolerance in transgenic tomatoes. *J. Integr. Plant Biol.* **2014**, *56*, 63–74. [CrossRef]
128. Che, X.; Zhang, Z.; Jin, L.; Liu, M.; Li, Y.; Gao, H.; Zhao, S. Effect of reducing nitric oxide in rumex K-1 leaves on the photoprotection of photosystem II under high temperature with strong light. *J. Plant Growth Regul.* **2016**, *35*, 1118–1125. [CrossRef]

7

# Mutation of YL Results in a Yellow Leaf with Chloroplast RNA Editing Defect in Soybean

**Xiaowei Zhu [1,2], Yi Pan [1], Zhi Liu [1], Yucheng Liu [1], Deyi Zhong [1,3], Zongbiao Duan [1], Zhixi Tian [1,4], Baoge Zhu [1,*] and Guoan Zhou [1,*]**

1. State Key Laboratory of Plant Cell and Chromosome Engineering, Institute of Genetics and Developmental Biology, Innovative Academy of Seed Design, Chinese Academy of Sciences, Beijing 100101, China; xwzhu@saas.sh.cn (X.Z.); ypan@genetics.ac.cn (Y.P.); zliu@genetics.ac.cn (Z.L.); ychliu@genetics.ac.cn (Y.L.); zhongdeyi@caas.cn (D.Z.); zbduan@genetics.ac.cn (Z.D.); zxtian@genetics.ac.cn (Z.T.)
2. Horticulture Research Institute, Shanghai Academy of Agricultural Sciences, Shanghai 201403, China
3. Agricultural Genomics Institute, Chinese Academy of Agricultural Sciences, Shenzhen 518120, China
4. College of Advanced Agricultural Sciences, University of Chinese Academy of Sciences, Beijing 100049, China

* Correspondence: bgzhu@genetics.ac.cn (B.Z.); guoanzhou@genetics.ac.cn (G.Z.)

**Abstract:** RNA editing plays a key role in organelle gene expression. Little is known about how RNA editing factors influence soybean plant development. Here, we report the isolation and characterization of a soybean *yl* (*yellow leaf*) mutant. The *yl* plants showed decreased chlorophyll accumulation, lower PS II activity, an impaired net photosynthesis rate, and an altered chloroplast ultrastructure. Fine mapping of *YL* uncovered a point mutation in *Glyma.20G187000*, which encodes a chloroplast-localized protein homologous to *Arabidopsis thaliana* (*Arabidopsis*) ORRM1. *YL* is mainly expressed in trifoliate leaves, and its deficiency affects the editing of multiple chloroplast RNA sites, leading to inferior photosynthesis in soybean. Taken together, these results demonstrate the importance of the soybean YL protein in chloroplast RNA editing and photosynthesis.

**Keywords:** soybean; *yellow leaf* mutant; photosynthesis; chloroplast RNA editing

## 1. Introduction

Photosynthesis is a vital process in which plants convert light energy into chemical energy [1,2]. In higher plants, chlorophyll *a* and *b* are the two major pigments located in the thylakoid membrane of photosynthetic organisms [2]. These chlorophyll molecules play essential roles in harvesting light energy and transferring that energy to reaction centers of the photosystem [2,3]. In the past few decades, tremendous research has been conducted on chlorophyll-deficient mutants in model plants, but the molecular genetic mechanisms underlying these soybean chlorophyll-deficiency mutations are not well understood.

For soybean, 15 chlorophyll-deficient mutant genes were reported to be mapped to chromosomes [4]. Of these genes, six lethal yellow mutant genes, including *Y11*, *Y18/Y18_1*, *Y18_2*, *YL_PR350*, *PsbP*, and *CD-5*, were mapped to chromosomes 13, 14, 17, 15, 3 and 15, respectively [5–9]. The other nine viable yellow mutant genes, including *Y9*, *Y10*, *Y12*, *Y13*, *Y17*, *Y20*, *Y23*, *Tic110*, and *Cd1*, were mapped to chromosomes 15, 3, 6, 13, 15, 12, 13, 2, and 10, respectively [8,10–16]. However, only the function of *Y11* and *Y9* genes in yellow foliage was validated by complementary analysis [6,9].

RNA editing is a post-transcriptional modification process that changes the sequence of RNA molecules so that the information in the mature RNA differs from that defined in the genome [17]. In land plants, RNA editing occurs in transcripts of chloroplasts and mitochondria. There are 20–60 chloroplasts and over 300 RNA mitochondrial editing sites in most flowering plants [17,18]. RNA editing is performed by an editosome that is assembled via protein-protein/RNA interactions [19]. Several organelle RNA recognition motif-containing (ORRM) proteins are known to be essential RNA editing factors. ORRM1 controls 62% of chloroplast editing sites in *Arabidopsis* and 81% of editing sites in maize, with the *Zm-orrm1* mutant exhibiting a pale green phenotype [20]. ORRM6 is primarily required for editing *psbF*-C77 and *accD*-C794 sites in *Arabidopsis* chloroplasts [21]. In addition, ORRM2, ORRM3, ORRM4 and ORRM5 are mitochondrial RNA editing factors [22–24]. Moreover, pentatricopeptide repeat (PPR) proteins, multiple organellar RNA editing factor (MORF)/RNA editing factor interacting proteins (RIPs), organelle zinc finger 1 (OZ1), protoporphyrinogen oxidase 1 (PPO1), and genomes uncoupled 1 (GUN1) have been identified as components of the plant RNA editosome [19,25–30].

Here, we report the characterization of a soybean *yellow leaf* (*yl*) mutant with chlorophyll deficiency and impaired photosynthesis. Fine mapping and DNA sequencing showed that *YL* encodes a GmORRM1. We showed that multiple chloroplast RNA editing sites were changed in *yl* mutants. Our findings provide new insight into the function of YL in soybean photosynthesis.

## 2. Results

### 2.1. *yl* Plants Show Yellow Leaves and Abnormal Photosynthesis at the Seedling Stage

The *yl* mutant was identified from the soybean cultivar Jindou 23 mutagenized with ethyl methane sulfonate (EMS). The *yl* mutant plants exhibited conspicuous yellow leaves (Figure 1A). Chlorophyll (Chl) analysis revealed that the Chl *a*, *b* and total Chl contents of wild type (Jindou 23) leaves were 1.15 mg/g, 0.89 mg/g and 2.04 mg/g, respectively (Figure 1B). In contrast, the Chl *a*, *b* and total Chl contents of *yl* leaves were 0.86 mg/g, 0.34 mg/g and 1.20 mg/g, respectively (Figure 1B). In addition, the Chl *a*/*b* ratio of *yl* leaves was significantly increased compared to that of the wild type (Figure 1C). To evaluate whether the photosynthesis of *yl* leaves was affected, we measured the maximum photochemical efficiency of photosystem II (PSII), defined as the Fv/Fm and net photosynthesis rate. The *yl* leaves displayed obviously decreased Fv/Fm values (wild type, 0.73; *yl*, 0.62) (Figure 1D). Moreover, the net photosynthesis rate of *yl* leaves was 9.03 μmol $CO_2$ $m^{-2}s^{-1}$, which was only 60% that of wild type leaves (14.93 μmol $CO_2$ $m^{-2}s^{-1}$) (Figure 1E). Therefore, the mutation in *yl* causes not only a reduction in chlorophyll contents but also abnormalities in photosynthesis.

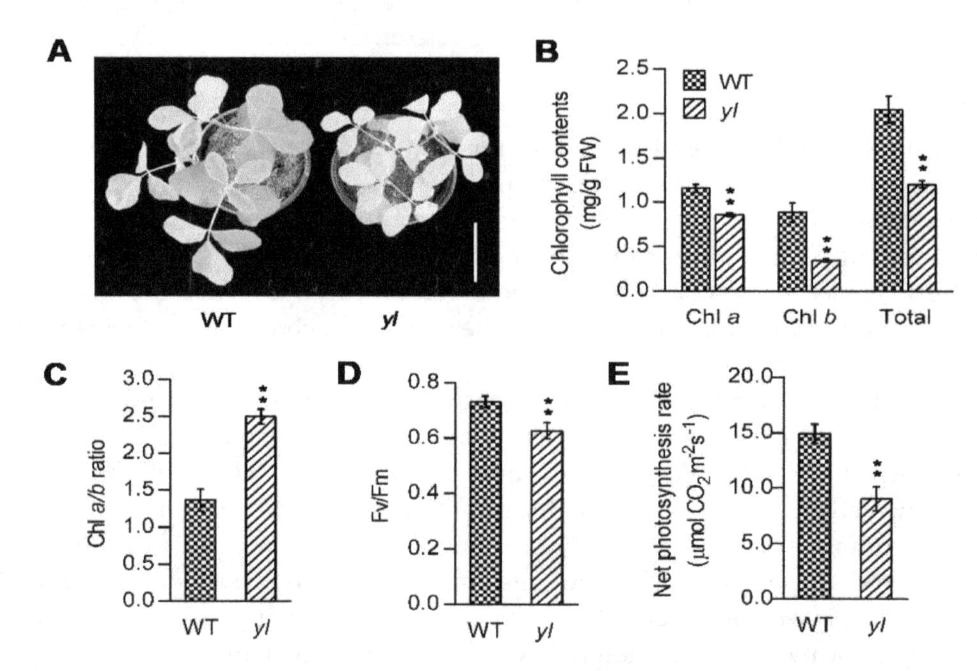


**Figure 1.** Phenotypic characterization of the wild type and *yl* mutant plants. (**A**) Two-week-old wild type (left) and *yl* (right) seedlings. (**B**) Chlorophyll contents of the third leaf of the wild type (left) and *yl* (right) plants. (**C**) The Chl *a*/*b* ratio of the wild type (left) and *yl* (right) leaves. (**D**) The maximum photochemical efficiency of PSII measured by the Fv/Fm chlorophyll fluorescence ratio in the wild type

(left) and *yl* (right) leaves. (**E**) The net photosynthesis rate of the wild type (left) and *yl* (right) plants. Error bars (**B–E**) represent the mean ± SD ($n \geq 3$); ** indicates a significant difference at the 0.01 level. The *p* values were calculated by Student's t-test. Bar = 5 cm in (**A**).

We further performed transmission electron microscopy (TEM) to compare the chloroplast ultrastructure between wild type and *yl* leaves. No differences in the number of chloroplasts per cell or chloroplast size were detected between the wild type and *yl*. However, the number of grana (stacked thylakoids) were slightly reduced in the *yl* chloroplasts (Figure 2A–D), indicating that *yl* chloroplasts were probably less functional than wild type chloroplasts.

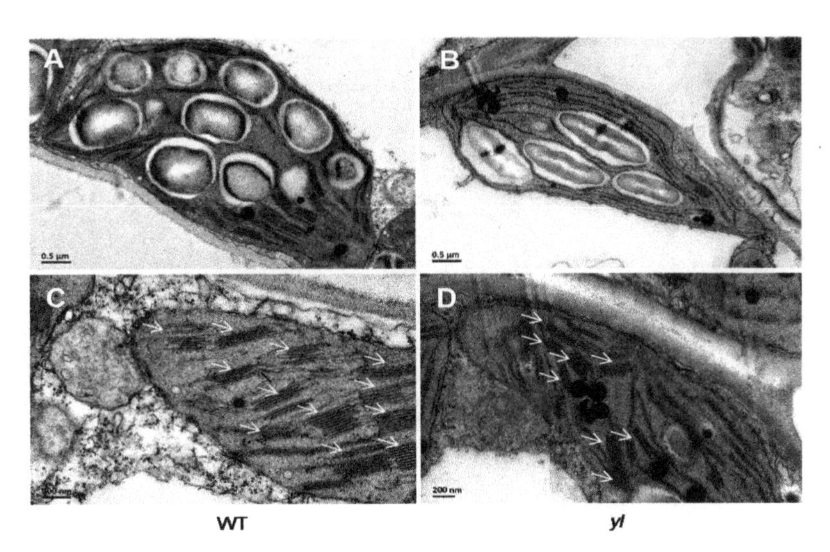


**Figure 2.** Chloroplast ultrastructure in the wild type and *yl* leaves. (**A**,**B**) Transmission electron micrographs (TEM) of chloroplasts in the wild type (**A**) and *yl* (**B**) leaves. (**C**,**D**) Magnified images of a chloroplast in the wild type (**C**) and *yl* (**D**) leaves. Arrows show grana. Bar = 0.5 μm in (**A**,**B**) and 200 nm in (**C**,**D**).

*2.2. Fine Mapping of the YL Gene*

To unravel the molecular mechanism of the *yl* phenotype, we performed genetic mapping to isolate the *YL* gene. The *yl* mutant was crossed to two soybean cultivars with normal green leaves, Williams 82 and Zhonghuang 13, and three $F_2$ mapping populations were produced by reciprocal crosses, *yl* × Williams 82, *yl* × Zhonghuang 13 and Zhonghuang 13 × *yl*. The leaf color of all $F_1$ plants was normal green. The segregation ratio of green to yellow leaves appeared to be 3:1 in $F_2$ populations, indicating that the *yl* phenotype is controlled by a single recessive nuclear gene (Table 1).

**Table 1.** Segregation pattern and chi-square tests for green/yellow leaves of $F_2$ progeny from the crosses between *yl* and Williams 82 or Zhonghuang 13.

| Crosses | $F_1$ | Number of $F_2$ Plants | | | $\chi^2$ (3:1) | *P* |
|---|---|---|---|---|---|---|
| | | Wild Type | Mutant | Total | | |
| *yl* × Williams 82 | Normal | 679 | 227 | 906 | 0 | >0.9 |
| *yl* × Zhonghuang13 | Normal | 440 | 163 | 603 | 1.221 | 0.50~0.25 |
| Zhonghuang13 × *yl* | Normal | 448 | 132 | 580 | 1.437 | 0.25~0.10 |

The *yl* mutation was primarily mapped on chromosome 20 between microsatellite markers Satt162 and Sat_155, which were 9.9 centimorgans (cM) and 1.1 cM from *yl*, respectively (Figure 3A). The *YL* locus was ultimately fine mapped into a 28-kb interval between single nucleotide polymorphism (SNP) markers S3 and S7-3 (Figure 3A). According to the soybean gene annotation database (www.phytozome.net) [31], there were three putative open reading frames (ORFs) within this 28-kb region

(Figure 3A). We sequenced the 28-kb sequences between the wild type and *yl* mutant and found a C to A transition at the eighth exon of *Glyma.20G187000* (Figure 3B).

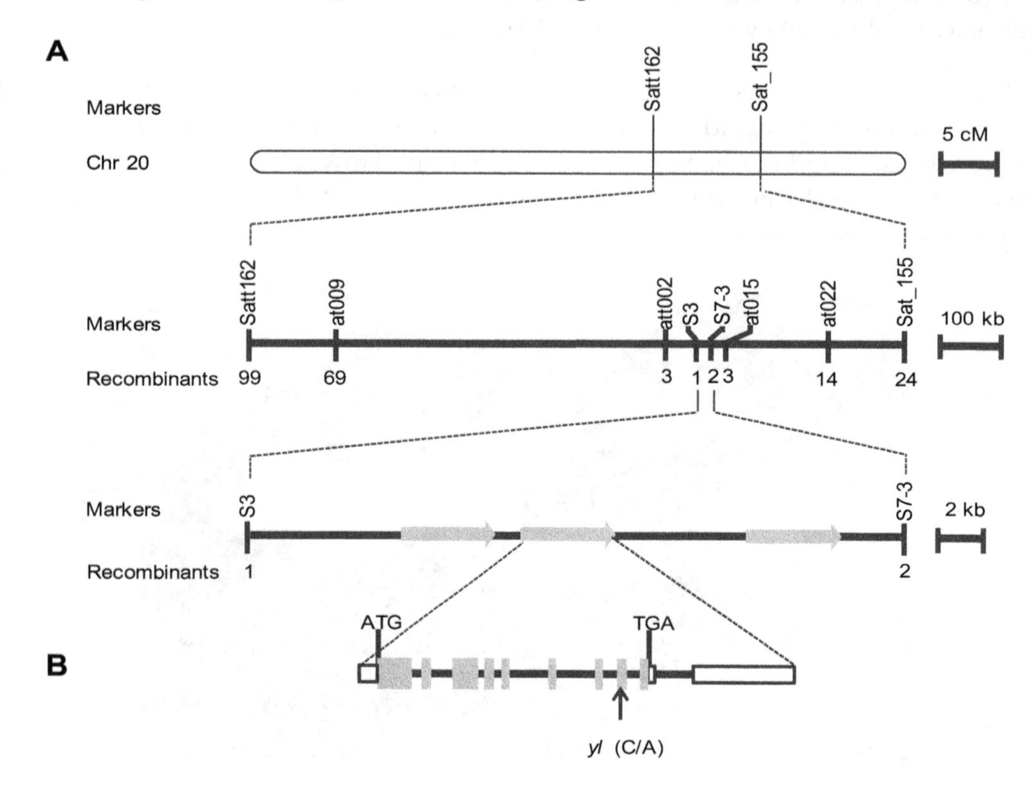


**Figure 3.** Fine mapping of the *YL* gene. (**A**) The *YL* locus was initially mapped to a region between markers Satt162 and Sat_155 on soybean chromosome 20. The gene was finally delimited to a 28 kb region between markers S3 and S7-3. Three predicted open reading frames (ORFs) were within this region. (**B**) *YL* (*Glyma.20G187000*) structure indicating nine exons (gray boxes), eight introns (line segments between the exons), and 5′ and 3′ untranslated regions (white boxes with black frame). Start (ATG) and stop (TGA) codons are marked. The *yl* mutation in the *YL* gene is shown.

### 2.3. *YL Encodes an Organelle RNA Recognition Motif-Containing Protein 1 (GmORRM1)*

The *YL* gene (*Glyma.20G187000*) encodes a protein of 390 amino acids with a C-terminal RNA recognition motif (Figure 4A). The *yl* mutation occurred in the RNA recognition motif and resulted in the substitution of an alanine with a glutamic acid (Figure 4A). The alignment of amino acid sequences homologous to YL from several plant species showed that this alanine residue in the RNA recognition motif was highly conserved (Figure 4B). In addition, the YL sequence shared 62.3% similarity with *Arabidopsis* ORRM1 (AtORRM1, At3G20930) and 58.2% similarity with maize ORRM1 (ZmORRM1, GRMZM5G899787), suggesting that YL (GmORRM1) and its homologs might have conserved functions.

### 2.4. *Expression Pattern and Protein Subcellular Localization of YL*

We performed quantitative real-time RT-PCR (qRT-PCR) to examine the expression of *YL* among various tissues. *YL* was expressed in all tissues tested, with the highest expression in trifoliate leaves, intermediate expression in meristem, nodules, expanded leaves, cotyledons, and young pods and weak expression in roots, stems, flowers, and young seeds (Figure 5A). The expression pattern of *YL* was consistent with the function of *YL* in affecting leaf color.

Using the Predotar and TargetP programs, the YL protein was predicted to localize in chloroplasts [32,33]. To obtain experimental verification of this localization, we expressed a YL-GFP fusion protein under the 35S promoter in *Nicotiana benthamiana* leaves. The green fluorescence signals of the YL-GFP fusion protein were observed in the chloroplasts of epidermal cells (Figure 5B,C).

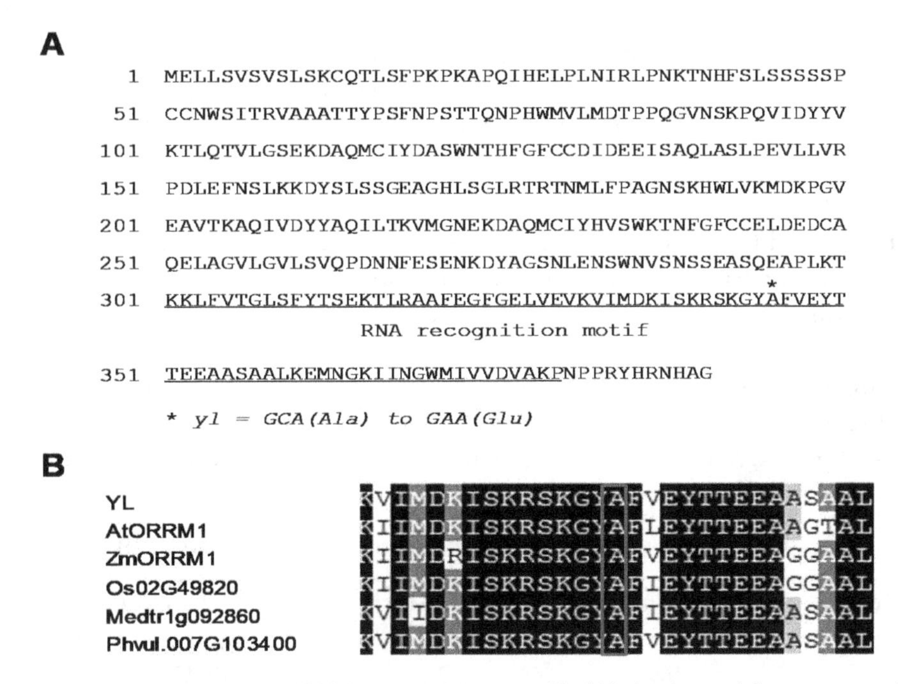


**Figure 4.** The YL protein contains an RNA recognition motif. (**A**) Amino acid sequence of YL. Position of the *yl* (asterisk) mutation is exhibited. The RNA recognition motif is underlined. (**B**) Partial alignment of the amino acid sequences of YL and its homologous proteins. The aligned sequences are from *Glycine max* (YL, Glyma.20G187000), *Arabidopsis thaliana* (AtORRM1, At3G20930), *Zea mays* (ZmORRM1, GRMZM5G899787), *Oryza sativa* (Os02g49820), *Medicago truncatula* (Medtr1g092860) and *Phaseolus vulgaris* (Phvul.007G103400). The red rectangle represents the conserved alanine residue in the RNA recognition motif.

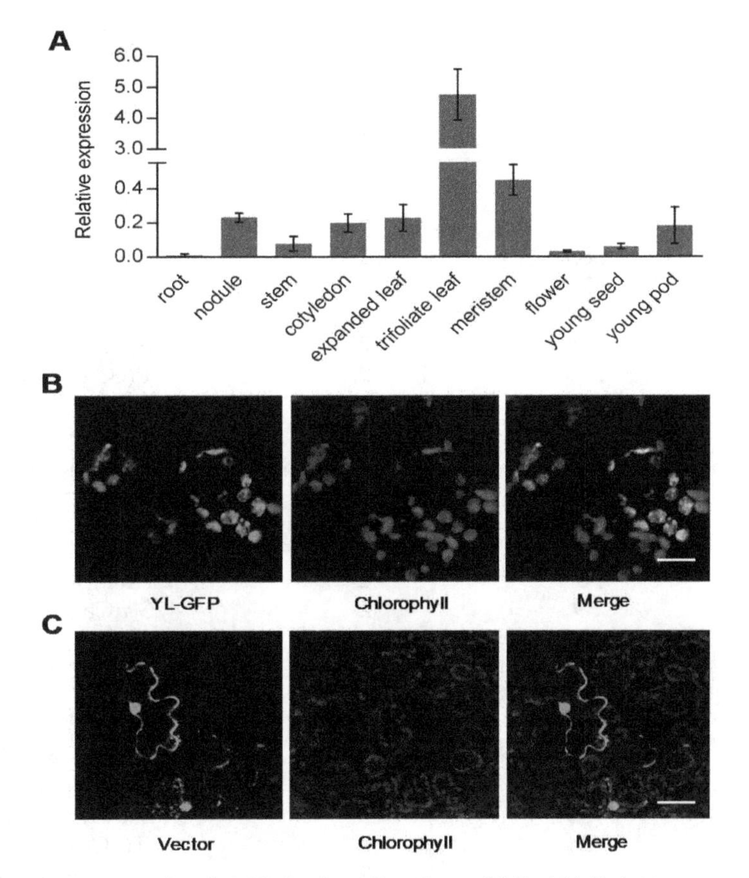


**Figure 5.** Expression pattern and subcellular localization of *YL*. (**A**) Relative expression levels of the *YL* gene in various soybean tissues. The real-time RT-PCR assays were performed in three biological

replicates. qRT-PCR values for *YL* are normalized with the actin and ATP synthase genes. Error bars represent the SD. (**B**,**C**) Subcellular localization of the YL-GFP fusion protein (**B**) and GFP (**C**) in epidermal cells of *Nicotiana benthamiana* leaves. Bar = 20 μm in (**B**) and 50 μm in (**C**).

### *2.5. Dramatic Defects of Chloroplast RNA Editing in yl*

To further understand the function of *YL*, we performed DNA resequencing and RNA sequencing (RNA-seq) for the wild type to screen out soybean chloroplast editing sites, obtaining 44 predicted sites from 22 chloroplast transcripts (Table 2, Tables S1 and S2). In addition, the *rpl23*-89 site was identified through comparative analyses with *Arabidopsis* chloroplast editing sites (Table 2). Most of these sites were verified by direct sequencing of PCR products of transcripts or the corresponding genomic DNA carrying them (Table S2). Among the 45 editing sites, 44 sites were C-to-U conversions, 43 sites were in the coding regions of genes, and most of them caused alteration of the encoded amino acids (Table 2).

**Table 2.** RNA editing sites in soybean chloroplast transcripts.

| Gene | Position [a] | Codon Site [b] | Conversion |
|---|---|---|---|
| *accD* | 617 | 2 | S(uCg)→L(uUg) |
| *atpF* | 92 | 2 | P(cCa)→L(cUa) |
| *clpP* | 559 | 1 | H(Cau)→Y(Uau) |
| *ndhA* | 341 | 2 | S(uCa)→L(uUa) |
| | 1073 | 2 | S(uCu)→F(uUu) |
| *ndhB* | 9 | 3 | W(ugG)→stop codon (ugA) |
| | 149 | 2 | S(uCa)→L(uUa) |
| | 542 | 2 | T(aCg)→M(aUg) |
| | 586 | 1 | H(Cau)→Y(Uau) |
| | 737 | 2 | P(cCa)→L(cUa) |
| | 746 | 2 | S(uCu)→F(uUu) |
| | 830 | 2 | S(uCa)→L(uUa) |
| | 836 | 2 | S(uCa)→L(uUa) |
| | 1112 | 2 | S(uCa)→L(uUa) |
| | 1255 | 1 | H(Cau)→Y(Uau) |
| | 1481 | 2 | P(cCa)→L(cUa) |
| *ndhC* | 323 | 2 | S(uCa)→L(uUa) |
| *ndhD* | 2 | 2 | T(aCg)→M(aUg) |
| | 383 | 2 | T(aCa)→I(aUa) |
| | 674 | 2 | S(uCa)→L(uUa) |
| | 878 | 2 | S(uCa)→L(uUa) |
| | 1298 | 2 | S(uCa)→L(uUa) |
| *ndhE* | 233 | 2 | P(cCg)→L(cUg) |
| *ndhF* | 290 | 2 | S(uCa)→L(uUa) |
| *petB* | 611 | 2 | S(uCa)→L(uUa) |
| *psaI* | 79 | 1 | H(Cau)→Y(Uau) |
| *psbE* | 214 | 2 | S(uCc)→F(uUc) |
| *psbF* | 6 | 3 | T(acC)→T(acU) |
| | 77 | 2 | S(uCu)→F(uUu) |
| *psbL* | 2 | 2 | T(aCg)→M(aUg) |
| *rpl23* | 89 | 2 | S(uCa)→L(uUa) |
| *rpoA* | 200 | 2 | S(uCa)→L(uUa) |
| *rpoB* | 338 | 2 | S(uCu)→F(uUu) |
| | 551 | 2 | S(uCa)→L(uUa) |
| | 566 | 2 | S(uCg)→L(uUg) |
| | 2000 | 2 | S(uCu)→F(uUu) |
| *rpoC1* | 41 | 2 | S(uCa)→L(uUa) |
| | 488 | 2 | S(uCa)→L(uUa) |
| *rps2* | 134 | 2 | T(aCa)→I(aUa) |
| | 248 | 2 | S(uCa)→L(uUa) |

**Table 2.** *Cont.*

| Gene | Position [a] | Codon Site [b] | Conversion |
|---|---|---|---|
| *rps12* | 554 [c] | | |
| *rps14* | 80 | 2 | S(uCa)→L(uUa) |
| *rps16* | 499 [c] | | |
| | 212 | 2 | S(uCa)→L(uUa) |
| *rps18* | 221 | 2 | S(uCg)→L(uUg) |

[a] Position is given with respect to the initiation codon of each chloroplast transcript. [b] Codon site is the order in the amino acid codon. [c] RNA editing sites are in introns of the chloroplast gene. The position here is given with respect to the initiation codon of each gene.

The RNA-seq method was combined with direct sequencing of PCR products of transcripts carrying chloroplast RNA editing sites to compare the RNA editing between wild type and *yl* leaves. The results showed that the editing was completely abolished for *ndhB*-737, *ndhD*-674 and *rpoB*-551 in the *yl* mutant (Figure 6A,B, Tables S1 and S2). These deficiencies caused changes in the encoded amino acid residues from Leu, Leu and Leu in the wild type to Pro, Ser and Ser in the *yl* mutant, respectively. In addition, the editing of 14 sites was decreased by 10% to 90% in the *yl* mutant (Figure 6A, Tables S1 and S2). However, the *ndhF*-290 and *rpoB*-566 sites exhibited higher editing levels in the *yl* mutant compared with the wild type (Figure 6A, Tables S1 and S2).

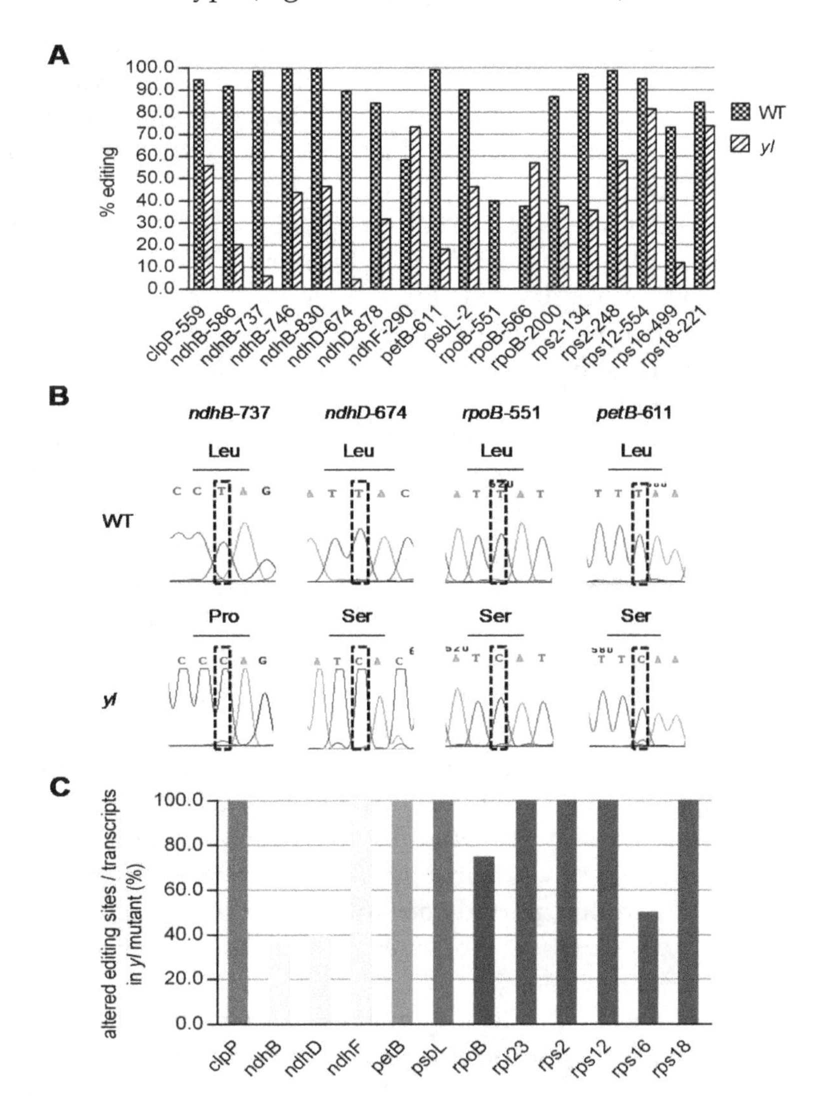


**Figure 6.** RNA editing at multiple chloroplast sites is impaired in the *yl* mutant. (**A**) Eighteen sites exhibit a significant alteration in editing of more than 10% in the *yl* mutant through RNA-seq analysis.

(**B**) The three abolished editing sites and *petB*-611 in the *yl* mutant. Rectangular frames indicate defective editing sites. The corresponding amino acids are underlined. (**C**) Percentage of altered editing sites/transcripts in the *yl* mutant. Each bar shows a transcript color-coded in accordance with the complex to which it belongs.

We then investigated the distribution of the affected editing sites in the *yl* mutant. The 19 affected editing sites were distributed in 12 chloroplast transcripts encoding components of the Clp protease proteolytic subunit, NDH complex, cytochrome $b_6f$ complex, PSII complex, RNA polymerase or ribosomal proteins. As shown in Figure 6C, the percentage of altered editing sites per transcript varied from 36.4% to 100%, suggesting that the effect of the *YL* mutation on editing was site specific but not transcript specific.

## 3. Discussion

In plants, characterization of yellow foliar mutants will present an interesting opportunity to understand the complex photosynthesis process. The *yl* mutant reported here showed reduced Chl *a*, *b* and total Chl contents (Figure 1B). Our study identified a point mutation in the *YL* gene responsible for the mutant phenotype (Figure 3B). In addition, we discovered decreased PSII activity and a decreased net photosynthesis rate of *yl* leaves (Figure 1D,E). These results reveal that *YL* plays an important role in the photosynthetic process. This conclusion is supported by two pieces of evidence: (1) Strong expression of *YL* was detected in the trifoliate leaves; (2) The *yl* mutant lacked stacked thylakoids in the chloroplast, where photosynthesis takes place. The YL protein was homologous to maize ORRM1. In addition, the *Mu* transposon insertional mutant in *ZmORRM1* showed defects in the major photosynthetic enzyme complexes [20]. However, it is not known whether YL could influence similar photosynthetic proteins.

Based on bioinformatic analysis and protein subcellular localization experiments, we conclude that *YL* encodes a 390-amino-acid chloroplast-localized protein (Figures 4A and 5B,C). Identification of the C-terminal RNA recognition motif of YL and the high level of homology with AtORRM1 and ZmORRM1 strongly suggest that YL functions in RNA editing. ORRM1 controlled the extent of editing in 62% of the chloroplast sites in *Arabidopsis* and 81% of sites in maize [20]. In our study, YL was required for the editing of 42% of chloroplast sites in 12 chloroplast transcripts (Figure 6A, Tables S1 and S2). Moreover, we compared RNA editing alterations in soybean, *Arabidopsis* and maize *orrm1* mutants. Notably, the three completely lost editing sites, *ndhB*-737, *ndhD*-674 and *rpoB*-551, in the *yl* mutant also exhibited a pronounced reduction in editing in *Arabidopsis* or maize *orrm1* mutants (Figure 6A,B, Tables S1 and S2). In contrast, the editing of *ndhF*-290 exhibited no change in the *Arabidopsis orrm1* mutant, whereas we observed a slight increase in the editing of *ndhF*-290 in the *yl* mutant (Figure 6A, Table S1). Consequently, it seems that ORRM1 showed species-specific functions in plants. These results suggest that YL may be involved in chloroplast RNA editing.

Twelve chloroplast transcripts, which encode components of the Clp protease proteolytic subunit, NDH complex, cytochrome $b_6f$ complex, PSII complex, RNA polymerase or ribosomal proteins, could not be edited in the *yl* mutant (Figure 6C). The RNA editing defect in transcripts encoding the components of the Clp protease proteolytic subunit, NDH complex, and RNA polymerase or ribosomal proteins may not be the main reason for the *yl* mutant phenotype. Although the *Arabidopsis orrm1* mutant exhibited severe editing defects in transcripts encoding these proteins, it did not show any phenotypic deficiencies [20]. In addition, partial knockout of *clpP* in tobacco resulted in an asymmetric, slender leaf shape but normal leaf color [34]. Moreover, no obvious phenotype was observed in tobacco mutants with disrupted NDH complexes [35].

The reduction in *petB*-611 and *psbL*-2 editing may have large contributions to the *yl* phenotype. The *petB* gene encodes cytochrome $b_6$, which is one of the major subunits of the cytochrome $b_6f$ complex mediating electron transfer between PSII and I [36–38]. In the *yl* mutant, defective editing at *petB*-611 resulted in a change from the wild type residue at position 204 (Leu) to Ser (Figure 6B). The $Leu^{204}$

residue belongs to the D helix span, which is involved in heme binding [37,39]. $Leu^{204}$ to $Ser^{204}$ alteration likely disrupts the assembly of cytochrome $b_6f$ complexes, as suggested by the behavior of *petB* mutants in *Chlamydomonas reinhardtii* [37]. The *yl* mutant is phenotypically similar to a tobacco *petB* mutant in which the portion of the *petB* coding region was replaced with an *aadA* cassette [38]. Both *yl* and *petB* mutants display yellow (pale green) leaf color and reduced chlorophyll contents and grana stacks. The *psbL* gene encodes a conserved low molecular weight protein of PS II [40]. In wild type soybean, the initiator codon (AUG) of *psbL* is formed by a C to U editing of the ACG codon. In the present *yl* mutant, defective editing at *psbL*-2 may partially influence the formation of the initiator codon (AUG). The lack of *PsbL* in tobacco has been implicated as impairing the assembly of PSII [41,42]. The reduced Fv/Fm in the *yl* mutant is similar to the behavior of *Arabidopsis* mutants deficient in RNA editing sites in plastid transcripts encoding PSII proteins, including *orrm6* mutants [21].

The cytochrome $b_6f$ and PSII complexes belong to the main components of the photosynthetic electron transfer chain [2]. The *yl* mutant, exhibiting RNA editing deficiency at *petB*-611 and *psbL*-2, may be unable to assemble functional cytochrome $b_6f$ and PSII complexes. Since the localization of cytochrome $b_6f$ and PSII complexes is predominantly in the grana regions of the thylakoid membrane system, this inference may account for lacked grana in *yl* chloroplast. In soybean, some mutants with defects in photosynthetic electron carrier proteins display a yellow phenotype [4,8]. For instance, the mutation in the PSII extrinsic protein GmpsbP leads to a lethal-yellow phenotype, extremely low Fv/Fm, and failure of proplastid differentiation into normal chloroplasts with grana [8]. The *yl* mutant showed reduced Fv/Fm, indicating that the *yl* mutant cannot utilize the absorbed light in photochemistry as effectively as the wild type. When more light energy is absorbed than is converted in plants, the photosynthetic organism is subjected to photooxidative stress, known as photoinhibition, leading to pigment bleaching, inactivation of electron transport and damage of the reaction center [43,44]. A recent study has shown that a leaf yellowing mutant phenotype in soybean may be largely due to the abnormal light absorption in the photosynthesis process [45]. Thus, photoinhibition may also be the cause of decreased Chl contents and net photosynthesis rate in the *yl* mutant.

## 4. Materials and Methods

### 4.1. Plant Materials and Growth Conditions

Seeds of soybean cultivar Jindou 23 (wildtype) were mutagenized with ethyl methane sulfonate (EMS). The phenotypes of the second mutant generation ($M_2$ progeny) plants were observed and the *yl* mutant with yellow foliage was identified from the $M_2$ progeny. The *yl* mutant has been continuously self-pollinated and selected based on the yellow foliar phenotype. For phenotype characterization of the *yl* mutant, the simultaneous field trial consisted of three replicates per genotype, planted in 0.50 m row spacing and 0.10 m spacing in the rows. To analyze the inheritance pattern of the *yl* mutant and fine-mapping of the *YL* gene, three $F_2$ populations were generated from reciprocal crosses between the *yl* mutant and two soybean cultivars, Williams 82 and Zhonghuang 13 (*yl* × Williams 82, *yl* × Zhonghuang 13 and Zhonghuang 13 × *yl*). All soybean materials were planted in the experimental field at the Institute of Genetics and Developmental Biology, Chinese Academy of Sciences (Beijing, China), during the natural growing seasons (May to October). The common meteorological conditions during soybean growing seasons in Beijing were as follows: (1) mean daily temperatures: about 24 °C. (2) overall precipitation: 400 ~ 600 mm.

### 4.2. Chl Contents, Net Photosynthetic Rate and Photochemical Efficiency Analysis

The third true leaves from the top of five-week-old soybean plants were used to estimate Chl contents, net photosynthetic rate and photochemical efficiency. The total Chl, Chl *a* and Chl *b* contents were determined as previously described [46]. Fresh leaves (200 mg) were immersed into 10 mL of 95% ethanol for pigments extraction under dark conditions. Then the absorbance of supernatant was quantified spectrophotometrically at 665 and 649 nm. Chlorophyll contents were calculated using the following formulas [46]: Chl *a* (mg/g) = [(13.95$OD_{665}$ − 6.88$OD_{649}$) × 10]/(tissue fresh weight (g) × 1000); Chl *b* (mg/g) = [(24.96$OD_{649}$ − 7.32$OD_{665}$) × 10]/(tissue fresh weight (g) × 1000); Total Chl (mg/g) = Chl *a* (mg/g) + Chl *b* (mg/g). Net photosynthetic rate was measured using an Li-6400 instrument (LI-COR, Lincoln, NE, USA) under a $CO_2$ concentration of 250 μmol $mol^{-1}$ and 1000 μmol $m^{-2}s^{-1}$ photosynthetically active radiation (PAR) between 9:00 and 11:00 in the morning. The maximum photochemical efficiency of photosystem II was measured as the Fv/Fm using an IMAGING-PAM Chlorophyll Fluorometer (Heinz Walz, Effeltrich, Germany).

Three to eight individual plants were examined for each material and the examination were repeated at least three times. The experimental data for each material are reported as mean ± standard deviation (SD). Statistical significance of differences between *yl* mutant and wildtype plants were tested using independent sample Student's t test algorithm. All statistical analysis were performed in SPSS Statistics 17.0.

### *4.3. Transmission Electron Microscopy*

The third leaves of 30-day-old wild type and *yl* plants were collected and fixed for at least 2 h at 4 °C in 0.1 M phosphate buffer, pH 7.4, with 2.5% glutaraldehyde and washed with the same buffer. The samples were fixed in 1% osmium tetroxide for 2 h and then dehydrated through an ethanol gradient, infiltrated and embedded in epoxy resin. Ultra-thin sections were obtained using a diamond knife and mounted on copper grids. Then the grids were stained with uranyl acetate and lead citrate, and examined using a JEM-1400 (JEOL, Tokyo, Japan) transmission electron microscope.

### *4.4. Fine Mapping of the YL Gene*

Three $F_2$ populations generated from reciprocal crosses between the *yl* mutant and two soybean cultivars, Williams 82 and Zhonghuang 13 (*yl* × Williams 82, *yl* × Zhonghuang 13 and Zhonghuang 13 × *yl*), were used for genetic analysis. In the $F_2$ populations, genomic DNA was isolated from selected etiolated seedlings exhibiting mutant phenotype for gene mapping. The cetyl trimethyl ammonium bromide (CTAB) method was used for genomic DNA extraction from the fresh young trifoliate leaf tissue and for all subsequent DNA extractions using this method unless otherwise stated. Briefly, frozen tissue was powdered and dispersed in 2 × CTAB extraction buffer, and incubated at 65 °C for about 60 min. Chloroform/Tris-phenol, 1:1 (vol/vol), was added and mixed to form an emulsion that was centrifuged for 10 min. The upper aqueous phase was transferred to a new tube, and 2/3 vol of isopropanol was added for DNA precipitation. The extracted DNA was treated with RNase to remove RNA contamination.

For primary mapping, 92 mutant individuals were selected from two $F_2$ populations between *yl* and Williams 82 or Zhonghuang 13. The markers used for primary mapping were 71 published SSR markers (http://soybase.org/) [47]. A total of 770 mutant individuals selected from the $F_2$ population between the *yl* mutant and Williams 82 were used for fine mapping. New molecular markers were developed for fine mapping (Table S3). To identify the candidate gene, the corresponding DNA fragments within the fine mapping region were amplified from the wild type and *yl* mutant using special primers (Table S3) and sequenced.

### *4.5. RNA Extraction and Quantitative Real-Time RT-PCR*

For analysis of the expression of *YL* in various soybean tissues, total RNA was isolated from root, nodule, stem, cotyledon, expanded leaf, trifoliate leaf, meristem, flower, young seed and pod collected from three individuals. For the analysis of chloroplast RNA editing in the *yl* mutant and wild type, total RNA was isolated from trifoliate leaves of five-week-old soybean plants. Fresh tissue was frozen immediately in liquid nitrogen and ground to powder using a mortar and pestle. Total RNA was extracted using 1 mL TRIzol reagent (Invitrogen Life Technologies, Carlsbad, CA, USA), according to the manufacturer's instructions. After ethanol precipitation, the RNA was dissolved in RNase-free water.

Approximately 2 μg of total RNA were reverse transcribed using the Prime Script RT reagent Kit with gDNA Eraser (TaKaRa, Beijing, China), according to the manufacturer's instructions. Quantitative real-time RT-PCR was performed in a 20 μL reaction mix containing 50 ng of cDNA, 1 × Light Cycler 480 SYBR Green I Master (Roche, Mannheim, Germany), 0.5 μM of each primer on a Lightcycler 480 (Roche, Mannheim, Germany) machine using the following PCR profile: 2 min at 94 °C, followed by 40 cycles of 15 s at 94 °C, 15 s at 60 °C, and 30 s at 72 °C. The dissociation curve analysis was conducted to verify the PCR specificity. Three biological replicates with three technical replicates were analyzed to quantify the levels of gene expression. The soybean actin and ATP synthase genes were used as internal standards to normalize the expression of *YL* using the $2^{-\Delta Ct}$ method. The primers used for quantitative real-time RT-PCR were designed using the Primer Premier 5 software (Premier, San Francisco, CA, USA) and listed in Table S3.

### 4.6. Subcellular Localization

To determine its subcellular localization, the coding sequence of *YL* lacking its stop codon was amplified by PCR in a 50 μL volume that contained 200 ng cDNA from wild type, 1 × PCR buffer, 0.4 mM dNTPs, 1 U KOD FX Neo (Toyobo, Osaka, Japan) and 0.3 μM of both forward and reverse primers of *YL-GFP* listed in Table S3. The sample was heated to 94 °C for 2 min, followed by 45 cycles of denaturation at 94 °C for 15 s, annealing at 61 °C for 30 s, elongation at 68 °C for 1 min. The amplified product was cloned into the pCAMBIA 1302 vector between the cauliflower mosaic virus 35S promoter and the GFP-coding sequence. The construct of YL-GFP vector was sequencing confirmed by Sangon Biotech (Shanghai, China).

For transformation of *Agrobacterium*, 1 μg purified plasmid DNA was added to competent cells of thawing *Agrobacterium tumefaciens* strain *GV3101* on ice. After ice bath for 30 min, the cell/DNA mix was immersed in liquid nitrogen for 1 min and subsequently incubated at 37 °C for 5 min, then ice bath for 2 min. By adding 900 μL liquid growth medium (no antibiotics), the cell/DNA mix was shocked for at least 120 min at 28 °C. After centrifugation, the cells were resuspended and plated on an agar plate containing kanamycin (50 mg/L) for selection of transformants.

For infiltration of *Nicotiana benthamiana*, the *Agrobacterium tumefaciens* strains carrying the YL-GFP, pCAMBIA 1302 and p19 of tomato bushy stunt virus plasmids were grown at 28 °C in liquid growth medium with kanamycin (50 mg/L), rifampicin (50 mg/L) and gentamicin (25 mg/L) until $OD_{600}$ reached 1.0. The *Agrobacterium tumefaciens* strains containing the YL-GFP and p19 or pCAMBIA 1302 and p19 plasmids were mixed. After centrifugation of the strain mixtures, the harvested cells were resuspended in 10 mM MES buffer containing 10 mM $MgCl_2$ and 100 mM acetosyringone to a final $OD_{600}$ of 1.0, followed by incubation at room temperature for 120 min. Strain mixtures were infiltrated into the abaxial surface of leaves of four-week-old *Nicotiana benthamiana* plants using a 1 mL syringe. The transformed epidermal cells were detected using a Zeiss LSM710 confocal microscope (Carl Zeiss Microscopy GmbH, Jena, Germany).

### 4.7. Chloroplast RNA Editing Analysis

For chloroplast RNA editing analysis, total DNA and RNA were isolated from wild type and *yl* leaves as described in Sections 4.4 and 4.5. Next, the DNA and RNA were submitted to Berry Genomics (Beijing) for DNA resequencing and rRNA-depleted strand-specific RNA-seq, respectively. For RNA-seq analysis, four cDNA libraries (two biological replicates per genotype) were constructed. All sequencing data used in this research were deposited in the National Center for Biotechnology Information (NCBI) Sequence Read Archive under BioProject ID PRJNA616185.

Clean reads were quality checked using FastQC (version 0.11.3) (http://www.bioinformatics.babraham.ac.uk/projects/fastqc/) [48]. After quality control, reads were aligned to the soybean chloroplast genome (ncbi.nlm.nih.gov/nuccore/DQ312375.1) [49] using HISAT (version 2.0.0) [50]. SNP calling was performed by GATK combined with Samtools [51,52]. SNPs with read numbers ≥ 20 from the RNA-seq and the corresponding DNA resequencing data were compared and the different chloroplast allele bases between the two data sets were considered as the candidate RNA editing sites. Because C-to-U editing in plastid and mitochondrial mRNAs appear to be ubiquitous in land plants, the editing efficiency of each editing site was calculated using the following equation: Editing (%) = $U/(C + U) \times 100$. U represents the read number of SNPs that are different from DNAs in plastid mRNAs. C represents the read number of SNPs that are identical as DNAs in the same plastid editing site.

To experimentally validate the RNA editing sites derived from RNA-seq, the genomic and transcript regions surrounding these sites were amplified from another wild type biological replicate and *yl* mutant biological replicate using KOD FX High-Fidelity DNA polymerase (Toyobo, Osaka, Japan). The PCR products were sequenced and compared to identify SNP changes resulting from RNA editing. The RNA editing extent was estimated by the relative heights of the peaks of the nucleotide in the sequence analyzed. The primer sequences are listed in Table S3.

## 5. Conclusions

In this study, we characterized one *yellow leaf* mutant and identified *YL*, a soybean homolog of *Arabidopsis ORRM1* that is involved in RNA editing. The *yl* mutant displayed yellow leaves, reduced chlorophyll contents, impaired photosynthesis and an altered chloroplast ultrastructure. Through fine mapping, the *yl* mutation was narrowed down to a 28-kb genomic region in chromosome 20 between markers S3 and S7-3. DNA sequencing revealed that the *YL* mutation in the *yl* mutant is a C to A transition in *Glyma.20G187000*, which causes an amino acid alteration. Further function analysis of *YL* uncovered that the point mutation in *yl* influenced the editing extent in 42% of the chloroplast sites surveyed. However, we speculate that only the decreased level of conversion of cytidine to uridine at *petB*-611 and *psbL*-2 may affect normal photosynthesis in *yl* leaves. Together, our findings indicate that soybean YL protein influences photosynthesis, possibly via its function in chloroplast RNA editing.

**Supplementary Materials:** Supplementary materials can be found at Table S1. Comparison of the extent of RNA editing of chloroplast sites between the wild type and *yl* mutant by RNA-seq. Table S2. Verification of the extent of chloroplast RNA editing by Sanger sequencing. Table S3. Primers used in this study.

**Author Contributions:** B.Z., G.Z. and Z.T. conceived and designed the experiments. X.Z. carried out the experiments and wrote the manuscript. Y.P., Z.L., D.Z. and Z.D. assisted with doing the experiments. Y.L. participated in the statistical analysis. B.Z. and G.Z. and Z.T. revised the manuscript. All authors have read

**Acknowledgments:** This work is dedicated to the memory of Baoge Zhu, who devoted his life to soybean breeding. We would like to thank Jingnan Liang (Institute of Microbiology, Chinese Academy of Sciences) for assistance with ultrathin sectioning and transmission electron microscopy.

## References

1. Stirbet, A.; Lazar, D.; Guo, Y.; Govindjee, G. Photosynthesis: Basics, history, and modeling. *Ann. Bot.* **2019**, *XX*, 1–27. [CrossRef] [PubMed]
2. Johnson, M.P. Photosynthesis. *Essays Biochem.* **2016**, *60*, 255–273. [CrossRef] [PubMed]
3. Von Wettstein, D.; Gough, S.; Kannangara, C.G. Chlorophyll biosynthesis. *Plant Cell* **1995**, *7*, 1039–1047. [CrossRef] [PubMed]
4. Sandhu, D.; Coleman, Z.; Atkinson, T.; Rai, K.M.; Mendu, V. Genetics and physiology of the nuclearly inherited yellow foliar mutants in soybean. *Front. Plant Sci.* **2018**, *9*, 471. [CrossRef] [PubMed]
5. Kato, K.K.; Palmer, R.G. Duplicate chlorophyll-deficient loci in soybean. *Genome* **2014**, *47*, 190–198. [CrossRef] [PubMed]
6. Campbell, B.W.; Mani, D.; Curtin, S.J.; Slattery, R.A.; Michno, J.M.; Ort, D.R.; Schaus, P.J.; Palmer, R.G.; Orf, J.H.; Stupar, R.M. Identical substitutions in magnesium chelatase paralogs result in chlorophyll-deficient soybean mutants. *G3 Genes Genom. Genet.* **2014**, *5*, 123–131. [CrossRef] [PubMed]
7. Reed, S.; Atkinson, T.; Gorecki, C.; Espinosa, K.; Przybylski, S.; Goggi, A.S.; Palmer, R.G.; Sandhu, D. Candidate gene identification for a lethal chlorophyll-deficient mutant in soybean. *Agronomy* **2014**, *4*, 462–469. [CrossRef]

8. Sandhu, D.; Atkinson, T.; Noll, A.; Johnson, C.; Espinosa, K.; Boelter, J.; Abel, S.; Dhatt, B.K.; Barta, T.; Singsaas, E.; et al. Soybean proteins GmTic110 and GmPsbP are crucial for chloroplast development and function. *Plant Sci.* **2016**, *252*, 76–87. [CrossRef]
9. Li, Q.; Fang, C.; Duan, Z.B.; Liu, Y.C.; Qin, H.; Zhang, J.X.; Sun, P.; Li, W.B.; Wang, G.D.; Tian, Z.X. Functional conservation and divergence of *GmCHLI* genes in polyploid soybean. *Plant J.* **2016**, *88*, 584–596. [CrossRef]
10. Weiss, M.G. Genetic linkage in soybeans: Linkage group I. *Crop Sci.* **1970**, *10*, 69–72. [CrossRef]
11. Weiss, M.G. Genetic linkage in soybeans: Linkage group VII. *Crop Sci.* **1970**, *10*, 627–629. [CrossRef]
12. Palmer, R.G.; Nelson, R.L.; Bernard, R.L.; Stelly, D.M. Genetics and linkage of three chlorophyll-deficient mutants in soybean: *y19*, *y22*, and *y23*. *J. Hered.* **1990**, *81*, 404–406.
13. Zou, J.J.; Singh, R.J.; Hymowitz, T. Association of the *yellow leaf* (*y10*) mutant to soybean chromosome 3. *J. Hered.* **2003**, *94*, 352–355. [CrossRef] [PubMed]
14. Palmer, R.G.; Xu, M. Positioning 3 qualitative trait loci on soybean molecular linkage group E. *J. Hered.* **2008**, *99*, 674–678. [CrossRef] [PubMed]
15. Zhang, H.; Zhang, D.; Han, S.; Zhang, X.; Yu, D. Identification and gene mapping of a soybean chlorophyll-deficient mutant. *Plant Breed.* **2011**, *130*, 133–138. [CrossRef]
16. Sandhu, D.; Ghosh, J.; Johnson, C.; Baumbach, J.; Baumert, E.; Cina, T.; Grant, D.; Palmer, R.G.; Bhattacharyya, M.K. The endogenous transposable element *Tgm9* is suitable for generating knockout mutants for functional analyses of soybean genes and genetic improvement in soybean. *PLoS ONE* **2017**, *12*, e0180732. [CrossRef]
17. Takenaka, M.; Zehrmann, A.; Verbitskiy, D.; Hartel, B.; Brennicke, A. RNA editing in plants and its evolution. *Annu. Rev. Genet.* **2013**, *47*, 335–352. [CrossRef]
18. Ichinose, M.; Sugita, M. RNA editing and its molecular mechanism in plant organelles. *Genes* **2016**, *8*, 5. [CrossRef]
19. Sun, T.; Bentolila, S.; Hanson, M.R. The unexpected diversity of plant organelle RNA editosomes. *Trends Plant Sci.* **2016**, *21*, 962–973. [CrossRef]
20. Sun, T.; Germain, A.; Giloteaux, L.; Hammani, K.; Barkan, A.; Hanson, M.R.; Bentolila, S. An RNA recognition motif-containing protein is required for plastid RNA editing in *Arabidopsis* and maize. *Proc. Natl. Acad. Sci. USA* **2013**, *110*, E1169–E1178. [CrossRef] [PubMed]
21. Hackett, J.B.; Shi, X.; Kobylarz, A.T.; Lucas, M.K.; Wessendorf, R.L.; Hines, K.M.; Bentolila, S.; Hanson, M.R.; Lu, Y. An organelle RNA recognition motif protein is required for photosystem II subunit *psbF* transcript editing. *Plant Physiol.* **2017**, *173*, 2278–2293. [PubMed]
22. Shi, X.; Hanson, M.R.; Bentolila, S. Two RNA recognition motif-containing proteins are plant mitochondrial editing factors. *Nucleic Acids Res.* **2015**, *43*, 3814–3825. [CrossRef] [PubMed]
23. Shi, X.; Germain, A.; Hanson, M.R.; Bentolila, S. RNA recognition motif-containing protein ORRM4 broadly affects mitochondrial RNA editing and impacts plant development and flowering. *Plant Physiol.* **2016**, *170*, 294–309. [CrossRef] [PubMed]
24. Shi, X.; Castandet, B.; Germain, A.; Hanson, M.R.; Bentolila, S. ORRM5, an RNA recognition motif-containing protein, has a unique effect on mitochondrial RNA editing. *J. Exp. Bot.* **2017**, *68*, 2833–2847. [CrossRef] [PubMed]
25. Bentolila, S.; Heller, W.P.; Sun, T.; Babina, A.M.; Friso, G.; van Wijk, K.J.; Hanson, M.R. RIP1, a member of an *Arabidopsis* protein family, interacts with the protein RARE1 and broadly affects RNA editing. *Proc. Natl. Acad. Sci. USA* **2012**, *109*, E1453–E1461. [CrossRef] [PubMed]
26. Takenaka, M.; Zehrmann, A.; Verbitskiy, D.; Kugelmann, M.; Hartel, B.; Brennicke, A. Multiple organellar RNA editing factor (MORF) family proteins are required for RNA editing in mitochondria and plastids of plants. *Proc. Natl. Acad. Sci. USA* **2012**, *109*, 5104–5109. [CrossRef]
27. Barkan, A.; Small, I. Pentatricopeptide repeat proteins in plants. *Annu. Rev. Plant Biol.* **2014**, *65*, 415–442. [CrossRef]
28. Zhang, F.; Tang, W.J.; Hedtke, B.; Zhong, L.L.; Liu, L.; Peng, L.W.; Lu, C.M.; Grimm, B.; Lin, R.C. Tetrapyrrole biosynthetic enzyme protoporphyrinogen IX oxidase 1 is required for plastid RNA editing. *Proc. Natl. Acad. Sci. USA* **2014**, *111*, 2023–2028. [CrossRef]
29. Sun, T.; Shi, X.; Friso, G.; Van Wijk, K.; Bentolila, S.; Hanson, M.R. A zinc finger motif-containing protein is essential for chloroplast RNA editing. *PLoS Genet.* **2015**, *11*, e1005028. [CrossRef]

30. Zhao, X.B.; Huang, J.Y.; Chory, J. GUN1 interacts with MORF2 to regulate plastid RNA editing during retrograde signaling. *Proc. Natl. Acad. Sci. USA* **2019**, *116*, 10162–10167. [CrossRef]
31. Phytozome. Available online: www.phytozome.net (accessed on 14 June 2020).
32. Emanuelsson, O.; Nielsen, H.; Brunak, S.; von Heijne, G. Predicting subcellular localization of proteins based on their N-terminal amino acid sequence. *J. Mol. Biol.* **2000**, *300*, 1005–1016. [CrossRef] [PubMed]
33. Small, I.; Peeters, N.; Legeai, F.; Lurin, C. Predotar: A tool for rapidly screening proteomes for N-terminal targeting sequences. *Proteomics* **2004**, *4*, 1581–1590. [CrossRef] [PubMed]
34. Shikanai, T.; Shimizu, K.; Ueda, K.; Nishimura, Y.; Kuroiwa, T.; Hashimoto, T. The chloroplast *clpP* gene, encoding a proteolytic subunit of ATP-dependent protease, is indispensable for chloroplast development in tobacco. *Plant Cell Physiol.* **2001**, *42*, 264–273. [CrossRef] [PubMed]
35. Burrows, P.A.; Sazanov, L.A.; Svab, Z.; Maliga, P.; Nixon, P.J. Identification of a functional respiratory complex in chloroplasts through analysis of tobacco mutants containing disrupted plastid *ndh* genes. *EMBO J.* **1998**, *17*, 868–876. [CrossRef]
36. Voelker, R.; Barkan, A. Nuclear genes required for post-translational steps in the biogenesis of the chloroplast cytochrome $b_6f$ complex in maize. *Mol. Gen. Genet.* **1995**, *249*, 507–514. [CrossRef]
37. Zito, F.; Kuras, R.; Choquet, Y.; Kossel, H.; Wollman, F.A. Mutations of *cytochrome* $b_6$ in *Chlamydomonas reinhardtii* disclose the functional significance for a proline to leucine conversion by *petB* editing in maize and tobacco. *Plant Mol. Biol.* **1997**, *33*, 79–86. [CrossRef]
38. Monde, R.A.; Zito, F.; Olive, J.; Wollman, F.A.; Stern, D.B. Post-transcriptional defects in tobacco chloroplast mutants lacking the cytochrome $b_6/f$ complex. *Plant J.* **2000**, *21*, 61–72. [CrossRef]
39. Szczepaniak, A.; Black, M.T.; Cramer, W.A. Topography of the chloroplast cytochrome $b_6$: Orientation of the cytochrome and accessibility of the lumen-side interhelix loops. *Z. Naturforsch. C J. Biosci.* **1989**, *44*, 453–461. [CrossRef]
40. Hankamer, B.; Morris, E.; Nield, J.; Carne, A.; Barber, J. Subunit positioning and transmembrane helix organisation in the core dimer of photosystem II. *FEBS Lett.* **2001**, *504*, 142–151. [CrossRef]
41. Swiatek, M.; Regel, R.E.; Meurer, J.; Wanner, G.; Pakrasi, H.B.; Ohad, I.; Herrmann, R.G. Effects of selective inactivation of individual genes for low-molecular-mass subunits on the assembly of photosystem II, as revealed by chloroplast transformation: The *psbEFLJ* operon in Nicotiana tabacum. *Mol. Genet. Genom.* **2003**, *268*, 699–710. [CrossRef]
42. Suorsa, M.; Regel, R.E.; Paakkarinen, V.; Battchikova, N.; Herrmann, R.G.; Aro, E.M. Protein assembly of photosystem II and accumulation of subcomplexes in the absence of low molecular mass subunits PsbL and PsbJ. *Eur. J. Biochem.* **2004**, *271*, 96–107. [CrossRef]
43. Krieger-Liszkay, A. Singlet oxygen production in photosynthesis. *J. Exp. Bot.* **2005**, *56*, 337–346. [CrossRef] [PubMed]
44. Eva-Mari, A.; Ivar, V.; Bertil, A. Photoinhibition of Photosystem II. Inactivation, protein damage and turnover. *Biochim. Biophys. Acta* **1993**, *1143*, 113–134.
45. Liu, M.; Wang, Y.; Nie, Z.; Gai, J.; Bhat Akhter, J.; Kong, J.; Zhao, T. Double mutation of two homologous genes *YL1* and *YL2* results in a leaf yellowing phenotype in soybean [*Glycine max* (L.) Merr]. *Plant Mol. Biol.* **2020**, online. [CrossRef] [PubMed]
46. Lichtenthaler, H.K. Chlorophylls and carotenoids: Pigments of photosynthetic biomembranes. *Method Enzymol.* **1987**, *148*, 350–382.
47. SoyBase. Available online: http://soybase.org/ (accessed on 14 March 2014).
48. FastQC. Available online: http://www.bioinformatics.babraham.ac.uk/projects/fastqc/ (accessed on 15 September 2015).
49. Glycine Max Cultivar PI 437654 Chloroplast, Complete Genome. Available online: ncbi.nlm.nih.gov/nuccore/DQ312375.1 (accessed on 12 January 2016).
50. Kim, D.; Langmead, B.; Salzberg, S.L. HISAT: A fast spliced aligner with low memory requirements. *Nat. Methods* **2015**, *12*, 357–360. [CrossRef] [PubMed]
51. Li, H.; Handsaker, B.; Wysoker, A.; Fennell, T.; Ruan, J.; Homer, N.; Marth, G.; Abecasis, G.; Durbin, R. 1000 Genome Project Data Processing Subgroup. The sequence alignment/map format and SAMtools. *Bioinformatics* **2009**, *25*, 2078–2079. [CrossRef]
52. McKenna, A.; Hanna, M.; Banks, E.; Sivachenko, A.; Cibulskis, K.; Kernytsky, A.; Garimella, K.; Altshuler, D.; Gabriel, S.; Daly, M.; et al. The Genome Analysis Toolkit: A MapReduce framework for analyzing next-generation DNA sequencing data. *Genome Res.* **2010**, *20*, 1297–1303. [CrossRef]

# 8

# Characterization of the Chloroplast Genome of *Trentepohlia odorata* (Trentepohliales, Chlorophyta) and Discussion of its Taxonomy

**Huan Zhu [1], Yuxin Hu [1], Feng Liu [2], Zhengyu Hu [3] and Guoxiang Liu [1,*]**

[1] Key Laboratory of Algal Biology, Institute of Hydrobiology, Chinese Academy of Sciences, Wuhan 430072, China; huanzhu@ihb.ac.cn (H.Z.); ssshyx@163.com (Y.H.)
[2] Key Laboratory of Marine Ecology and Environmental Sciences, Institute of Oceanology, Chinese Academy of Sciences, Qingdao 266071, China; liufeng@qdio.ac.cn
[3] State Key Laboratory of Freshwater Ecology and Biotechnology, Institute of Hydrobiology, Chinese Academy of Sciences, Wuhan 430072, China; huzy@ihb.ac.cn
* Correspondence: liugx@ihb.ac.cn

**Abstract:** Trentepohliales is an aerial order of Chlorophyta with approximately 80 species distributed mainly in tropical and subtropical regions. The taxonomy of this genus is quite difficult and presents a challenge for many phycologists. Although plentiful molecular data is available, most of the sequences are not identified at the species level. In the present study, we described a new specimen with detailed morphological data and identified it as *Trentepohlia odorata*. A phylogenetic analysis showed *T. odorata* as a novel lineage in Trentepohliales. *T. odorata* has the closest relationship with *T. annulata*, which is expected since sporangia of both species are without stalk cell and with dorsal pore. Species with such morphological characteristics may represent deep lineages in Trentepohliales. Although an increasing number of chloroplast genomes of Ulvophyceae have been reported in recent years, the whole plastome of Trentepohliales has not yet been reported. Thus, the chloroplast genome of *Trentepohlia odorata* was reported in the present study. The whole plastome was 399,372 bp in length, with 63 predicted protein-coding genes, 31 tRNAs, and 3 rRNAs. Additionally, we annotated 95 free-standing open reading frames, of which seven were annotated with plastid origins, 16 with eukaryotic genome origins, and 33 with bacterial genome origins. Four rpo genes (*rpoA*, *rpoB*, *rpoC1*, and *rpoC2*) were annotated within ORF clusters. These four genes were fragmented into several (partial) ORFs by in-frame stop codons. Additionally, we detected a frame shift mutation in the *rpoB* gene. The phylogenetic analysis supported that Trentepohliales clustered with Dasycladales and nested into the BDT clade (Bryopsidales, Dasycladales and Trentepohliales). Our results present the first whole chloroplast genome of a species of Trentepohliales and provided new data for understanding the evolution of the chloroplast genome in Ulvophyceae.

**Keywords:** chloroplast genome; free-standing ORFs; introns; phylogenetic analysis; taxonomic study; *Trentepohlia odorata*; Trentepohliales

---

## 1. Introduction

Algae in the order Trentepohliales are characterized by uniseriate and branched filaments. The order contains five genera and approximately 80 species that are primarily found in tropical and subtropical areas, and many of them form masses of a striking color (such as orange, yellow, and red) on tree trunks, walls, and stones [1]. Since the recorded species are completely aerial or subaerial, Trentepohliales is a special group in Ulvophyceae, of which the other groups are mainly composed of seaweeds or freshwater algae. The order can be easily distinguished from other ulvophycean algae by its specific sporangiate-lateral. Other special morphological and ultrastructural features include the absence of pyrenoid in the net-like chloroplasts, plasmodesmata in the septa, and a phragmoplast similar to Streptophyta visible during cytokinesis [2]. As the type genus of Trentepohliaceae, *Trentepohlia* Martius is the earliest known and most studied. Several studies have confirmed that *Trentepohlia* are polyphyletic groups [3–6]. There are more than 46 valid *Trentepohlia* species recorded in AlgaeBase [1]. Among all the *Trentepohlia* species, *T. odorata* (F.H. Wiggers) Wittrock is a controversial species since several phycologists considered this species synonymous with *T. umbrina* or *T. iolithus*, whereas most phycologists treated *T. odorata* as valid separate species [7–10]. Moreover, the current molecular data (i.e., 18S rDNA, *rbcL* cpDNA) available has shown that several species in *Trentepohlia* are also polyphyletic, especially several widespread and common species, i.e., *T. arborum* and *T. umbrina* [5,6]. Additionally, the morphology and phylogenetic position of *Trentepohlia odorata* has rarely been reported.

Recent studies based on genome-wide data or multigene chloroplast data revealed that the Ulvophyceae is not monophyletic and was recovered as two or more distinct clades [11–13]. However, there are only several different chloroplast coding genes available in the database (i.e., GenBank), and no whole chloroplast genome of Trentepohliales has yet been reported. Chloroplasts in species of Ignatiales, Ulotrichales, Oltmannsiellopsidales, and Ulvales have been reported to share a similar structure with most Viridiplantae, of which the circular plastome was composed by two inverted repeat regions (IRa and IRb) and two single copy regions (LSC and SSC) [12,14–16]. The sequenced chloroplast genome of the BCDT clade (Bryopsidales, Cladophorales, Dasycladales and Trentepohliales) have unique plastome structures. For example, the chloroplast genome of a species of Cladophorales was reported to consist of 34 small hairpin chromosomes and lost many genes [17]. All available chloroplast genomes of Bryopsidales lack a large inverted repeat region [18,19]. In comparison with the relatives of Trentepohiales, there is little known about the evolution and chloroplast genome of the order. Thus, it is important to sequence the entire chloroplast genome of a species within Trentepohliales to increase our knowledge about the order.

During 2010–2016, we collected lots of Trentepohliacean specimens from China, as reported in previous study [5]. In the present study, we identified one corticolous specimen as *T. odorata* based on morphological evidence. Short- and long-read high-throughput sequencing data of this isolate were obtained and assembled. The aims of our study were: (1) to present the complete chloroplast genome of *Trentepohlia odorata*, (2) to study the taxonomy of *Trentepohlia odorata*, (3) to reconstruct phylogenetic relationship between *Trentepohlia odorata* and other species.

## 2. Results

### *2.1. Morphological Observation*

*Trentepohlia odorata* (F.H. Wiggers) Wittrock 1880

Description: The alga formed a dense mat over tree bark (Figure 1A,B). The thallus mainly consisted of abundant erect parts and poorly prostrate filaments. Most of the erect filaments were nearly parallel, consisting of 3–10 cells, rarely branched and with a tapering end (Figure 1C). Apical vegetative cells of erect filaments were substantially longer than basal vegetative cells (Figure 1C,D). Cells of erect filaments were approximately 7–16 μm in width and 14–21 μm in length, with a length/width ratio of about 1.1–1.9. Prostrate filaments were often branched and form compact patches, the cells of which were mostly globose to ellipsoid or with other irregular shapes (Figure 1D,E). The size of the vegetative cells was approximately 8–13 μm in width and 12–18 μm in length, with a length/width ratio about 1.0–1.5. Presumptive zoosporangia were apical, intercalary, or lateral and mainly produced on erected filaments that are globose, ellipsoid, and obovate (Figure 1C–E). Most intercalary zoosporangia were globes and smaller than apical and lateral ones. The lateral zoosporangia were often clustered and are mainly obovate with an obvious dorsal pore (Figure 1E). The intercalary zoosporangia were approximately 8–12 μm in width and 9–16 μm in length. The lateral and apical zoosporangia were approximately 10–24 μm in width and 18–28 μm in length.

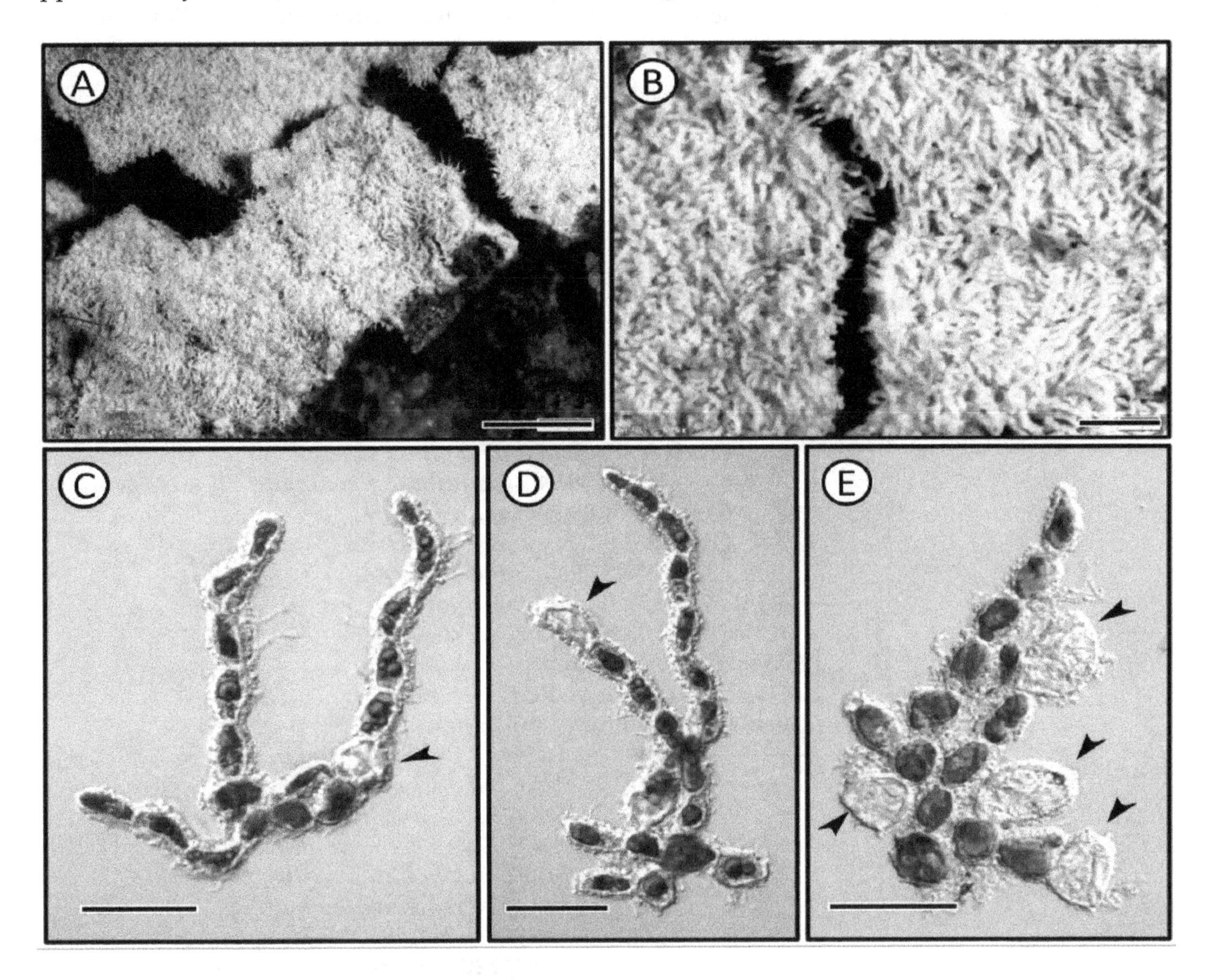


**Figure 1.** Morphology of *Trentepohlia odorata*. (**A**,**B**), steroscopic view of *Trentepohlia odorata* specimen (preserved in Formol acetic alcohol solution) on surface of *Fagus longipetiolata*. (**C**), microscopic view of heterotrichous thallus and its intercalary zoosporangium (arrow). (**D**), the apical zoosporangium with a dorsal pore (arrow). (**E**), the lateral zoosporangia with dorsal pore in cluster (arrows) at the basal part of thallus. Scale bars, 1 mm in (**A**), 200 μm in (**B**), 30 μm in (**C**–**E**).

This specimen was collected from a tropical botanical garden and formed abundant red growth on the surface of *Fagus longipetiolata* trunk. The morphology of specimen DZ1317 was completely consistent with primary description by Wittrock (1880) and Printz (1939) [7,10]. Despite being unable to observe the holotype of *Trentepohlia odorata* from its type locality (Fiona, Denmark), there is little morphological difference between our specimen and the holotype according to its primary description.

### 2.2. Phylogenetic Analysis

A phylogenetic analysis based on 73 sequences recovered from species in the Trentepohliales consisted of two main clades. One main clade was composed of several small clades including the *Cephaleuros* clade, *T. aurea* clade, *T. arborum*, and *Printzina lagenifera* clade. The other main clade was composed by *Phycopeltis*, *Printzina bosseae*, *T. umbrina*, *T. iolithus*, and *T. annulata* (Figure S1). According to the 18S rDNA phylogeny, *T. odorata* had the closest relationship with *T. annulata* (KM020077) and

*T.* cf. *umbrina* (KX586916) (Figure 2). According to the phylogeny based upon *rbc*L matrix, *T. odorata* clustered with Trentepohliales sp. (GU549443) and formed a robust clade with *T. annulata* (MH940266) and *T. abietina* (MH940276) (Figure 3). Such topology was consistent with previous studies. All of the analyses recovered *T. odorata* as a well resolved lineage.

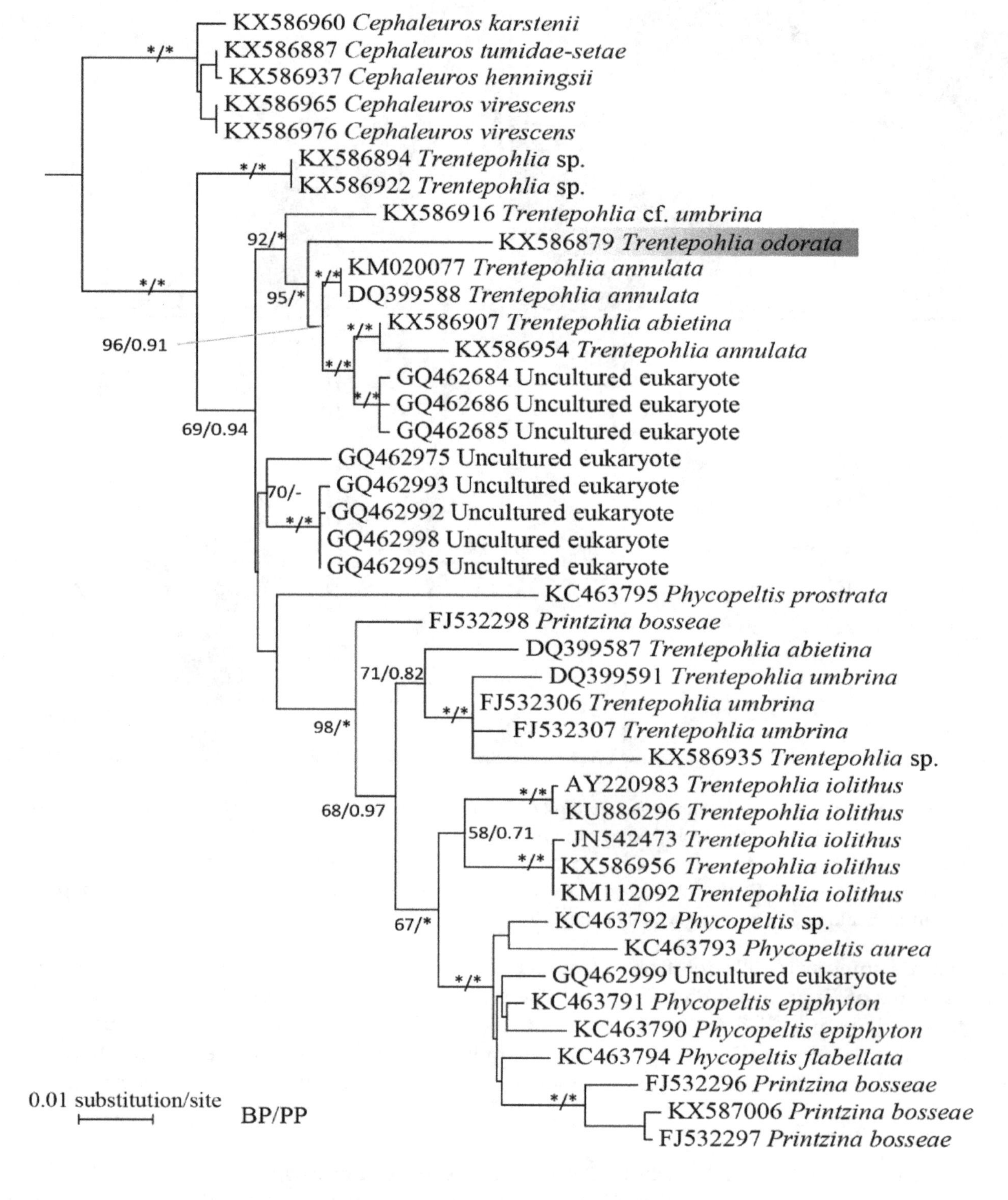


**Figure 2.** The maximum likelihood phylogram inferred from 18S rDNA sequences of Trentepohliales. The monophyletic genus *Cephaleuros* was used as outgroup. Maximum likelihood bootstrap values (1000 replicates) and Bayesian posterior probabilities (PPs) are given near the nodes, and only bootstrap proportion (BP) and PP values above 50 and 0.50 are shown. The full statistical support (100/1.00) is marked with an asterisk. The *Trentepohlia odorata* was shaded in grey.

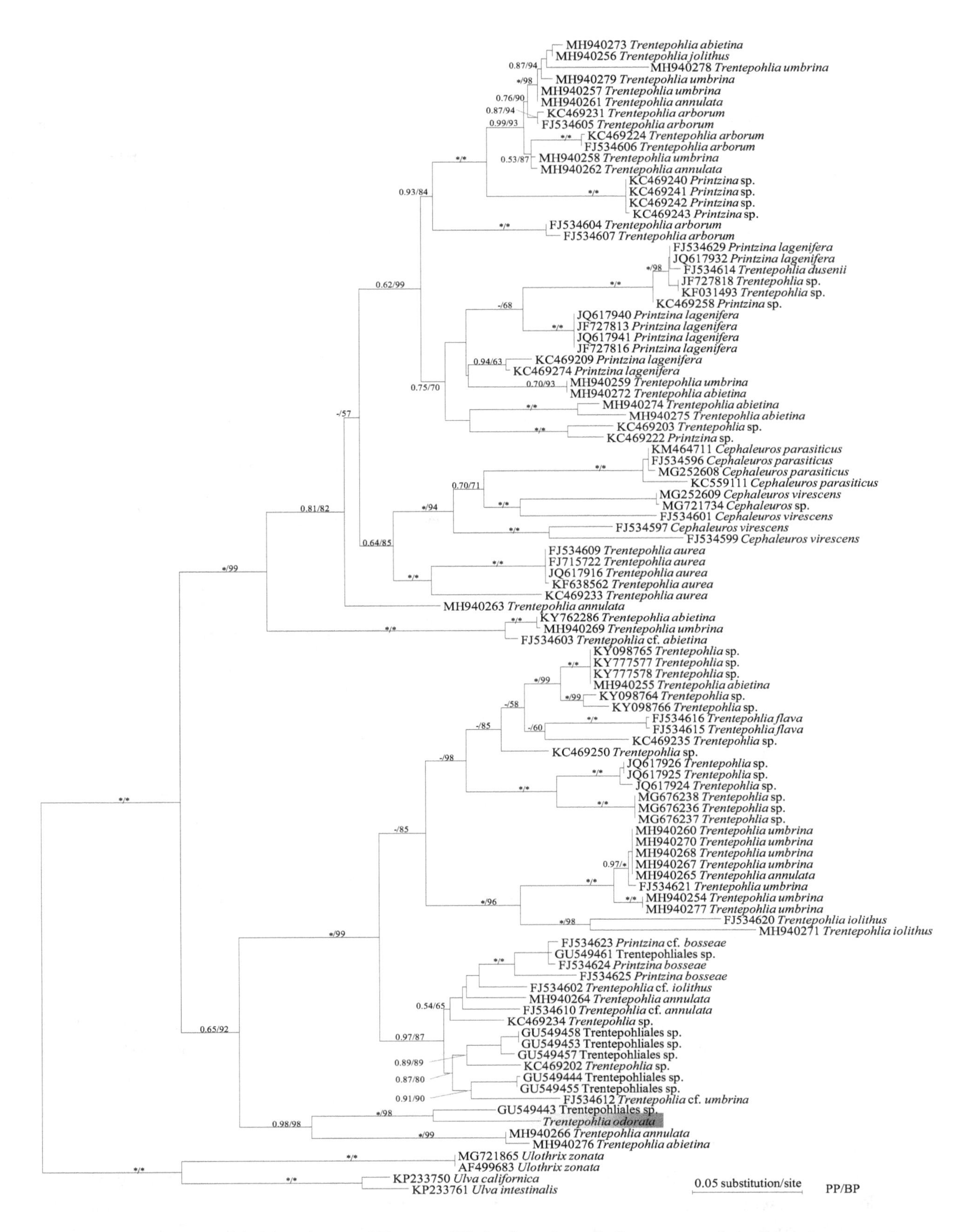


**Figure 3.** Maximum likelihood tree of Trentepohliales based on rbcL sequence data. Bayesian posterior probability (pp $\geq$ 0.50) and maximum likelihood (ML $\geq$ 50) bootstrap values are shown near the branches. The asterisks represent full statistical support (1.00/100). The *Trentepohlia odorata* was shaded in grey.

Our phylogenetic analysis based on chloroplast genome showed that both Trebouxiophyceae and Ulvophyceae were paraphyletic due to Chlorellaceae and the Bryopsidales, Trentepohliales, and Dasycladales clade (BDT clade) did not fall into core Trebouxiophyceae and core Ulvophyceae, respectively (Figure 4). Both maximum likelihood analysis and Bayesian analysis strongly supported that Trentepohliales clustered with Dasycladales. In the maximum likelihood analysis, the BDT clade was supported by a bootstrap value of 60; however, such topology did not occur in the Bayesian inference analysis (Figure S2). Both maximum likelihood analysis and Bayesian inference supported that Trentepohliales may have the closest relationship with Dasycladales.

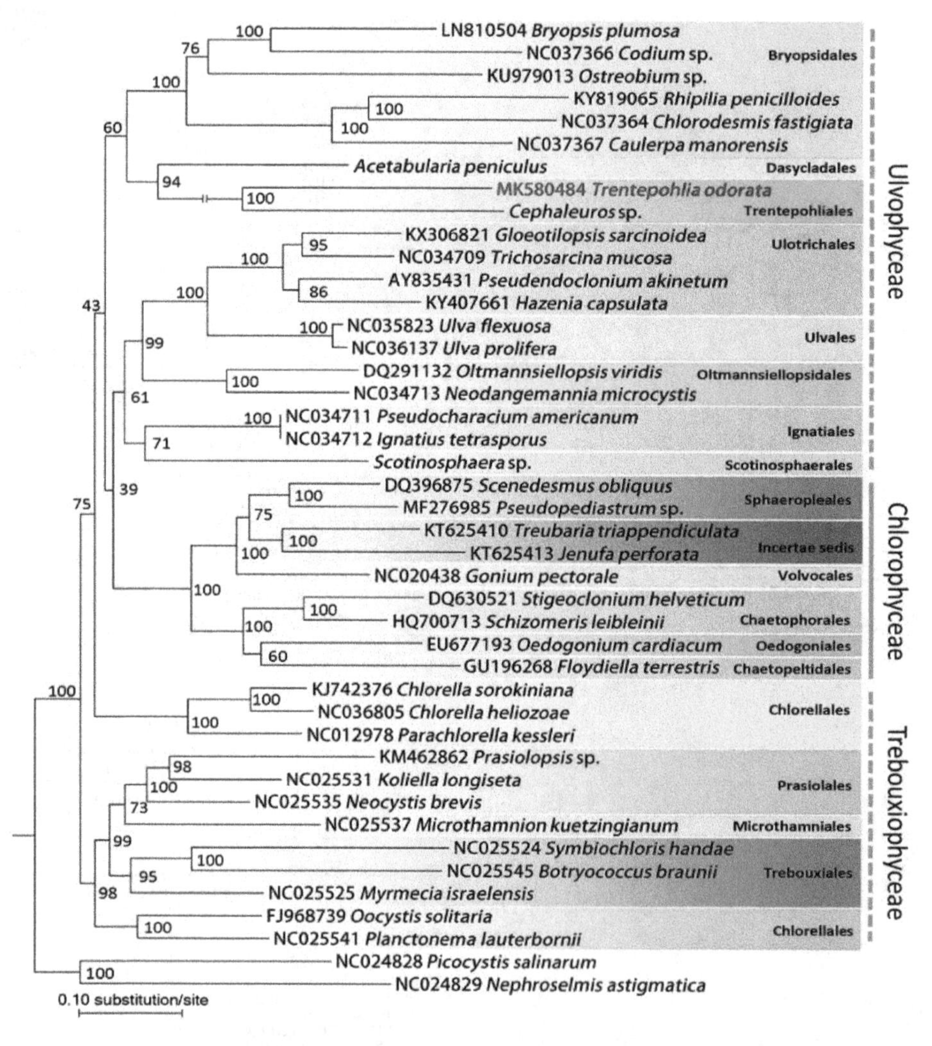


**Figure 4.** Maximum likelihood tree inferred from a dataset consisted of 31 common cpDNA-encoded genes of 43 core Chlorophytes. The genera *Picocystis* (NC024828) and *Nephroselmis* (NC024829) were selected as outgroup. Maximum likelihood bootstrap values (1000 replicates) are given near the nodes. Taxonomic arrangement follows AlgaeBase, and *Trentepohlia odorata* is in red.

### *2.3. Chloroplast Genome Analysis*

We obtained 21 contigs with a total of 355,893 bp and one circular molecule with a length 399,372 bp from short-read sequencing data and long-read sequencing data, respectively. The dot plot showed a high-level congruence between the 21 contigs and one circular molecule (Figure S3), indicating that the plastome of *T. odorata* was circular and had a length of up to 399,372 bp. Our annotation results showed that the plastome possesses the typical quadripartite structure. Two inverted repeat regions were 26,700 bp and 26,778 bp, respectively. The large single copy region (LSC) was 178,629 bp, while the small single copy region (SSC) was 167,265 bp (Figure 5). The overall G + C content of the circular cpDNA was calculated to be 29.75%. The analysis revealed that the cpDNA encodes 97 genes (Table 1).

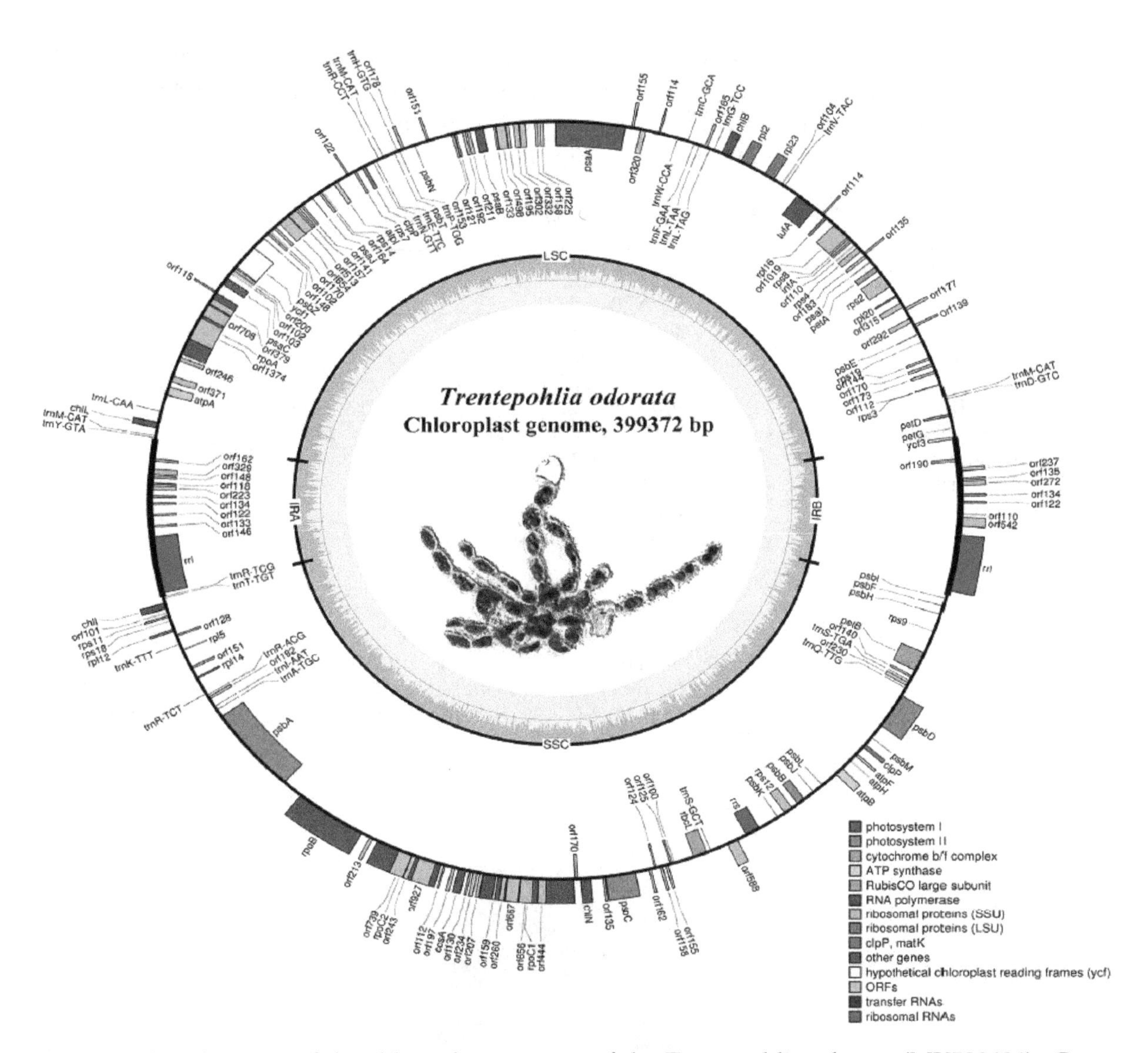


**Figure 5.** Circular map of the chloroplast genome of the Trentepohlia odorata (MK580484). Genes are color coded according to the functional categories listed in the index below the map. The GC content and inverted repeats (IRA and IRB) which separate the genome into two single copy regions are indicated on the inner circle. Genes on the inside of the outside circle are transcribed in a clockwise direction; those on the outside of the map are transcribed counterclockwise.

The genes were grouped into two major categories, coding genes and non-coding genes. The coding genes consisted of 63 predicted protein-coding genes, including five *atp* genes, four *chl* genes, four *pet* genes, five *psa* genes, 15 *psb* genes, seven *rpl* genes, four *rpo* genes, 11 *rps* genes, two *ycf* genes, and six other genes (*ccsA*, two *clpP*, *infA*, *rbcL*, and *tufA*). The non-coding gene category included 31 tRNAs and three rRNAs (Table 1). We annotated a total of 49 introns, of which 39 group I introns were present in eight genes (*rrl*-IRa (8), *rrl*-IRb (8), *rrs* (2), *psaA* (2), *psbA* (3), *psbC* (5), *psbD* (3), *petB* (4), and *rbcL* (2)), nine group II introns in eight genes (*rpl*2, *rps*12, *psaA* (3), *psbA*, *psbD*, *psaC*, *petB*, and *tufA*), and one unidentified type intron (*rpoB*). The IR regions contained only two of the same rRNA gene (*rrl*) and no other genes. Additionally, we detected that the *rrs* gene was not located in IR regions but had only one copy in the SSC region. The *ycf3* gene was located across LSC region and IRb region; however, there was only a partial sequence of this gene detected in the IRa region. We annotated two copies of clpP in the LSC region and SSC region, respectively. Additionally, there were 95 free-standing ORFs (length >100 aa) annotated in the intergenic region, with a total length up to 72,720 bp (18.21%) (Figure 6). Among the free-standing ORFs, seven were annotated with plastid origins (POP), 16 with eukaryotic genome origins (EOP), and 33 with bacterial genome origins (BOP) (Figure 7). Four genes (*rpoA*, *rpoB*, *rpoC1*, and *rpoC2*) were annotated within four ORF clusters (including partial ORFs). All four genes were fragmented into several ORFs by in-frame stop codons (Figure 8, red asterisks). There were three fragments annotated in *rpoC1*, three in *rpoC2*, eight as *rpoB*, and two as *rpoA*. In the *rpoB* ORF cluster, we detected a frame shift mutation (Figure 8, arrow).

**Table 1.** Genes encoded by *Trentepohlia odorata* chloroplast genome.

| Gene Products | Genes |
|---|---|
| ATP synthase | *atp*A, B, F, H, I |
| Chlorophyll biosynthesis | *chl*B, I, L, N |
| Cytochrome b6/f | *pet*A, B, D, G |
| Photosystem I | *psa*A, B, C, I, J |
| Photosystem II | *psb*A, B, C, D, E, F, H, I, J, K, L, M, N, T, Z |
| Large subunit ribosomal proteins | *rpl*2, 5, 12, 14, 16, 20, 23 |
| RNA polymerase | *rpo*A, B, C1, C2 |
| Small subunit ribosomal proteins | *rps*2, 3, 4, 7, 8, 9, 11, 12, 14, 18, 19 |
| Unknown function proteins | *ycf*1, 3 |
| Other proteins | *clpp* (2 copies), *ccs*A |
| Rubisco | *rbc*L |
| Translation factors | *tuf*A, *inf*A |
| Ribosomal RNAs | *rrl* (2 copies), *rrs* |
| Transfer RNAs | *trn*A(TGC), C(GCA), D(GTC), E(TTC), F(GAA), G(TCC), H(GTG), I(AAT), K(TTT), L(CAA), L(TAA), L(TAG), M(CAT) (5 copies), N(GTT), P(TGG), Q(TTG), R(ACG), R(CCT), R(TCG), R(TCT), S(GCT), S(TGA), T(TGT), V(TAC), W(CCA), Y(GTA) |

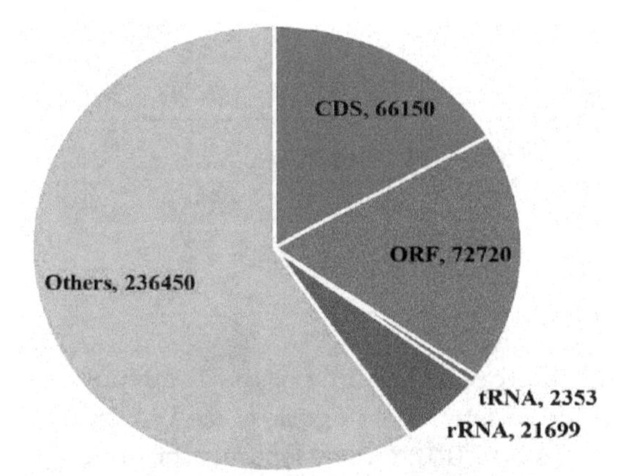


**Figure 6.** The size of CDS (protein coding regions), ORF (open reading frames, >100 aa), tRNA, rRNA and other regions in *Trentepohlia odorata* chloroplast genome. The number represents its size (bp).

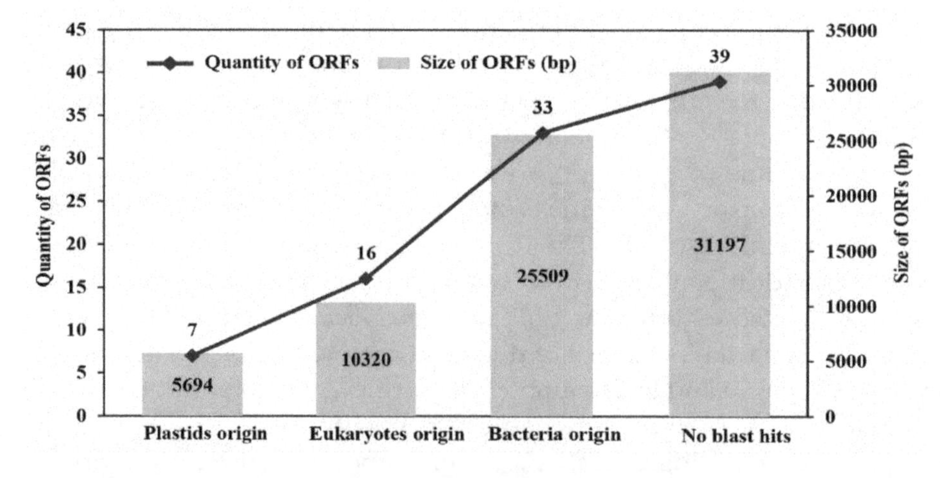


**Figure 7.** Blast result of free-standing ORFs in *Trentepohlia odorata* chloroplast genome.

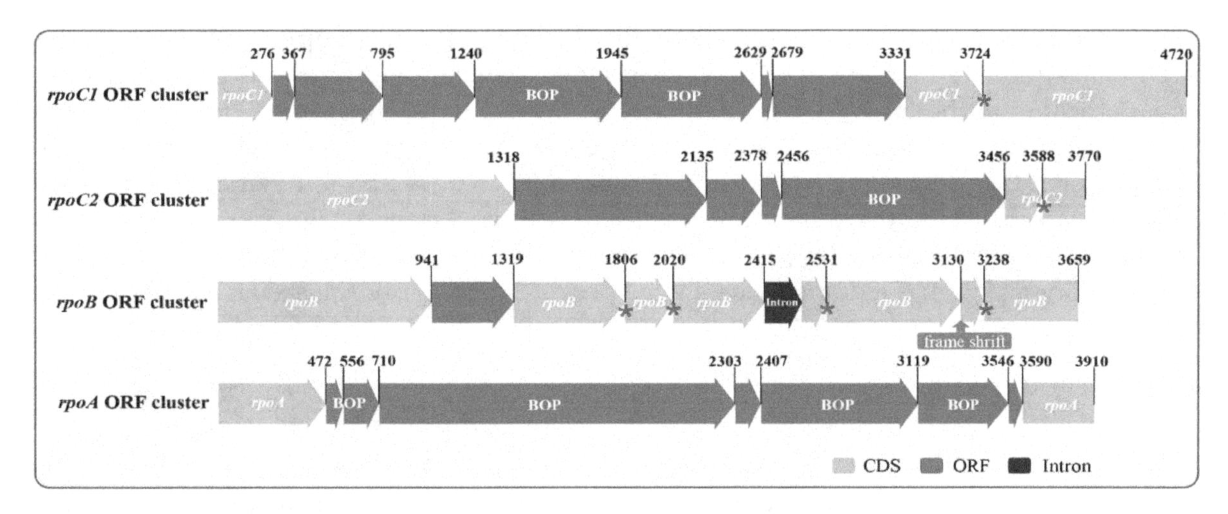


**Figure 8.** The fragmentation of *rpo* gene cluster. The red asterisk represents a detected in-frame stop codon; the shift mutation was labelled in a blue arrow. The BOP, EOP and POP represent ORF may have a bacteria, eukaryotic nucleus, and chloroplast origin respectively. ORF without BOP, EOP or POP label means no blast hit.

## 3. Discussion

The taxonomic controversy related to *T. odorata* primarily focused on whether or not the species was synonymous with *T. umbrina* or *T. iolithus* [7,8]. The main differences between *T. iolithus* and *T. odorata* was their substratum. *Trentepohlia iolithus* was only found on exposed stones or concrete, and *T. odorata* was found on tree bark [7,9,20]. A previous study reported *T. odorata* on other substratum, such as concrete, which was not consistent with the original description [21]. Although *T. odorata* has a very similar vegetative morphology with *T. iolithus* var. *yajiagengensis*, the morphology of sporangia and the phylogenetic position suggests that they are different species [22]. *Trentepohlia umbrina* is a paraphyletic species as sequences from *T. umbrina* clustered into several small clades in many studies. There was an obvious morphological difference between the two species. According to the original description, the thallus of *T. odorata* is heterotrichous and the thallus of *T. umbrina* is prostrate. The vegetative cells of *T. odorata* have a greater length/width ratio than that of *T. umbrina*. In the present study, our observation was consistent with the original description and the Printz description, thus we support Printz in that *T. odorata* is a morphologically distinct species, rather than Hariot [7,10]. The phylogenetic result shows that *Trentepohlia odorata* has the closest relationship with *Trentepohlia annulata*. One possible explanation for their close relationship is that both algae seem not to possess sporangiate-lateral. *Trentepohlia* species with lateral or intercalary sporangia and dorsal pore sporangia may represent several deep lineages in Trentepohliales. Additionally, there are few images regarding *Trentepohlia odorata* in previous studies, and our study provided new morphological evidence to compare those *Trentepohlia* species.

Although a considerable number of published plastomes are available, there are many gaps in Chlorophytes plastomes, especially in several orders of Ulvophyceae [12]. A recent study reported that chloroplast genomes in Cladophorales are fragmented into many small hairpin chromosomes [17]. The chloroplast genome in Bryopsidales are circular but lack a large inverted repeat [18]. Our study reported the first whole plastome of Trentepohliales, with a size up to 399,372 bp, which is the largest currently identified within Ulvophyceae. The plastome of *Trentepohlia odorata* presented a quadripartite structure, which differs from its close relatives, Cladophorales and Bryopsidales. We found several free-standing ORFs of bacterial origin and fragmentation by in-frame stop-codons in *rpoA*, *rpoB*, *rpoC1*, and *rpoC2* genes, which is similar with Bryopsidales [18]. The *rrf* gene was not detected in this study. We detected that the *rrl* gene located in IR region, and the *rrs* gene located in the SSC region. Fragmentation by introns were found in the *rrl* gene, with eight group-I introns. Similar cases were also found in *Caulerpa manorensis*, *Jenufa perforata*, *Schizomeris leibleinii*, and *Floydiella terrestris*, with seven, five, seven, and eight introns, respectively [18,19,23,24]. Two copies of the *clpP* gene were annotated. This gene duplication is very common in the nuclear genome and is usually caused by two repeats located at the two sides of the genes in the organellar genome. However, we did not detect repeat sequences at the sides of the two *clpP* genes. Our phylogenetic analysis using chloroplast genomes indicated that both Ulvophyceae and Trebouxiophyceae are paraphyletic, which is consistent with previous studies [11,13]. Trentepohliales clustered with Dasycladales in present study, which is also reported by previous studies [11,13]. However, we cannot rule out that Trentepohliales have a closest relationship with Cladophorales since species in Cladophorales was not included in our phylogenomic analysis.

## 4. Materials and Methods

### 4.1. Morphology and Cultivation

The specimen of *T. odorata* was derived from samples collected from the tree bark of *Fagus longipetiolata* located in the Hainan Tropical Botany Garden (19°30′49″ N, 109°30′14″ E) in December 2013, with the voucher number DZ1317. The specimen was primarily examined under a stereoscope microscope (Zeiss model KL1500 LCD; Carl Zeiss, Göttingen, Germany). Filaments were isolated using forceps and dissecting needles. We selected clean and complete filaments, placed them in culture dishes, and then spread the branches while using a stereoscope. Morphology was observed under differential interference contrast microscopy using a Leica DM5000B microscope (Leica Microsystems, Wetzlar, Germany). Micrographs were captured using a Leica DFC320 digital camera. Voucher specimens were deposited in the Freshwater Algal Herbarium (IHB), Institute of Hydrobiology, Chinese Academy of Sciences. The unialgal strain was aerially cultivated in a photo-reactor described by Chen et al. [25].

### 4.2. DNA Extraction, PCR, and Phylogenetic Reconstruction

The genomic DNA was extracted using a Axygen Universal DNA Isolation Kit (Axygen, Suzhou, China). Partial 18S rDNA was obtained as described in Zhu et al. (2017). The PCR products were purified and then sent to Tsingke Biotech Company, Inc. (Wuhan, China) for sequencing. Additional 18S rDNA and *rbcL* sequences from Trentepohliales species were downloaded from GenBank for analyses. Sequence matrices for phylogenetic analysis were initially aligned with MAFFT 7.0 and refined manually using Seaview [26,27]. ModelFinder was utilized to select the best-fitting evolutionary models for each marker according to Bayesian information criterion calculations [28]. IQ-TREE and MrBayes3.2 were used to infer the phylogeny [29,30]. We performed two phylogenetic analyses using the 18S rDNA matrix. First, Cladophorales was used as the outgroup to infer the topology of Trentepohliales. Based upon the results of the first phylogenetic analysis, we selected *Cephaleuros* as the outgroup to perform the second analysis. In the phylogenetic analysis using the *rbc*L matrix, Ulotricales was selected as the outgroup.

### 4.3. Chloroplast Genome Assembly and Annotation

A paired-end Illumina sequencing library was prepared from total DNA using the NEBNext Ultra DNA Library Prep Kit (E7370S). The libraries were sequenced using an Illumina NovaSeq 6000 with 150 bp insertion fragments (Illumina, San Diego, CA, USA). High-throughput sequencing data was sequentially analyzed by SOAPnuke v1.3.0 and SPAdes v3.10.0 [31,32]. A 1D genomic DNA by ligation (SQK-LSK108) kit was used to construct a long-reads library according to the manufacturer's instructions. The prepared library was loaded on Oxford Nanopore GridION X5 platform and sequenced. Only reads with mean scores >7 were retained. Long-read sequencing data were assembled using Canu v.1.6.0 [33]. The contigs were used to screen the chloroplast genome using the Blast program [34]. The selected chloroplast genome contig was assembled using Sequencher 4.10. Following this, we used Geneious 8.1 to map all the reads to the spliced genome sequence to verify that the contig was concatenated [35]. Finally, we obtained 21 chloroplast contigs from short-read sequencing data and one circular molecule from long-read sequencing data. A multicollinear dot plot was performed using Genome Pair Rapid Dotter v1.40 to detect the homology between 21 contigs and one circular molecule [36].

The chloroplast genome was primarily annotated using the online program DOGMA (Wyman et al., 2004) [37] and MAKER [38]. All open reading frames (ORFs) (with length >300 bp) were extracted by ORFfinder (https://www.ncbi.nlm.nih.gov/orffinder/), and then BLASTn and BLASTp (http://blast.ncbi.nlm.nih.gov/) were used with the e value set to 1e-10 to annotate the free-standing ORFs. Transfer RNA and ribosomal RNA genes were identified using tRNAscan-SE v1.23 and RNAmmer, respectively [39,40]. Intron boundaries were determined by modeling intron secondary structures and by comparing intron-containing genes with intronless homologs [41,42]. The graphical gene map was designed with Organellar Genome DRAW program (https://chlorobox.mpimp-golm.mpg.de/OGDraw.html) [43]. The annotated chloroplast genome was submitted to GenBank under the accession number MK580484.

We obtained a nucleoid dataset of 16,359 unambiguously aligned positions consisting of 31 common cpDNA-encoded genes of 43 Chlorophytes from Genbank (https://www.ncbi.nlm.nih.gov/genbank/), of which *Picocystis salinarum* and *Nephroselmis astigmatica* were used as the outgroup taxa. Most the genome accession numbers were presented before species name in Figure 8. Accession numbers of *Cephaleuros* sp., *Acetabularia peniculus* and *Scotinosphaera* sp. chloroplast genome sequences were MG721699-MG721754, MH545187-MH545222, and MG721898-MG721961 respectively. The data partition and best-fit models were selected using ModelFinder according to Bayesian inference criteria [28]. We used IQtree v1.7 and MrBayes 3.2 to perform maximum-likelihood analysis and Bayesian inference, respectively [29,30]. Additionally, because genes in the Cladophorales plastome were unique and there were fewer genes than the chloroplast genomes of other green algae, we did not include it in our analysis.

## 5. Conclusions

The morphological observation and phylogenetic analysis recovered *Trentepohlia odorata* as a separate species. According to phylogenetic results, *Trentepohlia odorata* has the closest relationship with *Trentepohlia annulata*, which could be explained by the shared morphological evidence, i.e., the absent sporangiate-lateral. The chloroplast genome of *Trentepohlia odorata* is 399,372 bp, with 97 genes and 95 free-standing ORFs. The typical quadripartite structure of *Trentepohlia odorata* is different from Cladophorales and Bryopsidales. However, the fragmention of *rpo* gene clusters by in-frame stop codons and the frame shift mutation detected in the present study is similar to Bryopsidales. Our study describes the first chloroplast genome of Trentepohliales and provides new data for understanding the evolution of the chloroplast genome in BCDT clade. Since Trentepohliles is an exclusively aerial order in core Chlorophytes, future study focused on plastomes of the other Trentepohliaceaen groups such as *Trentepohlia aurea*, *Trentepohlia bosseae* and *Cephaleuros* might have more interesting findings.

**Supplementary Materials:** Supplementary materials can be found at Figure S1. Maximum likelihood phylogram inferred from 18S rDNA sequences. Cladophorales was used as outgroup. Bootstrap values (1000 replicates) were given near the nodes. Figure S2. Bayesian tree inferred from a dataset consisted of 31 common cpDNA-encoded genes from 43 core Chlorophytes. Figure S3. The dot plot between the 21 contigs and one circular molecule.

**Author Contributions:** H.Z., G.L. and Z.H. conceived the study. H.Z., Y.H. and F.L. analyzed the data. H.Z. wrote the manuscript. All authors have read and approved the manuscript.

**Acknowledgments:** We are grateful to two anonymous reviewers, for their insightful and helpful comments and editorial corrections to improve our manuscript.

## References

1. Guiry, M.D.; Guiry, G.M. AlgaeBase. World-Wide Electronic Publication, National University of Ireland, Galway. 2019. Available online: http://www.algaebase.org (accessed on 5 March 2019).

2. López-Bautista, J.M.; Rindi, F.; Guiry, M.D. Molecular systematics of the subaerial green algal order Trentepohliales: An assessment based on morphological and molecular data. *Int. J. Syst. Evol. Microbiol.* **2006**, *56*, 1709–1715. [CrossRef]
3. Rindi, F.; Lam, D.W.; López-Bautista, J.M. Phylogenetic relationships and species circumscription in *Trentepohlia* and *Printzina* (Trentepohliales, Chlorophyta). *Mol. Phylogenetics Evol.* **2009**, *52*, 329–339. [CrossRef]
4. Nelsen, M.P.; Plata, E.R.; Andrew, C.J.; Lücking, R.; Lumbsch, H.T. Phylogenetic diversity of Trentepohlialean algae Associated with lichen-forming fungi. *J. Phycol.* **2011**, *47*, 282–290. [CrossRef] [PubMed]
5. Zhu, H.; Hu, Z.; Liu, G. Morphology and molecular phylogeny of Trentepohliales (Chlorophyta) from China. *Eur. J. Phycol.* **2017**, *52*, 330–341. [CrossRef]
6. Klimešová, M.; Rindi, F.; Škaloud, P. DNA cloning demonstrates high genetic heterogeneity in populations of the subaerial green alga *Trentepohlia* (Trentepohliales, Chlorophyta). *J. Phycol.* **2018**, *55*, 224–235. [CrossRef] [PubMed]
7. Hariot, M.P. Notes sur le genre Trentepohlia Martius (Suite). *J. Bot.* **1889**, *3*, 345–405.
8. Cribb, A.B. A revision of some species of *Trentepohlia* especially from Queensland. *Proc. R. Soc. Qld.* **1958**, *82*, 17–34.
9. Sarma, P. The freshwater Chaetophorales of New Zealand. *Nova Hedwigia* **1986**, *58*, 1–169.
10. Printz, H. Vorarbeiten zu einer Monographie der Trentepohliaceae. *Nytt Magazin für Naturvidenskaberne* **1939**, *80*, 137–210.
11. Fučíková, K.; Leliaert, F.; Cooper, E.D.; Škaloud, P.; Clerck, O.D.; Gurgel, C.F.D.; Lewis, L.A.; Lewis, P.O.; López-Bautistaet, J.M.; Delwiche, C.F.; et al. New phylogenetic hypotheses for the core Chlorophyta based on chloroplast sequence data. *Front. Ecol. Evol.* **2014**, *2*, 63.
12. Turmel, M.; Lemieux, C. Evolution of the plastid genome in green algae. *Adv. Bot. Res.* **2017**, *85*, 157–193.
13. Fang, L.; Leliaert, F.; Novis, P.M.; Zhang, Z.; Zhu, H.; Liu, G.; Penny, D.; Zhong, B. Improving phylogenetic inference of core Chlorophyta using chloroplast sequences with strong phylogenetic signals and heterogeneous models. *Mol. Phylogenetics Evol.* **2018**, *127*, 248–255. [CrossRef]
14. Pombert, J.F.; Lemieux, C.; Turmel, M. The complete chloroplast DNA sequence of the green alga *Oltmannsiellopsis viridis* reveals a distinctive quadripartite architecture in the chloroplast genome of early diverging ulvophytes. *BMC Biol.* **2006**, *4*, 3–4. [CrossRef]
15. Melton, J.T., III; Leliaert, F.; Tronholm, A.; López-Bautistaet, J.M. The complete chloroplast and mitochondrial genomes of the green macroalga *Ulva* sp. UNA00071828 (Ulvophyceae, Chlorophyta). *PLoS ONE* **2015**, *10*, e0121020. [CrossRef]
16. Turmel, M.; Otis, C.; Lemieux, C. Mitochondrion-to-chloroplast DNA transfers and intragenomic proliferation of chloroplast group II introns in *Gloeotilopsis* green algae (Ulotrichales, Ulvophyceae). *Genome Biol. Evol.* **2016**, *8*, 2789–2805. [CrossRef]
17. Cortona, A.D.; Leliaert, F.; Bogaert, K.A.; Turmel, M.; Boedeker, C.; Janouškovec, J.; López-Bautistaet, J.M.; Verbruggen, H.; Vandepoele, K.; Clerck, O.D. The plastid genome in Cladophorales green algae is encoded by hairpin chromosomes. *Curr. Biol.* **2017**, *27*, 3771–3782. [CrossRef]
18. Leliaert, F.; López-Bautistaet, J.M. The chloroplast genomes of *Bryopsis plumosa* and *Tydemania expeditiones* (Bryopsidales, Chlorophyta): Compact genomes and genes of bacterial origin. *BMC Genom.* **2015**, *16*, 204. [CrossRef]
19. Mcm, C.; Leliaert, F.; Marcelino, V.R.; Verbruggen, H. Large diversity of nonstandard genes and dynamic evolution of chloroplast genomes in Siphonous green algae (Bryopsidales, Chlorophyta). *Genome Biol. Evol.* **2018**, *10*, 1048–1061.
20. Rindi, F.; Guiry, M.D. Diversity, life history, and ecology of *Trentepohlia* and *Printzina* (Trentepohliales, Chlorophyta) in urban habitats in Western Ireland. *J. Phycol.* **2002**, *38*, 39–54. [CrossRef]
21. Ong, B.L.; Lim, M.; Wee, Y.C. Effects of desiccation and illumination on photosynthesis and pigmentation of an edaphic population of *Trentepohlia odorata* (Chlorophyta). *J. Phycol.* **1992**, *28*, 768–772. [CrossRef]
22. Liu, G.X.; Zhang, Q.; Zhu, H.; Hu, Z.Y. Massive Trentepohlia-Bloom in a glacier valley of Mt. Gongga, China, and a new variety of *Trentepohlia* (Chlorophyta). *PLoS ONE* **2012**, *7*, e37725. [CrossRef]
23. Brouard, J.S.; Otis, C.; Lemieux, C.; Turmel, M. The exceptionally large chloroplast genome of the green alga *Floydiella terrestris* illuminates the evolutionary history of the Chlorophyceae. *Genome Biol. Evol.* **2010**, *2*, 240–256. [CrossRef]

24. Brouard, J.S.; Otis, C.; Lemieux, C.; Turmel, M. The Chloroplast genome of the green alga *Schizomeris leibleinii* (Chlorophyceae) provides evidence for bidirectional DNA replication from a single origin in the Chaetophorales. *Genome Biol. Evol.* **2011**, *3*, 505–515. [CrossRef]
25. Chen, L.; Zhang, L.; Zhang, W.; Liu, T. Comparative analysis of growth and carotenoid accumulation of *Trentepohlia arborum* in aerial, subaerial, and aquatic cultivation. *J. Appl. Phycol.* **2015**, *27*, 1079–1087. [CrossRef]
26. Katoh, K.; Standley, D.M. MAFFT multiple sequence alignment software version 7: Improvements in performance and usability. *Mol. Biol. Evol.* **2013**, *30*, 772–780. [CrossRef]
27. Gouy, M.; Guindon, S.; Gascuel, O. SeaView version 4: A multiplatform graphical user interface for sequence alignment and phylogenetic tree building. *Mol. Biol. Evol.* **2010**, *27*, 221–224. [CrossRef]
28. Kalyaanamoorthy, S.; Minh, B.Q.; Wong, T.K.F.; Haeseler, A.V.; Jermiin, L.S. ModelFinder: Fast model selection for accurate phylogenetic estimates. *Nat. Methods* **2017**, *14*, 587–589. [CrossRef]
29. Lam-Tung, N.; Schmidt, H.A.; Arndt, V.H.; Quang, M.B. IQ-TREE: A fast and effective stochastic algorithm for estimating maximum-likelihood phylogenies. *Mol. Biol. Evol.* **2015**, *32*, 268–274.
30. Ronquist, F.; Teslenko, M.; Paul, V.D.M.; Ayres, D.L.; Darling, A.; Höhna, S.; LArget, B.; Liu, L.; Suchard, M.A.; Huelsenbeck, J.P. MrBayes 3.2: Efficient Bayesian phylogenetic inference and model choice across a large model space. *Syst. Biol.* **2012**, *61*, 539–542. [CrossRef]
31. Bankevich, A.; Nurk, S.; Antipov, D.; Gurevich, A.A.; Dvorkin, M.; Kulikov, A.S.; Lesin, V.M.; Nikolenko, S.I.; Pham, S.; Prjibelski, A.D.; et al. SPAdes: A new genome assembly algorithm and its applications to single-cell sequencing. *J. Comput. Biol.* **2012**, *19*, 455–477. [CrossRef]
32. Luo, R.; Liu, B.; Xie, Y.; Li, Z.; Huang, W.; Yuan, J.; He, G.; Chen, Y.; Pan, Q.; Liu, Y.; et al. SOAPdenovo2: An empirically improved memory-efficient short-read de novo assembler. *Gigascience* **2012**, *1*, 18. [CrossRef]
33. Koren, S.; Walenz, B.P.; Berlin, K.; Miller, J.R.; Bergman, N.H.; Phillippy, A.M. Canu: Scalable and accurate long-read assembly via adaptive k-mer weighting and repeat separation. *Genome Res.* **2017**, *27*, 722–736. [CrossRef]
34. Altschul, S.F.; Madden, T.L.; Schaffer, A.A.; Zhang, J.H.; Zhang, Z.; Miller, W.; Lipman, D.J. Gapped BLAST and PSI-BLAST: A new generation of protein database search programs. *Nucleic Acids Res.* **1997**, *25*, 3389–3402. [CrossRef]
35. Kearse, M.; Moir, R.; Wilson, A.; Stones-Havas, S.; Cheung, M.; Sturrock, S.; Buxton, S.; Cooper, A.; Markowitz, S.; Duran, C.; et al. Geneious Basic: An integrated and extendable desktop software platform for the organization and analysis of sequence data. *Bioinformatics* **2012**, *28*, 1647–1649. [CrossRef]
36. Krumsiek, J.; Arnold, R.; Rattei, T. Gepard: A rapid and sensitive tool for creating dotplots on genome scale. *Bioinformatics* **2007**, *23*, 1026–1028. [CrossRef]
37. Wyman, S.K.; Jansen, R.K.; Boore, J.L. Automatic annotation of organellar genomes with DOGMA. *Bioinformatics* **2004**, *20*, 3252–3255. [CrossRef]
38. Cantarel, B.L.; Korf, I.; Robb, S.M.C.; Parra, G.; Ross, E.; Moore, B.; Holt, C.; Alvarado, A.S.; Yandell, M. MAKER: An easy-to-use annotation pipeline designed for emerging model organism genomes. *Genome Res.* **2008**, *18*, 188–196. [CrossRef]
39. Lowe, T.M.; Eddy, S.R. tRNAscan-SE: A program for improved detection of transfer RNA genes in genomic sequence. *Nucleic Acids Res.* **1997**, *25*, 955–964. [CrossRef]
40. Lagesen, K.; Hallin, P.; Rødland, E.A.; Stærfeldt, H.H.; Rognes, T.; Ussery, D.W. RNAmmer: Consistent and rapid annotation of ribosomal RNA genes. *Nucleic Acids Res.* **2007**, *35*, 3100–3108. [CrossRef]
41. Michel, F.; Umesono, K.; Ozeki, H. Comparative and functional anatomy of group II catalytic introns—A review. *Gene* **1989**, *82*, 5–30. [CrossRef]
42. Michel, F.; Westhof, E. Modelling of the three-dimensional architecture of group I catalytic introns based on comparative sequence analysis. *J. Mol. Biol.* **1990**, *216*, 585–610. [CrossRef]
43. Lohse, M.; Drechsel, O.; Bock, R. OrganellarGenomeDRAW (OGDRAW): A tool for the easy generation of high-quality custom graphical maps of plastid and mitochondrial genomes. *Curr. Genet.* **2007**, *52*, 267–274. [CrossRef]

# 9

# Transcriptional and Post-Transcriptional Regulation of Organellar Gene Expression (OGE) and its Roles in Plant Salt Tolerance

**Pedro Robles and Víctor Quesada ***

Instituto de Bioingeniería, Universidad Miguel Hernández, Campus de Elche, 03202 Elche, Spain; probles@umh.es
* Correspondence: vquesada@umh.es

**Abstract:** Given their endosymbiotic origin, chloroplasts and mitochondria genomes harbor only between 100 and 200 genes that encode the proteins involved in organellar gene expression (OGE), photosynthesis, and the electron transport chain. However, as the activity of these organelles also needs a few thousand proteins encoded by the nuclear genome, a close coordination of the gene expression between the nucleus and organelles must exist. In line with this, OGE regulation is crucial for plant growth and development, and is achieved mainly through post-transcriptional mechanisms performed by nuclear genes. In this way, the nucleus controls the activity of organelles and these, in turn, transmit information about their functional state to the nucleus by modulating nuclear expression according to the organelles' physiological requirements. This adjusts organelle function to plant physiological, developmental, or growth demands. Therefore, OGE must appropriately respond to both the endogenous signals and exogenous environmental cues that can jeopardize plant survival. As sessile organisms, plants have to respond to adverse conditions to acclimate and adapt to them. Salinity is a major abiotic stress that negatively affects plant development and growth, disrupts chloroplast and mitochondria function, and leads to reduced yields. Information on the effects that the disturbance of the OGE function has on plant tolerance to salinity is still quite fragmented. Nonetheless, many plant mutants which display altered responses to salinity have been characterized in recent years, and interestingly, several are affected in nuclear genes encoding organelle-localized proteins that regulate the expression of organelle genes. These results strongly support a link between OGE and plant salt tolerance, likely through retrograde signaling. Our review analyzes recent findings on the OGE functions required by plants to respond and tolerate salinity, and highlights the fundamental role that chloroplast and mitochondrion homeostasis plays in plant adaptation to salt stress.

**Keywords:** Salt stress; organellar gene expression (OGE); Arabidopsis; rice; mitochondrial transcription termination factors (mTERFs); pentatricopeptide repeat (PPR) proteins; DEAD-box RNA helicases (RHs)-containing proteins; RNA-recognition motifs (RRMs)-containing proteins; SIGMA FACTOR 5; PLASTID-SPECIFIC RIBOSOMAL PROTEIN 2

## 1. Introduction

Soil salinity severely affects plant development and growth and leads to yield losses. The high concentration of salts causes ionic imbalance and osmotic stress. Accordingly, salinity alters the ionic homeostasis of plant cells by causing toxic excess of sodium ($Na^+$) and a deficiency of ions, such as potassium ($K^+$) and $Ca^{+2}$ because $Na^+$ in excess disturbs the uptake of these cationic nutrients [1]. High $Na^+$ content in the cytosol can also inhibit enzymatic activity, and can even interfere with protein surface charges by destabilizing molecular interactions. Apart from $Na^+$, high levels of other ions, such as $Cl^-$, $Mg^{2+}$, $SO_4{}^{2-}$, or $HCO^{3-}$ in the cytoplasm, can also contribute to salt toxicity [1]. To deal with the ionic and osmotic stresses imposed by salinity, plants have developed throughout their evolution several adaptive strategies [2].

A bigger world population and its associated food demand have meant having to increase the area dedicated to agricultural use and to develop ever-growing intensive and productive agriculture. Along these lines, the world's human population is expected to reach more than 9 billion by 2050, and according to FAO 2011, global food production will need to increase by about 70% to match this population growth [1]. To face these demands, salinization of arable soils poses a serious threat. Accordingly, losses in the world's agricultural production caused by the salinization of arable land are estimated to be around 12 trillion dollars per year, with more than 800 million hectares with high salinity levels [3]. This accounts for approximately 6% of the world's land and affects one third of the world's cultivated land area [4]. Therefore, it is necessary to advance in improving tolerance to crop salinity. One of the main strategies developed to avoid yields being lost to soil salinization is the selection of crop varieties that display enhanced tolerance to salinity. The development of these more halotolerant varieties requires unraveling the cellular, physiological, genetic, and molecular mechanisms that underlie plant salt tolerance. This is a difficult task, as plant salt tolerance is usually a genetically complex trait that is frequently modulated by different biosynthetic and signaling pathways. Crosstalks among these pathways have been reported under salinity conditions [5].

Abiotic stresses can be sensed by different plant cellular compartments, including mitochondria and chloroplasts [6], which are organelles with double membranes and with their own genomes. Cellular respiration in mitochondria (ATP production through oxidative phosphorylation) and photosynthesis in chloroplasts (the process by which plants produce organic substances from carbon dioxide and water using light energy from the Sun) are two vital processes to maintain life on our planet. The synthesis of essential biological molecules, such as nucleotides, amino acids, lipids, and vitamins, also takes place in these organelles [7–10]. Therefore, perturbation of chloroplast and mitochondrion homeostasis by endogenous or environmental cues may severely compromise plant growth and development. In line with this, chloroplasts are one of the organelles principally affected by salinity, which leads to lower carbon fixation rates and increased reactive oxygen species (ROS) levels [11]. An impaired plastid or mitochondrial function is communicated to the nucleus through retrograde signaling, and the activity of these organelles adjusts to cope with adverse environmental conditions. One way of achieving this is by modifying the activity of the nuclear genes that encode the chloroplast or mitochondrial proteins required for plants to adapt to stress [12,13]. Nonetheless, very little is known about the role that chloroplasts and mitochondria play in the response to abiotic stress in general, and to salinity in particular.

Chloroplast and mitochondrial genomes are the relics of those free-living prokaryotic organisms that invaded or were swallowed by a primitive eukaryotic cell with which they established an endosymbiotic relationship [14–16]. In the course of evolution, the number of genes in the genomes of these endosymbionts was drastically lowered, most of which were transferred to the nuclear genome. In this way, the gene functions needed for these organelles to properly function do not exclusively lie in their genomes because they depend largely on other functions located in the nucleus. This implies that a close coordination of the gene expression between nuclear and organellar genomes must exist. Along these lines, the vast majority of the proteins located in chloroplasts and mitochondria are translated in the cytosol and imported into these organelles as they are encoded by the nuclear genome. Accordingly, most studies about the effect of salinity on gene expression have focused on analyzing nuclear genes [5]. Proteomic and bioinformatic studies have indicated that chloroplast and plant mitochondrion proteomes, respectively, contain some 3000 and 2000 proteins [17,18]. In stark contrast, the genomes of these organelles harbor only between 100 and 200 genes that encode proteins involved in organellar gene expression (OGE, including transcription, RNA processing, and translation), photosynthesis in chloroplasts, and ATP production through oxidative phosphorylation in mitochondria. OGE regulation is crucial for plant growth and development. In chloroplasts, this is achieved mainly at the post-transcriptional level through the control of RNA translation, processing, splicing, decay, or editing [19]. Therefore, OGE must appropriately respond to changes that occur via development and environmental cues, including abiotic stresses, such as salinity. However, information about the effects that OGE disturbance has on plant tolerance to salinity is still limited. Notwithstanding, several works have recently reported thorough phenotypic and molecular characterizations of plant mutants, and to a lesser extent, of transgenic overexpression (OE) lines affected in nuclear genes involved in OGE in plastids or mitochondria, which exhibit altered sensitivity to salt stress. These genes regulate the expression of organellar genes at both the transcriptional and (mainly) post-transcriptional levels. Our review focuses on the results taken from analyzing these mutants and OE lines, and consequently, on OGE functions required for plant salt tolerance and response. Together, they highlight the important role of chloroplast and mitochondrion homeostasis in plant adaptation to salinity.

## 2. Effects of Perturbed OGE on Plant Tolerance to Salinity

As many of the genes involved in OGE reside in the nuclear genome, their mRNAs must be translated into the cytoplasm, and the proteins they code must be imported to chloroplasts or mitochondria. In this way, the nucleus can control the activity of these organelles, while the latter transmit information about their functional state to the nucleus by modulating nuclear activity according to these organelles' physiological requirements [20]. Accordingly, perturbed plastid gene expression is signaled to the nucleus, and affects the expression of photosynthesis-associated genes. In line with this, OGE is one of the several plant retrograde signaling routes proposed [21–23].

In the last few years, dozens of salt-responsive nuclear genes encoding chloroplast- or mitochondrial-targeted proteins from different plant species have been reported, some of which are involved in gene expression in these organelles (reviewed in [24,25]). The analysis of several plant mutants, mainly in *Arabidopsis thaliana* (hereafter Arabidopsis) and rice, has revealed a connection between chloroplast and mitochondrial functions and the response to salt stress. Some of these mutants are affected in nuclear genes involved in OGE and display altered responses to different abiotic stress conditions, including salinity (reviewed in [20]). Examples of such are genes that encode: (a) mitochondrial transcription termination factors (mTERFs), pentatricopeptide repeat (PPR), and other RNA-binding proteins that post-transcriptionally regulate OGE through RNA metabolism (e.g., DEAD-box RNA helicases (RHs) and RNA-recognition motifs (RRMs)-containing proteins); (b) the SIGMA FACTOR 5 (SIG5) required for plastid gene transcription; and (c) PLASTID-SPECIFIC RIBOSOMAL PROTEIN 2 (PSRP2).

### *2.1. Post-Transcriptional Regulation of OGE and Salt Tolerance*

#### 2.1.1. Defective mTERF Mutants Show Altered Responses to Salinity

One of the gene families involved in OGE regulation is mTERF, for which several Arabidopsis mutants showing altered sensitivity to salt have been reported [26,27]. The mTERF proteins are targeted to chloroplasts or mitochondria, and contain a variable number of repeats of a motif dubbed "mTERF", which is about 30 amino acids long. Plant genomes, mainly those of higher plants, contain larger numbers of *mTERF* genes than animals. Accordingly, 35 and 45 mTERFs have been reported in Arabidopsis and rice, respectively [28,29], and larger numbers have been proposed for other plants (e.g., 55, 56, and 62 for *Populous trichocarpa*, *Glycine max*, and *Malus domestica*, respectively [29]). This contrasts with the four mTERF (MTERF1 to 4) genes present in the nuclear genomes of metazoans. Both the expansion and diversification of plant mTERFs have been related to the tolerance and acclimation of plants to abiotic stress [27]. In vertebrates, mTERFs participate in mitochondria transcription termination, initiation, translation, and likely in mtDNA replication [30,31]. In plants, information about mTERF functions is rather limited. To date, very few mTERFs have been molecularly characterized; they all post-transcriptionally regulate the expression of chloroplast or mitochondria genes, and can bind DNA or RNA. Accordingly, Arabidopsis BELAYA SMERT/RUGOSA2 (BSM/RUG2) [28,32], mTERF15 [33], and *Zea mays* ZmTERF4 [34] are involved in organellar intron II splicing; mTERF6 has been proposed to participate in the maturation of the chloroplast isoleucine tRNA (*trnI.2*) gene [35] and the transcription termination of the plastid *rpoA* polycistron, which is important for transcription and translation in this organelle [36]. Along these lines, the first mTERF characterized in a photosynthetic organism, the *MOC1* gene of the unicellular green alga *Chlamydomonas reinhardtii*, is also involved in transcription termination in mitochondria [37,38]. Recently, Sun et al. [39] reported that mTERF4/COE1 (also known as BSM/RUG2) cooperates with the PPR protein GUN1 (GENOMES UNCOUPLED1) in plastid gene expression and retrograde signaling. Taken together, these results strongly support a role for plant mTERFs in regulating OGE.

As far as we know, a phenotype of altered sensitivity to salinity has been reported for *mda1* (*mTERF defective in Arabidopsis1*), *mterf6*, *mterf9*, *mterf10*, and *mterf11* mutants. Indeed, mutants *mda1*, affected in the *mTERF5* gene [40], and *mterf9* [41] are less sensitive to NaCl than the wild type, whereas *mterf6-2*, *mterf6-5* [42], *mterf10*, and *mterf11* [43] are salt-hypersensitive (Table 1). This suggests that mTERF5 and mTERF9 would negatively regulate Arabidopsis salt tolerance, whereas mTERF6, mTERF10, and mTERF11 would function as positive regulators of such tolerance [42]. Consistent with all this, the *mTERF10* and *mTER11* overexpression lines are more insensitive to NaCl than the wild type [43]. Interestingly, some of these mutants also exhibit an altered response to the abscisic acid (ABA) hormone, which plays a central role in the response and adaptation of plants to different abiotic stress conditions [44]. Accordingly, mutants *mda1* and *mterf9* are less sensitive to ABA than the wild type, whereas mutants *mterf6* and *mterf11* are ABA hypersensitive (Table 1) [40–43]. These results suggest that the enhanced or reduced tolerance of these mutants to salinity might be due, at least in part, to altered ABA signaling. The phenotype of an altered ABA response has also been reported for other mutants that exhibit enhanced or reduced sensitivity to abiotic stresses, besides salinity, which reinforces the fundamental role of this hormone in plant tolerance to adverse environmental conditions [20]. All the *mterf* mutants reported to show a salt stress phenotype are affected in mTERFs targeted to chloroplasts, and belong to either the "chloroplast cluster" (mTERF5, mTERF6, and mTERF9) or the "chloroplast associated-cluster" (mTERF10 and mTERF11) of proteins, which reveals the importance of chloroplast homeostasis for plant response to salinity [29].

**Table 1.** Mutants affected in OGE displaying a salt stress phenotype.

| Mutant | Species | Organelle | Affected Gene (AGI code) | Mutant Stress Phenotype | Molecular Function | Reference |
|---|---|---|---|---|---|---|
| *mda1* | *Arabidopsis thaliana* | Chloroplast | *mTERF5* (AT4G14605) | Reduced sensitivity to ABA, salt, and osmotic stress; altered sugar responses. | | [40] |
| *mterf6* | *Arabidopsis thaliana* | Chloroplast | *mTERF6* (AT4G38160) | Salt- and ABA-hypersensitive. | Maturation of the chloroplast isoleucine tRNA (*trnI.2*) gene; transcription termination of the plastid *rpoA* polycistron | [35,36,42] |
| *mterf9* | *Arabidopsis thaliana* | Chloroplast | *mTERF9* (AT5G55580) | Reduced sensitivity to ABA, salt, and osmotic stress; altered sugar responses. | | [41] |
| *mterf10* | *Arabidopsis thaliana* | Chloroplast | *mTERF10* (AT2G34620) | Hypersensitive to salt stress; *mTERF10* overexpression leads to enhanced salt and ABA tolerance. | | [43] |
| *mterf11* | *Arabidopsis thaliana* | Chloroplast | *mTERF11* (AT3G18870) | Hypersensitive to ABA and salt stress. *mTERF11* overexpression leads to enhanced salt tolerance and increased ABA sensitivity. | | [43] |
| *ppr40* | *Arabidopsis thaliana* | Mitochondria | *PPR40* (AT3G16890) | Enhanced sensitivity to salt, ABA, and oxidative stress. *PPR40* overexpression improves salt tolerance and reduces oxidative stress. | | [45,46] |
| *pgn* | *Arabidopsis thaliana* | Mitochondria | *PGN* (AT1G56570) | Hypersensitive to salt, ABA, and glucose. Enhanced ROS levels under salt stress. | | [47] |
| *ahg11* | *Arabidopsis thaliana* | Mitochondria | *AHG11* (AT2G44880) | Hypersensitive to ABA, salt, and osmotic stress. Enhanced ROS levels. | *nad4* mitochondrial RNA editing | [48] |
| *slg1* | *Arabidopsis thaliana* | Mitochondria | *SLG1* (AT5G08490) | Hypersensitive to ABA, salt, and osmotic stress. Enhanced tolerance to drought in adult plants. | *nad3* mitochondrial RNA editing | [49] |
| *slo2* | *Arabidopsis thaliana* | Mitochondria | *SLO2* (AT2G13600) | Hypersensitive to ABA, salt, and osmotic stress. Insensitive to ethylene. Enhanced tolerance to drought and salt in adult plants. Enhanced ROS levels. | *nad4L*, *nad7*, *mttB*, and *nad1* mitochondrial RNA editing | [50,51] |
| *wsl* | *Oryza sativa* | Chloroplast | *WSL* (LOC_Os01g37870) | Hypersensitive to ABA, salinity, and sugar. Enhanced $H_2O_2$ levels. | Splicing of chloroplast transcript *rpl2* | [52] |
| *ppr96* | *Arabidopsis thaliana* | Mitochondria | *PPR96* (AT2G03380) | Reduced sensitivity to ABA, salt, and oxidative stress. | | [53] |
| *atrh3* | *Arabidopsis thaliana* | Chloroplast | *AtRH3* (AT5G26742) | ABA-deficient and salt-sensitive. | Splicing of most intron-containing chloroplast genes | [54–56] |
| *psrp2* | *Arabidopsis thaliana* | Chloroplast | *PSRP2* (AT3G52150) | Enhanced tolerance to salt, osmotic, and cold stress. *PSRP2* overexpression leads to reduced tolerance to salt, dehydration, and cold stress. | | [57] |
| *sig5* | *Arabidopsis thaliana* | Chloroplast | *SIG5* (AT5G24120) | Hypersensitive to salt stress. | Repair of stress-damaged PSII through the transcriptional activation of the *psbD* blue light receptor | [58] |

In short, the analysis of the *mterf* mutants clearly indicates the involvement of mTERFs in the tolerance and acclimation of plants to adverse environmental conditions, especially to salt stress.

Apart from the above-reported results, Zhao et al. [59] made a systematic characterization of maize *mTERF* genes by bioinformatics and molecular biological approaches. By quantitative RT-PCR (qRT-PCR), these authors analyzed the expression profile of several maize *mTERF* genes under light and dark, and with salt and phytohormone treatments. Interestingly, the transcript levels of the maize *mTERF12* gene, the ortholog of Arabidopsis *mTERF6*, as well as those of maize *mTERF13* and *mTERF28*, changed after exposure to NaCl, $AlCl_3$, or ABA compared to those of the untreated plants. This scenario supports the proposed function of *mTERF* genes in response to salinity [59].

### 2.1.2. PPRs and Tolerance to Salt Stress

Like mTERFs, the PPR family of proteins participates in plant OGE, as PPRs regulate the processing of chloroplast and mitochondria RNAs, including splicing, stability, editing, and even translation [60]. As for mTERFs, the plant PPR family has also expanded, but to a higher level. Indeed, PPR is one of the largest families in land plants [61]. Accordingly, lycophyte Selaginella, rice, and Arabidopsis genomes, respectively, harbor more than 800, 650, and 450 *PPR* genes [62]. In contrast, metazoan genomes typically encode fewer than 10 PPR proteins [60]. PPRs are grouped into two subfamilies, namely P and PLS, according to the pattern of tandem repeats of a degenerate 35-amino acid repeat [63]. The PPR proteins of the P subfamily contain the canonical P motif, whereas those of the PLS subfamily contain, apart from the P motif, two variants that derive from the P motif: the large (L) and short (S) motifs [63]. The vast majority of PPR proteins are predicted to be targeted to mitochondria or chloroplasts. They can specifically bind RNA or DNA, and control different biological processes, such as embryogenesis, chloroplast, and seed development, as well as plastid retrograde signaling [47].

Despite the many PPR identified proteins in higher plants, very few have been reported to be involved in stress response: GENOMES UNCOUPLED 1 (GUN1) [64], PPR40 [45,46], ABA OVERLY-SENSITIVE 5 (ABO5) [65], LOVASTATIN INSENSITIVE 1 (LOI1) [66], PENTATRICOPEPTIDE REPEAT PROTEIN FOR GERMINATION ON NaCl (PGN) [47], SLOW GROWTH 1 (SLG1) [49], ABA HYPERSENSITIVE GERMINATION 11 (AHG11) [48], SLOW GROWTH 2 (SLO2) [50] and PPR96 [53] in Arabidopsis, and WHITE STRIPE LEAF (WSL) in rice [52]. All these PPR proteins are located in mitochondria, except for GUN1 and WSL, which are targeted to chloroplasts. Among these proteins, a role in salinity tolerance has been described for Arabidopsis PPR40, PGN, SLO2, PPR96, AHG11, and SLG1, and for rice WSL. Mitochondrial mutants *ppr40*, *pgn*, *ahg11*, *slg1*, and *slo2* and chloroplast mutant *wsl* display a similar response to ABA, salt, and osmotic stresses. They are all hypersensitive to these adverse environmental conditions during germination and early seedling growth. Most of these mutants exhibit changes in the endogenous levels of ROS, probably due to a perturbed mitochondrial function, which is the consequence of defects in the regulation of different mitochondrial RNA editing events in some cases.

The first PPR protein reported to be involved in salt tolerance was Arabidopsis PPR40. As mentioned above, the *ppr40-1* mutant is hypersensitive to ABA, salinity, and osmotic stress, accumulates high ROS levels, and shows an altered expression of stress-responsive genes and semidwarf growth (Table 1). No defects have been found in mitochondria RNA editing in the *ppr40-1* mutant [45]. A second mutant allele, *ppr40-2*, displays less severe alterations of developmental and stress responses, which suggests that is a weak loss-of-function allele of the *PPR40* gene [45]. The overexpression of the *PPR40* gene improves salt tolerance, likely because of the mitochondrial electron transport's enhanced stability and reduced oxidative damage when faced with this adverse condition [46].

The Arabidopsis PGN protein is also targeted to mitochondria, and similarly to *ppr40* mutant seedlings, *pgn* mutants exhibit high ROS levels in response to salinity, as well as the down-regulation of different stress responsive genes (Table 1) [47]. Loss of PGN function results in altered responses to biotic and abiotic stresses. Thus, *pgn* mutants display susceptibility to necrotrophic fungal pathogens, as well as enhanced sensitivity to salinity, ABA, and glucose. These results suggest a role for PGN in the mitochondria homeostasis of ROS in response to biotic or abiotic stresses, including salinity [47].

The mutant *ahg11* was identified in a genetic screening, conducted to isolate Arabidopsis mutants with an altered response to ABA. To avoid the isolation of mutants affected in ABA synthesis or catabolism, EMS mutagenesis was performed in an *aba2-1* genetic background, which displayed low ABA levels. Mutants *ahg11* showed enhanced sensitivity to ABA, NaCl, and mannitol during seed germination, but not later during development, which indicates that *AHG11* functions in the stress response in both seeds and young seedlings [48]. RNA editing of the mitochondrial *nad4* transcript is lacking in mutant *ahg11*, which results in an amino acid change in the NAD4 protein of complex I (Table 1). Murayama et al. [48] suggest that this editing defect might presumably affect the activity of complex I and lead to a redox imbalance in mutant *ahg11* by causing altered responses to stress.

Like *ahg11*, the Arabidopsis *slg1* mutant shows similar ABA, salt, and osmotic stress phenotypes during germination and in early growth stages (Table 1). Unlike the AHG11 protein, SLG1 is also involved in stress tolerance after germination [49]. In this way, *slg1* adult plants are more tolerant to drought stress than the wild type, likely due to a more rapid stomata response in the mutant, which is consistent with greater sensitivity to ABA, as this hormone plays a key role in the response to drought stress by controlling stomata closure. Similarly to *ahg11*, the *slg1* mutant also shows defects in mitochondrial RNA editing, specifically in the *nad3* transcript, by bringing about a change from serine to proline in the NAD3 protein that strongly affects the structure of the protein, and concomitantly, the activity of complex I (Table 1) [49].

The *slo2* mutants are defective in an RNA editing factor that plays a key role in Arabidopsis growth through energy metabolism regulation [51]. Loss of *SLO2* function causes several RNA editing defects, and affects different mitochondrial transcripts and results in amino acid changes in four proteins belonging to complex I of the electron transport chain (Table 1). Interestingly, the levels of complexes I, III, and IV are substantially lower in *slo2* mutants. These results led Zhu et al. [51] to propose that RNA editing defects result in the dysfunction of mitochondrial electron transfer chain complexes, and contribute to the *slo2* mutant phenotype. In a later work, Zhu et al. [50] investigated the involvement of SLO2 in tolerance to stress. They found that *slo2* mutants display hypersensitivity to ABA, salt, and osmotic stress, and insensitivity to ethylene during germination and in early seedling stages (Table 1). On the contrary, adult *slo2* plants proved more drought- and salinity-tolerant. The *slo2* mutants also accumulate high $H_2O_2$ levels and are more susceptible to infection by the pathogenic fungus *Botrytis cinerea* (Table 1). Taken together, these results indicate that SLO2 is required for proper sensitivity to ABA, ethylene, biotic, and abiotic stress [50].

Recently, Liu and colleagues [53] characterized the Arabidopsis *PPR96* gene encoding a mitochondria-located protein, which is involved in the response to salt, ABA, and oxidative stress. The *ppr96* knockout mutants are more tolerant than the wild type to NaCl, ABA, and the oxidative stress caused by hydrogen peroxide (Table 1). *PPR96* expression was up-regulated in response to salt and oxidative stress treatments, according to a microarray in silico analysis, and was experimentally confirmed by qRT-PCR [53]. Nevertheless, Liu et al. [53] did not test whether the mutant displayed defects in the editing of mitochondrial transcripts.

Altogether, the phenotypic and molecular characterizations of mutants *ahg11*, *slg1*, and *slo2* strongly support a connection between OGE regulation in mitochondria at the level of RNA editing and plant tolerance to ABA and abiotic stresses, such as salinity.

A function performed by a chloroplast PPR protein in the abiotic stress response has been reported in rice [52]. The *wsl* mutant displays enhanced sensitivity to NaCl and ABA during germination, and accumulates higher ROS levels than the wild type (Table 1). Interestingly, plastid rRNAs and proteins accumulate in *wsl* plants at lower levels than in the wild type. This is likely the result of the defective plastid translation caused by the inefficient splicing of the chloroplast *rpl2* gene, which encodes the ribosomal protein RPL2 of this organelle [52]. Unlike the mitochondrial cases discussed above, Tan et al. [52] were unable to detect defects in chloroplast RNA editing in the *wsl* mutant, which indicates a role for this PPR protein in RNA regulation at the splicing rather than the editing level.

Very recently, the importance of PPR proteins in abiotic stresses tolerance has been further supported by a genome-wide transcriptomic analysis of the PPR family in poplar, which identified 154 *PtrPPR* genes induced by biotic and abiotic treatments, including salinity [67].

### 2.1.3. Roles of Plant Organellar DEAD-Box RHs in Salinity Response

Another group of proteins involved in the control of OGE, which are hence experimentally related to the stress responses mediated by chloroplasts and plant mitochondria, is that of DEAD-box RNA helicases (DEAD-box RHs), the largest known RNA helicase subfamily. The helicase core of DEAD-box RHs is composed of the N and C terminal domains, which are structurally similar to recombination protein RecA, in which nine conserved motifs can be identified. Motif II of the N-terminal domain contains the amino acid sequence Asp-Glu-Ala-Asp (D-E-A-D), which gave a name to the subfamily. Variations in motif II determine three related RNA helicase subgroups, namely DEAD, DEAH, and Ski2, whose members are referred to as DExD/H-box proteins [68,69].

DEAD-box RHs are present in some viruses, many prokaryotes, and all eukaryotes, and they usually possess ATP-dependent RNA helicase activity, which allows them the localized unwinding of RNAs, and hence, to participate in different RNA metabolism processes, such as ribosome biogenesis, translation initiation, RNA splicing, turnover, and decay [70]. DEAD-box RHs are also involved in OGE in both chloroplasts and mitochondria. In fact, one of the first RH family members to be identified was Mss116, which is required for mitochondrial splicing in yeast [71].

Like mTERFs and PPRs, this family also extends in plants. Accordingly, Arabidopsis and rice show a similar number of presumed DEAD-box RHs, 58 and 60 [24], respectively, while *Drosophila melanogaster* and *Caenorhabditis elegans* show nearly half these numbers with 30 and 34 [72], respectively. By an in silico analysis, Nawaz and Kang [24] found that the numbers of chloroplastic DEAD-box RHs ranged from 7 to 12, and the mitochondrial ones from 4 to 7, in four different plant species.

Nowadays, growing evidence reveals the implication of DEAD-box RHs in plant responses to different kinds of stresses [24]. Regarding chloroplasts and mitochondria, it is known that some nuclear genes that encode organellar-localized DEAD-box RHs are induced by different stresses, including salinity. *HVD1* (*Hordeum vulgare* DEAD box protein1) was the first salt-responsive DEAD-box RH gene to be described encoding a chloroplast protein [73]. The expression of the *HVD1* gene is highly induced under salt and cold stresses, with subsequent recovery after exposure to salinity stress. In a genome-wide in silico study performed later, Umate et al. [74] found that several Arabidopsis DEAD-box RHs located in chloroplasts or mitochondria regulate transcription in response to various abiotic stresses (reviewed in [24]). In addition, the overexpression of some organellar-localized nuclear-encoded DEAD-box RHs has been described to confer resistance to salinity. Accordingly, the overexpression of OsSUV3 (*Oryza sativa* SUPPRESSOR OF VAR3), a mitochondrially-localized DExH/D-box-related DNA/RNA helicase, confers salt tolerance in rice by maintaining photosynthesis and antioxidant machinery [69]. Under salt stress, the levels of the plant hormones $GA_3$ (gibberellic acid 3), zeatin, and IAA (indole-3-acetic acid) increase in the SUV3 overexpressing lines, which, in turn, may induce the expression of genes to help cope with stress [75]. Recently, the chloroplast-targeted DEAD-box RH BrRH22 of *Brassica rapa* has been described to be induced by several abiotic stresses, including salinity and ABA treatment. Likewise, the chloroplast-localized DEAD-box RH OsRH58 of rice is up-regulated by salinity, dehydration, and heat, but not ABA, unlike BrRH22 [76]. Transgenic Arabidopsis OsRH58-expressing plants show increased growth and yield seeds under no stress conditions. The heterologous expression of BrRH22 or OsRH58 in Arabidopsis improves germination and plant growth under salinity conditions. Both proteins may possess RNA chaperone activity, which may, in turn, influence plastid translation under stress conditions, as the Arabidopsis transgenic plants expressing BrRH22 or OsRH58 show larger amounts of some chloroplast genome-encoded proteins when treated with either NaCl or mannitol [76,77].

To our knowledge, Arabidopsis *AtRH3* is the only gene to encode an organellar-localized DEAD-box RH, whose involvement in the response to salinity has been reported by the isolation and characterization of loss-of-function mutant alleles. The impaired function of chloroplast targeted AtRH3 causes a range of effects, from embryonic lethality to delayed growth, as well as reduced greening of vegetative tissues, depending on the mutant allele's strength [54]. Lee and colleagues [55] showed that young viable *atrh3-4* mutant seedlings were impaired in chloroplast biogenesis as these are smaller, and include fewer and smaller chlororibosomes than the wild type. The defective chloroplast development noted in *atrh3-4* seedlings gives rise to ABA-deficient and salt- and cold-sensitive phenotypes (Table 1) [55,56]. The function of AtRH3 is needed for the splicing of most intron-containing chloroplast genes [54,56] and the spliced/unspliced transcript ratio is lower in *atrh3* mutants (Table 1) [56]. Stress conditions do not affect this RNA processing defect, except for genes *ndhA* and *ndhB*, as the spliced/unspliced transcript ratio in *atrh3* mutants is significantly lower, especially under salt or cold stress, but not after dehydration stress [56]. The correct splicing of chloroplast transcripts is probably mediated by the RNA chaperone activity of AtRH3 [56]. More recently, loss of function mutations of *AtRH50*, another Arabidopsis gene that codes for a chloroplast-localized DEAD-box RH, has been shown to increase cold sensitivity as a result of the defective processing of chloroplastic rRNA, and hence inefficient plastid translation [78]. The response of *ath50* mutants to other abiotic stress, such as salinity, remains to be tested.

#### 2.1.4. Mutations in Plastid Ribosomal Protein PSRP2 Alter Salinity Tolerance

The previous sections noticeably illustrate the relationship between changes in the activity of nuclear genes involved in the post-transcriptional regulation of OGE at the RNA level and salt stress phenotypes. The involvement of plastidial translational machinery in plant salt responses has also been investigated. Accordingly, salt stress causes a transient suppression of the de novo synthesis of proteins in Arabidopsis suspension cell cultures [79]. Furthermore, Omidbakhshfard and colleagues [80] tested the expression by qRT-PCR of 170 genes related to protein synthesis in Arabidopsis leaves after NaCl exposure at different time points. Some genes showed up-regulation under salinity stress, and coded for chloroplast-located translation-related proteins, such as ribosomal proteins L11 (PRPL11) and L9-1 (also known as PIGGYBACK 2), ATAB2, an A/U-rich RNA-binding protein, which likely functions as an activator of translation, and PDF1B, a peptide deformylase required to remove the N-formyl group from nascent peptides. Interestingly, these genes have been previously described as being important for chloroplast development, and have been proposed to represent potential biotechnological targets for plant salt tolerance optimization [80].

As far as we know, only a salt-stress related phenotype has been described for a mutant affected in a plastid translational protein. Accordingly, the Arabidopsis *psrp2* mutant defective in PLASTID-SPECIFIC RIBOSOMAL PROTEIN 2 (PSRP2) of the 30S ribosomal subunit, one of the six Arabidopsis PSRP proteins, shows enhanced seedling growth under salinity stress (Table 1) [57]. On the contrary, the transgenic plants overexpressing PSRP2 display delayed germination and reduced seedling growth in response to salinity compared to the wild type (Table 1). These results suggest that PSRP2 functions as a negative regulator of germination and seedling growth under salinity conditions. Interestingly, PSRP2 contains two RNA recognition motifs (RRM), it can bind RNA and ssDNA, and it possesses RNA chaperone activity. All this suggests a connection between plastid translation and regulation of RNA metabolism [57].

### *2.2. Transcriptional Regulation of OGE and Plant Salt Tolerance*

#### 2.2.1. Arabidopsis SIG5 Protects Chloroplasts from Abiotic Stress Damage

Transcriptional regulation is also fundamental for OGE in plants, and mutations in the genes involved in this process can lead to a salt stress phenotype. Along these lines, a small family of nuclear genes, called *RpoT* (*RNA polymerase T7 phage-type*), encoding monomeric RNA polymerases, is responsible for the transcription of chloroplast or mitochondria genes [81]. Furthermore, the transcription of plastid genes requires a second type of DNA-dependent RNA polymerase, similar to those of prokaryotes, which is encoded by the organellar genome and dubbed PEP (plastid-encoded RNA polymerase). PEP comprises four catalytic core subunits (α, β, β′, and β″, respectively, encoded by plastid genes *rpoA, rpoB, rpoC1,* and *rpoC2*) and requires nucleus-encoded sigma factors (SIG) to recognize specific promoter sequences to initiate plastid gene transcription [82].

One of the six *SIG* Arabidopsis genes, *SIG5,* is up-regulated in response to different stress conditions, including salinity [58]. SIG5 is required for the transcriptional activation of the blue light-responsive promoter of the *psbD* gene that encodes photosystem II (PSII) reaction center protein D2, which binds the essential redox co-factors needed for photosynthetic electron transfer [58]. Besides, seed germination and recovery from damage of PSII after salt stress are delayed in loss-of-function mutant *sig5-2* compared to the wild type (Table 1). This led Nagashima et al. [58] to propose that SIG5 enhances the repair of the PSII reaction center under stress by protecting chloroplasts, and hence, the plants exposed to adverse environmental stress conditions. Interestingly, the expression of the orthologous gene from the liverwort *Marchantia polymorpha* (*MpSIG5*) is induced by blue-light irradiation under several stress conditions, which suggests the conservation of the responsible mechanism [83]. Nevertheless, the expression of the *psbD* gene of Marchantia did not occur in conjunction with *MpSIG5* induction, which indicates that SIG5 might play a divergent physiological role in different plant phyla [83].

Recently, Zhao et al. [84] obtained Arabidopsis *SIG5* overexpression lines, which are more tolerant to salt stress than the wild type. These results are in agreement with those from Nagashima et al. [58] mentioned above. Furthermore, Zhao and colleagues [84] also identified an upstream regulator of *SIG5,* ATHB17 (ARABIDOPSIS THALIANA HOMEOBOX 17), an Arabidopsis HD-Zip transcription factor that binds to the *cis*-elements present in the *SIG5* promoter. Similarly to *SIG5, ATHB17* loss-of-function reduced plant salt tolerance, whereas ATHB17 overexpression enhanced it. Interestingly, these phenotypes were at least partially dependent on SIG5. Accordingly, *ATHB17* overexpression in a *sig5-1* mutant background showed tolerance to a salinity intermediate between that of the *ATHB17* overexpression lines and the *sig5-1* mutant [84]. Under salt stress conditions, various plastid genes regulated by SIG5 were down-regulated in the *ATHB17* overexpression lines, but were up-regulated in the *ATHB17* knockout lines, which suggests the positive regulation of these genes through SIG5 [84].

## 3. Conclusions and Future Perspectives

The results compiled in this review reveal that accurate OGE regulation in chloroplasts and mitochondria is fundamental for plants to tolerate and adapt to adverse environmental conditions, such as salt stress (Figure 1). Consequently, perturbed OGE homeostasis affects plant responses to salinity, which can be detrimental for their survival. However, we are still far from fully understanding the role that OGE regulation plays in adapting plants to salinity. To make progress, it is necessary to know more details of the molecular functions of several genes involved in the control of OGE, whose disturbance results in stress phenotypes. This is especially relevant for the *mTERF* genes related to stress tolerance, because the mechanistic insight for most of them is lacking. Furthermore, unraveling why particular stressors specifically affect the expression of some OGE-related genes would shed light on their functional roles in plant tolerance to abiotic stresses. If we consider that a stress mutant phenotype has been reported for only a few *PPR, mTERF,* or *DEAD-box RH* genes, it is fundamental to identify and characterize mutants in plant model systems and crops affected in novel genes that transcriptionally or post-transcriptionally regulate OGE in plastids and mitochondria. This information is expected to contribute to the comprehensive understanding of the mechanisms regulating OGE in plants, and to provide insights to improve stress tolerance by identifying potential biotechnological targets for better plant growth and crop yields under salinity conditions.

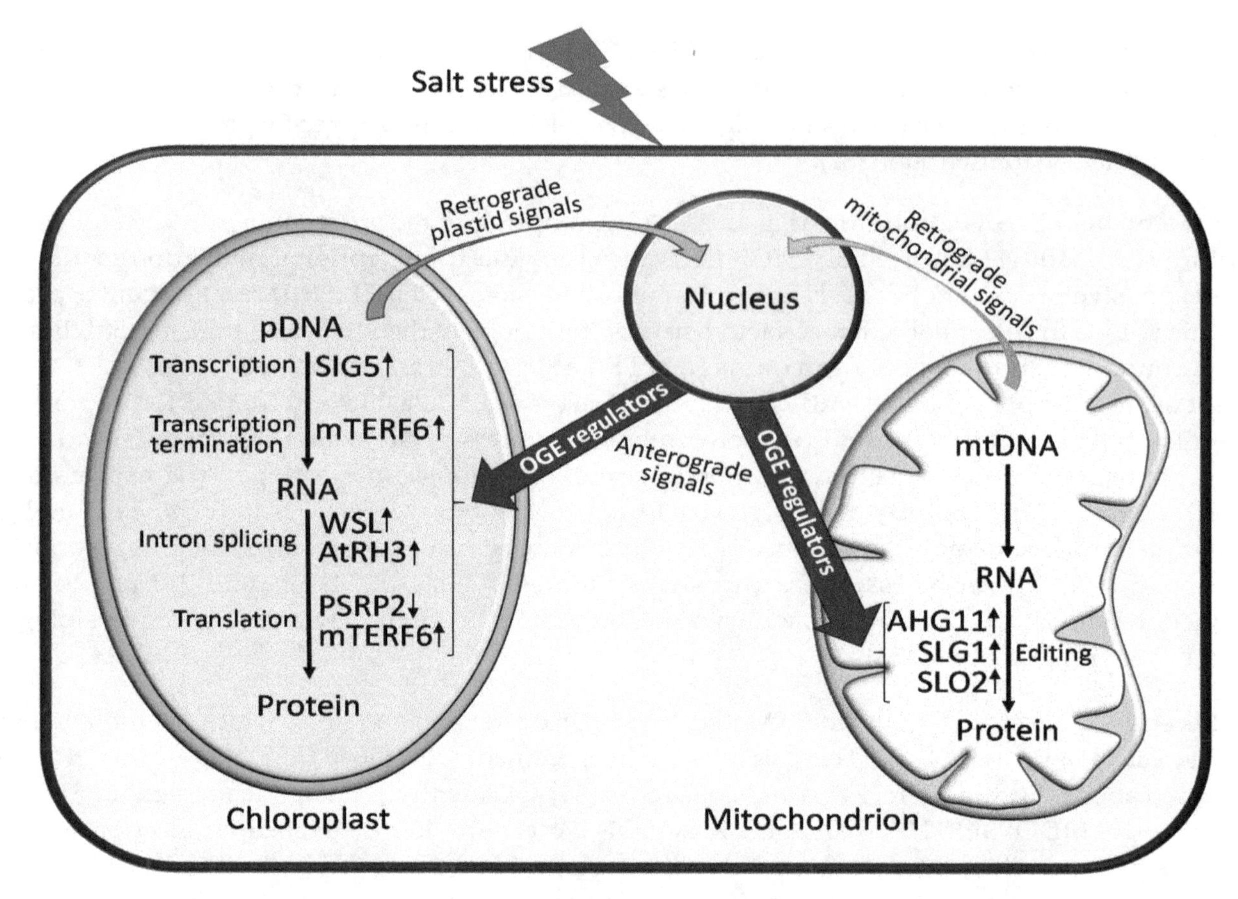


**Figure 1.** Schematic representation of the cellular functions of OGE regulators in plant response to salinity. Salt stress can perturb chloroplast or mitochondrion homeostasis and this would be communicated to the nucleus through retrograde signals, leading to changes in nuclear gene expression. In turn, this would activate the expression of OGE regulators (e.g., mTERFs, PPRs or DEAD-box RHs proteins), which would result in anterograde signalling responses to adjust organellar function to salinity. In the figure, inside a mitochondrion (purple) and chloroplast (green), only those molecularly-characterized OGE regulators whose mutations lead to altered responses to salinity are depicted. The OGE processes affected in these mutants, as well as their enhanced (↑) or reduced (↓) sensitivity to salt stress, are shown (see also Table 1 for further information).

## Abbreviations

| | |
|---|---|
| ABA | abscisic acid |
| DEAD-box RHs | DEAD-box RNA helicases |
| mTERF | mitochondrial transcription termination factor |
| OGE | organellar gene expression |
| PPR | pentatricopeptide repeat proteins |
| PSRP2 | PLASTID-SPECIFIC RIBOSOMAL PROTEIN 2 |
| RRM | RNA-recognition motifs (RRMs) |
| SIG5 | SIGMA FACTOR 5 |

## References

1. Zörb, C.; Geilfus, C.M.; Dietz, K.J. Salinity and crop yield. *Plant Biol.* **2019**, *21* (Suppl. 1), 31–38. [CrossRef] [PubMed]
2. Quesada, V.; Ponce, M.R.; Micol, J.L. Genetic analysis of salt-tolerant mutants in *Arabidopsis thaliana*. *Genetics* **2000**, *154*, 421–436. [PubMed]
3. Flowers, T.; Gaur, P.M.; Gowda, C.L.; Krishnamurthy, L.; Samineni, S.; Siddique, K.H.; Turner, N.C.; Vadez, V.; Varshney, R.K.; Colmer, T.D. Salt sensitivity in chickpea. *Plant Cell Environ.* **2010**, *33*, 490–509. [CrossRef] [PubMed]

4. Mian, A.A.; Senadheera, P.; Maathuis, F.J.M. Improving crop salt tolerante: Anion and cation transporters as genetic engineering targets. *Plant Stress* **2009**, *5*, 64–72.
5. Jamil, A.; Riaz, S.; Ashraf, M.; Foolad, M. Gene Expression Profiling of Plants under Salt Stress. *CRC Crit. Rev. Plant Sci.* **2011**, *30*, 435–458. [CrossRef]
6. Zhu, J.K. Abiotic stress signalling and responses in plants. *Cell* **2016**, *167*, 313–324. [CrossRef] [PubMed]
7. Rébeillé, F.; Macherel, D.; Mouillon, J.M.; Garin, J.; Douce, R. Folate biosynthesis in higher plants: Purification and molecular cloning of a bifunctional 6-hydroxymethyl-7,8-dihydropterin pyrophosphokinase/7,8-dihydropteroate synthase localized in mitochondria. *EMBO J.* **1997**, *16*, 947–957. [CrossRef] [PubMed]
8. Bartoli, C.G.; Pastori, G.M.; Foyer, C.H. Ascorbate biosynthesis in mitochondria is linked to the electron transport chain between complexes III and IV. *Plant Physiol.* **2000**, *123*, 335–344. [CrossRef] [PubMed]
9. Gueguen, V.; Macherel, D.; Jaquinod, M.; Douce, R.; Bourguignon, J. Fatty acid and lipoic acid biosynthesis in higher plant mitochondria. *J. Biol. Chem.* **2000**, *275*, 5016–5025. [CrossRef] [PubMed]
10. Sandelius, A.S.; Aronsson, H. *The Chloroplast. Interactions with the Environment. Plant Cell Monographs*, 1st ed.; Springer: Berlin/Heidelberg, Germany, 2009.
11. Peharec Štefanić, P.; Koffler, T.; Adler, G.; Bar-Zvi, D. Chloroplasts of salt-grown Arabidopsis seedlings are impaired in structure, genome copy number and transcript levels. *PLoS ONE* **2013**, *8*, e82548. [CrossRef] [PubMed]
12. Woodson, J.D.; Chory, J. Coordination of gene expression between organellar and nuclear genomes. *Nat. Rev. Genet.* **2008**, *9*, 383–395. [CrossRef] [PubMed]
13. Chan, K.X.; Crisp, P.A.; Estavillo, G.M.; Pogson, B.J. Chloroplast-to-nucleus communication: Current knowledge, experimental strategies and relationship to drought stress signaling. *Plant Signal. Behav.* **2010**, *5*, 1575–1582. [CrossRef] [PubMed]
14. Gray, M.W. Evolution of organellar genomes. *Curr. Opin. Genet. Dev.* **1999**, *9*, 678–687. [CrossRef]
15. Lang, B.F.; Gray, M.W.; Burger, G. Mitochondrial genome evolution and the origin of eukaryotes. *Annu. Rev. Genet.* **1999**, *33*, 351–397. [CrossRef] [PubMed]
16. Race, H.L.; Hermann, R.G.; Martin, W. Why have organelles retained genomes? *Trends Genet.* **1999**, *15*, 364–370. [CrossRef]
17. Huang, M.; Friso, G.; Nishimura, K.; Qu, X.; Olinares, P.D.; Majeran, W.; Sun, Q.; van Wijk, K.J. Construction of plastid reference proteomes for maize and Arabidopsis and evaluation of their orthologous relationships; the concept of orthoproteomics. *J. Proteome Res.* **2013**, *12*, 491–504. [CrossRef] [PubMed]
18. Lee, C.P.; Taylor, N.L.; Millar, A.H. Recent advances in the composition and heterogeneity of the Arabidopsis mitochondrial proteome. *Front. Plant Sci.* **2013**, *4*, 4. [CrossRef] [PubMed]
19. Jung, H.J.; Park, S.J.; Kang, H. Regulation of RNA metabolism in plant development and stress responses. *J. Plant Biol.* **2013**, *56*, 123–129. [CrossRef]
20. Leister, D.; Liangsheng, W.; Tatjana, K. Organellar Gene Expression and Acclimation of Plants to Environmental Stress. *Front. Plant Sci.* **2017**, *8*, 387. [CrossRef] [PubMed]
21. Pesaresi, P.; Masiero, S.; Eubel, H.; Braun, H.P.; Bhushan, S.; Glaser, E.; Salamini, F.; Leister, D. Nuclear photosynthetic gene expression is synergistically modulated by rates of protein synthesis in chloroplasts and mitochondria. *Plant Cell* **2006**, *18*, 970–991. [CrossRef] [PubMed]
22. Pesaresi, P.; Schneider, A.; Kleine, T.; Leister, D. Interorganellar communication. *Curr. Opin. Plant Biol.* **2007**, *10*, 600–606. [CrossRef] [PubMed]
23. Kleine, T.; Voigt, C.; Leister, D. Plastid signalling to the nucleus: Messengers still lost in the mists? *Trends Genet.* **2009**, *25*, 185–192. [CrossRef] [PubMed]
24. Nawaz, G.; Kang, H. Chloroplast- or mitochondria-targeted DEAD-box RNA helicases play essential roles in organellar RNA metabolism and abiotic stress responses. *Front. Plant Sci.* **2017**, *8*, 871. [CrossRef] [PubMed]
25. Suo, J.; Zhao, Q.; David, L.; Chen, S.; Dai, S. Salinity Response in Chloroplasts: Insights from Gene Characterization. *Int. J. Mol. Sci.* **2017**, *18*, 1011. [CrossRef] [PubMed]
26. Kleine, T.; Leister, D. Emerging functions of mammalian and plant mTERFs. *Biochim. Biophys. Acta* **2015**, *1847*, 786–797. [CrossRef] [PubMed]
27. Quesada, V. The roles of mitochondrial transcription termination factors (MTERFs) in plants. *Physiol. Plant.* **2016**, *157*, 389–399. [CrossRef] [PubMed]

28. Babiychuk, E.; Vandepoele, K.; Wissing, J.; Garcia-Diaz, M.; De Rycke, R.; Akbari, H.; Joubès, J.; Beeckman, T.; Jänsch, L.; Frentzen, M.; et al. Plastid gene expression and plant development require a plastidic protein of the mitochondrial transcription termination factor family. *Proc. Natl. Acad. Sci. USA* **2011**, *108*, 6674–6679. [CrossRef] [PubMed]
29. Kleine, T. Arabidopsis thaliana mTERF proteins: Evolution and functional classification. *Front. Plant Sci.* **2012**, *3*, 233. [CrossRef] [PubMed]
30. Roberti, M.; Polosa, P.L.; Bruni, F.; Manzari, C.; Deceglie, S.; Gadaleta, M.N.; Cantatore, P. The MTERF family proteins: Mitochondrial transcription regulators and beyond. *Biochim. Biophys. Acta* **2009**, *1787*, 303–311. [CrossRef] [PubMed]
31. Terzioglu, M.; Ruzzenente, B.; Harmel, J.; Mourier, A.; Jemt, E.; López, M.D.; Kukat, C.; Stewart, J.B.; Wibom, R.; Meharg, C.; et al. MTERF1 binds mtDNA to prevent transcriptional interference at the light-strand promoter but is dispensable for rRNA gene transcription regulation. *Cell Metab.* **2013**, *17*, 618–626. [CrossRef] [PubMed]
32. Quesada, V.; Sarmiento-Mañús, R.; González-Bayón, R.; Hricová, A.; Pérez-Marcos, R.; Graciá-Martínez, E.; Medina-Ruiz, L.; Leyva-Díaz, E.; Ponce, M.R.; Micol, J.L. Arabidopsis *RUGOSA2* encodes an mTERF family member required for mitochondrion, chloroplast and leaf development. *Plant J.* **2011**, *68*, 738–753. [CrossRef] [PubMed]
33. Hsu, Y.W.; Wang, H.J.; Hsieh, M.H.; Hsieh, H.L.; Jauh, G.Y. Arabidopsis mTERF15 is required for mitochondrial *nad2* intron 3 splicing and functional complex I activity. *PLoS ONE* **2014**, *9*, e112360. [CrossRef] [PubMed]
34. Hammani, K.; Barkan, A. An mTERF domain protein functions in group II intron splicing in maize chloroplasts. *Nucleic Acids Res.* **2014**, *42*, 5033–5042. [CrossRef] [PubMed]
35. Romani, I.; Manavski, N.; Morosetti, A.; Tadini, L.; Maier, S.; Kühn, K.; Ruwe, H.; Schmitz-Linneweber, C.; Wanner, G.; Leister, D.; et al. A Member of the Arabidopsis Mitochondrial Transcription Termination Factor Family Is Required for Maturation of Chloroplast Transfer RNAIle(GAU). *Plant Physiol.* **2015**, *169*, 627–646. [CrossRef] [PubMed]
36. Zhang, Y.; Cui, Y.L.; Zhang, X.L.; Yu, Q.B.; Wang, X.; Yuan, X.B.; Qin, X.M.; He, X.F.; Huang, C.; Yang, Z.N. A nuclear-encoded protein, mTERF6, mediates transcription termination of rpoA polycistron for plastid-encoded RNA polymerase-dependent chloroplast gene expression and chloroplast development. *Sci. Rep.* **2018**, *8*, 11929. [CrossRef] [PubMed]
37. Schonfeld, C.; Wobbe, L.; Borgstadt, R.; Kienast, A.; Nixon, P.J.; Kruse, O. The nucleus-encoded protein MOC1 is essential for mitochondrial light acclimation in *Chlamydomonas reinhardtii*. *J. Biol. Chem.* **2004**, *279*, 50366–50374. [CrossRef] [PubMed]
38. Wobbe, L.; Nixon, P.J. The mTERF protein MOC1 terminates mitochondrial DNA transcription in the unicellular green alga *Chlamydomonas reinhardtii*. *Nucleic Acids Res.* **2013**, *41*, 6553–6567. [CrossRef] [PubMed]
39. Sun, X.; Xu, D.; Liu, Z.; Kleine, T.; Leister, D. Functional relationship between mTERF4 and GUN1 in retrograde signaling. *J. Exp. Bot.* **2016**, *67*, 3909–3924. [CrossRef] [PubMed]
40. Robles, P.; Micol, J.L.; Quesada, V. Arabidopsis MDA1, a nuclear-encoded protein, functions in chloroplast development and abiotic stress responses. *PLoS ONE* **2012**, *7*, e42924. [CrossRef] [PubMed]
41. Robles, P.; Micol, J.L.; Quesada, V. Mutations in the plant-conserved MTERF9 alter chloroplast gene expression, development and tolerance to abiotic stress in *Arabidopsis thaliana*. *Physiol. Plant.* **2015**, *154*, 297–313. [CrossRef] [PubMed]
42. Robles, P.; Navarro-Cartagena, S.; Ferrández-Ayela, A.; Núñez-Delegido, E.; Quesada, V. The characterization of Arabidopsis *mterf6* mutants reveals a new role for mTERF6 in tolerance to abiotic stress. *Int. J. Mol. Sci.* **2018**, *19*, 2388. [CrossRef] [PubMed]
43. Xu, D.; Leister, D.; Kleine, T. *Arabidopsis thaliana* mTERF10 and mTERF11, but not mTERF12, are involved in the response to salt stress. *Front. Plant Sci.* **2017**, *8*, 1213. [CrossRef] [PubMed]
44. Zhu, J.K. Salt and drought stress signal transduction in plants. *Annu. Rev. Plant Biol.* **2002**, *53*, 247–273. [CrossRef] [PubMed]
45. Zsigmond, L.; Rigó, G.; Szarka, A.; Székely, G.; Otvös, K.; Darula, Z.; Medzihradszky, K.F.; Koncz, C.; Koncz, Z.; Szabados, L. Arabidopsis PPR40 connects abiotic stress responses to mitochondrial electron transport. *Plant Physiol.* **2008**, *146*, 1721–1737. [CrossRef] [PubMed]

46. Zsigmond, L.; Szepesi, A.; Tari, I.; Rigó, G.; Király, A.; Szabados, L. Overexpression of the mitochondrial *PPR40* gene improves salt tolerance in Arabidopsis. *Plant Sci.* **2012**, *182*, 87–93. [CrossRef] [PubMed]
47. Laluk, K.; Abuqamar, S.; Mengiste, T. The Arabidopsis mitochondria-localized pentatricopeptide repeat protein PGN functions in defense against necrotrophic fungi and abiotic stress tolerance. *Plant Physiol.* **2011**, *156*, 2053–2068. [CrossRef] [PubMed]
48. Murayama, M.; Hayashi, S.; Nishimura, N.; Ishide, M.; Kobayashi, K.; Yagi, Y.; Asami, T.; Nakamura, T.; Shinozaki, K.; Hirayama, T. Isolation of Arabidopsis *ahg11*, a weak ABA hypersensitive mutant defective in *nad4* RNA editing. *J. Exp. Bot.* **2012**, *63*, 5301–5310. [CrossRef] [PubMed]
49. Yuan, H.; Liu, D. Functional disruption of the pentatricopeptide protein SLG1 affects mitochondrial RNA editing, plant development, and responses to abiotic stresses in Arabidopsis. *Plant J.* **2012**, *70*, 432–444. [CrossRef] [PubMed]
50. Zhu, Q.; Dugardeyn, J.; Zhang, C.; Mühlenbock, P.; Eastmond, P.J.; Valcke, R.; De Coninck, B.; Oden, S.; Karampelias, M.; Cammue, B.P.; et al. The Arabidopsis thaliana RNA editing factor SLO2, which affects the mitochondrial electron transport chain, participates in multiple stress and hormone responses. *Mol. Plant* **2014**, *7*, 290–310. [CrossRef] [PubMed]
51. Zhu, Q.; Dugardeyn, J.; Zhang, C.; Takenaka, M.; Kühn, K.; Craddock, C.; Smalle, J.; Karampelias, M.; Denecke, J.; Peters, J.; et al. SLO2, a mitochondrial pentatricopeptide repeat protein affecting several RNA editing sites, is required for energy metabolism. *Plant J.* **2012**, *71*, 836–849. [CrossRef] [PubMed]
52. Tan, J.; Tan, Z.; Wu, F.; Sheng, P.; Heng, Y.; Wang, X.; Ren, Y.; Wang, J.; Guo, X.; Zhang, X.; et al. A novel chloroplast-localized pentatricopeptide repeat protein involved in splicing affects chloroplast development and abiotic stress response in rice. *Mol. Plant* **2014**, *7*, 1329–1349. [CrossRef] [PubMed]
53. Liu, J.M.; Zhao, J.Y.; Lu, P.P.; Chen, M.; Guo, C.H.; Xu, Z.S.; Ma, Y.Z. The E-Subgroup Pentatricopeptide Repeat Protein Family in Arabidopsis thaliana and Confirmation of the Responsiveness PPR96 to Abiotic Stresses. *Front. Plant Sci.* **2016**, *7*, 1825. [CrossRef] [PubMed]
54. Asakura, Y.; Galarneau, E.; Watkins, K.P.; Barkan, A.; van Wijk, K.J. Chloroplast RH3 DEAD box RNA helicases in maize and Arabidopsis function in splicing of specific group II introns and affect chloroplast ribosome biogenesis. *Plant Physiol.* **2012**, *159*, 961–974. [CrossRef] [PubMed]
55. Lee, K.H.; Park, J.; Williams, D.S.; Xiong, Y.; Hwang, I.; Kang, B.H. Defective chloroplast development inhibits maintenance of normal levels of abscisic acid in a mutant of the Arabidopsis RH3 DEAD-box protein during early postgermination growth. *Plant J.* **2013**, *73*, 720–732. [CrossRef] [PubMed]
56. Gu, L.; Xu, T.; Lee, K.; Lee, K.H.; Kang, H. A chloroplast-localized DEAD-box RNA helicase AtRH3 is essential for intron splicing and plays an important role in the growth and stress response in *Arabidopsis thaliana*. *Plant Physiol. Biochem.* **2014**, *82*, 309–318. [CrossRef] [PubMed]
57. Xu, T.; Lee, K.; Gu, L.; Kim, J.I.; Kang, H. Functional characterization of a plastid-specific ribosomal protein PSRP2 in Arabidopsis thaliana under abiotic stress conditions. *Plant Physiol. Biochem.* **2013**, *73*, 405–411. [CrossRef] [PubMed]
58. Nagashima, A.; Hanaoka, M.; Shikanai, T.; Fujiwara, M.; Kanamaru, K.; Takahashi, H.; Tanaka, K. The multiple-stress responsive plastid sigma factor, SIG5, directs activation of the psbD blue light-responsive promoter (BLRP) in *Arabidopsis thaliana*. *Plant Cell Physiol.* **2004**, *45*, 357–368. [CrossRef] [PubMed]
59. Zhao, Y.; Cai, M.; Zhang, X.; Li, Y.; Zhang, J.; Zhao, H.; Kong, F.; Zheng, Y.; Qiu, F. Genome-Wide identification, evolution and expression analysis of mTERF gene family in maize. *PLoS ONE* **2014**, *9*, e94126. [CrossRef] [PubMed]
60. Schmitz-Linneweber, C.; Small, I. Pentatricopeptide repeat proteins: A socket set for organelle gene expression. *Trends Plant Sci.* **2008**, *13*, 663–670. [CrossRef] [PubMed]
61. Barkan, A.; Small, I. Pentatricopeptide repeat proteins in plants. *Annu. Rev. Plant Biol.* **2014**, *65*, 415–442. [CrossRef] [PubMed]
62. Grotewold, E.; Chappell, J.; Kellogg, E.A. Genomes of organelles. In *Plant Genes, Genomes and Genetics*; John Wiley & Sons, Ltd.: Hoboken, NJ, USA, 2015.
63. Lurin, C.; Andres, C.; Aubourg, S.; Bellaoui, M.; Bitton, F.; Bruyere, C.; Caboche, M.; Debast, C.; Gualberto, J.; Hoffmann, B.; et al. Genome-wide analysis of Arabidopsis pentatricopeptide repeat proteins reveals their essential role in organelle biogenesis. *Plant Cell* **2004**, *16*, 2089–2103. [CrossRef] [PubMed]
64. Koussevitzky, S.; Nott, A.; Mockler, T.C.; Hong, F.; Sachetto-Martins, G.; Surpin, M.; Lim, J.; Mittler, R.; Chory, J. Signals from chloroplasts converge to regulate nuclear gene expression. *Science* **2007**, *316*, 715–719. [CrossRef]

65. Liu, Y.; He, J.; Chen, Z.; Ren, X.; Hong, X.; Gong, Z. ABA overly-sensitive 5 (ABO5), encoding a pentatricopeptide repeat protein required for *cis*-splicing of mitochondrial *nad2* intron 3, is involved in the abscisic acid response in Arabidopsis. *Plant J.* **2010**, *63*, 749–765. [CrossRef] [PubMed]
66. Tang, J.; Kobayashi, K.; Suzuki, M.; Matsumoto, S.; Muranaka, T. The mitochondrial PPR protein LOVASTATIN INSENSITIVE 1 plays regulatory roles in cytosolic and plastidial isoprenoid biosynthesis through RNA editing. *Plant J.* **2010**, *61*, 456–466. [CrossRef] [PubMed]
67. Xing, H.; Fu, X.; Yang, C.; Tang, X.; Guo, L.; Li, C.; Xu, C.; Luo, K. Genome-wide investigation of pentatricopeptide repeat gene family in poplar and their expression analysis in response to biotic and abiotic stresses. *Sci. Rep.* **2018**, *8*, 2817. [CrossRef] [PubMed]
68. Tuteja, N.; Tuteja, R. Helicases as molecular motors: An insight. *Physica A* **2006**, *372*, 70–83. [CrossRef]
69. Tuteja, N.; Sahoo, R.K.; Garg, B.; Tuteja, R. OsSUV3 dual helicase functions in salinity stress tolerance by maintaining photosynthesis and antioxidant machinery in rice (*Oryza sativa* L. cv. IR64). *Plant J.* **2013**, *76*, 115–127. [CrossRef] [PubMed]
70. Linder, P.; Fuller-Pace, F. Happy birthday: 25 years of DEAD-box proteins. *Methods Mol. Biol.* **2015**, *1259*, 17–33. [PubMed]
71. Seraphin, B.; Simon, M.; Boulet, A.; Faye, G. Mitochondrial splicing requires a protein from a novel helicase family. *Nature* **1989**, *337*, 84–87. [CrossRef] [PubMed]
72. Boudet, N.; Aubourg, S.; Toffano-Nioche, C.; Kreis, M.; Lecharny, A. Evolution of intron/exon structure of DEAD helicase family genes in Arabidopsis, Caenorhabditis, and Drosophila. *Genome Res.* **2001**, *11*, 2101–2114. [CrossRef] [PubMed]
73. Nakamura, T.; Muramoto, Y.; Yokota, S.; Ueda, A.; Takabe, T. Structural and transcriptional characterization of a salt-responsive gene encoding putative ATP dependent helicase in barley. *Plant Sci.* **2004**, *167*, 63–70. [CrossRef]
74. Umate, P.; Tuteja, R.; Tuteja, N. Genome-wide analysis of helicase gene family from rice and Arabidopsis: A comparison with yeast and human. *Plant Mol. Biol.* **2010**, *73*, 449–465. [CrossRef] [PubMed]
75. Sahoo, R.K.; Ansari, M.W.; Tuteja, R.; Tuteja, N. OsSUV3 transgenic rice maintains higher endogenous levels of plant hormones that mitigates adverse effects of salinity and sustains crop productivity. *Rice* **2014**, *7*, 17. [CrossRef] [PubMed]
76. Nawaz, G.; Kang, H. Rice OsRH58, a chloroplast DEAD-box RNA helicase, improves salt or drought stress tolerance in Arabidopsis by affecting chloroplast translation. *BMC Plant Biol.* **2019**, *19*, 17. [CrossRef] [PubMed]
77. Nawaz, G.; Lee, K.; Park, S.J.; Kim, Y.O.; Kang, H. A chloroplast-targeted cabbage DEAD-box RNA helicase BrRH22 confers abiotic stress tolerance to transgenic Arabidopsis plants by affecting translation of chloroplast transcripts. *Plant Physiol. Biochem.* **2018**, *127*, 336–342. [CrossRef] [PubMed]
78. Paieri, F.; Tadini, L.; Manavski, N.; Kleine, T.; Ferrari, R.; Morandini, P.; Pesaresi, P.; Meurer, J.; Leister, D. The DEAD-box RNA helicase RH50 is a 23S-4.5S rRNA maturation factor that functionally overlaps with the plastid signaling factor GUN1. *Plant Physiol.* **2018**, *176*, 634–648. [CrossRef] [PubMed]
79. Ndimba, B.K.; Chivasa, S.; Simon, W.J.; Slabas, A.R. Identification of Arabidopsis salt and osmotic stress responsive proteins using two-dimensional difference gel electrophoresis and mass spectrometry. *Proteomics* **2005**, *5*, 4185–4196. [CrossRef] [PubMed]
80. Omidbakhshfard, M.A.; Omranian, N.; Ahmadi, F.S.; Nikoloski, Z.; Mueller-Roeber, B. Effect of salt stress on genes encoding translation-associated proteins in Arabidopsis thaliana. *Plant Signal. Behav.* **2012**, *7*, 1095–1102. [CrossRef] [PubMed]
81. Börner, T.; Aleynikova, A.Y.; Zubo, Y.O.; Kusnetsov, V.V. Chloroplast RNA polymerases: Role in chloroplast biogenesis. *Biochim. Biophys. Acta* **2015**, *1847*, 761–769. [CrossRef] [PubMed]
82. Kanamaru, K.; Tanaka, K. Roles of chloroplast RNA polymerase sigma factors in chloroplast development and stress response in higher plants. *Biosci. Biotechnol. Biochem.* **2004**, *68*, 2215–2223. [CrossRef] [PubMed]
83. Kanazawa, T.; Ishizaki, K.; Kohchi, T.; Hanaoka, M.; Tanaka, K. Characterization of four nuclear-encoded plastid RNA polymerase sigma factor genes in the liverwort *Marchantia polymorpha*: Blue-light- and multiple stress-responsive SIG5 was acquired early in the emergence of terrestrial plants. *Plant Cell Physiol.* **2013**, *54*, 1736–1748. [CrossRef] [PubMed]
84. Zhao, P.; Cui, R.; Xu, P.; Wu, J.; Mao, J.L.; Chen, Y.; Zhou, C.Z.; Yu, L.H.; Xiang, C.B. ATHB17 enhances stress tolerance by coordinating photosynthesis associated nuclear gene and *ATSIG5* expression in response to abiotic stress. *Sci. Rep.* **2017**, *7*, 45492. [CrossRef] [PubMed]

# 10

# Plastid Transformation: How does it Work? Can it be Applied to Crops? What can it Offer?

**Yihe Yu [1,†], Po-Cheng Yu [2,†], Wan-Jung Chang [3], Keke Yu [1] and Choun-Sea Lin [4,*]**

1 College of Forestry, Henan University of Science and Technology, Luoyang 471023, China; yuyihe@haust.edu.cn (Y.Y.); yukeke918@163.com (K.Y.)
2 Department of Chemistry, Temple University, Philadelphia, PA 19122, USA; tug55568@temple.edu
3 Perelman School of Medicine, University of Pennsylvania, Philadelphia, PA 19104, USA; wanjungc@upenn.edu
4 Agricultural Biotechnology Research Center, Academia Sinica, Taipei 115, Taiwan
* Correspondence: cslin99@gate.sinica.edu.tw
† These authors contributed equally to this work.

**Abstract:** In recent years, plant genetic engineering has advanced agriculture in terms of crop improvement, stress and disease resistance, and pharmaceutical biosynthesis. Cells from land plants and algae contain three organelles that harbor DNA: the nucleus, plastid, and mitochondria. Although the most common approach for many plant species is the introduction of foreign DNA into the nucleus (nuclear transformation) via Agrobacterium- or biolistics-mediated delivery of transgenes, plastid transformation offers an alternative means for plant transformation. Since there are many copies of the chloroplast genome in each cell, higher levels of protein accumulation can often be achieved from transgenes inserted in the chloroplast genome compared to the nuclear genome. Chloroplasts are therefore becoming attractive hosts for the introduction of new agronomic traits, as well as for the biosynthesis of high-value pharmaceuticals, biomaterials and industrial enzymes. This review provides a comprehensive historical and biological perspective on plastid transformation, with a focus on current and emerging approaches such as the use of single-walled carbon nanotubes (SWNTs) as DNA delivery vehicles, overexpressing morphogenic regulators to enhance regeneration ability, applying genome editing techniques to accelerate double-stranded break formation, and reconsidering protoplasts as a viable material for plastid genome engineering, even in transformation-recalcitrant species.

**Keywords:** chloroplast; SWNTs; morphogenic regulators; CRISPR; protoplast regeneration

---

## 1. Introduction

The plastids of algae and land plants are semi-autonomous organelles with their own genomes, complete with transcription and translation machinery [1] that arose from endosymbiosis between a eukaryotic ancestor and a photosynthetic cyanobacterium [2]. The plastid genome is a circular double-stranded DNA molecule that encodes 100–250 genes and is maternally inherited in angiosperms [3]. The size of chloroplast genomes varies between species, ranging from 107 kbp (Cathay silver fir, *Cathaya argyrophylla*) to 218 kbp (Geranium, *Pelargonium* spp.). The plastid genome is present in many copies inside the organelle and is protected from gene silencing pathways that typically lower transgene expression, thereby allowing high levels of foreign protein to accumulate, achieving 5–40% total soluble protein (TSP) [4] and up to 70% of total soluble protein in tobacco (*Nicotiana tabacum*) [5–7]. An even higher protein yield (>75% TSP) was recently achieved in tobacco transplastomic plants expressing a hyper-thermostable form of β-glucosidase cloned from the hyperthermophilic bacterium *Pyrococcus furiosus* [8]. Compared to nuclear transformation, plastid transformation has many advantages.

First, the maternal inheritance of the chloroplast genome prevents genes from escaping through pollen grains in most plants and thus reduces the spread of transgenes in the environment and avoids issues with Genetically Modified Organisms (GMOs) [9]. Second, the multiple copies of the plastid genome contained within the organelle results in the high expression of foreign genes and the accumulation of the encoded proteins [10]. Third, since plastid transformation involves homologous recombination (HR) of the transgene into a so-called neutral site, the method largely avoids gene silencing caused by position effects [11]. Fourth, the chloroplast employs a prokaryotic gene expression system and allows the easy and simultaneous expression of polycistronic genes [12]. Last, the stacking of multiple transgenes can be performed in a single and highly efficient transformation event [13]. Plastid transformation is, therefore, gaining traction [14].

The main crop species used for plastid transformation are in the Solanaceae, including tobacco, potato (*Solanum tuberosum*), tomato (*S. lycopersicum*), eggplant (*S. melongena*), and pepper (*Capsicum annuum*). Plastid transformation has also been reported in other important crop species, including soybean (*Glycine max*, Fabaceae), lettuce (*Lactuca sativa*, Asteraceae), cauliflower (*Brassica oleracea* var. *botrytis*, Brassicaceae), cotton (*Gossypium* spp., Malvaceae), carrot (*Daucus carota*, Umbelliferae), bitter squash (*Momordica charantia*, Cucurbitaceae) and rice (*Oryza sativa*, Poaceae) (Table 1). To date, the plastids of over 20 flowering plants have been transformed [15] (Table 1). In addition to the crops mentioned above, recent successes in plastid transformation have been reported in the plant species bitter melon [16], and the medicinal plant sweet wormwood (*Artemisia annua*) [17] and licorice weed (*Scoparia dulcis*) [18,19] (Table 1).

**Table 1.** Species in which plastid transformation has been demonstrated.

| Family | Scientific Name | Common Name | Selection Marker | Resistance | Method | Reference |
|---|---|---|---|---|---|---|
| Chlamydomonadaceae | *Chlamydomonas reindhartii* | Chlamydomonas | *aphA6* | Kan [1] | Biolistic | [20] |
| Euglenaceae | *Euglena gracilis* | Euglena | *aadA* | Spec [2]/Strep [3] | Biolistic | [21] |
| Funariaceae | *Physcomitrella patens* | moss | *aadA* | Spec | PEG [4] | [22] |
| Asteraceae | *Lactuca sativa* | lettuce | *aadA* | Spec | Biolistic | [23] |
| Amaranthaceae | *Beta vulgaris* | sugarbeet | *aadA* | Spec | Biolistic | [24] |
| Asteraceae | *Artemisia annua* | sweet wormwood | *aadA* | Spec | Biolistic | [17] |
| Brassicaceae | *Arabidopsis thaliana* | Arabidopsis | *aadA* | Spec | Biolistic | [25] |
| Brassicaceae | *Brassica capitate* | cabbage | *aadA* | Spec/Strep | Biolistic | [26] |
| Brassicaceae | *Brassica napus* | oilseed rape | *aadA* | Spec | Biolistic | [27] |
| Brassicaceae | *Brassica oleracea* var. *botrytis* | cauliflower | *aadA* | Spec | PEG | [28] |
| Brassicaceae | *Lesquerella fendleri* | popweed | *aadA/GFP* | Spec/Strep | Biolistic | [29] |
| Cucurbitaceae | *Momordica charantia* | bitter squash | *aadA* | Spec | Biolistic | [16] |
| Fabaceae | *Glycine max* | soybean | *aadA* | Spec | Biolistic | [30] |
| Malvaceae | *Gossypium* spp. | cotton | *aphA6/nptII* | $KNO_3$/Kan | Biolistic | [31] |
| Poaceae | *Oryza sativa* | rice | *hpt* | Hygromycin | Biolistic | [32] |
| Salicaceae | *Populus alba* | poplar | *aadA* | Spec | Biolistic | [33] |
| Scrophulariaceae | *Scoparia dulcis* | licorice weed | *aadA* | Spec | Biolistic | [19] |
| Solanaceae | *Capsicum annuum* | pepper | *aadA* | Spec | Biolistic | [34] |
| Solanaceae | *Nicotiana tabacum* | tobacco | *aadA* | Spec | Biolistic | [8] |
| Solanaceae | *Solanum lycopersicum* | tomato | *aadA* | Spec | Biolistic | [35] |
| Solanaceae | *Scoparia melongena* | eggplant | *aadA* | Spec | Biolistic | [36] |
| Solanaceae | *Solanum tuberosum* | potato | *aadA* | Spec/Strep | Biolistic | [37] |
| Solanaceae | *Petunia xhybrida* | petunia | *aadA* | Spec/Strep | Biolistic | [38] |
| Umbelliferae | *Daucus carota* | carrot | *aadA* | Spec | Biolistic | [39] |

[1] Kanamycin; [2] Spectinomycin; [3] Streptomycin; [4] PEG-mediated transformations.

Based on these successful cases, plastid transformation should be applicable to many plant families, whether they are monocots or dicots. However, plastid transformation remains much more challenging than nuclear transformation and is not as widespread in plant research. Furthermore, Bock [7] raised the issue of reproducibility of plastid transformation, as the plastids of fewer than ten species have been demonstrably transformed in at least two independent reports since 1988, the year the first example of plastid transformation was published in the unicellular green alga Chlamydomonas (*Chlamydomonas reinhardtii*) [7,40]. Major critical points limiting current plastid transformation attempts are 1) the method of DNA delivery, 2) homologous recombination efficiency and 3) methods for efficient selection and robust regeneration of transformants. In this review, we will address each of these issues and will highlight recent innovative technologies and strategies in plastid transformation. We hope to give readers a new perspective on the potential of plastid transformation with concrete examples, setting the stage for plastid transformation in more plant species in the future.

## 2. How to Transform a Chloroplast in Three Steps

Plastid transformation can be divided into three steps: first, foreign DNA is delivered to cells of an explant [41]. Second, the foreign DNA is inserted into the chloroplast genome through homologous recombination at a predetermined and precise location. Third, candidate transformants are repeatedly screened on selection medium until the wild-type genome is eliminated (a state known as homoplasmy) [9]. Positive explants are then regenerated into stable transgenic plants.

### *2.1. Robust Methods for DNA Delivery into the Chloroplast*

The two most common methods for introducing foreign DNA into chloroplasts are biolistic transformation [9] and polyethylene glycol (PEG)-mediated transfection [42]. Biolistics delivers particles coated with DNA into plant cells by high-speed bombardment through a gene gun or a particle delivery system. This method can be applied to various plants by adjusting bombardment parameters such as distance to the target tissue, chamber vacuum pressure, particle size and DNA: particle ratios, to accommodate variations in leaf texture. The PEG-mediated plastid transformation method works on plant cells from which the cell wall has been removed (protoplasts). The co-culture of protoplasts in the presence of PEG vesicles loaded with plasmid DNA allows DNA uptake by protoplasts, leading to the integration of foreign DNA into the plastid genome [42]. Although the PEG method requires the enzymatic digestion of tissues to release protoplasts, it is a more economical procedure because it does not rely on a specialized and expensive delivery system.

A new strategy for plastid transformation via nanoparticles was recently introduced [43]. This method allows DNA to be delivered to chloroplasts simply through single-walled carbon nanotubes (SWNTs) without the need for additional instruments or protoplast isolation, or the aid of chemical reagents. Nanocarriers for transformation consisted of chitosan-complexed single-walled carbon nanotubes (CS–SWNTs). These nanotubes are positively charged and can, therefore, carry negatively charged plasmid DNA via electrostatic interactions; the resulting DNA-SWNT conjugate may easily enter leaf mesophyll cells by infiltration using a syringe from stomatal pores. Once the DNA-SWNT conjugate passes through the leaf surface and enters the mesophyll, it will eventually be trafficked to the chloroplast bilayers by way of lipid exchange envelope penetration [43].

The power of this delivery strategy lies in the fact that the DNA can be selectively released in the chloroplast due to differences in pH within the cell. The acidic cytosol (around pH 5.5) leaves the DNA tightly bound to chitosan; by contrast, the carriers tend to unload the DNA inside the chloroplast due to its weakly alkaline environment (~pH 8.0). This preferential release accomplishes the selective release of DNA at its intended target site. A similar strategy may also be applied to protoplasts by effectively replacing PEG with nanotubes [43]. Indeed, DNA alone does not have the ability to penetrate plant cells in the absence of the SWNT carrier, as demonstrated with protoplasts co-cultured with DNA only. Notably, the efficiency of DNA entrance depends on the zeta potential and hydrodynamic radius of SWNTs rather than the concentration of plasmid DNA. Using a reporter construct carrying the *Yellow Fluorescent Protein* gene (*YFP*), YFP fluorescence was detected after 24 hr incubation of protoplasts with a DNA:SWNT mixture in a 1:6 ratio, and the YFP signal largely coincided with chloroplasts. However, no fluorescence was detected when the DNA:SWNT ratio was increased to 1:1. Excess DNA, bound to SWNTs, will neutralize their surface charge, which will not only reduce the entry of DNA-SWNT conjugates but also increase their instability, resulting in a partial or complete loss of membrane crossing [43].

Using the same YFP reporter, the method was further tested on living plants, in the form here of four-week old arugula (*Eruca sativa*) plants. The authors tested different DNA:SWNT ratios (1:1, 1:3 and 1:6); not surprisingly, based on the reasoning provided above, only ratios of 1:3 and 1:6 expressed YFP. Transient YFP expression reached its peak 48 h after the initial infiltration. An impressive aspect of the study was the low amounts of DNA (~20 ng) required for successful delivery and expression of plasmid DNA to chloroplasts, far less than for PEG-mediated transformation (20–50 μg) or biolistics (5 μg) [44–46].

This article demonstrated the delivery and expression of a transgene to chloroplasts in five different plant species, including four mature living plants and isolated protoplasts, which indicates the potential of this nanoparticle delivery method for widespread application [43]. Recently, another method by Santana et al. used a chloroplast signal peptide as a guide to delivering nanomaterials loaded with chemicals into Arabidopsis chloroplasts [47]. Tagged with a 14-amino acid guide peptide, the cargo was sent to the translocons at the outer and inner membrane of chloroplasts (TOC/TIC) and was transported into the chloroplast stroma. This, therefore, provides a targeted way to deliver materials into plastids via a biorecognition motif [47]. In addition to biolistics and PEG-mediated delivery methods, carbon nanotubes and biorecognition peptides now join the team of plastid transformation tools and combine the advantages of low DNA amounts and high delivery accuracy.

However, all SWNT tests were based on transient expression using a single plasmid carrying flanking regions from switchgrass (*Panicum virgatum*) [43]. Although the YFP signal indeed originated from the chloroplast and thus demonstrated the trafficking function of SWNTs to the organelle, homologous recombination within the chloroplast genome remains to be confirmed. In the case of the signal peptide delivery technique [47], the chemical cargo was the subject of the test and not DNA. Therefore, the ability and the efficiency of stable transformation through these two methods remain to be determined.

### *2.2. Harnessing Homologous Recombination in Chloroplasts*

Homologous recombination is a crucial step after DNA delivery into the chloroplast that determines the subsequent success of the transformation. The frequency of HR events is highest when the foreign DNA carries a sequence of at least 121 bp that is identical to the target integration site [48]. Careful consideration should be given to the choice of promoter and regulatory elements (5′ and 3′ untranslated regions), as well as the insertion site in the plastid genome, to maximize transformation efficiency, as shown in Table 2. The *psbA* promoter, from the plastid-encoded photosystem II protein D1 precursor, was first used over 30 years ago in Chlamydomonas and still appears to be the best position in which to insert a target gene, as the *psbA* gene product is the most highly translated plastid protein [49]. Boynton, Gillham, and colleagues first achieved plastid transformation of Chlamydomonas with this plastid promoter in 1988 [40]; Svab, Hajukiewicz and Maliga followed, in 1990, with the first report of plastid transformation in tobacco, although they used a different fragment of chloroplast DNA in their target plasmid [50]. Plastid transformation has since been reported in many flowering plants, such as tobacco [50–52], Arabidopsis [53,54], potato [55], rice [56], rapeseed [57], and tomato [58]. A protocol for plastid transformation of an elite rapeseed cultivar (*B. napus* L.) has been developed [27]. The highest protein yields were accomplished by using the *rrn* promoter from the plastid rRNA operon, as reported in several reviews [7,8,59,60].

**Table 2.** Promoters, untranslated regions, and insertion sites commonly used for plastid transformation.

| Promoters | 5′-UTRs | 3′-UTRs | Insertion Sites |
|---|---|---|---|
| *psbA* | *ggagg* | *psbA* | *trnI/trnA* |
| *rrn* | T7g10 | *rps16* | *rbcL/accD* |
| *rbcL* | *rbcL* | *rbcL* | *trnfM-trnG* |
| | *psbA* | *petD* | *trnV/rps12* |
| | *atpB* | | *trnN-trnR* |
| | | | *ycf3-trnS* |

### 2.3. Selection Methods and Regeneration Protocols for Transplastomic Cells and Plants

Ideally, plant species commonly used for plastid transformation, such as tobacco, should have a well-developed tissue culture system and high regeneration potential. It is therefore generally considered that plastid transformation is confined to crops that fulfill these two criteria. Different plant species, or even different cultivars from the same species, require specific tissue culture conditions. Protocols for in vitro culture often take time to set up, including the optimization of the growth temperature, humidity, the composition of the culture medium, to name a few variables. Plastid transformation studies and their applications are thus still largely restricted to certain plant species like rice, tobacco, and lettuce.

Although plant tissue culture conditions vary extensively, the selection procedure for screening homoplasmic plants are quite similar and have not changed much over the past decades [60]. Several antibiotics have been used in plastid transformation [15,61]. For example, the expression of *Neomycin Phosphotransferase* (*nptII*) or *Aminoglycoside 3′-Phosphotransferase* (*aphA6*) confers resistance to kanamycin [31,62], while the expression of *Bialaphos Resistance* (*bar*) provides resistance to the herbicide glyphosate in transgenic plants. However, the gene *Streptomycin 3′-Adenylyltransferase* (*aadA*), a spectinomycin and streptomycin resistance gene, remains the most commonly used marker for plastid transformation [15,60]. Spectinomycin inhibits plastid protein translation by binding to chloroplast ribosomes [63,64]. At least in Chlamydomonas and tobacco, mutations in the plastid ribosomal subunit targeted by spectinomycin confer resistance to the antibiotic. In fact, the first report of plastid transformation in tobacco used a plastid DNA fragment carrying such a mutation to demonstrate transformation potential and act as a selectable marker at the same time [50].

Parker et al. reported that one spectinomycin tolerance strategy in Arabidopsis involves Acetyl-Coa Carboxylase 2 (ACC2) [65,66]. Indeed, seedlings cannot develop beyond the cotyledon stage under spectinomycin selection when they have a functional copy of the nuclear *ACC2* gene, thereby limiting selection efficiency in Arabidopsis plastid transformation. Based on these results, Yu et al. went on to use an *acc2* loss of function Arabidopsis mutant (SALK_148966C) to test the effectiveness of particle bombardment with a plasmid encoding an aadA-GFP fusion [25]. Transformation efficiency increased around 100-fold in the *acc2* mutant background. Ruf et al. later used clustered regularly interspaced short palindromic repeats (CRISPR)/Cas9-mediated genome editing to inactivate the *ACC2* locus and create a recipient line for plastid transformation. Transformants in this *acc2* mutant background grew to maturity and produced seeds, as the loss of ACC2 function is not accompanied by visible growth phenotypes [67]. These reports may, therefore, open an avenue for routine engineering of the plastid genome in Arabidopsis, and provide valuable information for plastid transformation in other, more recalcitrant species.

## 3. Can Plastid Transformation Work in Crops? CRISPR-Cas, Morphogenic Regulators, and Protoplast Regeneration Can Help

Although tobacco is an excellent species for plastid transformation, its leaves are not edible, so any recombinant protein produced in the chloroplast needs to be purified before use. By contrast, plastid transformation of vegetable and fruit plants would offer significant advantages for the production of edible vaccines. Fruits are easier to store and transport relative to leafy vegetables like lettuce. The establishment of plastid transformation in species that bear fruit suitable for human consumption is a hot research topic.

For plastid transformation, the formation of double-strand DNA breaks (DSBs) in the chloroplast genome is a critical factor for HR [56], which might be stimulated at the chosen editing site by the use of the CRISPR nuclease Cas9 or transcription activator-like effector nucleases (TALENs). The CRISPR/Cas9 system allows for precise genome editing by guide RNAs (gRNAs) that direct the Cas9 nuclease to a target site. To increase plastid transformation efficiency, Yoo et al. [68] applied genome editing in Chlamydomonas chloroplasts by introducing two plasmids: one plasmid carried both a gRNA and a Cas9 expression cassette, while the other plasmid bore the donor DNA fragment for integration at the DSB site created by the action of Cas9 and the gRNA. Both plasmids were transformed into the alga by biolistics; cell lysates from individual transformants were screened by PCR 28 days after bombardment. When Cas9/gRNA was placed under the control of the strong chloroplast *psaA* promoter, two transformants out of 20 had the donor DNA at the intended integration site. By contrast, no HR events were observed when Cas9/gRNA were expressed from the weaker plastid *psbD* promoter, or when the first plasmid carried the gRNA but not Cas9. These results indicate that DNA breaks indeed promote donor DNA integration. Although the CRISPR-Cas approach has not yet been applied to plastid transformation in plants, a sequential transformation method was used to generate nuclear HR transgenic lines in Arabidopsis [69]. Two transgenic lines expressing Cas9 were used as parental lines for stable transformation with a second construct carrying the sgRNA and donor DNA. Using this sequential transformation approach, transgenic plants with HR in the target site of the nuclear genome were successfully identified [69]. Similarly, incorporating Cas9 and gRNAs into plastid transformation protocols should accelerate DSB formation and raise the frequency of HR, as high expression of Cas9 would enhance the likelihood of HR in transgenic plants.

TALENs constitute another potential strategy for site-specific gene modification and have been widely used in algae and flowering plants to generate transformants with desired traits [70–72]. For example, Li et al. [73] co-bombarded a plasmid carrying a TALEN construct and a plasmid carrying an HR DNA fragment in rice using biolistics. The transformation efficiency of the TALEN-assisted group was twice as high as that of a group transformed only with the HR fragment-containing plasmid. Together, these recent results provide a convincing foundation for the application of genome editing by Cas9 or TALENs in plastid transformation.

Another limiting factor for any plastid transformation effort is the regeneration efficiency during tissue culture. Harnessing growth-related genes to promote plant growth is not a new concept: in fact, the idea of affecting plant morphogenesis to recover transformants can be traced back to the 1980s [74]. Previous studies have shown that overexpressing morphogenic genes can increase nuclear transformation rates and enhance regeneration ability [75,76]. To date, several morphogenic genes (*Baby boom* (*BBM*) and *Wuschel2* (*WUS2*)) have been successfully used in nuclear transformation [76–78]. This strategy has not yet been applied to plastid transformation and is worth exploring as an approach to support plastid transformation in more species, especially those plants that are recalcitrant to transformation or are only marginally transformable.

Thanks to the advances in transformation technology, research facilities and purity of chemicals, protoplasts might be an option worth (re)-considering for plants that are not amenable to transformation by biolistics or when morphogenic regulators are ineffective. Protoplast regeneration was reported as early as the 1970s [79]. Unfortunately, protoplast regeneration is perceived by many as being hard to establish and labor-intensive and is always considered as a last resort, when it is even considered. This impression might stem from the poor quality of chemicals and laboratory conditions in the early days. With the development of more effective tissue culture protocols and the evolution of laboratory facilities, protoplast regeneration may be established in any plant tissue culture laboratory with minimal effort [44,80]. This technique, in fact, offers many advantages: there is no need for expensive instruments and consumables cost is low. Protoplast transformation may even be considered as being more efficient than other transformation techniques for flowering plants. Multiple rounds of particle bombardment are typically required on hundreds of leaves to obtain transformants, while protoplasts derived from only two leaves and transformed with PEG vesicles loaded with target DNA can produce positive clones in a single experiment [42]. Numerous articles have reported success in PEG-mediated protoplast transformation and regeneration [42,80–82]; some studies also demonstrated that protoplast transformation supported the transformation of non-transformable plants and increased the expression of a foreign (target) gene. In cauliflower, no transformants were obtained by biolistics, but DNA uptake was achieved by protoplasts in the presence of PEG, although the transformation efficiency was not high [82]. In potato, high activity levels were detected for a reporter carrying the *beta-Glucuronidase* (*GUS*) gene transformed into protoplasts using the PEG method. By contrast, transgenic plants obtained by biolistics showed variable GUS activity levels, and truncated RNA species were detected in plants with low GUS activity [81]. This phenomenon of transgene silencing in lines generated by biolistics is frequently reported, in contrast to Agrobacterium- or PEG-mediated transformation in rice and barley [83,84].

To date, our laboratory has successfully transformed protoplasts with the PEG-mediated method and regenerated whole plants for tobacco and *N. benthamiana*, tomato, wild tomato (*S. peruvianum*), rapid-cycling Brassica and Arabidopsis. Taking tobacco as an example, it only took our laboratory about two years to establish the entire procedure, from protoplast isolation and CRISPR/Cas9 transformation by PEG-mediated transfection to protoplast regeneration [80]; by no means a short time frame, but definitely manageable and attainable. With the incorporation of the latest tools such as SWNTs delivery or morphogenic regulators, we believe that our platform can attain a wider application by reaching more crops or when applied to plastid transformation.

## 4. What Needs Can Plastid Transformation Fill?

Plastid transformation was developed over three decades ago, and numerous transplastomic algae and flowering plants have been created successfully. What can these transplastomic plants do? Here, we review the plastid transformation literature to give our readers an understanding of the potential applications.

Engineering chloroplasts with desired agronomic traits has garnered interest in recent years. For instance, expressing a bacterial *4-Hydroxyphenylpyruvate Dioxygenase* (*HPPD*) gene in tobacco or soybean chloroplasts conferred enhanced herbicide resistance [30], while expression of *Betaine Aldehyde Dehydrogenase* (*BADH*) in carrot chloroplasts provided strong salt tolerance [39]. As mentioned earlier, tobacco remains by far the most suitable species for plastid transformation, although the technique has also been successfully applied to other species such as tomato [85,86], potato [37], maize [87], sugar beet [24], cotton [31] and wheat [88]. For example, the simultaneous expression of protease inhibitors and chitinase in transplastomic tobacco plants conferred resistance to multiple biotic and abiotic stresses [89]. Multiple economic and agronomic traits of interest have been engineered into chloroplasts, including resistance to cold, drought, insects or herbicides as well as salt tolerance [30,90–92]. Herbicide resistance is perhaps one of the most notable traits in plastid transformation. Plants resistant to the herbicide glyphosate (commercialized as Roundup) were generated by introducing the *5-Enolpyruvylshikimate-3-Phosphate Synthase* (*EPSPS*) gene into the tobacco plastid genome, which encodes an enzyme that detoxifies glyphosate [4]. Agronomic characters obtained by the engineering of the chloroplast genome are listed in Table 3; most transformations relied on biolistics, although these largely predate the publication of the SWNTs tool, which may provide a new opportunity for scientists thanks to its low cost and ease of use.

**Table 3.** Agronomic traits engineered into crops by plastid transformation.

| Integration Site | Regulatory Sequence Promoter/Terminator | Transgene | Efficiency of Expression | Enhanced Trait(s) | References |
|---|---|---|---|---|---|
| *rbcL/accD* | *Prrn/rbcL 3′* | *panD* | >4-fold β-alanine | Photosynthesis and biomass production in response to high temperature stress | [90] |
| *trnI/trnA* | *T7g10* or *PpsbA* | *RbcS* | >150-fold RbcS transcript | Photosynthetic performance | [93] |
| *trnI/trnA* | *PpsbA/TpsbA* | *AQP1, TicAQP1* | 16-fold transcript | Photosynthetic performance | [94] |
| *trnV/orf708* | *PpsbA/TpsbA* | *bicA* | ~0.1% TLP | Photosynthetic performance | [95] |
| *trnV/rps12* | *Prrn/T7g10/Trps16* | *Trx f, Trx m* | 700% leaf starch increased | Carbohydrate/starch content | [96] |
| *trnI/trnA* | *PpsbA/TpsbA* | *bgl-1* | >160-fold enzyme | Resistance to whitefly and aphids | [97] |
| *trnI/trnA* | *Prrn/ggagg/psbA 3′* | *tps1* | >169-fold transcript | Drought tolerance | [98] |
| *trnI/trnA* | *PpsbA/T7g10/Trps16* | *badh* | 93–101 μmol/g DW | Salt tolerance (up to 400 mM NaCl) | [39] |
| *rbcL/accD* | *PpsbA/rbcL 3′* | *hppd* | 5% TSP | Resistance to herbicide | [30] |
| *rbcL/accD* | *Prrn/TpsbA* | *EPSPS* | NR | Resistance to the herbicide glyphosate (>5 mM) | [99] |

**Table 3.** *Cont.*

| Integration Site | Regulatory Sequence Promoter/Terminator | Transgene | Efficiency of Expression | Enhanced Trait(s) | References |
|---|---|---|---|---|---|
| *rps7,12/trnV* | *Prrn/T7g10/Trps16* | *EPSPS* | >10% TSP | Resistance to the herbicide glyphosate | [92] |
| *trnV/rps12,7* | *Prrn/TrbcL* | *bar* | >7% in TSCP | Resistance to the herbicide phosphinothricin | [100] |
| *trnfM/trnG* | *PatpI/Trps16* | *Lycopene β-cyclase, Phytoene synthase* | NR | Herbicide resistance and carotenoid biosynthesis | [85] |
| *trnI/trnA* | *Prrn/T7g10/Trps16* | *mt1* | NR | Phytoremediation capability on mercury accumulation | [101] |
| *trnI/trnA* | *Prrn/ggagg/TpsbA* | *merA, merB* | NR | Phytoremediation capability on mercury accumulation | [102] |
| *trnI/trnA* | *PpsbA/TpsbA* | *RC101, PG1* | 32–38% TSP; 17–26% TSP | Resistance to viral and bacterial infections | [103] |
| *trnI/trnA* | *Prrn/TpsbA* | *Bt-cry2Aa2* | 45.3% TSP | Insecticidal protein content | [104] |
| *trnI/trnA* | *Prrn/T7g10/Trps16* | *MSI-99* | 89.75 μg/g FW | Resistance to rice blast fungus | [105] |
| *trnV/rps12,7* | *Prrn/T7g10/TrbcL* | *cry1Ab* | NR | Resistance to caterpillar (*Anticarsia gemmatalis*) | [106] |
| *trnI/trnA* | *TrbcL* | *Bt-cry9Aa2* | ~10% of TSP | Resistance to potato tuber moth (*Phthorimaea operculella*) | [107] |
| *rbcL/accD* | *Prrn/TpsbA* | *cry2Aa2* | 2–3% TSP | Resistance to moth (*Heliothis virescens, Helicoverpa zea,* and *Spodoptera exigua*) | [108] |
| *rbcL/accD* | *PpsbA/SD/Trsp16* | *TC, γ-TMT* | 3.05 nmol $h^{-1}mg^{-1}$ protein | Vitamin E content in tobacco and lettuce | [109] |
| *trnfM/trnG* | *Prrn/TrbcL* | *HPT, TCY, TMT* | NR | Vitamin E content in fruit; cold-stress tolerance | [35] |
| *trnI/trnA* | *Prrn/TpsbA* | *sporamin. CeCPI, chitinase* | 0.85–1% TSP | Resistance to phytopathogens and insects | [89] |
| *trnI/trnA* | *PpsbA/TpsbA* | *cpo* | 15-fold increased | Resistance to fungal infection (*Fusarium verticillioides, Verticillium dahliae* and *Alternaria alternata*) | [110] |
| *trnI/trnA* | *Prrn/T7g10/TpsbA* | *γ-TMT* | 7.7% TLP | α-tocopherol content to regulate abiotic stress resistance | [91] |
| *trnI/trnA* | *PpsbA/TpsbA* | *PelB, PelD* | 2.42 U/mg; 2.31 U/mg | Resistance to *Erwinia* soft rot | [111] |
| *trnI/trnA* | *PpsbA/TpsbA* | *pta* | 5.16–9.27% TSP | Resistance to aphid, whitefly, Lepidopteran insects, and bacterial and viral pathogens | [112] |
| *trnI/trnA* | *PpsbA/TpsbA* | *phaA* | 14.71 U/mg plant protein | Capacity for cytoplasmic male sterility engineering | [113] |

DW: dry weight; FW: fresh weight; NR: not recorded; SD: Shine-Dalgarno sequence; TLP: total leaf protein; TSCP: total soluble cellular protein; TSP: total soluble protein.

### *4.1. Antigen Vaccines and Protein-based Drugs*

Globally, the number of individuals suffering from diabetes is expected to rise from 170 million in 2000 to a projected 366 million by 2030 [114]. More than 90% of the global population cannot afford the cost of insulin [3,115]. Protein-based drugs such as insulin are expensive because they are produced in yeast fermentation systems and later kept in cold storage, but the final pure product still has a short shelf-life [116]. However, protein-based drugs produced in transplastomic plants may solve many of the associated issues without raising costs or compromising drug efficacy [3]. Many vaccine antigens and biopharmaceuticals have been successfully produced from the chloroplasts of flowering plants.

The induction of insulin production in human subjects is an attractive alternative to daily insulin injections. Exendin-4 (EX4), an analog of the peptide hormone Glucagon-like peptide, was expressed in tobacco chloroplasts, fused to the Cholera toxin B (CTB) subunit to facilitate delivery by crossing the intestinal epithelium. Lyophilized tobacco leaf extracts increased insulin production levels in mice without inducing hypoglycemia, even when a 5000-fold excess dose of CTB-EX4 was delivered orally [117]. In addition, the accumulation of human interferon-gamma in tobacco chloroplasts reached 0.42% of total soluble protein [118]. Unlike microorganisms, plant chloroplasts can perform post-translational modifications of protein-based drugs and promote their proper folding: phosphorylation, amidation, and disulfide bond formation [119].

Human papillomavirus (HPV) is a cause of cervical cancer, which kills over 250,000 women each year. Protein E7 from HPV type 16 (HPV-16 E7) is an attractive anti-cancer vaccine antigen that has been expressed in tobacco via plastid transformation or transient expression [120–122]. The plant-produced proteins successfully induced an immune response and mediated tumor regression in the murine model. Using the transient transfection system with Agrobacterium LBA4404 and the pBIN-NSs vector containing the TSWV NSs silencing suppressor gene, the E7 protein fused with Zera®was expressed only at levels ranging from 0.1–6 g/kg [121]. Via plastid transformation, E7 could reach 0.1% TSP in transplastomic plants [122]. Notably, E7-potato virus X coat protein fusion proteins accumulated to levels around five times higher than the unfused E7 [120,122].

Human coagulation factors made from plants have also been shown to improve immune tolerance in hemophilia murine and canine models [23,123]. In addition, high-level expression of vaccine antigens and therapeutic proteins has been achieved in plant chloroplasts (leaves and roots) or chromoplasts (fruits) for antigens associated with the plague, tetanus, human immunodeficiency virus (HIV), cholera, malaria, Alzheimer's disease and hemophilia [123–130]. Table 4 provides a partial list of vaccine antigens and drug proteins expressed in the chloroplast. Although high levels of protein expression are desirable for chloroplast production of protein-based drugs, excessive expression of foreign proteins may poison host plants. However, the chloroplast of the unicellular green alga Chlamydomonas largely possesses the same machinery necessary for folding and assembling complex eukaryotic proteins, as that of flowering plants and tolerates the accumulation of eukaryotic toxins [20]. Protein-based drugs and vaccine antigens produced in Chlamydomonas chloroplasts are shown in Table 5 [20,131–137].

**Table 4.** Vaccine antigens and protein-based drugs produced by chloroplasts.

| Trait | Protein Being Expressed | Expression | Host Plant | References |
|---|---|---|---|---|
| Insulin | EX4 | 14.3% TSP | tobacco | [117] |
| Hemophilia B | FIX | 1.79 mg/g DW in lettuce; 3.8% TSP in tobacco | lettuce; tobacco | [123] [124] |
| Hemophilia A | FVIII | 852 μg/g DW in lettuce; 370 mg/g FW in tobacco | lettuce; tobacco | [23] [138] |
| Malaria | PMK, MVK, MDD, AACT, HMGS, HMGRt; IPP, FPP, ADS, CYP71AV1, AACPR | 0.1 mg/g FW | tobacco | [125] |
| HIV | Pr55gag | 78–% TSP | tobacco | [139] |
| HPV | E7 | 3–8% TSP | tobacco | [120] |
| Human cytokine | IFNα2b | 3 mg/g FW | tobacco | [126] |
| Human cytokine | IFN-γ | 6% TSP | tobacco | [140] |
| Human cytokine | hCT-1 | 5% TSP | tobacco | [141] |
| Cholera | AMA1 | 7.3 % TSP in tobacco; 13.2 % TSP in lettuce | tobacco; lettuce | [127] |
| Tuberculosis | Mtb72F and ESAT6 | 1.2–7.5% TSP | tobacco | [142] |
| Tuberculosis | CFP10, ESTA6 and dIFN | >0.035% TSP | carrot | [143] |
| Dengue virus | EDIII | 0.8–1.6% TSP | tobacco | [144] |

DW: dry weight; FW: fresh weight; NR: not recorded; TLP: total leaf protein; TSP: total soluble protein.

**Table 5.** Protein-based drugs produced in *Chlamydomonas reinhardtii* chloroplasts.

| Therapeutic Protein | Expression | References |
|---|---|---|
| αCD22HCH23PE40, αCD22PE40 -Targets and kills B cell tumor | 0.2–0.3% TSP | [20] |
| Tumor necrosis factor-related apoptosis-inducing ligand (TRAIL) -Leads to the apoptosis of cancer cells | 0.43–0.67% TSP | [131] |
| GBSS-AMA1, GBSS-MSP1 -Anti-malarial | 0.2–1.0 μg/mg Starch | [132] |
| Human glutamic acid decarboxylase (hGAD65) -For the treatment of Type I diabetes | 0.25–0.3% TSP | [133] |
| Protein VP1 of Foot-and-mouth disease virus (FMDV-VP1) -Mucosal vaccine | 3% TSP | [134] |
| Bovine mammary-associated serum amyloid (M-SAA) -Mucin induction | 3–5% TSP | [135] |
| Protein E2 of classical swine fever virus (CSFV-E2) -Prevents classical swine fever | 1.5–2% TSP | [136] |
| Protein VP2 of Infectious burial disease virus (IBDV-VP2) -Prevents IBDV infection | 0.8–4% TCP | [137] |

DW: dry weight; FW: fresh weight; NR: not recorded; TCP: total cellular protein; TSP: total soluble protein.

### 4.2. Industrial Enzymes and Biomaterials

The chloroplast genome has been repeatedly engineered to produce industrial enzymes and biomaterials. Polyhydroxyalkanoates (PHAs) are a large class of biodegradable polyesters biopolymers naturally synthesized by many microorganisms that can be used as an alternative to petroleum-based plastics [137]. The first described and most well-studied PHA is polyhydroxybutyrate (PHB). Various systems have been adapted for the production of PHB, including microbial cells and various plant tissues. To date, however, the highest level of PHB accumulation was achieved in tobacco plastids, with levels of 18.8% of dry weight (DW). The tobacco system was based on an operon extension strategy to synthesize high PHB levels by introducing a bacterial operon, consisting of three genes encoding enzymes necessary for PHB biosynthesis, into the tobacco chloroplast genome [145]. The high amounts of PHB produced in this system stems from the high flux of the PHB biosynthetic precursor acetyl-CoA released during fatty acid biosynthesis [146]. Typical examples of industrial enzymes and biomaterials obtained through plastid transformation are given in Table 6 [8,97,145,147–151].

**Table 6.** Industrial enzymes and biomaterials obtained via chloroplast production in tobacco.

| Products | Gene(s) | Expression | References |
|---|---|---|---|
| β-Glucosidase | *bgl-1* | 44.4 U/g FW | [97] |
| β-Glucosidase, Cellulases | *bgl1, celA, celB* | 9.9–58.2 U/mg of TSP | [147] |
| Cellulases, Xylanase | *endo, celB, xyn* | 0.38–75.6% TSP | [8] |
| Cell wall-degrading enzyme | *bgl1C, cel6B, cel9A, xeg74* | 5–40% TSP | [148] |
| β-Mannanase | *manI* | 25 U/g FW | [149] |
| Xylanase | *xynA, xyn10A, xyn11B* | 0.2–6% TSP | [150] |
| p-Hydroxybenzoic acid | *UbiC* | 25% DW | [151] |
| Polyhydroxybutyrate | PHB pathway genes | 18.8% TSP | [145] |

DW: dry weight; FW: fresh weight; NR: not recorded; TCP: total cellular protein; TSP: total soluble protein.

### 4.3. Phytoremediation

Mercury (Hg), especially in its organic form, is a highly toxic pollutant that affects humans, animals, and plants alike. At present, phytoremediation is a cost-effective method to remove heavy metals from contaminated soils by using plants to clean up contaminated environments by taking up the desired pollutant [152]. In plants, Hg mainly targets chloroplasts, where it impairs electron transport and photosynthesis. Therefore, chloroplasts would be ideal sites in which to increase resistance to organic and inorganic Hg and repair damage resulting from Hg exposure [153].

Transgenic tobacco plants with engineered chloroplasts exhibited enhanced uptake of inorganic Hg, accumulating about 100-fold more than in untransformed plants [154]. An operon containing the bacterial genes *merA* (mercuric ion reductase gene) and *merB* (organomercurial lyase gene), expressed in tobacco chloroplasts, significantly improved plant tolerance to organic compounds [102]. Integrating the murine *Metallothionein* gene (*MT1*) into the tobacco chloroplast genome allowed high Hg accumulation within tobacco cells. These transplastomic lines were resistant up to 20 μM Hg and remained healthy with normal chlorophyll content and biomass [101]. Plastid transformation may also increase tolerance to high concentrations of copper, and sustain higher growth rates [91].

### 4.4. Biofuels Production

At present, biofuels research mainly focuses on the production of liquid fuel using sugars and lignocellulose from cassava (*Manihot esculenta*), sweet sorghum and other starchy or sugary non-grain crops as raw materials. The most important step in biofuels production is the hydrolysis of lignocellulose [155], with enzymatic digestion being the most efficient and environmentally friendly method, although fungi or bacteria producing the necessary cellulase make the process less efficient and more expensive. A chloroplast-based cellulolytic enzyme has been applied as an industrially pretreated feedstock (*Arundo donax*) for biofuel production [8]. The high levels and compartmentalization of toxic proteins possible within chloroplasts can however protect transgenic plants from multidirectional effects, turning the many chloroplasts within each cell into ideal bioreactors for industrial enzyme production [156]. Enzymes from various fungi and bacteria have been successfully produced in plant chloroplasts: for example, β-glucosidase [8,157], β-1,4-endoglucanase [158], cutinase, exoglucanase, pectinase, xylanase, lipase and acetyl xylan esterase were expressed in tobacco chloroplasts to produce fermentable sugars [111,159–162].

The enzymes derived from transplastomic plants have high activities for further applications. The β-glucosidase BglC and the endoglucanase Cel6A from *Thermobifida fusca* were highly active against synthetic test substrates when expressed in tobacco chloroplasts [157,163]. Treatment of cotton fiber with chloroplast-derived cutinase resulted in enlarged segments and the irreversible unwinding of intertwined inner fibers due to the expansion activity of cutinase. Transgenic plants accumulating cutinase also exhibited esterase and lipase activities [164]. A cocktail of these enzymes efficiently promoted sugar release from filter paper, pine wood and citrus peel [111]. The β-1,4-endoglucanase EGPh from the Archaeon *P. horikoshiiwere*, expressed in tobacco chloroplasts, can hydrolyze carboxymethyl cellulose (CMC) equally well in dry and fresh leaves. Furthermore, the inactive form of EGPh in mature leaves is easily removed by heat treatment [158]. The expression of endo-β-mannanase from *Trichoderma reesei* reached 25 units per gram of leaf (fresh weight), and the activity of endo-β-mannanase from chloroplast extracts was 6–7 fold higher than in *Echerichia coli* extracts, while also having higher temperature stability (40 °C to 70 °C) and wider pH optimum (pH 3.0 to 7.0) [149]. These reports provide convincing evidence that chloroplast-made enzymes have better temperature stability and a wider pH optimum range than those made in other systems [8,111,162].

The contributions of different elements of plastid transformation vectors can be assessed by comparing different efforts to produce the same protein. Xylanase is an important enzyme for lignocellulosic biomass fermentation and sugar release that has been transgenically expressed in tobacco. The xylanase genes used to date have been selected from different fungi or bacteria, expressed using different promoters, and inserted to the different sites in the chloroplast genome (Table 7). A comparison shows that the *Prrn* promoter drives the highest accumulation among these *xyn* transplastomic tobacco plants, although the *Prrn*-driven xylanase still has variable accumulation levels that might arise from the insertion site or source of the gene [8,111,150,161,165]. Kolotilin et al. tested different expression cassettes with *Prrn* or *psbA* promoter to express xylanase in tobacco [150]. Using *Prrn* as the promoter did indeed generate the most transcript in tobacco leave; however, it caused growth retardation in the transgenic plants, and ultimately produced similar protein accumulation levels as the construct using *psbA* as the promoter. These results indicate that mRNA accumulation can be too high in some cases and might even be lethal to the plant.

Another factor to consider is that different sources of genes will have dissimilar codon usage. As the amount of foreign protein expressed in the chloroplast is related to both the promoter and its codon usage preference, codon optimization of the gene of interest provides an alternative way to increase protein expression [166].

**Table 7.** Xylanse produced by tobacco chloroplasts.

| Gene | Source | Expression Level | Promoter | Insertion Site | Reference |
|---|---|---|---|---|---|
| *xyn* | *Alicyclobacillus acidocaldarius* | 35.7% TSP | *Prrn* | rrn16/trnV–rps12/7 | [8] |
| *xynA* | *Bacillus subtilis strain NG-27* | 6% TSP | *Prrn* | rbcL-accD | [165] |
| *xyn2* | *Trichoderma reesei* | 421 U/mg TSP | *Prrn* | trnI-trnA | [111] |
| *xyl10B* | *Thermotoga maritima* | 13%TSP; 61.9 U/mg DW | *Prrn* | rbcL-accD | [161] |
| *xynA* | *Clostridium cellulovorans* | 0.5% TSP | *Prrn* or *PpsbA* | trnI-trnA | [150] |
| *xyn10A* | *Aspergillus niger* | 0.2% TSP | *PpsbA* | trnI-trnA | [150] |
| *xyn11B* | *Aspergillus niger* | 6% TSP | *PpsbA* | trnI-trnA | [150] |

DW: dry weight; TSP: total soluble protein.

### *4.5. Everything Looks Great, Right?*

There are still many challenges facing the widespread adoption of plastid transformation technologies. These include: low transformation efficiency; lack of efficient screening methods for homoplasmy in transgenic plants outside of tobacco, especially in important crops such as rice, corn and other monocotyledonous plants, due to the lack of suitable selection markers and regulatory elements [167]; lack of appropriate tissue-specific regulatory sequences [168]; degradation of foreign proteins [169]; and foreign protein expression sometimes causing male sterility, yellow leaves and stunting. Inducible systems such as the ethanol-induced T7 promoter system, IPTG-lac system, and theophylline-inducible riboswitch system were developed because of the damage caused by constitutive foreign protein expression in host plants, but even these systems suffer from drawbacks: they are complex, toxic and costly. Although plastid transformation has been applied to various fields, it will also be critical to raise awareness among the general public of the usefulness of marker-free transplastomic plants.

## 5. Conclusions and Prospects

There is no doubt that plastid transformation has afforded a new direction for plant genetic engineering and constitutes a research hotspot because of its many advantages over nuclear transformation. Indeed, high transgene expression and engineering of polygenic traits are not amenable to classical nuclear transformation. However, plastid transformation can meet human needs, it is cost-effective, environmentally friendly, safe and efficient. It can be used to modify agronomic traits, and for phytoremediation and biofuels production. More importantly for human health, antigen vaccines and protein-based drugs can be produced in chloroplasts. Although plastid transformation has been achieved in many crops, many others still remain recalcitrant to plastid transformation [137]. The protocols of plastid transformation in crops would offer significant advantages for the production of edible vaccines and medical proteins, biofuels and industrial enzymes, as well as enhanced agronomic traits. In this review, we have illustrated the potential advantages from new studies and technologies like CRISPR-Cas9 for introducing double-strand DNA breaks for HR and create new varieties/mutants (*acc2*) to increase selection efficiency during plastid transformation. No longer limited to biolistic methods, nanotubes can be an alternative material for DNA delivery to increase the donor DNA into chloroplasts. Although they have not yet been applied to plastid transformation, the morphogenic regulators (*BBM*, *WUS2*, and cytokinin biosynthesis genes) and the new chemicals/technology for tissue culture and regeneration can increase the regeneration of transformed cells. These new approaches should also entice researchers to reconsider protoplast-based strategies for plastid transformation. Combined with nuclear transformation and other methods, plastid transformation may allow the production of important proteins.

**Author Contributions:** C.S.L. conceived and designed the review. C.-S.L., P.-C.Y. and W.-J.C. prepared the plastid transformation section. Y.Y., P.-C.Y., W.-J.C., and K.Y. prepared the plastid transformation application section. P.-C.Y., W.-J.C. and C.-S.L. wrote the manuscript with input from all co-authors. All authors read and approved the final manuscript.

**Acknowledgments:** We thank Plant Editors for English editing.

## Abbreviations

| | |
|---|---|
| CMC | Carboxymethyl Cellulose |
| CRISPR | Clustered Regularly Interspaced Short Palindromic Repeats |
| CSFV | Classical Swine Fever Virus |
| DAP | Day After Pollination |
| DSB | Double Strand Breaks |
| DW | Dry Weight |
| FW | Fresh Weight |
| GMO | Genetically Modified Organisms |

| | |
|---|---|
| HIV | Human Immunodeficiency Virus |
| HR | Homologous Recombination |
| IBDV | Infectious Burial Disease Virus |
| PEG | Polyethylene Glycol |
| SWNT | Single-Walled Carbon Nanotubes |
| TALEN | Transcription Activator-Like Effector Nucleases |
| TLP | Total Leaf Protein |
| TRAIL | Tumor Necrosis Factor Related Apoptosis-Inducing Ligand |
| TSP | Total Soluble Protein |
| TSCP | Total Soluble Cellular Protein |
| YFP | Yellow Fluorescent Protein |

## References

1. Maliga, P. Plastid transformation in higher plants. *Annu. Rev. Plant Biol.* **2004**, *55*, 289–313. [CrossRef]
2. McFadden, G.I. Chloroplast origin and integration. *Plant Physiol.* **2001**, *125*, 50–53. [CrossRef] [PubMed]
3. Adem, M.; Beyene, D.; Feyissa, T. Recent achievements obtained by chloroplast transformation. *Plant Methods* **2017**, *13*, 30. [CrossRef] [PubMed]
4. Fathi Roudsari, M.; Salmanian, A.H.; Mousavi, A.; Hashemi Sohi, H.; Jafari, M. Regeneration of glyphosate-tolerant *Nicotiana tabacum* after plastid transformation with a mutated variant of bacterial aroA gene. *Iran. J. Biotechnol.* **2009**, *7*, 247–253.
5. Daniell, H.; Singh, N.D.; Mason, H.; Streatfield, S.J. Plant-made vaccine antigens and biopharmaceuticals. *Trends Plant Sci.* **2009**, *14*, 669–679. [CrossRef] [PubMed]
6. Bock, R.; Warzecha, H. Solar-powered factories for new vaccines and antibiotics. *Trends Biotechnol.* **2010**, *28*, 246–252. [CrossRef] [PubMed]
7. Bock, R. Engineering plastid genomes: Methods, tools, and applications in basic research and biotechnology. *Annu. Rev. Plant Biol.* **2015**, *66*, 211–241. [CrossRef]
8. Castiglia, D.; Sannino, L.; Marcolongo, L.; Ionata, E.; Tamburino, R.; De Stradis, A.; Cobucci-Ponzano, B.; Moracci, M.; La Cara, F.; Scotti, N. High-level expression of thermostable cellulolytic enzymes in tobacco transplastomic plants and their use in hydrolysis of an industrially pretreated *Arundo donax* L. biomass. *Biotechnol. Biofuels* **2016**, *9*, 154. [CrossRef]
9. Lu, X.M.; Yin, W.B.; Hu, Z.M. Chloroplast transformation. *Methods Mol. Biol.* **2006**, *318*, 285–303.
10. Cui, Y.; Qin, S.; Jiang, P. Chloroplast transformation of *Platymonas (Tetraselmis) subcordiformis* with the *bar* gene as selectable marker. *PLoS ONE* **2014**, *9*, e98607. [CrossRef]
11. Hanson, M.R.; Gray, B.N.; Ahner, B.A. Chloroplast transformation for engineering of photosynthesis. *J. Exp. Bot.* **2013**, *64*, 731–742. [CrossRef]
12. Fuentes, P.; Armarego-Marriott, T.; Bock, R. Plastid transformation and its application in metabolic engineering. *Curr. Opin. Biotechnol.* **2018**, *49*, 10–15. [CrossRef] [PubMed]
13. Sarwar Khan, M.; Mustafa, G.; Ahmad Joyia, F. Technical advances in chloroplast biotechnology. In *Transgenic Crops—Emerging Trends and Future Perspectives*; IntechOpen: London, UK, 2019.
14. Sánchez, E.A.E.; Castillo, J.; Cruz, Q.; García, S. *Biotechnological Applications of Plastid Foreign Gene Expression*; IntechOpen: London, UK, 2018.
15. Day, A.; Goldschmidt-Clermont, M. The chloroplast transformation toolbox: Selectable markers and marker removal. *Plant Biotechnol. J.* **2011**, *9*, 540–553. [CrossRef] [PubMed]
16. Narra, M.; Kota, S.; Velivela, Y.; Ellendula, R.; Allini, V.; Abbagani, S. Construction of chloroplast transformation vector and its functional evaluation in *Momordica charantia* L. *3 Biotech* **2018**, *8*, 140. [CrossRef] [PubMed]
17. Kaushal, C.; Abdin, M.Z.; Kumar, S. Chloroplast genome transformation of medicinal plant *Artemisia annua*. *Plant Biotechnol. J.* **2020**. [CrossRef] [PubMed]

18. Muralikrishna, N.; Srinivas, K.; Kumar, K.; Sadanandam, A. Stable plastid transformation in *Scoparia dulcis* L. *Physiol. Mol. Biol. Plants* **2016**, *22*, 575–581. [CrossRef] [PubMed]
19. Kota, S.; Hao, Q.; Narra, M.; Anumula, V.; Rao, A.V.; Hu, Z.; Abbagani, S. Improved plastid transformation efficiency in *Scoparia dulcis* L. *J. Plant Biotechnol.* **2019**, *46*, 323–330. [CrossRef]
20. Tran, M.; Van, C.; Barrera, D.J.; Pettersson, P.L.; Peinado, C.D.; Bui, J.; Mayfield, S.P. Production of unique immunotoxin cancer therapeutics in algal chloroplasts. *Proc. Natl. Acad. Sci. USA* **2013**, *110*, E15–E22. [CrossRef]
21. Doetsch, N.A.; Favreau, M.R.; Kuscuoglu, N.; Thompson, M.D.; Hallick, R.B. Chloroplast transformation in Euglena gracilis: Splicing of a group III twintron transcribed from a transgenic psbK operon. *Curr. Genet.* **2001**, *39*, 49–60. [CrossRef]
22. Sugiura, C.; Sugita, M. Plastid transformation reveals that moss tRNA(Arg)-CCG is not essential for plastid function. *Plant J.* **2004**, *40*, 314–321. [CrossRef]
23. Kwon, K.C.; Sherman, A.; Chang, W.J.; Kamesh, A.; Biswas, M.; Herzog, R.W.; Daniell, H. Expression and assembly of largest foreign protein in chloroplasts: Oral delivery of human FVIII made in lettuce chloroplasts robustly suppresses inhibitor formation in haemophilia A mice. *Plant Biotechnol. J.* **2018**, *16*, 1148–1160. [CrossRef] [PubMed]
24. De Marchis, F.; Bellucci, M. Plastid transformation in sugar beet: *Beta vulgaris*. *Methods Mol. Biol.* **2014**, *1132*, 367–373.
25. Yu, Q.G.; Lutz, K.A.; Maliga, P. Efficient plastid transformation in Arabidopsis. *Plant Physiol.* **2017**, *175*, 186–193. [CrossRef] [PubMed]
26. Liu, C.W.; Lin, C.C.; Chen, J.J.W.; Tseng, M.J. Stable chloroplast transformation in cabbage (*Brassica oleracea* L. var. capitata L.) by particle bombardment. *Plant Cell Rep.* **2007**, *26*, 1733–1744. [CrossRef]
27. Cheng, L.; Li, H.-P.; Qu, B.; Huang, T.; Tu, J.-X.; Fu, T.-D.; Liao, Y.-C. Chloroplast transformation of rapeseed (*Brassica napus*) by particle bombardment of cotyledons. *Plant Cell Rep.* **2010**, *29*, 371–381. [CrossRef] [PubMed]
28. Nugent, G.D.; Coyne, S.; Nguyen, T.T.; Kavanagh, T.A.; Dix, P.J. Nuclear and plastid transformation of *Brassica oleracea* var. botrytis (cauliflower) using PEG-mediated uptake of DNA into protoplasts. *Plant Sci.* **2006**, *170*, 135–142.
29. Skarjinskaia, M.; Svab, Z.; Maliga, P. Plastid transformation in *Lesquerella fendleri*, an oilseed Brassicacea. *Transgenic Res.* **2003**, *12*, 115–122. [CrossRef]
30. Dufourmantel, N.; Dubald, M.; Matringe, M.; Canard, H.; Garcon, F.; Job, C.; Kay, E.; Wisniewski, J.P.; Ferullo, J.M.; Pelissier, B.; et al. Generation and characterization of soybean and marker-free tobacco plastid transformants over-expressing a bacterial 4-hydroxyphenylpyruvate dioxygenase which provides strong herbicide tolerance. *Plant Biotechnol. J.* **2007**, *5*, 118–133. [CrossRef]
31. Kumar, S.; Dhingra, A.; Daniell, H. Stable transformation of the cotton plastid genome and maternal inheritance of transgenes. *Plant Mol. Biol.* **2004**, *56*, 203–216. [CrossRef]
32. Wang, Y.P.; Wei, Z.Y.; Xing, S.C. Stable plastid transformation of rice, a monocot cereal crop. *Biochem. Biophys. Res. Commun.* **2018**, *503*, 2376–2379. [CrossRef]
33. Okumura, S.; Sawada, M.; Park, Y.W.; Hayashi, T.; Shimamura, M.; Takase, H.; Tomizawa, K.I. Transformation of poplar (*Populus alba*) plastids and expression of foreign proteins in tree chloroplasts. *Transgenic Res.* **2006**, *15*, 637–646. [CrossRef] [PubMed]
34. Kota, S.; Lakkam, R.; Kasula, K.; Narra, M.; Qiang, H.; Allini, V.; Hu, Z.; Abbagani, S. Construction of a species-specific vector for improved plastid transformation efficiency in *Capsicum annuum* L. *3 Biotech* **2019**, *9*, 226. [CrossRef] [PubMed]
35. Lu, Y.H.; Rijzaani, H.; Karcher, D.; Ruf, S.; Bock, R. Efficient metabolic pathway engineering in transgenic tobacco and tomato plastids with synthetic multigene operons. *Proc. Natl. Acad. Sci. USA* **2013**, *110*, E623–E632. [CrossRef]
36. Singh, A.K.; Verma, S.S.; Bansal, K.C. Plastid transformation in eggplant (*Solanum melongena* L.). *Transgenic Res.* **2010**, *19*, 113–119. [CrossRef] [PubMed]
37. Valkov, V.T.; Gargano, D.; Scotti, N.; Cardi, T. Plastid transformation in potato: *Solanum tuberosum*. *Methods Mol. Biol.* **2014**, *1132*, 295–303.
38. Zubko, M.K.; Zubko, E.I.; van Zuilen, K.; Meyer, P.; Day, A. Stable transformation of petunia plastids. *Transgenic Res.* **2004**, *13*, 523–530. [CrossRef]

39. Kumar, S.; Dhingra, A.; Daniell, H. Plastid-expressed betaine aldehyde dehydrogenase gene in carrot cultured cells, roots, and leaves confers enhanced salt tolerance. *Plant Physiol.* **2004**, *136*, 2843–2854. [CrossRef] [PubMed]
40. Boynton, J.E.; Gillham, N.W.; Harris, E.H.; Hosler, J.P.; Johnson, A.M.; Jones, A.R.; Shark, K.B. Chloroplast transformation in *Chlamydomonas* with high velocity microprojectiles. *Science* **1988**, *240*, 1534–1538. [CrossRef] [PubMed]
41. Daniell, H. Transformation and foreign gene expression in plants mediated by microprojectile bombardment. In *Recombinant Gene Expression Protocols. Methods in Molecular Biology*; Tuan, R.S., Ed.; Humana Press: Totowa, NJ, USA, 1997; Volume 62, pp. 463–489.
42. O'neill, C.; Horvath, G.V.; Horvath, E.; Dix, P.J.; Medgyesy, P. Chloroplast transformation in plants: Polyethylene glycol (PEG) treatment of protoplasts is an alternative to biolistic delivery systems. *Plant J.* **1993**, *3*, 729–738. [CrossRef] [PubMed]
43. Kwak, S.Y.; Lew, T.T.S.; Sweeney, C.J.; Koman, V.B.; Wong, M.H.; Bohmert-Tatarev, K.; Snell, K.D.; Seo, J.S.; Chua, N.H.; Strano, M.S. Chloroplast-selective gene delivery and expression in planta using chitosan-complexed single-walled carbon nanotube carriers. *Nat. Nanotechnol.* **2019**, *14*, 447–455. [CrossRef]
44. Wu, F.H.; Shen, S.C.; Lee, L.Y.; Lee, S.H.; Chan, M.T.; Lin, C.S. Tape-Arabidopsis Sandwich—A simpler Arabidopsis protoplast isolation method. *Plant Methods* **2009**, *5*, 16. [CrossRef] [PubMed]
45. Verma, D.; Samson, N.P.; Koya, V.; Daniell, H. A protocol for expression of foreign genes in chloroplasts. *Nat. Protoc.* **2008**, *3*, 739. [CrossRef] [PubMed]
46. Scotti, N.; Cardi, T. Plastid transformation as an expression tool for plant-derived biopharmaceuticals. In *Transgenic Plants. Methods in Molecular Biology*; Dunwell, J.M., Wetten, A.C., Eds.; Humana Press: Totowa, NJ, USA, 2012; Volume 847, pp. 451–466.
47. Santana, I.; Wu, H.; Hu, P.; Giraldo, J.P. Targeted delivery of nanomaterials with chemical cargoes in plants enabled by a biorecognition motif. *Nat. Commun.* **2020**, *11*, 2045. [CrossRef] [PubMed]
48. Dauvillee, D.; Hilbig, L.; Preiss, S.; Johanningmeier, U. Minimal extent of sequence homology required for homologous recombination at the psbA locus in *Chlamydomonas reinhardtii* chloroplasts using PCR-generated DNA fragments. *Photosynth. Res.* **2004**, *79*, 219–224. [CrossRef] [PubMed]
49. Ruhlman, T.; Verma, D.; Samson, N.; Daniell, H. The role of heterologous chloroplast sequence elements in transgene integration and expression. *Plant Physiol.* **2010**, *152*, 2088–2104. [CrossRef] [PubMed]
50. Svab, Z.; Hajukiewicz, P.; Maliga, P. Stable transformation of plastids in higher plants. *Proc. Natl. Acad. Sci. USA* **1990**, *87*, 8526–8530. [CrossRef]
51. Carrer, H.; Hockenberry, T.N.; Svab, Z.; Maliga, P. Kanamycin resistance as a selectable marker for plastid transformation in tobacco. *Mol. Gen. Genet.* **1993**, *241*, 49–56. [CrossRef]
52. Staub, J.M.; Garcia, B.; Graves, J.; Hajdukiewicz, P.T.; Hunter, P.; Nehra, N.; Paradkar, V.; Schlittler, M.; Carroll, J.A.; Spatola, L. High-yield production of a human therapeutic protein in tobacco chloroplasts. *Nat. Biotechnol.* **2000**, *18*, 333–338. [CrossRef]
53. Maliga, P. Towards plastid transformation in flowering plants. *Trends Biotechnol.* **1993**, *11*, 101–107. [CrossRef]
54. Sikdar, S.; Serino, G.; Chaudhuri, S.; Maliga, P. Plastid transformation in *Arabidopsis thaliana*. *Plant Cell Rep.* **1998**, *18*, 20–24. [CrossRef]
55. Sidorov, V.A.; Kasten, D.; Pang, S.Z.; Hajdukiewicz, P.T.; Staub, J.M.; Nehra, N.S. Stable chloroplast transformation in potato: Use of green fluorescent protein as a plastid marker. *Plant J.* **1999**, *19*, 209–216. [CrossRef] [PubMed]
56. Khan, M.S.; Maliga, P. Fluorescent antibiotic resistance marker for tracking plastid transformation in higher plants. *Nat. Biotechnol.* **1999**, *17*, 910–915. [CrossRef]
57. Hou, B.K.; Zhou, Y.H.; Wan, L.H.; Zhang, Z.L.; Shen, G.F.; Chen, Z.H.; Hu, Z.M. Chloroplast transformation in oilseed rape. *Transgenic Res.* **2003**, *12*, 111–114. [CrossRef] [PubMed]
58. Ruf, S.; Hermann, M.; Berger, I.J.; Carrer, H.; Bock, R. Stable genetic transformation of tomato plastids and expression of a foreign protein in fruit. *Nat. Biotechnol.* **2001**, *19*, 870–875. [CrossRef] [PubMed]
59. Scotti, N.; Rigano, M.M.; Cardi, T. Production of foreign proteins using plastid transformation. *Biotechnol. Adv.* **2012**, *30*, 387–397. [CrossRef] [PubMed]
60. Bock, R. Genetic engineering of the chloroplast: Novel tools and new applications. *Curr. Opin. Biotechnol.* **2014**, *26*, 7–13. [CrossRef]

61. Rosellini, D. Selectable markers and reporter genes: A well furnished toolbox for plant science and genetic engineering. *CRC Crit. Rev. Plant Sci.* **2012**, *31*, 401–453. [CrossRef]
62. Huang, F.C.; Klaus, S.M.J.; Herz, S.; Zou, Z.; Koop, H.U.; Golds, T.J. Efficient plastid transformation in tobacco using the aphA-6 gene and kanamycin selection. *Mol. Genet. Genom.* **2002**, *268*, 19–27. [CrossRef]
63. Carter, A.P.; Clemons, W.M.; Brodersen, D.E.; Morgan-Warren, R.J.; Wimberly, B.T.; Ramakrishnan, V. Functional insights from the structure of the 30S ribosomal subunit and its interactions with antibiotics. *Nature* **2000**, *407*, 340–348. [CrossRef]
64. Borovinskaya, M.A.; Shoji, S.; Holton, J.M.; Fredric, K.; Cate, J.H.D. A steric block in translation caused by the antibiotic spectinomycin. *ACS Chem. Biol.* **2007**, *2*, 545–552. [CrossRef]
65. Parker, N.; Wang, Y.X.; Meinke, D. Natural variation in sensitivity to a loss of chloroplast translation in arabidopsis. *Plant Physiol.* **2014**, *166*, 2013–2027. [CrossRef] [PubMed]
66. Parker, N.; Wang, Y.X.; Meinke, D. Analysis of Arabidopsis accessions hypersensitive to a loss of chloroplast translation. *Plant Physiol.* **2016**, *172*, 1862–1875. [CrossRef] [PubMed]
67. Ruf, S.; Forner, J.; Hasse, C.; Kroop, X.; Seeger, S.; Schollbach, L.; Schadach, A.; Bock, R. High-efficiency generation of fertile transplastomic Arabidopsis plants. *Nat. Plants* **2019**, *5*, 282–289. [CrossRef]
68. Yoo, B.C.; Yadav, N.S.; Orozco, E.M.; Sakai, H. Cas9/gRNA-mediated genome editing of yeast mitochondria and Chlamydomonas chloroplasts. *PeerJ* **2020**, *8*, e8362. [CrossRef] [PubMed]
69. Miki, D.; Zhang, W.X.; Zeng, W.J.; Feng, Z.Y.; Zhu, J.K. CRISPR/Cas9-mediated gene targeting in Arabidopsis using sequential transformation. *Nat. Commun.* **2018**, *9*, 1967. [CrossRef] [PubMed]
70. Li, T.; Liu, B.; Spalding, M.H.; Weeks, D.P.; Yang, B. High-efficiency TALEN-based gene editing produces disease-resistant rice. *Nat. Biotechnol.* **2012**, *30*, 390–392. [CrossRef]
71. Ma, L.; Zhu, F.G.; Li, Z.W.; Zhang, J.F.; Li, X.; Dong, J.L.; Wang, T. TALEN-based mutagenesis of lipoxygenase LOX3 enhances the storage tolerance of rice (*Oryza sativa*) seeds. *PLoS ONE* **2015**, *10*, e0143877. [CrossRef] [PubMed]
72. Jung, J.H.; Altpeter, F. TALEN mediated targeted mutagenesis of the caffeic acid O-methyltransferase in highly polyploid sugarcane improves cell wall composition for production of bioethanol. *Plant Mol. Biol.* **2016**, *92*, 131–142. [CrossRef]
73. Li, D.; Tang, N.; Fang, Z.; Xia, Y.M.; Cao, M.L. Co-transfer of TALENs construct targeted for chloroplast genome and chloroplast transformation vector into rice using particle bombardment. *J. Nanosci. Nanotechnol.* **2016**, *16*, 12194–12201. [CrossRef]
74. Smigocki, A.C.; Owens, L.D. Cytokinin gene fused with a strong promoter enhances shoot organogenesis and zeatin levels in transformed plant-cells. *Proc. Natl. Acad. Sci. USA* **1988**, *85*, 5131–5135. [CrossRef]
75. Ebinuma, H.; Sugita, K.; Matsunaga, E.; Yamakado, M. Selection of marker-free transgenic plants using the isopentenyl transferase gene. *Proc. Natl. Acad. Sci. USA* **1997**, *94*, 2117–2121. [CrossRef] [PubMed]
76. Lowe, K.; Wu, E.; Wang, N.; Hoerster, G.; Hastings, C.; Cho, M.J.; Scelonge, C.; Lenderts, B.; Chamberlin, M.; Cushatt, J.; et al. Morphogenic regulators Baby boom and Wuschel improve monocot transformation. *Plant Cell* **2016**, *28*, 1998–2015. [CrossRef] [PubMed]
77. Atkins, P.A.; Voytas, D.F. Overcoming bottlenecks in plant gene editing. *Curr. Opin. Plant Biol.* **2020**, *54*, 79–84. [CrossRef]
78. Maher, M.F.; Nasti, R.A.; Vollbrecht, M.; Starker, C.G.; Clark, M.D.; Voytas, D.F. Plant gene editing through de novo induction of meristems. *Nat. Biotechnol.* **2020**, *38*, 84–89. [CrossRef]
79. Takebe, I.; Labib, G.; Melchers, G. Regeneration of whole plants from isolated mesophyll protoplasts of tobacco. *Naturwissenschaften* **1971**, *58*, 318–320. [CrossRef]
80. Lin, C.S.; Hsu, C.T.; Yang, L.H.; Lee, L.Y.; Fu, J.Y.; Cheng, Q.W.; Wu, F.H.; Hsiao, H.C.W.; Zhang, Y.S.; Zhang, R.; et al. Application of protoplast technology to CRISPR/Cas9 mutagenesis: From single-cell mutation detection to mutant plant regeneration. *Plant Biotechnol. J.* **2018**, *16*, 1295–1310. [CrossRef]
81. Craig, W.; Gargano, D.; Scotti, N.; Nguyen, T.T.; Lao, N.T.; Kavanagh, T.A.; Dix, P.J.; Cardi, T. Direct gene transfer in potato: A comparison of particle bombardment of leaf explants and PEG-mediated transformation of protoplasts. *Plant Cell Rep.* **2005**, *24*, 603–611. [CrossRef]
82. Nugent, G.D.; ten Have, M.; van der Gulik, A.; Dix, P.J.; Uijtewaal, B.A.; Mordhorst, A.P. Plastid transformants of tomato selected using mutations affecting ribosome structure. *Plant Cell Rep.* **2005**, *24*, 341–349. [CrossRef] [PubMed]

83. Kohli, A.; Gahakwa, D.; Vain, P.; Laurie, D.A.; Christou, P. Transgene expression in rice engineered through particle bombardment: Molecular factors controlling stable expression and transgene silencing. *Planta* **1999**, *208*, 88–97. [CrossRef]
84. Travella, S.; Ross, S.M.; Harden, J.; Everett, C.; Snape, J.W.; Harwood, W.A. A comparison of transgenic barley lines produced by particle bombardment and Agrobacterium-mediated techniques. *Plant Cell Rep.* **2005**, *23*, 780–789. [CrossRef] [PubMed]
85. Wurbs, D.; Ruf, S.; Bock, R. Contained metabolic engineering in tomatoes by expression of carotenoid biosynthesis genes from the plastid genome. *Plant J.* **2007**, *49*, 276–288. [CrossRef] [PubMed]
86. Ruf, S.; Bock, R. Plastid transformation in tomato. *Methods Mol. Biol.* **2014**, *1132*, 265–276.
87. Bock, R. Towards plastid transformation in maize. *News Rep.* **2007**, *6*, 2–5.
88. Cui, C.; Song, F.; Tan, Y.; Zhou, X.; Zhao, W.; Ma, F.; Liu, Y.; Hussain, J.; Wang, Y.; Yang, G.; et al. Stable chloroplast transformation of immature scutella and inflorescences in wheat (*Triticum aestivum* L.). *Acta Biochim. Biophys. Sin.* **2011**, *43*, 284–291. [CrossRef]
89. Chen, P.J.; Senthilkumar, R.; Jane, W.N.; He, Y.; Tian, Z.; Yeh, K.W. Transplastomic *Nicotiana benthamiana* plants expressing multiple defence genes encoding protease inhibitors and chitinase display broad-spectrum resistance against insects, pathogens and abiotic stresses. *Plant Biotechnol. J.* **2014**, *12*, 503–515. [CrossRef]
90. Fouad, W.M.; Altpeter, F. Transplastomic expression of bacterial L-aspartate-alpha-decarboxylase enhances photosynthesis and biomass production in response to high temperature stress. *Transgenic Res.* **2009**, *18*, 707–718. [CrossRef]
91. Jin, S.; Daniell, H. Expression of gamma-tocopherol methyltransferase in chloroplasts results in massive proliferation of the inner envelope membrane and decreases susceptibility to salt and metal-induced oxidative stresses by reducing reactive oxygen species. *Plant Biotechnol. J.* **2014**, *12*, 1274–1285. [CrossRef] [PubMed]
92. Ye, G.N.; Hajdukiewicz, P.T.; Broyles, D.; Rodriguez, D.; Xu, C.W.; Nehra, N.; Staub, J.M. Plastid-expressed 5-enolpyruvylshikimate-3-phosphate synthase genes provide high level glyphosate tolerance in tobacco. *Plant J.* **2001**, *25*, 261–270. [CrossRef] [PubMed]
93. Dhingra, A.; Portis, A.R.; Daniell, H. Enhanced translation of a chloroplast-expressed RbcS gene restores small subunit levels and photosynthesis in nuclear RbcS antisense plants. *Proc. Natl. Acad. Sci. USA* **2004**, *101*, 6315–6320. [CrossRef] [PubMed]
94. Fernandez-San Millan, A.; Aranjuelo, I.; Douthe, C.; Nadal, M.; Ancin, M.; Larraya, L.; Farran, I.; Flexas, J.; Veramendi, J. Physiological performance of transplastomic tobacco plants overexpressing aquaporin AQP1 in chloroplast membranes. *J. Exp. Bot.* **2018**, *69*, 3661–3673. [CrossRef]
95. Pengelly, J.; Förster, B.; von Caemmerer, S.; Badger, M.; Price, G.D.; Whitney, S. Transplastomic integration of a cyanobacterial bicarbonate transporter into tobacco chloroplasts. *J. Exp. Bot.* **2014**, *65*, 3071–3080. [CrossRef] [PubMed]
96. Sanz-Barrio, R.; Corral-Martinez, P.; Ancin, M.; Segui-Simarro, J.M.; Farran, I. Overexpression of plastidial thioredoxin f leads to enhanced starch accumulation in tobacco leaves. *Plant Biotechnol. J.* **2013**, *11*, 618–627. [CrossRef] [PubMed]
97. Jin, S.; Kanagaraj, A.; Verma, D.; Lange, T.; Daniell, H. Release of hormones from conjugates: Chloroplast expression of beta-glucosidase results in elevated phytohormone levels associated with significant increase in biomass and protection from aphids or whiteflies conferred by sucrose esters. *Plant Physiol.* **2011**, *155*, 222–235. [CrossRef] [PubMed]
98. Lee, S.-B.; Kwon, H.-B.; Kwon, S.-J.; Park, S.-C.; Jeong, M.-J.; Han, S.-E.; Byun, M.-O.; Daniell, H. Accumulation of trehalose within transgenic chloroplasts confers drought tolerance. *Mol. Breed.* **2003**, *11*, 1–13. [CrossRef]
99. Daniell, H.; Datta, R.; Varma, S.; Gray, S.; Lee, S.-B. Containment of herbicide resistance through genetic engineering of the chloroplast genome. *Nat. Biotechnol.* **1998**, *16*, 345–348. [CrossRef]
100. Lutz, K.A.; Knapp, J.E.; Maliga, P. Expression of bar in the plastid genome confers herbicide resistance. *Plant Physiol.* **2001**, *125*, 1585–1590. [CrossRef]
101. Ruiz, O.N.; Alvarez, D.; Torres, C.; Roman, L.; Daniell, H. Metallothionein expression in chloroplasts enhances mercury accumulation and phytoremediation capability. *Plant Biotechnol. J.* **2011**, *9*, 609–617. [CrossRef]
102. Ruiz, O.N.; Hussein, H.S.; Terry, N.; Daniell, H. Phytoremediation of organomercurial compounds via chloroplast genetic engineering. *Plant Physiol.* **2003**, *132*, 1344–1352. [CrossRef]

103. Lee, S.B.; Li, B.; Jin, S.; Daniell, H. Expression and characterization of antimicrobial peptides Retrocyclin-101 and Protegrin-1 in chloroplasts to control viral and bacterial infections. *Plant Biotechnol. J.* **2011**, *9*, 100–115. [CrossRef]
104. De Cosa, B.; Moar, W.; Lee, S.-B.; Miller, M.; Daniell, H. Overexpression of the Bt cry 2Aa2 operon in chloroplasts leads to formation of insecticidal crystals. *Nat. Biotechnol.* **2001**, *19*, 71–74. [CrossRef]
105. Wang, Y.-P.; Wei, Z.-Y.; Zhang, Y.-Y.; Lin, C.-J.; Zhong, X.-F.; Wang, Y.-L.; Ma, J.-Y.; Ma, J.; Xing, S.-C. Chloroplast-expressed MSI-99 in tobacco improves disease resistance and displays inhibitory effect against rice blast fungus. *Int. J. Mol. Sci.* **2015**, *16*, 4628–4641. [CrossRef] [PubMed]
106. Dufourmantel, N.; Tissot, G.; Goutorbe, F.; Garcon, F.; Muhr, C.; Jansens, S.; Pelissier, B.; Peltier, G.; Dubald, M. Generation and analysis of soybean plastid transformants expressing Bacillus thuringiensis Cry1Ab protoxin. *Plant Mol. Biol.* **2005**, *58*, 659–668. [CrossRef] [PubMed]
107. Chakrabarti, S.K.; Lutz, K.A.; Lertwiriyawong, B.; Svab, Z.; Maliga, P. Expression of the cry9Aa2 Bt gene in tobacco chloroplasts confers resistance to potato tuber moth. *Transgenic Res.* **2006**, *15*, 481. [CrossRef] [PubMed]
108. Kota, M.; Daniell, H.; Varma, S.; Garczynski, S.F.; Gould, F.; Moar, W.J. Overexpression of the *Bacillus thuringiensis* (Bt) Cry2Aa2 protein in chloroplasts confers resistance to plants against susceptible and Bt-resistant insects. *Proc. Natl. Acad. Sci. USA* **1999**, *96*, 1840–1845. [CrossRef]
109. Yabuta, Y.; Tanaka, H.; Yoshimura, S.; Suzuki, A.; Tamoi, M.; Maruta, T.; Shigeoka, S. Improvement of vitamin E quality and quantity in tobacco and lettuce by chloroplast genetic engineering. *Transgenic Res.* **2013**, *22*, 391–402. [CrossRef]
110. Ruhlman, T.A.; Rajasekaran, K.; Cary, J.W. Expression of chloroperoxidase from *Pseudomonas pyrrocinia* in tobacco plastids for fungal resistance. *Plant Sci.* **2014**, *228*, 98–106. [CrossRef]
111. Verma, D.; Kanagaraj, A.; Jin, S.; Singh, N.D.; Kolattukudy, P.E.; Daniell, H. Chloroplast-derived enzyme cocktails hydrolyse lignocellulosic biomass and release fermentable sugars. *Plant Biotechnol. J.* **2010**, *8*, 332–350. [CrossRef]
112. Jin, S.; Zhang, X.; Daniell, H. Pinellia ternata agglutinin expression in chloroplasts confers broad spectrum resistance against aphid, whitefly, lepidopteran insects, bacterial and viral pathogens. *Plant Biotechnol. J.* **2012**, *10*, 313–327. [CrossRef]
113. Ruiz, O.N.; Daniell, H. Engineering cytoplasmic male sterility via the chloroplast genome by expression of β-ketothiolase. *Plant Physiol.* **2005**, *138*, 1232–1246. [CrossRef]
114. Davidson, J.A. Advances in therapy for type 2 diabetes: GLP-1 receptor agonists and DPP-4 inhibitors. *Cleve Clin. J. Med.* **2009**, *76* (Suppl. 5), S28–S38. [CrossRef]
115. Daniell, H.; Lin, C.S.; Yu, M.; Chang, W.J. Chloroplast genomes: Diversity, evolution, and applications in genetic engineering. *Genome Biol.* **2016**, *17*, 134. [CrossRef]
116. Basaran, P.; Rodriguez-Cerezo, E. Plant molecular farming: Opportunities and challenges. *Crit. Rev. Biotechnol.* **2008**, *28*, 153–172. [CrossRef] [PubMed]
117. Kwon, K.C.; Nityanandam, R.; New, J.S.; Daniell, H. Oral delivery of bioencapsulated exendin-4 expressed in chloroplasts lowers blood glucose level in mice and stimulates insulin secretion in beta-TC6 cells. *Plant Biotechnol. J.* **2013**, *11*, 77–86. [CrossRef] [PubMed]
118. Razmi, S.; Jalali Javaran, M.; Bagheri, A.; Honari, H.; Soleimani Zadeh, M. Expression of human interferon gamma in tobacco chloroplasts. *Rom. Biotechnol. Lett.* **2019**, *24*, 208–215. [CrossRef]
119. Řepková, J. Potential of chloroplast genome in plant breeding. *Czech J. Genet. Plant Breed.* **2010**, *46*, 103–113. [CrossRef]
120. Morgenfeld, M.; Lentz, E.; Segretin, M.E.; Alfano, E.F.; Bravo-Almonacid, F. Translational fusion and redirection to thylakoid lumen as strategies to enhance accumulation of human papillomavirus E7 antigen in tobacco chloroplasts. *Mol. Biotechnol.* **2014**, *56*, 1021–1031. [CrossRef]
121. Whitehead, M.; Öhlschläger, P.; Almajhdi, F.N.; Alloza, L.; Marzábal, P.; Meyers, A.E.; Rybicki, E.P. Human papillomavirus (HPV) type 16 E7 protein bodies cause tumour regression in mice. *BMC Cancer* **2014**, *14*, 367. [CrossRef]
122. Morgenfeld, M.; Segretin, M.E.; Wirth, S.; Lentz, E.; Zelada, A.; Mentaberry, A.; Gissmann, L.; Bravo-Almonacid, F. Potato virus X coat protein fusion to human papillomavirus 16 E7 oncoprotein enhance antigen stability and accumulation in tobacco chloroplast. *Mol. Biotechnol.* **2009**, *43*, 243. [CrossRef]

123. Herzog, R.W.; Nichols, T.C.; Su, J.; Zhang, B.; Sherman, A.; Merricks, E.P.; Raymer, R.; Perrin, G.Q.; Hager, M.; Wiinberg, B.; et al. Oral tolerance induction in hemophilia B dogs fed with transplastomic lettuce. *Mol. Ther.* **2017**, *25*, 512–522. [CrossRef]
124. Verma, D.; Moghimi, B.; LoDuca, P.A.; Singh, H.D.; Hoffman, B.E.; Herzog, R.W.; Daniell, H. Oral delivery of bioencapsulated coagulation factor IX prevents inhibitor formation and fatal anaphylaxis in hemophilia B mice. *Proc. Natl. Acad. Sci. USA* **2010**, *107*, 7101–7106. [CrossRef]
125. Saxena, B.; Subramaniyan, M.; Malhotra, K.; Bhavesh, N.S.; Potlakayala, S.D.; Kumar, S. Metabolic engineering of chloroplasts for artemisinic acid biosynthesis and impact on plant growth. *J. Biosci.* **2014**, *39*, 33–41. [CrossRef] [PubMed]
126. Arlen, P.A.; Falconer, R.; Cherukumilli, S.; Cole, A.; Cole, A.M.; Oishi, K.K.; Daniell, H. Field production and functional evaluation of chloroplast-derived interferon-α2b. *Plant Biotechnol. J.* **2007**, *5*, 511–525. [CrossRef]
127. Davoodi-Semiromi, A.; Schreiber, M.; Nalapalli, S.; Verma, D.; Singh, N.D.; Banks, R.K.; Chakrabarti, D.; Daniell, H. Chloroplast-derived vaccine antigens confer dual immunity against cholera and malaria by oral or injectable delivery. *Plant Biotechnol. J.* **2010**, *8*, 223–242. [CrossRef] [PubMed]
128. Kohli, N.; Westerveld, D.R.; Ayache, A.C.; Verma, A.; Shil, P.; Prasad, T.; Zhu, P.; Chan, S.L.; Li, Q.; Daniell, H. Oral delivery of bioencapsulated proteins across blood–brain and blood–retinal barriers. *Mol. Ther.* **2014**, *22*, 535–546. [CrossRef] [PubMed]
129. Arlen, P.A.; Singleton, M.; Adamovicz, J.J.; Ding, Y.; Davoodi-Semiromi, A.; Daniell, H. Effective plague vaccination via oral delivery of plant cells expressing F1-V antigens in chloroplasts. *Infect. Immun.* **2008**, *76*, 3640–3650. [CrossRef] [PubMed]
130. Tregoning, J.S.; Nixon, P.; Kuroda, H.; Svab, Z.; Clare, S.; Bowe, F.; Fairweather, N.; Ytterberg, J.; van Wijk, K.J.; Dougan, G. Expression of tetanus toxin fragment C in tobacco chloroplasts. *Nucleic Acids Res.* **2003**, *31*, 1174–1179. [CrossRef]
131. Yang, Z.; Chen, F.; Li, D.; Zhang, Z.; Liu, Y.; Zheng, D.; Wang, Y.; Shen, G. Expression of human soluble TRAIL in *Chlamydomonas reinhardtii* chloroplast. *Chin. Sci. Bull.* **2006**, *51*, 1703–1709. [CrossRef]
132. Dauvillee, D.; Delhaye, S.; Gruyer, S.; Slomianny, C.; Moretz, S.E.; d'Hulst, C.; Long, C.A.; Ball, S.G.; Tomavo, S. Engineering the chloroplast targeted malarial vaccine antigens in Chlamydomonas starch granules. *PLoS ONE* **2010**, *5*, e15424. [CrossRef]
133. Wang, X.; Brandsma, M.; Tremblay, R.; Maxwell, D.; Jevnikar, A.M.; Huner, N.; Ma, S. A novel expression platform for the production of diabetes-associated autoantigen human glutamic acid decarboxylase (hGAD65). *BMC Biotechnol.* **2008**, *8*, 87. [CrossRef]
134. Sun, M.; Qian, K.; Su, N.; Chang, H.; Liu, J.; Shen, G. Foot-and-mouth disease virus VP1 protein fused with cholera toxin B subunit expressed in *Chlamydomonas reinhardtii* chloroplast. *Biotechnol. Lett.* **2003**, *25*, 1087–1092. [CrossRef]
135. Manuell, A.L.; Beligni, M.V.; Elder, J.H.; Siefker, D.T.; Tran, M.; Weber, A.; McDonald, T.L.; Mayfield, S.P. Robust expression of a bioactive mammalian protein in Chlamydomonas chloroplast. *Plant Biotechnol. J.* **2007**, *5*, 402–412. [CrossRef] [PubMed]
136. He, D.-M.; Qian, K.-X.; Shen, G.-F.; Zhang, Z.-F.; Yi-Nü, L.; Su, Z.-L.; Shao, H.-B. Recombination and expression of classical swine fever virus (CSFV) structural protein E2 gene in *Chlamydomonas reinhardtii* chroloplasts. *Colloids Surf. B Biointerfaces* **2007**, *55*, 26–30. [CrossRef]
137. Dobrogojski, J.; Spychalski, M.; Luciński, R.; Borek, S. Transgenic plants as a source of polyhydroxyalkanoates. *Acta Physiol. Plant* **2018**, *40*, 162. [CrossRef]
138. Sherman, A.; Su, J.; Lin, S.; Wang, X.; Herzog, R.W.; Daniell, H.B. The Journal of the American Society of Hematology, Suppression of inhibitor formation against FVIII in a murine model of hemophilia A by oral delivery of antigens bioencapsulated in plant cells. *Blood* **2014**, *124*, 1659–1668. [CrossRef] [PubMed]
139. Scotti, N.; Alagna, F.; Ferraiolo, E.; Formisano, G.; Sannino, L.; Buonaguro, L.; De Stradis, A.; Vitale, A.; Monti, L.; Grillo, S. High-level expression of the HIV-1 Pr55 gag polyprotein in transgenic tobacco chloroplasts. *Planta* **2009**, *229*, 1109–1122. [CrossRef] [PubMed]
140. Leelavathi, S.; Reddy, V.S. Chloroplast expression of His-tagged GUS-fusions: A general strategy to overproduce and purify foreign proteins using transplastomic plants as bioreactors. *Mol. Breed.* **2003**, *11*, 49–58. [CrossRef]

141. Farran, I.; Río-Manterola, F.; Íñiguez, M.; Gárate, S.; Prieto, J.; Mingo-Castel, A.M. High-density seedling expression system for the production of bioactive human cardiotrophin-1, a potential therapeutic cytokine, in transgenic tobacco chloroplasts. *Plant Biotechnol. J.* **2008**, *6*, 516–527. [CrossRef]
142. Lakshmi, P.S.; Verma, D.; Yang, X.; Lloyd, B.; Daniell, H. Low cost tuberculosis vaccine antigens in capsules: Expression in chloroplasts, bio-encapsulation, stability and functional evaluation in vitro. *PLoS ONE* **2013**, *8*, e54708. [CrossRef]
143. Permyakova, N.V.; Zagorskaya, A.A.; Belavin, P.A.; Uvarova, E.A.; Nosareva, O.V.; Nesterov, A.E.; Novikovskaya, A.A.; Zav'yalov, E.L.; Moshkin, M.P.; Deineko, E.V. Transgenic carrot expressing fusion protein comprising *M. tuberculosis* antigens induces immune response in mice. *Biomed. Res. Int.* **2015**, *2015*, 417565. [CrossRef]
144. Gottschamel, J.; Lössl, A.; Ruf, S.; Wang, Y.; Skaugen, M.; Bock, R.; Clarke, J.L. Production of dengue virus envelope protein domain III-based antigens in tobacco chloroplasts using inducible and constitutive expression systems. *Plant Mol. Biol.* **2016**, *91*, 497–512. [CrossRef]
145. Bohmert-Tatarev, K.; McAvoy, S.; Daughtry, S.; Peoples, O.P.; Snell, K.D. High levels of bioplastic are produced in fertile transplastomic tobacco plants engineered with a synthetic operon for the production of polyhydroxybutyrate. *Plant Physiol.* **2011**, *155*, 1690–1708. [CrossRef] [PubMed]
146. Snell, K.D.; Peoples, O.P. PHA bioplastic: A value-added coproduct for biomass biorefineries. *Biofuels Bioprod. Biorefining Innov. Sustain. Econ.* **2009**, *3*, 456–467. [CrossRef]
147. Espinoza-Sánchez, E.A.; Torres-Castillo, J.A.; Rascón-Cruz, Q.; Zavala-García, F.; Sinagawa-García, S.R. Production and characterization of fungal β-glucosidase and bacterial cellulases by tobacco chloroplast transformation. *Plant Biotechnol. Rep.* **2016**, *10*, 61–73. [CrossRef]
148. Petersen, K.; Bock, R. High-level expression of a suite of thermostable cell wall-degrading enzymes from the chloroplast genome. *Plant Mol. Biol.* **2011**, *76*, 311–321. [CrossRef]
149. Agrawal, P.; Verma, D.; Daniell, H. Expression of *Trichoderma reesei* beta-mannanase in tobacco chloroplasts and its utilization in lignocellulosic woody biomass hydrolysis. *PLoS ONE* **2011**, *6*, e29302. [CrossRef]
150. Kolotilin, I.; Kaldis, A.; Pereira, E.O.; Laberge, S.; Menassa, R. Optimization of transplastomic production of hemicellulases in tobacco: Effects of expression cassette configuration and tobacco cultivar used as production platform on recombinant protein yields. *Biotechnol. Biofuels* **2013**, *6*, 65. [CrossRef]
151. Viitanen, P.V.; Devine, A.L.; Khan, M.S.; Deuel, D.L.; Van Dyk, D.E.; Daniell, H. Metabolic engineering of the chloroplast genome using the *Echerichia coli* ubiC gene reveals that chorismate is a readily abundant plant precursor for p-hydroxybenzoic acid biosynthesis. *Plant Physiol.* **2004**, *136*, 4048–4060. [CrossRef]
152. Terry, N.; Zayed, A.M.; De Souza, M.P.; AS, T. Selenium in higher plants. *Annu. Rev. Plant Physiol. Plant Mol. Biol.* **2000**, *51*, 401–432. [CrossRef]
153. Pilon-Smits, E.; Pilon, M. Phytoremediation of metals using transgenic plants. *CRC Crit. Rev. Plant Sci.* **2002**, *21*, 439–456. [CrossRef]
154. Hussein, H.S.; Ruiz, O.N.; Terry, N.; Daniell, H. Phytoremediation of mercury and organomercurials in chloroplast transgenic plants-Enhanced root uptake, translocation to shoots, and volatilization. *Environ. Sci. Technol.* **2007**, *41*, 8439–8446. [CrossRef]
155. Zhang, Z.; Donaldson, A.A.; Ma, X. Advancements and future directions in enzyme technology for biomass conversion. *Biotechnol. Adv.* **2012**, *30*, 913–919. [CrossRef]
156. Jin, S.; Daniell, H. The engineered chloroplast genome just got smarter. *Trends Plant Sci.* **2015**, *20*, 622–640. [CrossRef] [PubMed]
157. Gray, B.N.; Yang, H.; Ahner, B.A.; Hanson, M.R. An efficient downstream box fusion allows high-level accumulation of active bacterial beta-glucosidase in tobacco chloroplasts. *Plant Mol. Biol.* **2011**, *76*, 345–355. [CrossRef] [PubMed]
158. Nakahira, Y.; Ishikawa, K.; Tanaka, K.; Tozawa, Y.; Shiina, T. Overproduction of hyperthermostable beta-1,4-endoglucanase from the archaeon *Pyrococcus horikoshii* by tobacco chloroplast engineering. *Biosci. Biotechnol. Biochem.* **2013**, *77*, 2140–2143. [CrossRef] [PubMed]
159. Pantaleoni, L.; Longoni, P.; Ferroni, L.; Baldisserotto, C.; Leelavathi, S.; Reddy, V.S.; Pancaldi, S.; Cella, R. Chloroplast molecular farming: Efficient production of a thermostable xylanase by *Nicotiana tabacum* plants and long-term conservation of the recombinant enzyme. *Protoplasma* **2014**, *251*, 639–648. [CrossRef]
160. Xue, X.-Y.; Mao, Y.-B.; Tao, X.-Y.; Huang, Y.-P.; Chen, X.-Y. New approaches to agricultural insect pest control based on RNA interference. *Adv. Insect. Phys.* **2012**, *42*, 73–117.

161. Kim, J.Y.; Kavas, M.; Fouad, W.M.; Nong, G.; Preston, J.F.; Altpeter, F. Production of hyperthermostable GH10 xylanase Xyl10B from Thermotoga maritima in transplastomic plants enables complete hydrolysis of methylglucuronoxylan to fermentable sugars for biofuel production. *Plant Mol. Biol.* **2011**, *76*, 357–369. [CrossRef] [PubMed]
162. Daniell, H.; Ribeiro, T.; Lin, S.N.; Saha, P.; McMichael, C.; Chowdhary, R.; Agarwal, A. Validation of leaf and microbial pectinases: Commercial launching of a new platform technology. *Plant Biotechnol. J.* **2019**, *17*, 1154–1166. [CrossRef]
163. Gray, B.N.; Ahner, B.A.; Hanson, M.R. High-level bacterial cellulase accumulation in chloroplast-transformed tobacco mediated by downstream box fusions. *Biotechnol. Bioeng.* **2009**, *102*, 1045–1054. [CrossRef]
164. Verma, D.; Jin, S.; Kanagaraj, A.; Singh, N.D.; Daniel, J.; Kolattukudy, P.E.; Miller, M.; Daniell, H. Expression of fungal cutinase and swollenin in tobacco chloroplasts reveals novel enzyme functions and/or substrates. *PLoS ONE* **2013**, *8*, e57187. [CrossRef]
165. Leelavathi, S.; Gupta, N.; Maiti, S.; Ghosh, A.; Reddy, V.S. Overproduction of an alkali- and thermo-stable xylanase in tobacco chloroplasts and efficient recovery of the enzyme. *Mol. Breed.* **2003**, *11*, 59–67. [CrossRef]
166. Kwon, K.C.; Chan, H.T.; Leon, I.R.; Williams-Carrier, R.; Barkan, A.; Daniell, H. Codon optimization to enhance expression yields insights into chloroplast translation. *Plant Physiol.* **2016**, *172*, 62–77. [CrossRef] [PubMed]
167. Zhang, J.; Li, Y.; Jin, J.; Chen, Q.; Xie, X.; Wang, Z.; Luo, Z.; Yang, J. Recent advances in tobacco chloroplast genetic engineering. *Tobacco Sci. Technol.* **2017**, *50*, 88–98.
168. Valkov, V.T.; Scotti, N.; Kahlau, S.; MacLean, D.; Grillo, S.; Gray, J.C.; Bock, R.; Cardi, T. Genome-wide analysis of plastid gene expression in potato leaf chloroplasts and tuber amyloplasts: Transcriptional and posttranscriptional control. *Plant Physiol.* **2009**, *150*, 2030–2044. [CrossRef]
169. Ahmad, N.; Michoux, F.; Lossl, A.G.; Nixon, P.J. Challenges and perspectives in commercializing plastid transformation technology. *J. Exp. Bot.* **2016**, *67*, 5945–5960. [CrossRef]

# Nuclear Integrants of Organellar DNA Contribute to Genome Structure and Evolution in Plants

**Guo-Jun Zhang [1,2], Ran Dong [1], Li-Na Lan [1], Shu-Fen Li [1], Wu-Jun Gao [1,*] and Hong-Xing Niu [1,*]**

[1] College of Life Sciences, Henan Normal University, Xinxiang 453007, China; zgj003@126.com (G.-J.Z.); randong0930@163.com (R.D.); lanlina281@163.com (L.-N.L.); lishufen83@163.com (S.-F.L.)
[2] School of Basic Medical Sciences, Xinxiang Medical University, Xinxiang 453003, China
* Correspondence: gaowujun@htu.cn (W.-J.G.); hongxingniu@htu.cn (H.-X.N.)

**Abstract:** The transfer of genetic material from the mitochondria and plastid to the nucleus gives rise to nuclear integrants of mitochondrial DNA (NUMTs) and nuclear integrants of plastid DNA (NUPTs). This frequently occurring DNA transfer is ongoing and has important evolutionary implications. In this review, based on previous studies and the analysis of NUMT/NUPT insertions of more than 200 sequenced plant genomes, we analyzed and summarized the general features of NUMTs/NUPTs and highlighted the genetic consequence of organellar DNA insertions. The statistics of organellar DNA integrants among various plant genomes revealed that organellar DNA-derived sequence content is positively correlated with the nuclear genome size. After integration, the nuclear organellar DNA could undergo different fates, including elimination, mutation, rearrangement, fragmentation, and proliferation. The integrated organellar DNAs play important roles in increasing genetic diversity, promoting gene and genome evolution, and are involved in sex chromosome evolution in dioecious plants. The integrating mechanisms, involving non-homologous end joining at double-strand breaks were also discussed.

**Keywords:** genome structure; genome evolution; DNA transfer; nuclear integrants of plastid DNA (NUPT); nuclear integrants of mitochondrial DNA (NUMT)

---

## 1. Introduction

Three genetic compartments coexist in the plant cells: the nucleus and two cytoplasmic organelles, namely, chloroplast (plastid) and mitochondria. Among these cellular parts, the nucleus harbors the majority of the genetic material, and chloroplast and mitochondria contain relatively little but essential genetic substance. It is widely accepted that chloroplasts and mitochondria in eukaryotic cells are descended from erstwhile free-living organisms (α-proteobacteria and cyanobacteria) by endosymbiosis more than a billion years ago [1–3]. When the three genetic compartments were involved in one cell, genetic flux among them occurred frequently [4–6]. In theory, there are six types of DNA transfer occur among the three genetic compartments. At least five types, including mitochondrion-to-nucleus, plastid-to-nucleus, plastid-to-mitochondrion, nucleus-to-mitochondrion, and mitochondrion-to-plastid DNA transfer, have been observed presently based on experimental and bioinformatics data [4]. The transfer frequency is remarked different among the different types of genetic flow. Mitochondrion and chloroplast DNA have been integrated to the nuclear genome with high frequency, whereas other types of transfer occur rarely. In fact, along with the evolutionary process of the plastid and mitochondrion, many genes have relocated from the organelle to host nuclear genomes. Such events combined with the deletion of dispensable organelle genes considerably reduced the genome size of the mitochondrion and plastid during the evolution from their progenitors [4,7]. Organellar DNA transfer into the nuclear genome not just occurred during the establishment of symbiosis; it still actively occurs continuously [8–13]. The organelle-derived DNAs in the nucleus are designated as nuclear integrants of plastid DNA (NUPTs) and nuclear integrants of mitochondrial DNA (NUMTs). The NUPTs and NUMTs discussed here refer to the nuclear organellar DNAs, which still have homologous copies in the organelles and do not include the relocated organellar genes that were previously reviewed [4,7,14].

With the rapid development of high-throughput sequencing technology, the plastid, mitochondrial, and nuclear genomes of a number of plant species have been deciphered. This makes the large-scale analysis of NUPTs and NUMTs in plants be possible. Currently, NUPTs and NUMTs in some plant genomes have been characterized [15–21]. The characterization of nuclear organellar DNAs in plants adds new insights into the role of NUPTs and NUMTs in genome structure and evolution. Growing pieces of evidence has revealed that DNA transfer from organelles contributes considerably to gene and genome evolution and provides a major source of genetic diversity [4,22–24]. In this review, we summarized recent advances in organelle-derived sequences and genome evolution in plants, with focus on the evolutionary fate of these insertions and the roles of NUPTs/NUMTs in the structure and evolution of plant genomes.

## 2. Characterization of Organellar DNA-Derived Sequences in Plants

The nuclear integrants of organellar DNA were first discovered in a study in which a mitochondrial ATPase subunit gene was found in the nuclear genome, as well as in the mitochondrial genome of *Neurospora crassa* [25]. Since then, organelle-derived sequences were examined in the nuclear genome of a number of animals [26–32] and plants [15–21]. The availability of a large amount of plant organelle and nuclear genome data has made it possible to investigate the prevalence and characteristics of NUMTs and NUPTs in plants. In addition to some plant genomes analyzed previously, we estimated the whole genome landscape of NUMTs and NUPTs in the majority of currently sequenced plant species. A dataset of NUPTs in 199 plant genomes and NUMTs in 91 plant genomes was obtained (Tables S1 and S2). The analysis methods were described in Supplemental File 1. It should be noted that the NUPT and NUMT insertions analyzed here are all contiguous fragments reflected from the BLAST results, and the rearrangements of NUPTs/NUMTs from different regions of chloroplast/mitochondrial genomes are not analyzed.

### 2.1. Number and Size Distribution

NUPTs and NUMTs were observed in all examined plant genomes. Their number and size distributions vary markedly among different species. The average size of NUPT ranges from 57 bp (*Chlorella variabilis*) to 3382 bp (*Porphyra umbilicalis*), and the total NUPT length ranges from 1038 bp (*Phaeodactylum tricomutum*) to 9.83 Mb (*Triticum urartu*). The NUPT proportion of these genomes varies from 0.004% to more than 1% in three genomes, including those of *Cucurbita maxima*, *Porphyridium purpureum*, and *Ziziphus jujuba*. The number of NUPTs is few in the majority of algae, and the genomes with NUPT number less than 200 are all from algae. By contrast, NUPTs are abundant in most of the flowering plants, with the highest number found in *Triticum urartu* (Table S1). Similarly, NUMTs are also highly varied among different plant species. The cumulative length within the examined plant genomes varies from 327 bp in *Cyanidioschyzon merolae* to 11.42 Mb in *Capsicum annuum*, and the NUMT proportion accounting for the nuclear genome is from 0.0002% to 2.08% (Table S2). Previous studies showed that the longest NUPT is a 131-kb integrant detected in rice [33,34]; a 620-kb NUMT insertion derived from partially duplicated mitochondrial DNA investigated in *Arabidopsis* is the largest NUMT examined to date [35]. Our analysis detected a 135-kb NUPT in *Gossypium hirsutum*; thus, it is the longest NUPT insertion known so far (Table S1). The NUPT/NUMT fraction usually accounts for less than 0.1%, which is a small fraction of the nuclear genome. However, it should be noted that "old" NUPT and NUMT sequences are usually difficult to detect because of constant mutation and rearrangement during the evolutionary process [29,36]. In addition, NUMTs and NUPTs with high sequence similarity to mitochondrial/chloroplast DNA sequences may be removed as organelle contamination when the nuclear genomes are assembled. Thus, even in "thoroughly sequenced" nuclear genomes, NUMTs/NUPTs may not be completely investigated. Therefore, the content of these organelle-derived sequences is usually underestimated using the standard BLAST method.

Different studies on the correlation between NUPT/NUMT abundance and nuclear genome size show conflicting findings [29,37]. Thus, correlation analysis between genome size and cumulative length or total number of NUPTs and NUMTs in more than 200 plant species was conducted to investigate whether nuclear genome size affect NUPT/NUMT content. A positive correlation existed between nuclear genome size and cumulative lengths of NUPTs/NUMTs, as well as the total number of NUPTs/NUMTs (Figure 1). Previous searches detected no such correlations, probably because of the smaller number of plant nuclear genomes analyzed. No correlations were detected between NUPT/NUMT content and chloroplast/mitochondrial genome size (Tables S1 and S2).

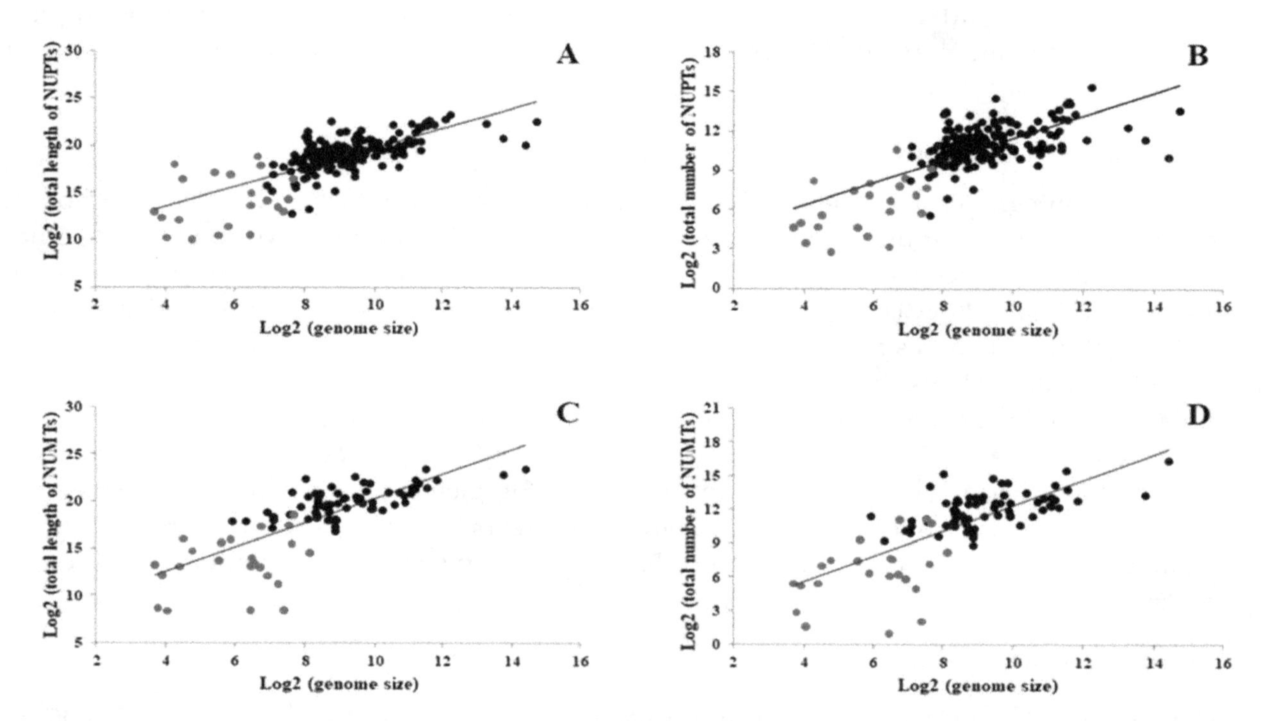


**Figure 1.** Correlation analysis between nuclear genome size and nuclear integrants of plastid DNA (NUPT)/nuclear integrants of mitochondrial DNA (NUMT) content or number in plants: (**A**) genome size versus cumulative length of NUPTs; (**B**) genome size versus total number of NUPTs; (**C**) genome size versus cumulative length of NUMTs; (**D**) genome size versus total number of NUMTs. This dot-plot was generated based on the data presented in Tables S1 and S2. The red dots indicate the results of algal plant analysis, whereas the black dots represent those of analysis involving other plants, most of which are flowering plants.

## 2.2. Organization and Distribution Patterns

NUPTs and NUMTs are frequently organized as clusters [21,38,39]. For example, we observed that approximately 45% of the 3155 NUPT insertions were organized in clusters in the genome of *Asparagus officinalis* [21]. In the model plant species *Arabidopsis* and rice, NUPTs and NUMTs are frequently nonrandomly arranged as loose clusters or tight clusters based on the physically linked degrees [39]. NUPTs/NUMTs are organized into three major patterns in plants: (a) continuous fragments of nuclear DNA collinear with mitochondrial or chloroplast DNA, (b) rearranged NUPTs/NUMTs originating from different regions of one organelle genome with non-uniform orientation, and (c) mosaics containing both NUPTs and NUMTs [40]. The presence of mosaic clusters containing NUPTs and NUMTs indicates that DNA fragments from different organelles might have concatemerized before insertion, or these nuclear regions are hotspots for integration [39,41]. Other organization patterns such as NUPTs/NUMTs with tandem duplications originating from one organelle fragment with the same orientation were occasionally observed [21]. These various organization patterns of NUPTs/NUMTs existing in plant genomes suggested that the origins and the evolutionary paths of the integrated regions may be different, and NUPTs/NUMTs are involved in shaping the plant genome via complicated mechanisms.

NUPTs/NUMTs are usually distributed unevenly in the analyzed plant genomes [20,21,39]. NUPTs and NUMTs are more preferred to distribute in centromeric and pericentromeric regions [36,39], which have few genes and a high level of heterochromatin content [42–44]. Such regions may offer a stable genomic environment for the maintenance of the alien organelle-originated DNA [16,36]. The integrations in these regions should be less harmful than those in other chromosomal regions [36]. For example, in rice and *A. officinalis*, large NUPTs are predominately distributed in the pericentromeric regions of the chromosomes [21,36]. In some species, such as *Arabidopsis* and sorghum, a considerable fraction of NUPTs and NUMTs is co-localized with transposable elements (TEs) [16]. These findings imply that recombination based on repetitive sequence can lead to the rearrangement of chromosome structure and contribute to the various organization patterns of organelle-derived sequences.

The chromatin state seems to be an essential factor that affects the successful insertion of organellar DNA into nucleus. The pre-insertion status of *Oryza sativa* subsp. *indica*-specific NUPTs suggests that the newly transferred organellar sequences are predominantly inserted into open chromatin. This phenomenon has also been observed in humans [45]. However, current existing NUPTs/NUMTs are often detected in heterochromatin regions. Such paradox can be explained by two reasons. One is that the accessibility of chromatin can be modified by external environment, such as stress [46], and/or by genetic crash, such as hybridization [47]. Alternatively, many new insertions in the open chromatin may not be retained because of selective pressure; for that, the insertion into exons of genes can damage gene function. Indeed, most NUPTs/NUMTs are located in introns or untranslated regions [23]. By contrast, the heterochromatin regions are more facilitated for the maintenance of organelle-derived sequences.

### 2.3. Modification Pattern

As alien gene materials of the nuclear genome, NUPTs/NUMTs can cause host genome instability; they may lead to genomic region reshuffling, genome size expansion, and heterochromatization [16,38,48]. Epigenetic regulation, mainly including DNA methylation and histone tail modifications, provides a defense measure for inhibiting the activity of mobile DNA and other types of extraneous DNA [49,50].

The mutation patterns of NUPTs of a number of plant species show biased mutations of cytosine (C) → thymime (T) on one DNA strand, and guanine (G) → adenine (A) substitutions on the opposite strand [4,33,38,51]. These biased substitutions might be due to the hypermethylation of cytosine residues with subsequent deamination [52]. These observations suggest that DNA methylation might play essential roles in regulating integrated organellar DNAs, mainly associated with the transcription suppression of integrated organelle DNA [53,54]. In fact, it has been revealed that a considerable number of NUPTs are methylated, and the DNA methylation intensity and level decrease over evolutionary time. The DNA methylation-modified NUPTs may maintain the stability of plant nuclear genomes against the insertion of organellar DNA sequences and play an important role in the symbiosis of nuclear and organelle genomes [51]. Further methylome data analysis of epigenetic mutants in *Arabidopsis* and rice show that organellar DNA sequences are methylated mainly via the maintenance methylation machinery, involving *DDM1*, *CMT3*, *CMT2*, and *SUVH4/KYP*. However, other mechanisms, such as RNA-directed DNA methylation and homology-dependent DNA methylation machinery may also play roles in the methylation level.

### 2.4. Transfer Rate of Organellar DNA to Nucleus

The frequency of organellar DNA transfer is positively correlated with the number of mitochondria or plastids in the cell [55–57]. Two independent laboratories established a similar screen experimental system to measure the rate of nuclear DNA transferred from plastids; this system was based on transplastomic lines containing a selectable marker gene, whose transfer to the nuclear can confer antibiotic resistance of the regenerated or outcrossed progenies [10,12]. De novo NUPT formation occurred once in approximately 16,000 pollen grains [10] or once in every five million somatic cells screened [12]; this result suggests the high frequency of DNA transfer from the chloroplast to the nucleus. Such high frequency of organellar DNA transfer to the nucleus was also observed in yeast, in which the transfer rate of the mitochondrial DNA to the nucleus is approximately $2 \times 10^{-5}$ per cell per generation [58]. Abiotic stresses, such as subtle heat or cold treatment, can even increase the transfer frequency [55,59]. It is noteworthy that in these experiments, the estimated frequency of DNA transfer from the organelle to the nucleus may be fewer than in reality, because the screen system only focused on the transfer of the selectable marker gene from the organelle genome; those events where the transferred organellar genome fragment did not contain the selective gene may have been missed.

The frequency of plastid-to-nucleus DNA transfer differs markedly among diverse tissues [10,12,60]. The transfer frequency in gametophytic tissue is higher than that in somatic cells. Specifically, the transfer frequency is much higher in male germlines (1 per 11,000 pollen grains) than that in female germlines (1 stable transposition in 273,000 ovules) [60]. The elevated frequency examined within male germline may be caused by the fact that the chloroplast DNA within pollens are more likely to degrade, which was examined in a number of plant species [61].

## 3. Genetic Consequence and Fate of NUPTs/NUMTs in Plants

Once the organellar DNA has integrated into the nuclear genome, the inserted sequences are placed into a new environment and will undergo gradual amelioration to adapt to the host genome. Most of the NUPTs/NUMTs begin to decay because they are not affected by selection pressure [33,38,40]. Analysis of the organization, dynamics, and interspecies variation of current NUPTs/NUMTs in plant genomes suggests the following five post-integration fates of organellar DNA: (a) elimination, (b) mutation, (c) fragmentation, (d) rearrangement, and (e) proliferation (Figure 2).

### *3.1. Elimination*

Considering the high transfer frequencies of organellar DNA to nuclear genome and the relatively low percentage of NUPTs/NUMTs in the plant genome, it is expected that a considerable number of these transferred organellar DNA fragments would be eliminated to prevent rapid genomic enlargement. The fate of the newly integrated chloroplast fragments (kanamycin resistance gene, *neo*) in tobacco reveals that four of nine lines show different level of kanamycin resistance instability, which is caused by the deletion of *neo*. The loss of the organellar transferred fragments usually occurs during mitosis, because intraspecific variation of NUPTs is observed frequently. However, it can also occur during meiosis occasionally [48,62]. A comprehensive analysis of the NUPTs existing in the rice genome showed that 80% of the NUPTs were eliminated within a million years following their integration. The average half-lives of these NUPTs are estimated to be 0.5 million years (Myr) for large insertions and 2.2 Myr for small insertions. Thus, the nuclear genome is in balance between continual integration and fast deletion of the organellar DNA [36]. The rapid elimination is a counterbalance mechanism of the frequent integration of plastid DNAs. Investigation of organellar DNA integrants in the nuclear genomes of *Arabidopsis* and rice have suggested that replication slippage is a mechanism for such deletion events, because short direct repeats are found frequently in the flanking regions of the deleted sequences [33,38].

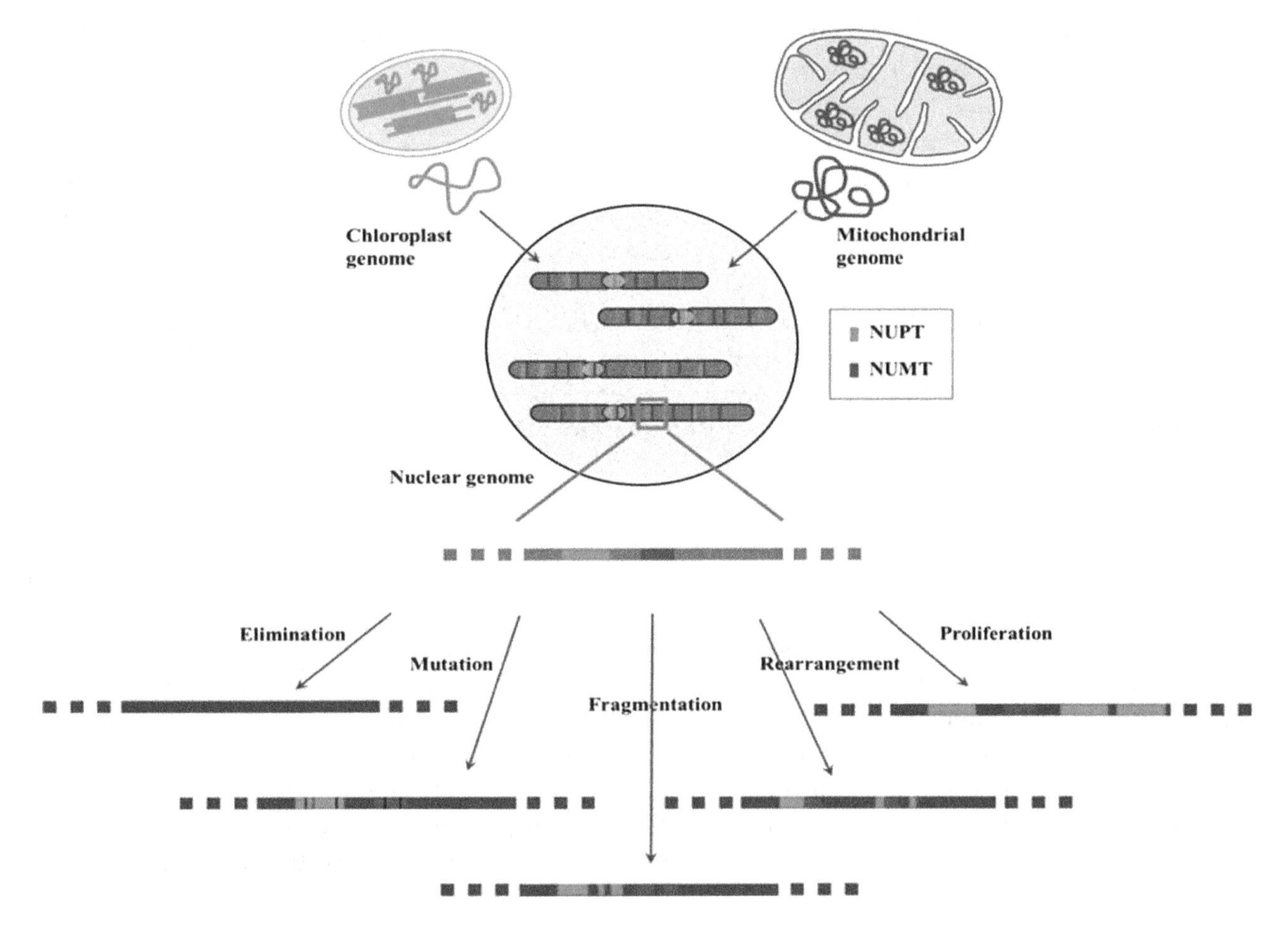


**Figure 2.** Diagrammatic sketch summarizing the fate of integrated organellar DNA in nuclear genome. See details in text.

### 3.2. Mutation

As an intrinsic property of DNA, mutation is a predominant post-integration fate of organelle DNA. Organelle-derived sequences in the nuclear genome are unavoidably subject to the evolutionary forces that act on nuclear DNA [4]. Given that mutation rates vary greatly among organellar and nuclear DNAs, the integrated sequences show different mutation rates compared with the organellar sequences that they derive. In most animals, the mutation rate of mitochondrial DNA is approximately ten times faster than that of the nuclear DNA [63–65]. Thus, NUMTs inserted into the nucleus reserve ancestral mitochondria information and become "molecular fossils" that can be useful phylogenetic markers to study the evolutionary route of related taxa [66,67]. By contrast, the mutation rate of organellar DNA in plant species is much slower than that in the nuclear genome [68]. Organellar DNA is evolved with approximately one or fewer mutations per kb per million years [69], and the mutation rate is approximately seven mutations per kb per million years for the nuclear genome [70]. Thus, the NUPTs/NUMTs inserted into the plant nuclear genome evolve much more rapidly. With the evolutionary process, NUPT/NUMT mutation accumulates until they are not recognized as NUPTs/NUMTs because of the large sequence divergence between the evolved integrated sequences and the ancestral organellar DNAs. A previous investigation of recently acquired insertions of nuclear organellar DNA in *Arabidopsis* and rice revealed that the C $\rightarrow$ T and G $\rightarrow$ A transition mutations have occurred far more frequently than other point mutations. The 5-methylcytosine hypermutation frequency is approximately 5.6-fold and 9.5-fold higher than other point mutations in *Arabidopsis* NUMTs and rice NUPTs, respectively. In addition, most of the short insertion and deletion mutations occurred at homopolymeric regions and were less frequent than point mutations [33,38].

Thus, the 5-methylcytosine hypermutations may be a major mechanism of mutational decay for the newly integrated organellar sequences.

### 3.3. Fragmentation

As discussed above, the sequence similarity between NUPTs/NUMTs and their original chloroplast/mitochondrial DNAs will decrease over evolutionary time. Such sequence similarity is correlated positively with the size of NUPTs/NUMTs. In addition, dot plot analysis showed many separate but tightly linked NUPTs, which are collinear with the chloroplast genome in *A. officinalis* [21]. These pieces of evidence indicate that the original insertions are large and then break into smaller fragments [36,39]. The insertion of TEs and other non-organelle DNA sequences into NUPTs and NUMTs plays a significant role in this fragmentation process [36,39]. Some NUPT/NUMT clusters were formed by such process of organellar DNA insertion and subsequent fragmentation accompanied by incorporation of non-organellar DNA.

### 3.4. Rearrangement

A considerable number of NUPTs/NUMTs are arranged in tandem arrays of sequences originating from various regions of the plastid or/and mitochondrion genome (s); this phenomenon suggests that the transfer of organellar sequences undergo rearrangement before and/or after their insertion into the nuclear genome. In rice, the fragmentation and reshuffling events could occur immediately after plastid DNAs are integrated into the nuclear genome [36]. A detailed analysis of the *A. thaliana* 620-kb large NUMT shows that at least four rearrangement events have occurred, among which, three were localized at the initial 76 kb of the insertion [36]. These rearrangements may have occurred after integration, but they can also occur before transfer in the organelle via homologous recombination between repetitive sequences [71] or in the nucleus either before or during integration [11,39,72,73].

### 3.5. Proliferation

The integrated organellar DNAs can amplify themselves to increase the copy number in the nuclear genome. The amplified fragment can be arranged in tandem arrays or distributed separately in the host genome. For example, in *A. officinalis*, we identified several NUPT clusters arranged by a number of nearly identical chloroplast DNA sequences [21]. In *Carica papaya*, a plastid-derived sequence containing *rsp15* gene has been proliferated 23 times in the hermaphrodite-specific regions of the $Y^h$ chromosome (HSY) [74]. The NUPTs/NUMTs themselves are not able to duplicate; thus, their proliferation in the nuclear genome is probably duplicated, co-amplified with neighboring retrotransposons, which can easily increase their copy number via "copy-and-paste" retrotransposition. Indeed, intact *Ty3-pypsy* retrotransposable elements are presented less than 1.5 kb upstream of most *rsp15* NUPT sequences, suggesting that the NUPT proliferation is probably mediated via retrotransposons [74].

We summarized the various fates of the organellar DNA integrated into the host nuclear genome. It is worth noting that the occurrence of the abovementioned events is not isolated and that these events usually occur integrated within one NUPT or NUMT sequence. Thus, after the organellar DNA integrated into the nuclear genome, they are fragmented, shuffled, and accompanied by nucleotide mutation. During this process, a considerable number of NUPTs/NUMTs are deleted from the nuclear genome. Thus, the current repertoire of NUPTs/NUMTs is in rapid variation of fragmentation, mutation, and rearrangement and in dynamic equilibrium between continuous integration and frequent deletion.

In addition, we only discuss the fate of the NUPTs/NUMTs with simple origin, that is, individual organellar fragment from chloroplast or mitochondrial genome transferred to the nucleus. However, the transferred sequence not only comes from one fragment of plastid or mitochondria but it can be derived from both plastids and mitochondria. In this case, the disparate fragments from plastids and mitochondria are combined and/or rearranged and then integrated into the nuclear genome [38,62]. Such integrants can undergo the abovementioned various evolutionary events and complicate the structure of NUPTs/NUMTs.

## 4. Effects of Organelle DNA-Derived Sequences on the Nuclear Genome Structure and Evolution

Organellar DNA integration provides novel source of plant genome evolution. In general, NUMT/NUPT insertions seem to expand the genome size due to the incorporation of novel DNA, although some of the newly integrated sequences can be lost through deletion. A recent study showed that the incorporated plastid DNA can be deleted and accompanied by the loss of adjacent non-plastid-derived DNA within one generation of transfer [62]. Thus, the integration of organellar DNA and its subsequent diverse variation events could contribute significantly to the dynamic nuclear genome evolution.

### *4.1. Contribution to the Genetic Diversity*

Organellar DNA integration is a substantial and ongoing process that generates genetic diversity. The NUMT/NUPT numbers, sizes, and densities show large divergence between different species. Even in the same species, different lines or ecotypes represent sequence variation of NUMTs and NUPTs. Among the 40 investigated NUMTs in humans, 12 are polymorphic [29]. In *Arabidopsis*, a 3.9-kb NUMT is presented in a polyubiquitin gene in four ecotypes, including Columbia, Eifel, Enkheim, and Hilversum [75,76]. However, this NUMT loci did not exist in five other ecotypes investigated [75]. Another 104-bp rearrangement NUMT derived from different mitochondrial DNA fragments is detected at the subtelomeric region of the short arm of chromosome 1 in 12 out of the 35 *Arabidopsis* accessions examined [77]. Similar results were found in rice and maize. A NUPT with length of 131 kb was detected on chromosome 10 of *O. sativa* subsp. *japonica*, but not detected in *O. sativa* subsp. *indica* or *O. rufipogon* nuclear genomes [33]. Fluorescence in situ hybridization (FISH) analyses using overlapping mitochondrial/chloroplast DNA fragments as probes revealed that the number and positions of detectable NUMTs/NUPTs varied markedly among different maize inbred lines [13,78]. For example, a mix of cosmids containing overlapping mitochondrial DNA fragments was probed, and only B73 showed a strong hybridization signal on 9L among the examined inbred lines; thus, this site may represent a recent mitochondrial DNA insertion [78]. Some NUMT sites are diverse even within the same maize inbred line collected from different laboratories (sources) [78]. Such diversity may be attributed to different mechanisms, such as integration of a new fragment of mitochondrial DNA, proliferation of an existing NUMT fragment, or deletion and degradation of NUMT sequences at a particular site [78]. The interspecific diversity in NUPT and NUMT accumulation might be affected by the number and stability of plastid and mitochondria in the germline; another factor is the species-specific mechanisms regulating nuclear DNA acquisition and deletion [37]. These results strongly suggested that the organelle DNA transfer is a frequently ongoing process, and the insertions of organelle-derived segments are components of the dynamic fraction of plant nuclear genomes.

### *4.2. Effects on Gene Structure and Evolution*

Certain NUPTs and NUMTs are related to genes. For example, two NUMTs are located within genes in the yeast genome [79]. In the *Arabidopsis* and rice genomes, approximately 25% of NUMTs and NUPTs are found within genes [39]. In some cases, NUMT/NUPT sequences that fuse to non-organelle-derived nuclear DNA are transcribed, suggesting that NUMTs/NUPTs can remodel genes and their products by providing novel exons [19,80,81]. The mosaic genes with one or more exons originated from organellar DNA were analyzed in detail in yeast, *Homo sapiens*, *Arabidopsis*, and rice. In that study, a total of 474 NUMTs and NUPTs were detected within, or adjacent to the annotated genes. Among these NUMTs and NUPTs, 45 insertions are involved in a total of 49 protein-coding exons belonged to 34 genes [23]. Moreover, considering that the sequence divergence between NUMT/NUPT and the donor organellar sequences was constantly enlarged with evolutionary time, some organelle-originated exon sequences may be difficult to directly detect through conventional methods. Thus, organelle-derived DNA integrations might contribute to many ancient functional exon acquisitions that are not found thus far [23]. In addition, the Ka/Ks ratios (non-synonymous substitutions/synonymous substitutions) of the nuclear protein-coding exon derived from organellar reading frames are higher than 1, thereby suggesting the non-neutral evolution of these integrants and their adaptation to their novel functions [23].

In addition to recruiting as exons, NUMTs/NUPTs can also be associated with the regulatory elements of genes. Some organellar DNA sequences are integrated into introns, as well as regions of 5′ and 3′ to nuclear genes; these integrated sequences might be associated with changes in the gene regulation [80]. Specifically, two organelle-derived sequences, namely, *Arabidopsis* enhancer 12-7 and 12-Q, make the reporter genes show cell-specific expression in transgenic tobacco [82]. However, the influence of the organellar insertions on gene expression still needs further investigation.

### *4.3. Roles Played in the Sex Chromosome Evolution*

Several reports revealed that NUPTs and NUMTs are strongly accumulated in the sex chromosomes of dioecious plants than other plants [74,83,84]. The first discovery of this phenomenon is in the model dioecious plant *Silene latifolia*, in which a bacterial artificial chromosome clone containing partial plastid genome sequences shows strong hybridization signals on the Y chromosome; however, the signals are very weak on the X chromosome and the autosomes. The large size of the Y chromosome in this species may be partially due to the accumulation of such NUPTs [83]. In another dioecious plant, *Carica papaya*, NUPTs are localized within the male-specific region of the Y chromosome (MSY) and HSY approximately 12 times the rate in the X chromosome counterpart and four times that of the genome wide average [74]. NUPTs/NUMTs accumulate in the sex chromosomes of other dioecious plants, such as *Rumex acetosa* [84] and *Coccinia grandis* [85]. In *A. officinalis*, the number and total length of NUPTs in Y chromosome were larger than those in other chromosomes. Sequence alignment and cytogenetic analysis show a 47-kb centromeric region in the Y chromosome with very high density of NUPTs, i.e., more than three-quarters of the sequences of this region originate from the plastid. However, the MSY and the nearby regions did not have NUPT integrations. Thus, NUPTs may play important roles in the centromere building of the sex chromosome of *A. officinalis* but are not implicated in MSY formation [21]. These studies reveal that the organellar DNA accumulation is strongly related to the evolution of plant sex chromosomes.

These integrated organellar DNA sequences are predicted as one of the evolutionary forces that drive the formation of sex chromosomes in dioecious plants. It is believed that the sex chromosomes originate from autosomes, and the evolutionary process involves some essential events, such as sex-determining gene emergence, recombination restriction, and Y chromosome degeneration [86]. In addition, sex chromosomes can recruit abundant TEs and organellar DNAs [74,85], and these novel sequences may benefit the structural differentiation and recombination restriction of sex chromosomes and contribute to Y chromosome degeneration [49]. The gathering of such novel sequences in sex chromosomes can promote recombination suppression of the sex chromosomes; when recombination is suppressed, such sequences can accumulate rapidly in the recombination-free region. The lack of recombination and release from purifying selection avoids their deletion from the recombination-suppressed region [87,88].

## 5. Mechanisms of Organellar DNA Integration

Organellar DNA sequences can be transferred to the nucleus through two distinct pathways, namely, direct DNA integration [89] and RNA-mediated DNA transfer via reverse transcription. Some experimental and bioinformatics studies suggest that many NUPTs/NUMTs are very large and include intergenic and/or non-coding regions of organellar DNA [11,37,38,89,90]. Thus, the migration of organellar DNA fragments into the nucleus may be mediated by DNA. However, the functional organellar gene relocated to the nuclear genome with splicing and RNA editing, and this event seems to have occurred before transfer; this phenomenon suggests the involvement of RNA intermediates [91–93]. Thus, the two pathways may both have important roles to play in the transfer process.

DNA escapes from organelles and its incorporation into the nucleus has been experimentally investigated in yeast [58,79] and tobacco [10,12]. Bioinformatics analysis and direct experimental evidence suggest that nonhomologous end-joining mechanism (NHEJ) is involved in organellar DNA incorporation during double-stranded break (DSB) repair. In yeast, fragments of mitochondrial DNA segments are captured into DSBs in the nuclear genome during the DSB healing process via NHEJ [79,94]. DSBs can be induced in vivo by exogenous and endogenous factors; they are the most potentially deleterious type of DNA damage [95–97]. DSBs can be usually repaired via the insertion of filler DNA into the lesion, whereas organellar DNA can be passively used as the filler DNA. The integration includes two DSB repair events, which are mediated by the short sequence microhomology (1–7 bp, "micro-identities") between the break chromosome ends and organellar DNA or by a blunt-end repair without homology [79,94,96,98]. A similar mechanism involving NHEJ at DSB sites was examined in tobacco [45,55,62]. The integrated fragments are usually complicated mixtures of multiple different segments of chloroplast DNA end joined together based on sequence microhomology [62]. These results suggested that linking of linear DNA fragments generates chimeric DNA molecules before or during insertion into the nuclear genome [38,62]. Sequence analysis revealed that such terminal micro-identities have also been presented for NUMT integrants in humans [99,100], indicating that organellar fragments integrating into the nuclear genome via NHEJ during DSB repair process might be a common mechanism conserved in eukaryotic organisms. A recent study has shown that several mosaic organellar integrants present signatures of long homology; thus, other mechanisms, such as homologous recombination may also contribute to organellar DNA integration [40].

## 6. Conclusions and Perspective

Organellar DNA integrated into the nuclear genome plays important functional roles in gene and genome evolution. The characterization and distribution patterns of organellar-derived sequences of the nuclear genome are crucial to the elucidation of the dynamics and evolution of the genome in plants and other eukaryotes. Nevertheless, many unanswered questions about organellar integrated nuclear sequences remain. For example, the precise mechanism of organellar DNA transfer, the extent and variety of the effects of organellar DNA transfer on gene activity regulation, and the genome instability and defense mechanism caused by organellar DNA transfer are still poorly understood. Thus, bioinformatics approaches, cytogenetic analysis, and experimental studies illustrating transfer events are necessary to answer these questions.

Currently, the experimental analysis of DNA transfer from organelle to nucleus has been performed only in tobacco and yeast. Studies that used these model organisms have presented considerable information on NUMT/NUPT insertion events, such as the transfer frequency, the size of transferred genome fragments, the environmental conditions favoring the transfer, and the mechanism of transfer. However, that such events are universal is not yet certain and should be further studied in other plant species. Experimental models involving other plant species should be established to provide more insights into the transfer event.

Similar to TE, organellar DNA integrated into nuclear genome is considered alien genetic material to the nuclear genome. Such sequences could result in nuclear genome instability, and the nuclear genome can initiate defense mechanisms [49,50]. A recent report showed that DNA methylation might play an important role in this process [51]. Future studies on DNA methylation and heterochromatization of organellar DNA insertions may improve our understanding of the organellar DNA integrants involved in dynamic genome evolution.

**Supplementary Materials:** Table S1, information of genome size, chloroplast genome size, and NUPT statistics results in sequenced plants. Table S2, information of genome size, mitochondrial genome size, and NUMT statistics results in sequenced plants.

**Author Contributions:** W.-J.G. and H.-X.N. conceived the whole project; G.-J.Z. and R.D. wrote the paper; L.-N.L. and S.-F.L. analyzed the NUPT and NUMT data. All authors have read and agreed to the published version of the manuscript.

## References

1. Douglas, S.E. Plastid evolution: Origins, diversity, trends. *Curr. Opin. Genet. Dev.* **1998**, *8*, 655–661. [CrossRef]
2. Gray, M.W. The evolutionary origins of organelles. *Trends Genet.* **1989**, *5*, 294–299. [CrossRef]

3. McBride, H.M. Mitochondria and endomembrane origins. *Curr. Biol.* **2018**, *28*, R367–R372. [CrossRef]
4. Kleine, T.; Maier, U.G.; Leister, D. DNA transfer from organelles to the nucleus: the idiosyncratic genetics of endosymbiosis. *Annu. Rev. Plant Biol.* **2009**, *60*, 115–138. [CrossRef]
5. Ma, P.F.; Zhang, Y.X.; Guo, Z.H.; Li, D.Z. Evidence for horizontal transfer of mitochondrial DNA to the plastid genome in a bamboo genus. *Sci. Rep.* **2015**, *5*, 11608. [CrossRef]
6. Pinard, D.; Myburg, A.A.; Mizrachi, E. The plastid and mitochondrial genomes of *Eucalyptus grandis*. *BMC Genom.* **2019**, *20*, 132. [CrossRef]
7. Timmis, J.N.; Ayliffe, M.A.; Huang, C.Y.; Martin, W.F. Endosymbiotic gene transfer: Organelle genomes forge eukaryotic chromosomes. *Nat. Rev. Genet.* **2004**, *5*, 123–135. [CrossRef]
8. Ayliffe, M.A.; Scott, N.S.; Timmis, J.N. Analysis of plastid DNA-like sequences within the nuclear genomes of higher plants. *Mol. Biol. Evol.* **1998**, *15*, 738–745. [CrossRef]
9. Yuan, Q.; Hill, J.; Hsiao, J.; Moffat, K.; Ouyang, S.; Cheng, Z.; Jiang, J.; Buell, C. Genome sequencing of a 239-kb region of rice chromosome 10L reveals a high frequency of gene duplication and a large chloroplast DNA insertion. *Mol. Genet. Genom.* **2002**, *267*, 713–720. [CrossRef]
10. Huang, C.Y.; Ayliffe, M.A.; Timmis, J.N. Direct measurement of the transfer rate of chloroplast DNA into the nucleus. *Nature* **2003**, *422*, 72–76. [CrossRef]
11. Huang, C.Y.; Ayliffe, M.A.; Timmis, J.N. Simple and complex nuclear loci created by newly transferred chloroplast DNA in tobacco. *Proc. Natl. Acad. Sci. USA* **2004**, *101*, 9710–9715. [CrossRef]
12. Stegemann, S.; Hartmann, S.; Ruf, S.; Bock, R. High-frequency gene transfer from the chloroplast genome to the nucleus. *Proc. Natl. Acad. Sci. USA* **2003**, *100*, 8828–8833. [CrossRef]
13. Roark, L.; Hui, A.; Donnelly, L.; Birchler, J.; Newton, K. Recent and frequent insertions of chloroplast DNA into maize nuclear chromosomes. *Cytogenet. Genome Res.* **2010**, *129*, 17–23. [CrossRef]
14. Sloan, D.B.; Warren, J.M.; Williams, A.M.; Wu, Z.; Abdel-Ghany, S.E.; Chicco, A.J.; Havird, J.C. Cytonuclear integration and co-evolution. *Nat. Rev. Genet.* **2018**, *19*, 635–648. [CrossRef]
15. Chen, H.Y.; Yu, Y.; Chen, X.L.; Zhang, Z.Z.; Gong, C.; Li, J.F.; Wang, A.X. Plastid DNA insertions in plant nuclear genomes: The sites, abundance and ages, and a predicted promoter analysis. *Funct. Integr. Genom.* **2015**, *15*, 131–139. [CrossRef]
16. Michalovova, M.; Vyskot, B.; Kejnovsky, E. Analysis of plastid and mitochondrial DNA insertions in the nucleus (NUPTs and NUMTs) of six plant species: Size, relative age and chromosomal localization. *Heredity* **2013**, *111*, 314–320. [CrossRef]
17. Yoshida, T.; Furihata, H.Y.; Kawabe, A. Analysis of nuclear mitochondrial DNAs and factors affecting patterns of integration in plant species. *Genes Genet. Syst.* **2017**, *92*, 27–33. [CrossRef]
18. Ko, Y.J.; Kim, S. Analysis of nuclear mitochondrial DNA segments of nine plant species: Size, distribution, and insertion loci. *Genom. Inform.* **2016**, *14*, 90–95. [CrossRef]
19. Shahmuradov, I.A.; Akbarova, Y.Y.; Solovyev, V.V.; Aliyev, J.A. Abundance of plastid DNA insertions in nuclear genomes of rice and *Arabidopsis*. *Plant Mol. Biol.* **2003**, *52*, 923–934. [CrossRef]
20. Kim, H.T.; Lee, J.M. Organellar genome analysis reveals endosymbiotic gene transfers in tomato. *PLoS ONE* **2018**, *13*, e0202279. [CrossRef]
21. Li, S.F.; Li, J.R.; Wang, J.; Dong, R.; Jia, K.L.; Zhu, H.W.; Li, N.; Yuan, J.H.; Deng, C.L.; Gao, W.J. Cytogenetic and genomic organization analyses of chloroplast DNA invasions in the nuclear genome of *Asparagus officinalis* L. provides signatures of evolutionary complexity and informativity in sex chromosome evolution. *BMC Plant Biol.* **2019**, *19*, 361. [CrossRef] [PubMed]
22. Martin, W.F.; Rujan, T.; Richly, E.; Hansen, A.; Cornelsen, S.; Lins, T.; Leister, D.; Stoebe, B.; Hasegawa, M.; Penny, D. Evolutionary analysis of Arabidopsis, cyanobacterial, and chloroplast genomes reveals plastid phylogeny and thousands of cyanobacterial genes in the nucleus. *Proc. Natl. Acad. Sci. USA* **2002**, *99*, 12246–12251. [CrossRef] [PubMed]
23. Noutsos, C.; Kleine, T.; Armbruster, U.; DalCorso, G.; Leister, D. Nuclear insertions of organellar DNA can create novel patches of functional exon sequences. *Trends Genet.* **2007**, *23*, 597–601. [CrossRef] [PubMed]
24. Knoop, V.; Brennicke, A. A mitochondrial intron sequence in the 5′-flanking region of a plant nuclear lectin gene. *Curr. Genet.* **1991**, *20*, 423–425. [CrossRef] [PubMed]
25. Boogaart, P.V.D.; Samallo, J.; Agsteribbe, E. Similar genes for a mitochondrial ATPase subunit in the nuclear and mitochondrial genomes of *Neurospora crassa*. *Nature* **1982**, *298*, 187–189. [CrossRef] [PubMed]

26. Lopez, J.V.; Yuhki, N.; Masuda, R.; Modi, W.; O'Brien, S.J. *Numt*, a recent transfer and tandem amplification of mitochondrial DNA to the nuclear genome of the domestic cat. *J. Mol. Evol.* **1994**, *39*, 174–190.
27. Mourier, T.; Hansen, A.J.; Willerslev, E.; Arctander, P. The Human Genome Project reveals a continuous transfer of large mitochondrial fragments to the nucleus. *Mol. Biol. Evol.* **2001**, *18*, 1833–1837. [CrossRef]
28. Hazkani-Covo, E.; Graur, D. A comparative analysis of numt evolution in human and chimpanzee. *Mol. Biol. Evol.* **2007**, *24*, 13–18. [CrossRef]
29. Hazkani-Covo, E.; Zeller, R.M.; Martin, W.F. Molecular poltergeists: Mitochondrial DNA copies (numts) in sequenced nuclear genomes. *PLoS Genet.* **2010**, *6*, e1000834. [CrossRef]
30. Ding, Y.R.; Li, B.; Zhang, Y.J.; Mao, Q.M.; Chen, B. Complete mitogenome of Anopheles sinensis and mitochondrial insertion segments in the nuclear genomes of 19 mosquito species. *PLoS ONE* **2018**, *13*, e0204667. [CrossRef]
31. Yan, Z.C.; Fang, Q.; Tian, Y.; Wang, F.; Chen, X.X.; Werren, J.H.; Ye, G.Y. Mitochondrial DNA and their nuclear copies in the parasitic wasp *Pteromalus puparum*: A comparative analysis in Chalcidoidea. *Int. J. Biol. Macromol.* **2019**, *121*, 572–579. [CrossRef]
32. Schiavo, G.; Strillacci, M.G.; Ribani, A.; Bovo, S.; Román-Ponce, S.I.; Cerolini, S.; Bertolini, F.; Bagnato, A.; Fontanesi, L. Few mitochondrial DNA sequences are inserted into the turkey (*Meleagris gallopavo*) nuclear genome: Evolutionary analyses and informativity in the domestic lineage. *Anim. Genet.* **2018**, *49*, 259–264. [CrossRef]
33. Huang, C.Y.; Grünheit, N.; Ahmadinejad, N.; Timmis, J.N.; Martin, W.F. Mutational decay and age of chloroplast and mitochondrial genomes transferred recently to angiosperm nuclear chromosomes. *Plant Physiol.* **2005**, *138*, 1723–1733. [CrossRef]
34. The Rice Chromosome 10 Sequencing Consortium. In-depth view of structure, activity, and evolution of rice chromosome 10. *Science* **2003**, *300*, 1566–1569. [CrossRef]
35. Stupar, R.M.; Lilly, J.W.; Town, C.D.; Cheng, Z.; Kaul, S.; Buell, C.R.; Jiang, J. Complex mtDNA constitutes an approximate 620-kb insertion on *Arabidopsis thaliana* chromosome 2: Implication of potential sequencing errors caused by large-unit repeats. *Proc. Natl. Acad. Sci. USA* **2001**, *98*, 5099–5103. [CrossRef]
36. Matsuo, M.; Ito, Y.; Yamauchi, R.; Obokata, J. The rice nuclear genome continuously integrates, shuffles, and eliminates the chloroplast genome to cause chloroplast–nuclear DNA flux. *Plant Cell* **2005**, *17*, 665–675. [CrossRef]
37. Richly, E.; Lartillot, N.; Philippe, H. NUMTs in sequenced eukaryotic genomes. *Mol. Biol. Evol.* **2004**, *21*, 1081–1084. [CrossRef]
38. Noutsos, C.; Richly, E.; Leister, D. Generation and evolutionary fate of insertions of organelle DNA in the nuclear genomes of flowering plants. *Genome Res.* **2005**, *15*, 616–628. [CrossRef]
39. Richly, E.; Leister, D. NUPTs in sequenced eukaryotes and their genomic organization in relation to NUMTs. *Mol. Biol. Evol.* **2004**, *21*, 1972–1980. [CrossRef]
40. Leister, D. Origin, evolution and genetic effects of nuclear insertions of organelle DNA. *Trends Genet.* **2005**, *21*, 655–663. [CrossRef]
41. Portugez, S.; Martin, W.F.; Hazkani-Covo, E. Mosaic mitochondrial-plastid insertions into the nuclear genome show evidence of both non-homologous end joining and homologous recombination. *BMC Evol. Biol.* **2018**, *18*, 162. [CrossRef]
42. Arabidopsis Genome Initiative. Analysis of the genome sequence of the flowering plant *Arabidopsis thaliana*. *Nature* **2000**, *408*, 796–815. [CrossRef]
43. Neumann, P.; Navrátilová, A.; Koblížková, A.; Kejnovsky, E.; Hribova, E.; Hobza, R.; Widmer, A.; Dolezel, J.; Macas, J. Plant centromeric retrotransposons: A structural and cytogenetic perspective. *Mob. DNA* **2011**, *2*, 4. [CrossRef]
44. Liao, Y.; Zhang, X.; Li, B.; Liu, T.; Chen, J.; Bai, Z.; Wang, M.; Shi, J.; Walling, J.G.; Wing, R.A.; et al. Comparison of *Oryza sativa* and *Oryza brachyantha* genomes reveals selection-driven gene escape from the centromeric regions. *Plant Cell* **2018**, *30*, 1729–1744. [CrossRef]
45. Wang, N.; Timmis, J.N. Cytoplasmic organelle DNA preferentially inserts into open chromatin. *Genome Biol. Evol.* **2013**, *5*, 1060–1064. [CrossRef]
46. Pecinka, A.; Dinh, H.Q.; Baubec, T.; Rosa, M.; Lettner, N.; Scheid, O.M. Epigenetic regulation of repetitive elements is attenuated by prolonged heat stress in *Arabidopsis*. *Plant Cell* **2010**, *22*, 3118–3129. [CrossRef]

47. Zhu, W.S.; Hu, B.; Becker, C.; Doğan, E.S.; Berendzen, K.W.; Weigel, D.; Liu, C. Altered chromatin compaction and histone methylation drive non-additive gene expression in an interspecific Arabidopsis hybrid. *Genome Biol.* **2017**, *18*, 157. [CrossRef]
48. Sheppard, A.E.; Timmis, J.N. Instability of plastid DNA in the nuclear genome. *PLoS Genet.* **2009**, *5*, e1000323. [CrossRef]
49. Li, S.F.; Zhang, G.J.; Yuan, J.H.; Deng, C.L.; Gao, W.J. Repetitive sequences and epigenetic modification: Inseparable partners play important roles in the evolution of plant sex chromosomes. *Planta* **2016**, *243*, 1083–1095. [CrossRef]
50. Galindo-González, L.; Sarmiento, F.; Quimbaya, M.A. Shaping plant adaptability, genome structure and gene expression through transposable element epigenetic control: Focus on methylation. *Agronomy* **2018**, *8*, 180. [CrossRef]
51. Yoshida, T.; Furihata, H.Y.; To, T.K.; Kakutani, T.; Kawabe, A. Genome defense against integrated organellar DNA fragments from plastids into plant nuclear genomes through DNA methylation. *Sci. Rep.* **2019**, *9*, 2060. [CrossRef]
52. Holliday, R.; Grigg, G. DNA methylation and mutation. *Mutat. Res.* **1993**, *285*, 61–67. [CrossRef]
53. Paszkowski, J.; Whitham, S.A. Gene silencing and DNA methylation processes. *Curr. Opin. Plant Biol.* **2001**, *4*, 123–129. [CrossRef]
54. Bender, J. Chromatin-based silencing mechanisms. *Curr. Opin. Plant Biol.* **2004**, *7*, 521–526. [CrossRef]
55. Wang, D.; Lloyd, A.H.; Timmis, J.N. Environmental stress increases the entry the cytoplasmic organellar DNA into the nucleus in plants. *Proc. Natl. Acad. Sci. USA* **2012**, *109*, 2444–2448. [CrossRef] [PubMed]
56. Wang, N.; Lloyd, A.H.; Timmis, J.N. Nuclear genome diversity in somatic cells is accelerated by environmental stress. *Plant Signal. Behav.* **2012**, *7*, 595–597. [CrossRef]
57. Smith, D.R.; Crosby, K.; Lee, R.W. Correlation between nuclear plastid DNA abundance and plastid number supports the limited transfer window hypothesis. *Genome Biol. Evol.* **2011**, *3*, 365–371. [CrossRef]
58. Thorsness, P.E.; Fox, T.D. Escape of DNA from mitochondria to the nucleus in *Saccharomyces cerevisiae*. *Nature* **1990**, *346*, 376–379. [CrossRef]
59. Ruf, S.; Hasse, C.; Stegemann, S.; Bock, R. *Plastid Transmission, Gene Transfer and the Impact of the Environment*; ISCGGE: Maynooth, Ireland, 2010.
60. Sheppard, A.E.; Ayliffe, M.A.; Blatch, L.; Day, A.; Delaney, S.K.; Khairul-Fahmy, N.; Li, Y.; Madesis, P.; Pryor, A.J.; Timmis, J.N. Transfer of plastid DNA to the nucleus is elevated during male gametogenesis in tobacco. *Plant Physiol.* **2008**, *148*, 328–336. [CrossRef]
61. Mogensen, H.L. The hows and whys of cytoplasmic inheritance in seed plants. *Am. J. Bot.* **1996**, *83*, 383–404. [CrossRef]
62. Wang, D.; Gu, J.B.; David, R.; Wang, Z.; Yang, S.T.; Searle, I.R.; Zhu, J.K.; Timmis, J.N. Experimental reconstruction of double-stranded break repair-mediated plastid DNA insertion into the tobacco nucleus. *Plant J.* **2018**, *93*, 227–234. [CrossRef] [PubMed]
63. Brown, W.M.; George, M.; Wilson, A.C. Rapid evolution of animal mitochondrial DNA. *Proc. Natl. Acad. Sci. USA* **1979**, *76*, 1967–1971. [CrossRef] [PubMed]
64. Brown, W.M.; Prager, E.M.; Wang, A.; Wilson, A.C. Mitochondrial DNA sequences of primates: Tempo and mode of evolution. *J. Mol. Evol.* **1982**, *18*, 225–239. [CrossRef] [PubMed]
65. Haag-Liautard, C.; Coffey, N.; Houle, D.; Lynch, M.; Charlesworth, B.; Keightley, P.D. Direct estimation of the mitochondrial DNA mutation rate in *Drosophila melanogaster*. *PLoS Biol.* **2008**, *6*, e204. [CrossRef]
66. Perna, N.T.; Kocher, T.D. Mitochondrial DNA: Molecular fossils in the nucleus. *Curr. Biol.* **1996**, *6*, 128–129. [CrossRef]
67. Zhang, D.X.; Hewitt, G.M. Nuclear integrations: Challenges for mitochondrial DNA markers. *Trends Ecol. Evol.* **1996**, *11*, 247–251. [CrossRef]
68. Wolfe, K.H.; Li, W.H.; Sharp, P.M. Rates of nucleotide substitution vary greatly among plant mitochondrial, chloroplast, and nuclear DNAs. *Proc. Natl. Acad. Sci. USA* **1987**, *84*, 9054–9058. [CrossRef]
69. Middleton, C.P.; Senerchia, N.; Stein, N.; Akhunov, E.D.; Keller, B.; Wicker, T.; Kilian, B. Sequencing of chloroplast genomes from wheat, barley, rye and their relatives provides a detailed insight into the evolution of the Triticeae tribe. *PLoS ONE* **2014**, *9*, e85761. [CrossRef]
70. Ossowski, S.; Schneeberger, K.; Lucas-Lledó, J.I.; Warthmann, N.; Clark, R.M.; Shaw, R.G.; Weigel, D.; Lynch, M. The rate and molecular spectrum of spontaneous mutations in *Arabidopsis thaliana*. *Science* **2010**, *327*, 92–94. [CrossRef]

71. Unseld, M.; Marienfeld, J.R.; Brandt, P.; Brennicke, A. The mitochondrial genome of *Arabidopsis thaliana* contains 57 genes in 366,924 nucleotides. *Nat. Genet.* **1997**, *15*, 57–61. [CrossRef]
72. Lin, X.; Kaul, S.; Rounsley, S.; Shea, T.P.; Benito, M.I.; Town, C.D.; Fujii, C.Y.; Mason, T.; Bowman, C.L.; Barnstead, M.; et al. Sequence and analysis of chromosome 2 of the plant *Arabidopsis thaliana*. *Nature* **1999**, *402*, 761–768. [CrossRef] [PubMed]
73. Hazkani-Covo, E.; Sorek, R.; Graur, D. Evolutionary dynamics of large Numts in the human genome: rarity of independent insertions and abundance of post-insertion duplications. *J. Mol. Evol.* **2003**, *56*, 169–174. [CrossRef]
74. VanBuren, R.; Ming, R. Organelle DNA accumulation in the recently evolved papaya sex chromosomes. *Mol. Genet. Genom.* **2013**, *288*, 277–284. [CrossRef] [PubMed]
75. Sun, C.W.; Callis, J. Recent stable insertion of mitochondrial-DNA into an Arabidopsis polyubiquitin gene by nonhomologous recombination. *Plant Cell* **1993**, *5*, 97–107.
76. Ullrich, H.; Lättig, K.; Brennicke, A.; Knoop, V. Mitochondrial DNA variations and nuclear RFLPs reflect different genetic similarities among 23 *Arabidopsis thaliana* ecotypes. *Plant Mol. Biol.* **1997**, *33*, 37–45. [CrossRef]
77. Kuo, H.F.; Olsen, K.M.; Richards, E.J. Natural variation in a subtelomeric region of Arabidopsis: Implications for the genomic dynamics of a chromosome end. *Genetics* **2005**, *173*, 401–417. [CrossRef]
78. Lough, A.N.; Roark, L.M.; Kato, A.; Ream, T.S.; Lamb, J.C.; Birchler, J.A.; Newton, K.J. Mitochondrial DNA transfer to the nucleus generates extensive insertion site variation in maize. *Genetics* **2008**, *178*, 47–55. [CrossRef]
79. Ricchetti, M.; Fairhead, C.; Dujon, B. Mitochondrial DNA repairs double-strand breaks in yeast chromosomes. *Nature* **1999**, *402*, 96–100. [CrossRef]
80. Blanchard, J.L.; Schmidt, G.W. Pervasive migration of organellar DNA to the nucleus in plants. *J. Mol. Evol.* **1995**, *41*, 397–406. [CrossRef]
81. Maiti, A.K.; Jorissen, M.; Bouvagnet, P. Isolation, in silico characterization and chromosomal localization of a group of cDNAs from ciliated epithelial cells after in vitro ciliogenesis. *Genome Biol.* **2001**, *2*. [CrossRef]
82. Ott, R.W.; Chua, N.H. Enhancer sequences from *Arabidopsis thaliana* obtained by library transformation of *Nicotiana tabacum*. *Mol. Genet. Genom.* **1990**, *223*, 169–179. [CrossRef] [PubMed]
83. Kejnovsky, E.; Kubat, Z.; Hobza, R.; Lengerova, M.; Sato, S.; Tabata, S.; Fukui, K.; Matsunaga, S.; Vyskot, B. Accumulation of chloroplast DNA sequences on the Y chromosome of *Silene latifolia*. *Genetica* **2006**, *128*, 167–175. [CrossRef] [PubMed]
84. Steflova, P.; Hobza, R.; Vyskot, B.; Kejnovsky, E. Strong accumulation of chloroplast DNA in the Y chromosomes of *Rumex acetosa* and *Silene latifolia*. *Cytogenet. Genome Res.* **2014**, *142*, 59–65. [CrossRef] [PubMed]
85. Sousa, A.; Bellot, S.; Fuchs, J.; Houben, A.; Renner, S.S. Analysis of transposable elements and organellar DNA in male and female genomes of a species with a huge Y chromosome reveals distinct Y centromeres. *Plant J.* **2016**, *88*, 387–396. [CrossRef]
86. Ming, R.; Bendahmane, A.; Renner, S.S. Sex chromosomes in land plants. *Annu. Rev. Plant Biol.* **2011**, *62*, 485–514. [CrossRef]
87. Bergero, R.; Charlesworth, D. The evolution of restricted recombination in sex chromosomes. *Trends Ecol. Evol.* **2009**, *24*, 94–102. [CrossRef]
88. Cioffi, M.B.; Kejnovsky, E.; Marquioni, V.; Poltronieri, J.; Molina, W.F.; Diniz, D.; Bertollo, L.A.C. The key role of repeated DNAs in sex chromosome evolution in two fish species with ZW sex chromosome system. *Mol. Cytogenet.* **2012**, *5*, 28. [CrossRef]
89. Woischnik, M.; Moraes, C.T. Pattern of organization of human mitochondrial pseudogenes in the nuclear genome. *Genome Res.* **2002**, *12*, 885–893. [CrossRef]
90. Nugent, J.M.; Palmer, J.D. RNA-mediated transfer of the gene coxII from the mitochondrion to the nucleus during flowering plant evolution. *Cell* **1991**, *66*, 473–481. [CrossRef]
91. Shafer, K.S.; Hanekamp, T.; White, K.H.; Thorsness, P.E. Mechanisms of mitochondrial DNA escape to the nucleus in the yeast *Saccharomyces cerevisiae*. *Curr. Genet.* **1999**, *36*, 183–194. [CrossRef]
92. Adams, K.L.; Daley, D.O.; Qiu, Y.L.; Whelan, J.; Palmer, J.D. Repeated, recent and diverse transfers of a mitochondrial gene to the nucleus in flowering plants. *Nature* **2000**, *408*, 354–357. [CrossRef] [PubMed]

93. Grohmann, L.; Brennicke, A.; Schuster, W. The mitochondrial gene encoding ribosomal protein S12 has been translocated to the nuclear genome in *Oenothera*. *Nucleic Acids Res.* **1992**, *20*, 5641–5646. [CrossRef] [PubMed]
94. Yu, X.; Gabriel, A. Patching broken chromosomes with extranuclear cellular DNA. *Mol. Cell* **1999**, *4*, 873–881. [CrossRef]
95. Shin, D.S.; Chahwan, C.; Huffman, J.L.; Tainer, J.A. Structure and function of the double-strand break repair machinery. *DNA Repair* **2004**, *3*, 863–873. [CrossRef]
96. Van Gent, D.C.; Hoeijmakers, J.H.J.; Kanaar, R. Chromosomal stability and the DNA double-stranded break connection. *Nat. Rev. Genet.* **2001**, *2*, 196–206. [CrossRef]
97. Puchta, H. The repair of double-strand breaks in plants: Mechanisms and consequences for genome evolution. *J. Exp. Bot.* **2005**, *56*, 1–14. [CrossRef]
98. Hazkani-Covo, E.; Covo, S. Numt-mediated double-strand break repair mitigates deletions during primate genome evolution. *PLoS Genet.* **2008**, *4*, e1000237. [CrossRef]
99. Blanchard, J.L.; Schmidt, G.W. Mitochondrial DNA migration events in yeast and humans: Integration by a common end-joining mechanism and alternative perspectives on nucleotide substitution patterns. *Mol. Biol. Evol.* **1996**, *13*, 537–548. [CrossRef]
100. Ricchetti, M.; Tekaia, F.; Dujon, B. Continued colonization of the human genome by mitochondrial DNA. *PLoS Biol.* **2004**, *2*, e273. [CrossRef]

# 12

# Comparative Mitogenome Analysis of the Genus *Trifolium* Reveals Independent Gene Fission of *ccmFn* and Intracellular Gene Transfers in Fabaceae

**In-Su Choi [1,*], Tracey A. Ruhlman [1] and Robert K. Jansen [1,2]**

[1] Department of Integrative Biology, University of Texas at Austin, Austin, TX 78712, USA; truhlman@austin.utexas.edu (T.A.R.); jansen@austin.utexas.edu (R.K.J.)

[2] Centre of Excellence in Bionanoscience Research, Department of Biological Sciences, Faculty of Science, King Abdulaziz University, Jeddah 21589, Saudi Arabia

* Correspondence: 86ischoi@gmail.com

**Abstract:** The genus *Trifolium* is the largest of the tribe Trifolieae in the subfamily Papilionoideae (Fabaceae). The paucity of mitochondrial genome (mitogenome) sequences has hindered comparative analyses among the three genomic compartments of the plant cell (nucleus, mitochondrion and plastid). We assembled four mitogenomes from the two subgenera (*Chronosemium* and *Trifolium*) of the genus. The four *Trifolium* mitogenomes were compact (294,911–348,724 bp in length) and contained limited repetitive (6.6–8.6%) DNA. Comparison of organelle repeat content highlighted the distinct evolutionary trajectory of plastid genomes in a subset of *Trifolium* species. Intracellular gene transfer (IGT) was analyzed among the three genomic compartments revealing functional transfer of mitochondrial *rps1* to nuclear genome along with other IGT events. Phylogenetic analysis based on mitochondrial and nuclear *rps1* sequences revealed that the functional transfer in Trifolieae was independent from the event that occurred in robinioid clade that includes genus *Lotus*. A novel, independent fission event of *ccmFn* in *Trifolium* was identified, caused by a 59 bp deletion. Fissions of this gene reported previously in land plants were reassessed and compared with *Trifolium*.

**Keywords:** legumes; clover; organelle genetics; mitochondria; endosymbiotic gene transfer; gene fission

## 1. Introduction

Plant cells comprise three genomic compartments (i.e., nucleus, mitochondrion and plastid). Unlike the typically conservative plastid genome (plastome) [1,2], plant mitochondrial genomes (mitogenome) display drastic evolutionary plasticity in size, content and structure, intracellular gene transfer (IGT) and interspecific horizontal gene transfer [3–8]. Substitution rates of mitochondrial protein coding genes, however, are the most conservative among the three genomic compartments [9]. In angiosperms, the relative rate of synonymous substitutions of mitogenome, plastome and nuclear genome is 1:3:16 [10].

Extensive gene loss and IGT of organelle DNA to the nucleus occurred in the early stages of endosymbiosis [11]. Nuclear genome sequences that originate from the mitogenome and plastome are referred to as nuclear mitochondrial DNA sequences (NUMTs) and nuclear plastid DNA sequences (NUPTs), respectively [12,13]. Transfer of mitochondrial DNA to the nuclear genome is an ongoing process in both of plants and animals but functional transfer of mitochondrial genes has almost ceased in animals [14]. Functional transfer of mitochondrial genes in plants has often involved ribosomal protein or succinate dehydrogenase genes [5]. Transfer of mitochondrial genes to the nuclear genome cannot substitute function of the original mitochondrial copy unless the nuclear copy acquires the appropriate expression and targeting signals [15]. Before the acquisition of regulatory signals, NUMTs must survive mutational decay in nuclear genome, which limits the lifespan of the nonfunctional sequences [16]. Mitochondrial IGT events may be successful or unsuccessful in terms of functionality and the phylogenetic distribution of pseudogenization and deletion of mitochondrial genes can be assessed in descendant lineages [17].

Following functional transfer, NUMTs attain higher substitution rates than their mitochondrial counterparts [14] because of substantial differences in the synonymous substitution rate between mitochondrial and nuclear genomes in plants [10]. Hence, functional transfer of mitochondrial genes into the nucleus is often detected by the presence of intact but highly diverged copies in nuclear genome compared to mitochondrial copies [18,19]. On rare occasions, functional transfers of mitochondrial genes exhibit an intriguing situation in which the nuclear and mitochondrial genomes contain different portions of the coding region resulting from mitochondrial gene fission and IGT (e.g., *rpl2* in many of eudicots) [20]. Szafranski [21] named this process "intercompartmental piecewise gene transfer." In plant mitogenome evolution, the protein that most commonly undergoes gene fission is cytochrome c maturation protein *ccmF* [22–25].

In *Escherichia coli*, the eight *ccm* genes (*ccmA-H*) are clustered in a single locus [26]. In most plants, three *ccm* genes (*ccmA*, *ccmE* and *ccmH*) have been transferred from the mitogenome to the nuclear genome, two (*ccmD* and *ccmG*) were lost and four (*ccmB*, *ccmC*, *ccmFc* and *ccmFn*) remain in the mitochondrion [27]. Since the fission of *ccmF* into *ccmFc* and *ccmFn* is shared by liverworts and seed plants [3], this event happened early in land plant evolution. In addition, there were independent fissions of *ccmF* in several lineages of land plants, including fission of *ccmFc* into *ccmFc1* and *ccmFc2* in *Marchantia* [22] and fissions of *ccmFn* into *ccmFn1* and *ccmFn2* in Brassicaceae [23,24] and *Allium* (Amaryllidaceae) [25].

Fabaceae are the third largest angiosperm family with approximately 20,000 species in six subfamilies [28]. Most species diversity occurs in subfamily Papilionoideae, which includes many economically important species [29]. The inverted repeat (IR) lacking clade (IRLC) is one of the major groups of Papilionoideae, which is defined by absence of the canonical plastome IR (~25 kb) [30]. Plastome studies of the IRLC elucidated several rare evolutionary phenomena, including high degree of genome rearrangement [31], localized hypermutation [32], genome size expansion with accumulation of dispersed repeats and unique sequences of unknown origin [33–35] and re-acquisition of a large IR [36]. However, mitogenome evolution in IRLC is poorly understood and represented by only two species, *Vicia faba* (tribe Fabeae) [37] and *Medicago truncatula* (tribe Trifolieae) [38]. In Trifolieae, a study of the mitochondrial *rps1* gene documented the existence of functional nuclear copies and putatively pseudogenized mitochondrial copies from three genera (*Medicago*, *Melilotus* and *Trigonella*) [39]. Deletion of mitochondrial *rps1* was also identified from another papilionoid species, *Lotus japonicus* [40]. The status of mitochondrial *rps1* across Trifolieae and related taxa has not been examined until recently. Parallel losses of several mitochondrial genes in Fabaceae were revealed in a previous study, however, whether the losses represent a single ancestral IGT or multiple IGTs was not determined [8].

*Trifolium* is the largest genus (ca. 250 species) of the tribe Trifolieae [41] and is divided into two subgenera (*Chronosemium* and *Trifolium*) [42]. Trifolieae belong to the IRLC and are closely related to Fabeae [30]. Several evolutionary studies of Trifolieae plastid [31,33,34] and nuclear [43–46] genomes have been conducted but mitogenome comparisons of *Trifolium* have been neglected. In this study, gene content, size and repeat structure of mitogenomes of four *Trifolium* species from the two subgenera *Chronosemium* (*T. aureum* and *T. grandiflorum*) and *Trifolium* (*T. meduseum* and *T. pratense*) were examined and compared to related papilionoid species.

## 2. Results

### 2.1. Mitogenome Features of Four Trifolium Species

For each of four *Trifolium*, a single chromosome was assembled that contained all expected mitochondrial coding sequences. The length of the four mitogenomes varied from to 294,911 to

348,724 bp (Table 1). The GC content was conserved among the species at 44.9–45.2 %. Gene content was identical with three rRNAs, 16 tRNAs and 32 protein coding genes while gene order was distinct for each species (Figure 1).

**Table 1.** Assembly information for four *Trifolium* mitogenomes.

| Species | Subgenus | Raw Reads | Mitogenome Coverage | Mitogenome Length (bp) | GC (%) | NCBI Accession |
|---|---|---|---|---|---|---|
| *T. aureum* | *Chronosemium* | 71,346,514 | 645 | 294,911 | 44.9 | MT039392 |
| *T. grandiflorum* | *Chronosemium* | 48,390,678 | 197 | 347,723 | 45.1 | MT039391 |
| *T. meduseum* | *Trifolium* | 68,712,286 | 207 | 348,724 | 45.0 | MT039390 |
| *T. pratense* | *Trifolium* | 47,909,108 | 212 | 301,823 | 45.2 | MT039389 |

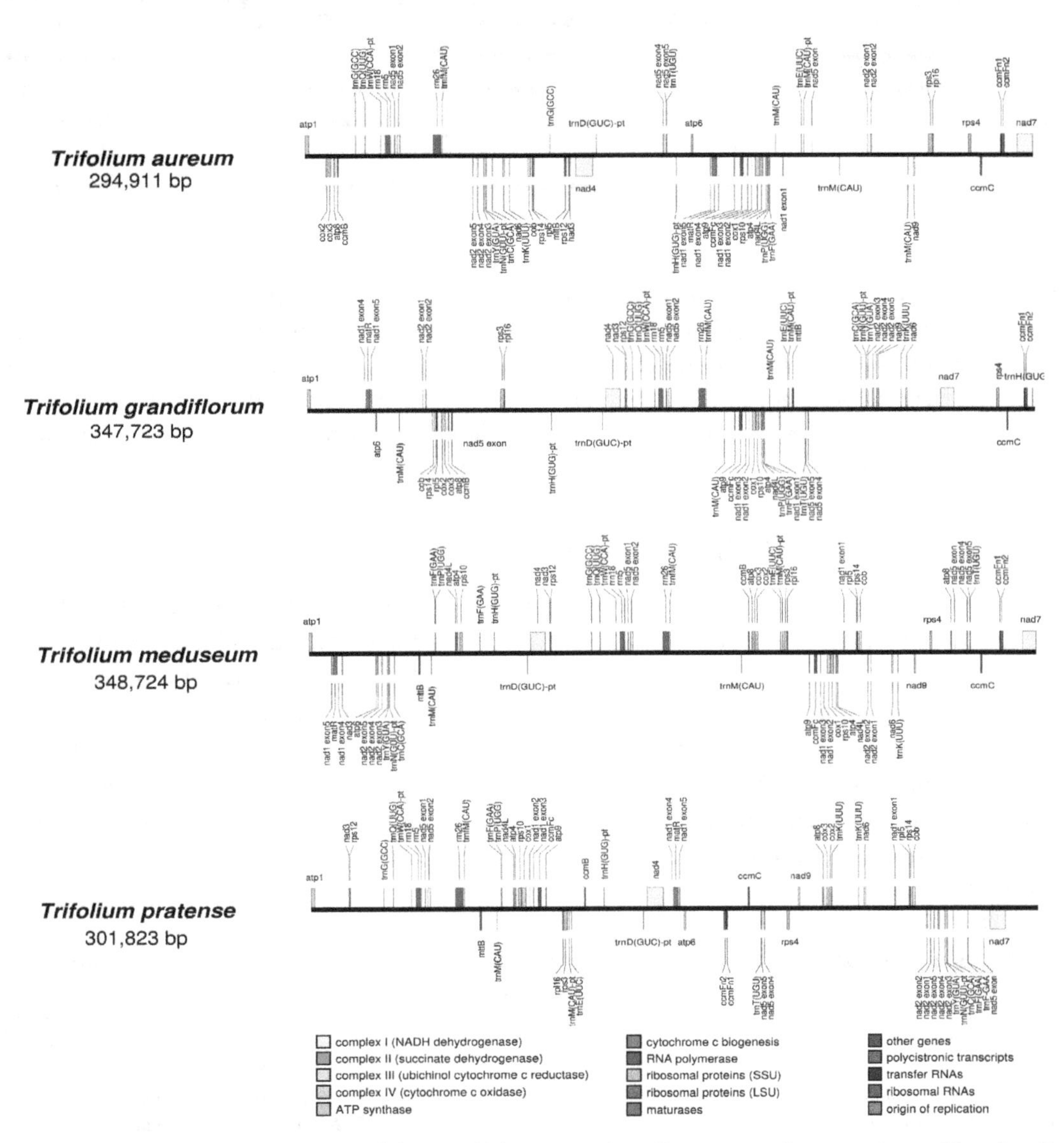


**Figure 1.** Linear mitogenome maps of four *Trifolium* species. Fragmented genes caused by duplication or pseudogenization are not depicted. pt indicates tRNAs of plastid origin.

Gene and intron content comparison with other published mitogenomes revealed one gene loss (*rps1*) (Figure S1), which was shared with *Lotus* and two cis-spliced intron losses (ccmFci829 and rps3i174) that were exclusive to *Trifolium* (Figure S2). Sequence alignment of *ccmFn* from *Trifolium* with other IRLC genera revealed a 59 bp deletion that resulted in a frame shift and premature stop codon (Figure 2). a putative downstream start codon for a second open reading frame (ORF) (*ccmFn2*) was also identified.

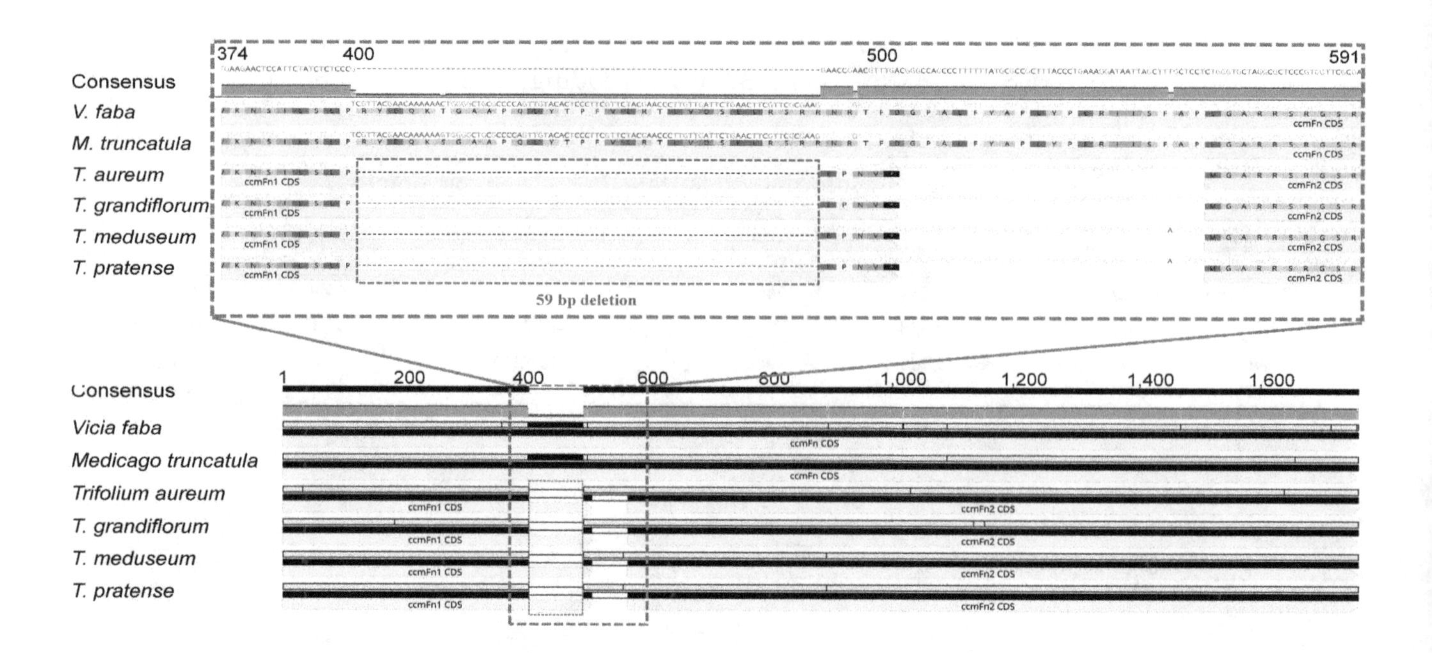


**Figure 2.** Fission of *ccmFn* in four *Trifolium*. Alignment of six *ccmFn* sequences of mitogenomes from species in inverted repeat lacking clade. The region (grey dashed rectangle, aligned positions 374-591) showing the 59 bp deletion (red dotted box) is enlarged above. Translated amino acid alignments are presented below corresponding nucleotide sequence alignments. Nucleotide coordinates are indicated above consensus of alignment. Sequence identity is shown below consensus (green = 100%, yellow-green = at least 30% and under 100%, red = below 30%).

## 2.2. *Repeat Composition of Organelle Genomes in Trifolium*

Repeat sequences were estimated four mitogenomes and thirteen plastomes (Table 2). The amount of repetitive sequences in mitogenomes was not highly variable (6.6~8.6 %). In contrast, the amount of repetitive DNA in plastomes was highly variable (4.4%~20.7%) and can be divided into two non-overlapping ranges that corresponded to two groups of two sections (subgen. *Chronosemium* sect. *Chronosemium* and subg. *Trifolium* sect. *Paramesus*, 4.4%~5.2 %) and five sects. of subg. *Trifolium* (*Lupinaster*, *Trichocephalum*, *Trifolium*, *Vesicastrum* and *Trifoliastrum*, 10.7%~20.7 %). The contrasting repeat composition between organelle genomes was particularly evident in *T. pratense*, which had smallest amount of repeat sequence in its mitogenome and the largest amount in its plastome (Figure 3; Table 2).

**Table 2.** Comparison of repeat percentage between organelle genomes in *Trifolium*.

| Species | Subgenus | Section | Mitogenome | | | Plastome | | |
|---|---|---|---|---|---|---|---|---|
| | | | Size (bp) | Repeat (%) | NCBI Accession | Size (bp) | Repeat (%) | NCBI Accession |
| *T. aureum* | *Chronosemium* | *Chronosemium* | 294,911 | 8.6 | MT039392 | 126,970 | 5.2 | NC_024035 |
| *T. grandiflorum* | *Chronosemium* | *Chronosemium* | 347,723 | 8.4 | MT039391 | 125,628 | 4.7 | NC_024034 |
| *T. boissieri* | *Chronosemium* | *Chronosemium* | | Not applicable | | 125,740 | 5.1 | NC_025743 |
| *T. strictum* | *Trifolium* | *Paramesus* | | Not applicable | | 125,834 | 4.4 | NC_025745 |
| *T. glanduliferum* | *Trifolium* | *Paramesus* | | Not applicable | | 126,149 | 4.8 | NC_025744 |
| *T. lupinaster* | *Trifolium* | *Lupinaster* | | Not applicable | | 135,049 | 10.9 | KJ788287 |
| *T. subterraneum* | *Trifolium* | *Trichocephalum* | | Not applicable | | 144,763 | 19.7 | NC_011828 |
| *T. meduseum* | *Trifolium* | *Trichocephalum* | 348,724 | 8.5 | MT039390 | 142,595 | 19.5 | NC_024166 |
| *T. pratense* | *Trifolium* | *Trifolium* | 301,823 | 6.6 | MT039389 | 146,573 | 20.7 | MT039393 |
| *T. hybridum* | *Trifolium* | *Vesicastrum* | | Not applicable | | 134,831 | 13.1 | KJ788286 |
| *T. semipilosum* | *Trifolium* | *Vesicastrum* | | Not applicable | | 138,194 | 15.8 | KJ788291 |
| *T. repens* | *Trifolium* | *Trifoliastrum* | | Not applicable | | 132,120 | 10.7 | NC_024036 |
| *T. occidentale* | *Trifolium* | *Trifoliastrum* | | Not applicable | | 133,806 | 11.1 | KJ788289 |

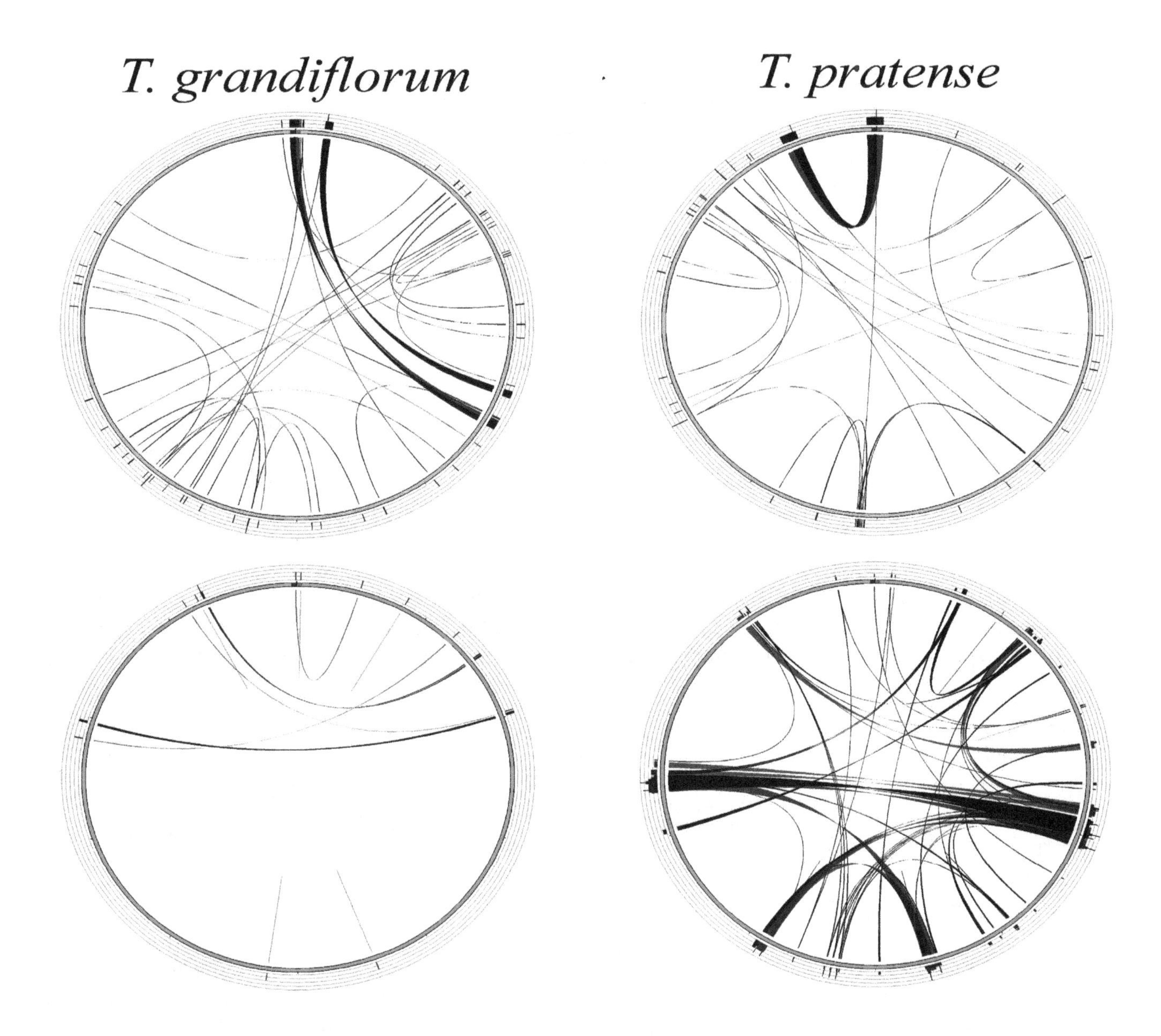


**Figure 3.** Distribution pattern of dispersed repeat sequences in circular representations of organelle genomes of two *Trifolium* species. Each ribbon represents a BLAST hit for a pair of dispersed repeats. Brown circles are mitogenomes and green circles are plastomes. Multiple hits in a single region are indicated by histograms in outer concentric rings. Data for repetitive sequences of all published organelle genomes of *Trifolium* is available in Table 2.

### *2.3. Intracellular Gene Transfer (IGT) in Trifolium*

The extent of IGT among the three genomic compartments was analyzed in *T. pratense* by BLAST (Figure 4; Table 3). The amount of DNA shared between the two organelle genomes was very low (0.3 kb). The organelle genomes shared considerable DNA with the nuclear genome and GC content of shared DNA reflected the compartment of origin (45.8% for mitogenome and 35.1% for plastome). In general, BLAST hits between nuclear and organelle genomes were very short and had high sequence identity (Table 3).

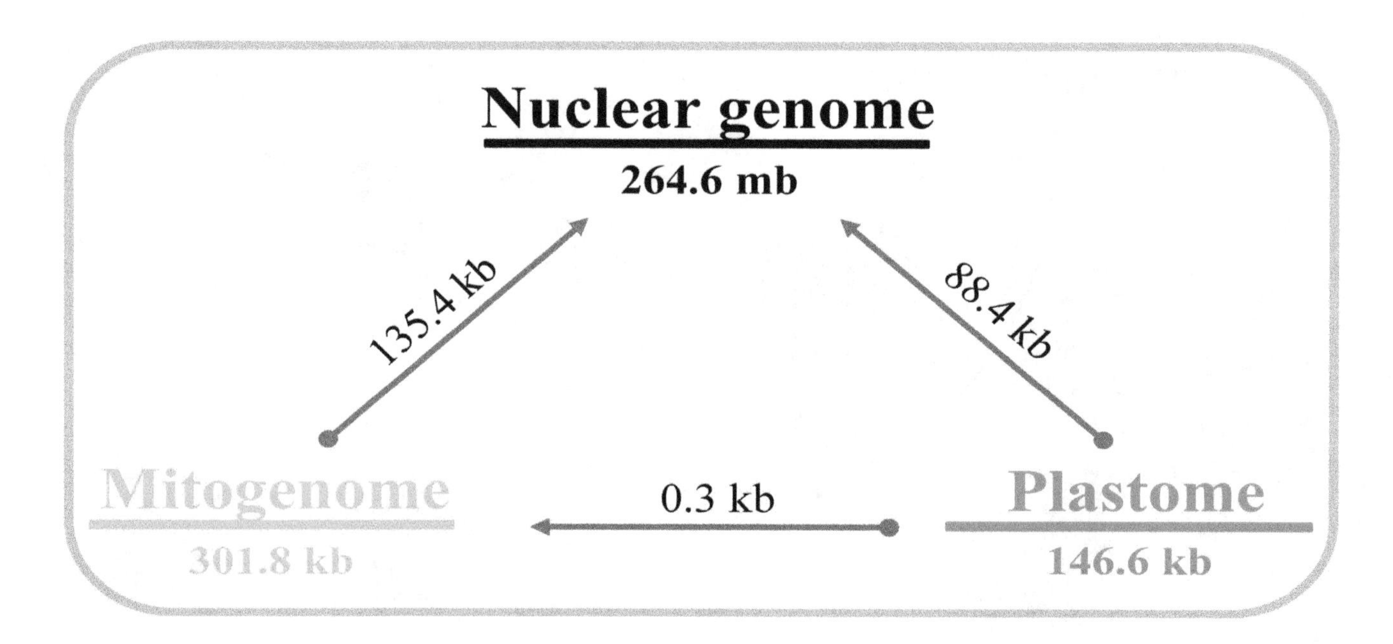


**Figure 4.** Shared DNA among three different genomic compartments in *Trifolium pratense*. Size of each genome is presented below the bar. Values on arrows represent the amount of shared DNA. Round end of arrow represents query sequence and pointed end represents subject sequence in BLAST analyses.

**Table 3.** BLAST hit statistics for shared DNA between nuclear and organelle genomes in *Trifolium pratense*.

| Comparison | Number | Average Identity | GC (%) | Length (bp) | | | |
|---|---|---|---|---|---|---|---|
| | | | | Min | Max | Mean | Median |
| Nuclear vs. Mitochondria | 1830 | 95.5 | 45.8 | 33 | 3950 | 121.8 | 93.5 |
| Nuclear vs. Plastid | 1086 | 95.7 | 35.1 | 34 | 2027 | 144.2 | 118.5 |

A long contiguous region (348.5 kb) was identified from chromosome 4 of *T. repens* (position: 72,476,623–72,825,180) that shared substantial DNA with the mitogenome of *T. meduseum* (Figure S3). This sequence had a high GC content (44.3%) compared to the entire chromosome 4 (33.2%).

### 2.4. Multiple Functional Transfers of Mitochondrial rps1 in Papilionoideae

A phylogenetic analysis of nuclear and mitochondrial copies of *rps1* for papilionoid legumes was conducted (Figure 5). Nuclear genomes of two *Trifolium* species (*T. pratense* and *T. repens*) (Table S1) included multiple *rps1* copies. Nuclear copies of *rps1* were placed in two separate positions, one that included *Lotus* sister to the taxa in the tribes Fabeae and Trifolieae and the second with four genera of the tribe Trifolieae (*Trigonella, Melilotus, Medicago* and *Trifolium*). Branch lengths for the nuclear copies of *rps1* were substantially longer than mitochondrial copies indicating accelerated substitution rates. The Trifolieae was monophyletic but the branch leading to the tribe was very short and the bootstrap value (BS = 43%) was low. In Trifolieae, the mitochondrial *rps1* sequences formed a paraphyletic grade sister to a monophyletic group of nuclear *rps1* (BS = 96%).

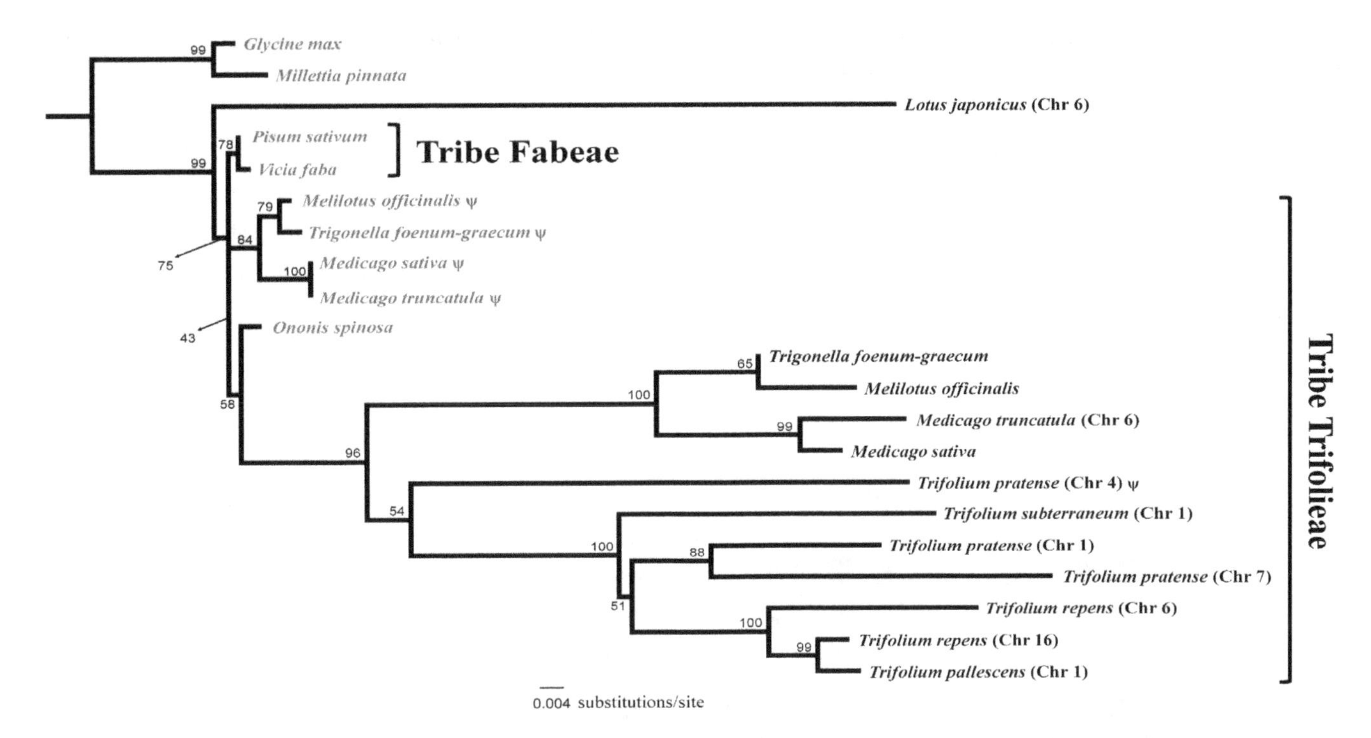


**Figure 5.** Maximum likelihood phylogeny of *rps1* gene. Bootstrap values are indicated at nodes. Sequences from mitogenome are indicated in brown. Sequences from nuclear genome are indicated as black and chromosome (Chr) numbers are specified in the parentheses when available.

### *2.5. Fission of ccmF in Land Plants*

To investigate the phylogenetic distribution of the fission of *ccmFn* and conservation of two ORFs *ccmFn1* and *ccmFn2* in *Trifolium*, mitochondrial *ccmF* sequences were assembled using available next-generation sequencing (NGS) reads (Table S2). The expanded taxon sampling confirmed the adjacency of the ORFs *ccmFn1* and *ccmFn2* and that the fission was restricted to *Trifolium*. All examined *Trifolium* species shared the *ccmFc* intron loss. Draft nuclear genome sequences of four species of *Trifolium* (*T. subterraneum*, *T. pratense*, *T. pallescens* and *T. repens*) were examined for intact copies of *ccmFn1* and *ccmFn2*. Fragments of sequences similar to *ccmFn1* and *ccmFn2* were identified in *T. subterraneum* and *T. pratense* but no intact copies were detected. However, intact copies both of *ccmFn1* and *ccmFn2* from *T. pallescens* (chromosome 4) and *T. repens* (chromosomes 4 and 9) were identified and were adjacent as in mitogenomes of *Trifolium*. Eleven *ccmFn* sequences (eight mitochondrial and three nuclear copies) were detected in *Trifolium* (Figure S4a). All nuclear copies were identical to their corresponding mitochondrial copy. Among mitochondrial copies, only three *Trifolium* species (*T. aureum*, *T. grandiflorum* and *T. pallescens*) showed unique sequence and the remaining sequences in the other five species were identical to each other in the coding region (Figure S4b).

Fission of *ccmFc* was analyzed in three species of *Marchantia* and two other genera of the Marchantiales. Sequence alignment revealed that a single nucleotide deletion caused *ccmFc* fission in one species of *Marchantia*, *M. paleacea* (Figure S5).

Examination of *ccmFn* fission in Brassicaceae included 17 taxa (Table S2). The *ccmF* genes were assembled from Cleomaceae (*Cleome violacea*), the sister family of Brassicaceae and two early diverging Brassicaceae genera (*Aethionema* and *Odontarrhena*). The fission of *ccmFn* was shared by all Brassicaceae except *Aethionema* and in all cases *ccmFn1* and *ccmFn2* were found in different loci. *Odontarrhena argentea* was the only member of the Brassicaceae that lost the *ccmFc* intron.

The phylogenetic position of *ccmFn* fission and separation in Fabaceae and Brassicaceae (Table S2), were plotted on cladograms of each of family (Figure 6a,b). The location of the breakpoint of *ccmFn* fission was also compared among the three families Fabaceae, Brassicaceae and Amaryllidaceae (Figure 6c). The fission occurred in different locations in the gene within each family and occurred in a more basal position in Brassicaceae than Fabaceae. The separation of *ccmFn1* and *ccmFn2* only occurred in Amaryllidaceae and Brassicaceae.

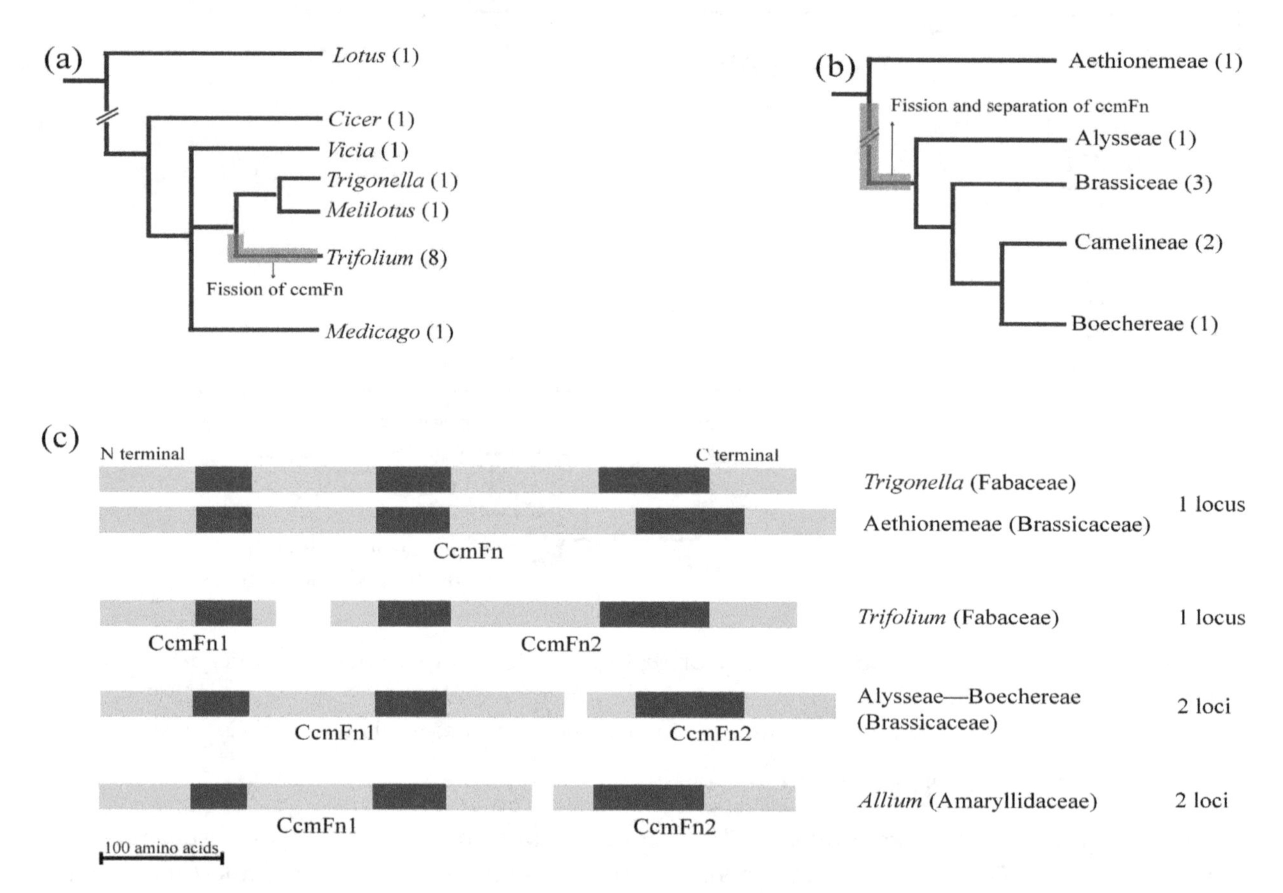


**Figure 6.** Fission events of *ccmFn* in angiosperms. (**a**) Examined genera of Fabaceae are plotted on cladogram from Ellison et al. [42]. Number of examined species is indicated in the parenthesis. The branch where the fission occurred is marked with grey bar. (**b**) Examined tribes of Brassicaceae are plotted on cladogram from Huang et al. [47]. Number of examined genera is indicated in the parenthesis. The branch where the fission and separation occurred is marked with grey bar. (**c**) *CcmFn* gene organization compared among various taxa of angiosperms. Dark brown box indicates conserved domains.

## 3. Discussion

### *3.1. Contrasting Evolutionary Trajectories of Trifolium Organelle Genomes*

*Trifolium* mitogenomes (294,911 to 348,724 bp) (Table 1) are similar in size to the other Trifolieae genus *Medicago* (271,618 bp), which has the smallest currently sequenced papilionoid mitogenome [8]. Mitogenomes of *Trifolium* have relatively little repetitive DNA (6.6–8.6%) (Table 2) compared to mitogenomes of other Papilionoideae species (2.9–60.6%) [8]. This low repeat content in the mitogenome is in contrast to the plastome of some *Trifolium* species. The acquisition of numerous, novel repeat sequences and drastic rearrangement in the plastome of *T. subterraneum* and related species has been reported [31,33,35]. Increased taxon sampling by Sveinsson and Cronk [34] revealed that plastome expansion is shared by five sections, referred to as the "refractory clade" in subgenus *Trifolium* (*Lupinaster*, *Trichocephalum*, *Trifolium*, *Vesicastrum* and *Trifoliastrum*). The distinct evolutionary trajectory of organelle genomes in the genus is particularly evident in *T. pratense*, which has the lowest percentage of repetitive DNA in the mitogenome and the highest in the plastome as well as the most highly rearranged structure (Table 2 and Figure 3). In plant mitogenomes, accumulation of repeats, genome expansions and rearrangements may be a consequence of error-prone DNA repair mechanisms such as nonhomologous end-joining or break-induced-replication [48–50]. In Geraniaceae, a correlation between nonsynonymous substitution rates for DNA replication, recombination and repair (DNA-RRR) genes and plastome complexity was reported [51]. The plastome-specific increase in repeat complexity in the *Trifolium* refractory clade may be the result of disruption of 'plastid specific' DNA-RRR-protein genes, some of which are targeted to both mitochondria and plastids [7]. More comprehensive taxon sampling that includes data from all three plant genomic compartments of *Trifolium* is required to test this hypothesis.

### 3.2. Multiple Functional Transfers of the Mitochondrial rps1 Gene to the Nucleus in Papilionoideae

An earlier investigation reported the functional transfer of mitochondrial *rps1* to the nucleus in three genera of Trifolieae (*Trigonella, Melilotus* and *Medicago*) [39]. In the current study, the complete deletion of *rps1* gene from mitogenomes of four *Trifolium* species was detected (Figure S1), which is shared by the distantly related genus *Lotus*, a member of the tribe Loteae (Figure S2). There are two possible explanations for the phylogenetic distribution of the loss/transfer. The loss of mitochondrial *rps1* could be due to a single IGT in a common ancestor with differential resolution in descendant lineages, that is, acquisition of functional signals (or not) to stabilize transfer. Alternatively, there may have been independent functional transfers from an ancestor in each of the two unrelated lineages. To examine these alternatives, a maximum likelihood (ML) analysis was conducted using expanded taxon sampling of nuclear and mitochondrial *rps1* sequences. The resulting tree (Figure 5) included some long branches, which may be affected by the well-known phenomenon of long-branch attraction [52]. Nuclear *rps1* from *Lotus* and Trifolieae species were split into two independent clades, with intact and pseudogenized mitochondrial *rps1* placed between them. This pattern supports the explanation that functional transfers of *rps1* occurred at least two times in Papilionoideae, once in *Lotus* and a separate event in the ancestor of the Trifolieae clade that includes *Trigonella, Melilotus, Medicago* and *Trifolium*. The timing of the functional transfer of *rps1* in Trifolieae would likely be after the divergence of *Ononis* (Figure 5), which only has a mitochondrial copy [39].

Despite the putative functional replacement by nuclear *rps1*, the mitochondrial *rps1* in three genera (*Trigonella, Melilotus* and *Medicago*) was retained with limited sequence divergence (Figure 5), whereas it is completely and precisely deleted in *Trifolium* (Figure S1). Coding regions of plant mitogenomes are conserved by an accurate long homology-based repair mechanism, while non-coding regions are not conserved and are repaired by error-prone mechanisms [50]. Differential selection on mitogenomic molecules, which reduces harmful mutations on coding regions after double strand breaks (DSBs), was proposed to explain this [48,49]. Pseudogenized copies of mitochondrial *rps1* in the three genera *Trigonella, Melilotus* and *Medicago* are located adjacent to *nad5* exon1 (ca. 200 bp apart) [39]. Mutations in 5′ region of *nad5* exon1 that do not disturb transcription or translation of the functional gene and only affect pseudogenized *rps1* can be inherited by selection after DSBs. So, the adjacent location of mitochondrial *rps1* to *nad5* exon1 may enable retention of high sequence identity after functional replacement by sharing the benefit of accurate repair. a similar situation is known for the *rps14* pseudogene that is adjacent to *rpl5* in grasses [53]. Conservation of non-coding regions adjacent to coding regions is also present in mitogenome-wide sequence divergence comparisons across Fabaceae [8].

### 3.3. Shared DNA Among Genomes of Trifolium

Comparative analyses of the three genomic compartments (nuclear, mitochondrial and plastid) in *T. pratense* revealed a substantial amount of shared DNA between nuclear and organelle genomes, most of which was short fragments (Figure 4, Table 3). The shared DNAs between nuclear and mitochondrial genome was 135.4 kb (Figure 4) and had GC content more similar to those of mitogenomes (Tables 1 and 3) suggesting that most IGT was unidirectional (i.e., mitochondrion to nucleus) and the nuclear genome of *T. pratense* includes numerous NUMTs. These NUMTs may integrate into the nuclear genome of *T. pratense* as short fragments. Alternatively, these short fragments may be the consequence of post-IGT mutational decay and rearrangement of longer NUMT sequences [54].

The discovery of a long stretch of NUMTs (spanning 348.5 kb; GC: 44.3%) in chromosome 4 of *T. repens* (Figure S3) supports a recent genomic scale IGT event. This type of large IGT was identified in *Arabidopsis thaliana* (Brassicaceae) in which ~270 kb of 367 kb mitogenome transferred to the nucleus [55] and covers an ~620 kb region of the nuclear genome [56]. To estimate the amount of NUMTs in *T. repens*, a mitogenome sequence from the same DNA source (white clover cv 'Crau' derivative) [46,57] is necessary. Large NUMTs were reported for animal nuclear genomes (little brown bat and fugu), however, these were later shown represent artifacts of genome assembly [58,59]. The nuclear genomes of *Trifolium* species are drafts with many gaps [43–46]. Verification of long putative NUMTs in *Trifolium* is needed to confirm genomic scale IGT events from the mitochondrial to nuclear genome.

### 3.4. Multiple Fissions of ccmF in Land Plants and a Novel Event in Trifolium

The first fission of mitochondrial *ccmF* dates back to the early evolution of land plants and split the gene into N-terminal (*ccmFn*) and C-terminal (*ccmFc*) coding regions [60]. In Marchantiales, the ORFs are closely adjacent (Figure S5). The mitogenome study of *Marchantia paleacea* (misidentified as *M. polymorpha* [61]) from the early 1990s [22] reported a fission of *ccmFc* (i.e., *ccmFc1* and *ccmFc2*) due to a single nucleotide deletion. This fission event was accepted in several subsequent papers [3,21,60], however, mitogenome sequences of two other *Marchantia* species (*M. inflexa* and *M. polymorpha* subsp. *ruderalis*) did not show the single nucleotide deletion, consistent with the other two available mitogenomes of Marchantiales (Figure S5). The initial report of a *ccmFc* fission in *Marchantia* should be re-examined to determine if it is specific to *M. paleacea* or the result of sequencing error.

In angiosperms, two independent fissions of *ccmFn* have been reported in *Allium* (Amaryllidaceae) [25] and Brassicaceae [24,62]. In both cases, *ccmFn1* and *ccmFn2* are distant from each other in the mitogenome and they share a similar breakpoint for the fission (Figure 6). The phylogenetic distribution of the fission in Amaryllidaceae was investigated by polymerase chain reaction using four genera in the family (*Narcissus, Tulbaghia, Ipheion* and *Allium*) and revealed that the separation of the two sequences is restricted to *Allium* [25]. However, the status of the other three genera without separation of *ccmFn* sequences does not necessarily guarantee that the gene is not split because there are cases of gene fission where the two new genes occupy a single locus, for example, fission of *ccmF* (into *ccmFn* and *ccmFc*) in Marchantiales (Figure S5) and *ccmFn* (into *ccmFn1* and *ccmFn2*) in *Trifolium* (Figure 2). The distribution and status of *ccmFn* fission in Amaryllidaceae needs further investigation including broad taxon sampling as well as confirmation with additional sequencing.

In Brassicaceae, it was argued that the fission is shared by all members of the family because it is present in five complete or draft mitochondrial genomes covering the earliest diverging genus (*Aethionema*) and other core genera (*Arabidopsis, Brassica, Raphanus*), whereas the mitogenome of the sister family Cleomaceae does not have the fission [62]. Further investigation, including additional published mitogenomes and assembled mitochondrial contigs for *ccmF* genes (Table S2), indicates that three species of *Aethionema* do not have the fission of *ccmFn* (Figure 6b). This discrepancy could be due an assembly error since the *Aethionema* data in the previous study was a draft mitogenome [62]. Whatever was the cause of discrepancy, it is clear that the fission of *ccmFn* is shared by many but not all Brassicaceae. The fission occurred after the divergence of *Aethionema* (Figure 6b); however, it is unknown if there was an intermediate stage that had experienced the fission but not physical separation of the *ccmFn1* and *ccmFn2*.

The independent fission of *ccmFn* in *Trifolium* represents a novel event. The fission was caused by a deletion of 59 bp resulting in a frame shift and premature stop codon (Figure 2). An alternative outcome of this deletion may be pseudogenization of the *ccmFn*. Mutational decay and deletion of pseudogenized mitochondrial genes can be delayed by proximity to functional genes (e.g., *rps1* in some

Trifolieae genera and *rps14* in grasses, see Section 3.2). However, the gene that is consistently adjacent to *ccmFn* (*ccmFn1* and *ccmFn2*) is *ccmC*, which is ca. 8kb away from *ccmFn* in the four *Trifolium* species (Figure 1). Moreover, the expanded *ccmFn* sequence sampling confirms that the two ORFs (*ccmFn1* and *ccmFn2*) are conserved in eight *Trifolium* species with only a limited amount of sequence variation in coding regions (Figure S4). The fission break point in *Trifolium* is different from other angiosperms that express cytochrome c maturation protein from two ORFs, yet the conserved domains of the product remain intact (Figure 6c). Hence, the two ORFs of *ccmFn* are regarded as functional. The fission occurred after the divergence of genera *Trigonella* and *Melilotus* in the Trifolieae. The conserved adjacency of the two ORFs (*ccmFn1* and *ccmFn2*) may represent an early stage of the fission as in *ccmFn* and *ccmFc* in Marchantiales (Figure S5).

The fission of *ccmFn* in *Trifolium* leads to another question: is this event related to "intercompartmental piecewise gene transfer" [21]? To explore this question, we searched for ORFs of *ccmFn* in draft nuclear genomes of four *Trifolium* species (*T. subterraneum*, *T. pratense*, *T. pallescens* and *T. repens*). Both *T. pallescens* and *T. repens* (Figure S4) contained the *ccmFn* NUMTs however these were not restricted to a single ORF but included a locus covering both ORFs (*ccmFn1* and *ccmFn2*) and their flanking regions. The NUMTs were identical to their counterpart in mitogenome suggesting that the transfer was a recent event (or artifact in nuclear genome assembly, see discussion Section 3.3). Furthermore, there was no post-IGT sequence modification to suggest a functional transfer. Evidence did not support a relationship between fission of the mitochondrial gene *ccmFn* and piecewise or functional transfer in *Trifolium* species.

## 4. Materials and Methods

### 4.1. Assembly of Trifolium Mitogenomes

Four species of *Trifolium* from the two subgenera *Chronosemium* (*T. aureum* and *T. grandiflorum*) and *Trifolium* (*T. meduseum* and *T. pratense*) were selected for mitogenome assembly. The 100 bp paired-end raw Illumina (San Diego, CA, US) reads (Table 1) for mitogenome assembly were from Sabir et al. [31]. Assembly and mapping were conducted in Geneious Prime (https://www.geneious.com) using Geneious assembler and mapper, respectively. To assemble mitogenomes, the methods in Choi et al. [8] were followed. First, raw reads from the plastome were excluded by mapping total raw reads to corresponding plastomes [*T. aureum* (NC_024035.1), *T. grandiflorum* (NC_024034.1), *T. meduseum* (NC_024166.1) and *T. pratense* (MT039393)]. De novo assembly was subsequently conducted for each with ~30 million plastome-filtered reads. Among the assembled contigs, mitochondrial contigs were selected by BLAST searches against reference Fabaceae mitogenome sequences at National Center for Biotechnology Information (NCBI) (https://www.ncbi.nlm.nih.gov/genome/organelle/) using BLASTN 2.8.0+ [63] with default options. Mitochondrial contigs were manually assembled as single chromosomes in Geneious. Finally, draft mitogenomes were refined by mapping total plastome-filtered reads.

### 4.2. Annotation and Genome Content Comparison of Mitogenomes

To compare gene and intron content of *Trifolium* mitogenomes with related taxa, five previously published mitogenomes were acquired—two from IRLC [*Vicia faba* (KC189947) and *Medicago truncatula* (NC_029641)], one from the robinioid clade [*Lotus japonicus* (NC_016743)], which is sister to IRLC; and two from millettioid *sensu lato* clade [*Millettia pinnata* (NC_016742)] and *Glycine max* (NC_020455)], which is sister to the hologalegina clade (robinioid + IRLC). Annotation of rRNAs, protein coding genes and introns was conducted based on a reference mitogenome of *Liriodendron tulipifera* (NC_021152) with a set of 41 conserved mitochondrial genes in Geseq [64]. Annotation for protein coding genes was manually corrected in Geneious to fit ORFs. The annotation for tRNAs was cross-checked by tRNAscan-SE v2.0 [65].

*4.3. Completion of the Trifolium Pratense Plastome*

Plastome drafts of *Trifolium pratense* were reported in two different studies [31,34] but these sequences contained a complex repeat structure. Since these previous assemblies were based on short insert size data only (400-800 bp), the *T. pratense* plastome was redone using sequences generated from one of the previous studies [31] as well as mapping data from mate-pair Illumina sequences (ERX946087) with long insert sizes (7 kb) [43]. The newly assembled plastome was annotated as described above but with MPI-MP chloroplast references in GeSeq [64].

*4.4. Repeat Estimation in Organelle Genomes*

Repeat content was estimated in four mitogenomes and 13 plastomes (Table 2). Tandem repeats were identified using Tandem Repeats Finder version 4.09 [66] with default options. Other repeats (larger than 30 bp) were analyzed by BLASTN [63] searches using each genome as both subject and query with a word size of 7 and an e-value of $1e^{-6}$ as described in Guo et al. [67]. All BLAST hits were retained. Sequence coordinate information for BLAST hits was transferred to each genome as an annotation in Geneious and overlapping regions between hits were excluded from the estimations for repetitive DNA content. The distribution of dispersed repeat sequences across the genomes was visualized by Circoletto [68].

*4.5. Shared DNA among Different Genomic Compartments*

Shared DNA was evaluated in *Trifolium pratense* because this is the only species examined with completed sequences from all three genomic compartments. The mitogenome (MT039389) and plastome (MT039393) in this study were utilized and the nuclear genome was available as a chromosome-scale reference draft (LT990601- LT990607) [43]. Shared DNA among the genomes was evaluated in MegaBLAST with a word size of 28 and an e-value of $1e^{-6}$. For nuclear and organelle genome comparisons, each organelle sequence was used as the query against a subject database comprising the nuclear genome. For the comparison of organelle genomes, the plastome was used as the query and the mitogenome was the subject. BLAST hits with sequence identity higher than 90% were retained. Overlapping regions between hits were excluded from the estimations of shared DNA.

To search for putative large-scale IGT (> 100 kb) events, shared DNA analysis was conducted as described above but in this case the largest mitogenome (*T. meduseum*) and other published nuclear genomes of *Trifolium* (Table S1) were utilized. BLAST hits between the mitogenome and a long stretch of the nuclear region of *T. repens* were visualized by Circoletto [68].

*4.6. Investigation on Status of rps1 in Nuclear and Mitochondrial Genome*

Nuclear and mitochondrial sequences of *rps1* generated for a previous study [39] were acquired from NCBI. Nuclear *rps1* sequences for other species were searched by MegaBLAST using the options described above. Mitochondrial *rps1* of *Vicia faba* was used to query nuclear genomes of *Lotus japonicus, Medicago truncatula, Trifolium subterraneum, T. pratense, T. pallescens* and *T. repens* (Table S1). Mitochondrial *rps1* sequences were also extracted from mitogenomes of *Glycine max, Millettia pinnata, Vicia faba* and *Medicago truncatula*. All *rps1* sequences were aligned with MAFFT v.7.017 [69] using default options. Nucleotide substitution models were evaluated in jModelTest v.2.1.6 [70] by Akaike information criterion. ML analysis (GTR +G with 1000 bootstrap replications) was conducted using *G. max* and *M. pinnata* as outgroups in RAxML v.8 [71] in the CIPRES Science Gateway [72].

The status of mitochondrial *rps1* in *Trifolium* was tested by sequence alignment of the mitochondrial locus containing *rps1* and *nad5* exon1 in *M. truncatula* and the corresponding regions in four mitogenomes of *Trifolium*. Sequences were aligned in MAFFT [69] using default options followed by manual adjustments to minimize gaps and maximize apparent homologous regions.

### *4.7. Investigation of ccmF Fissions in Selected Land Plants*

To investigate previously reported fission events of *ccmF* genes in land plants [22–25], all available sequences from published mitogenome sequences related to *Marchantia*, Brassicaceae, *Allium* (Amaryllidaceae) and Fabaceae were acquired. For *Marchantia*, published mitogenomes [22,73,74] of three species (*M. inflexa*, *M. polymorpha* subsp. *ruderalis* and *M. paleacea*) were examined: two from NCBI [*M. polymorpha* subsp. *ruderalis* (NC_037508.1) and *M. paleacea* (NC_001660.1)] and *M. inflexa*, which was downloaded from FigShare (https://figshare.com/articles/Marchantia_inflexa_mitochondrion_and_chloroplast_genomes/6639209/1). Two mitogenomes [*Dumortiera hirsuta* (NC_042873) and *Riccia fluitans* (NC_043906)], which are closely related to *Marchantia* in Marchantiales [75,76], were also included. For Amaryllidaceae, a single mitogenome [*Allium cepa* (NC_030100)] was available.

In addition to previously published and newly assembled mitogenomes, mitochondrial contigs were generated from available NGS reads for Brassicales and Fabaceae (Table S2). Raw sequences were mapped to reference *ccmF* sequences and the mapped reads were assembled in Geneious. The *ccmF* sequences of *Medicago truncatula* and *Batis maritima* were used as references for Fabaceae and Brassicales, respectively. Read depth of assembled *ccmF* genes (*ccmFn* and *ccmFc*) were compared to confirm that sequences originated from mitogenome rather than from other genomic compartments (i.e., nuclear and plastid genome). To search for nuclear copies of *ccmFn1* and *ccmFn2*, subject databases comprising four *Trifolium* nuclear genomes (Table S1) were queried with the mitochondrial *ccmFn* of *T. aureum* using MegaBLAST with default options. All sequences were aligned with MAFFT as described above. The status of *ccmFn* was plotted on cladograms from published phylogenetic studies of *Trifolium* [42] and Brassicaceae [47]. Conserved domains of *ccmFn* were detected using the Motif Scan of MyHits (http://myhits.isb-sib.ch/cgi-bin/motif_scan) [77,78].

## 5. Conclusions

The newly sequenced mitogenomes of *Trifolium* allowed comparative analyses of genome evolution for all three cellular compartments—mitochondrion, nucleus and plastid. Unlike many angiosperms, *Trifolium* lacks the highly repetitive genome organization of mitogenome. Some *Trifolium* plastomes has a much more complex organization and has accumulated more repeat contents than the mitogenome. a substantial amount of organellar DNA was detected in nuclear genomes of *Trifolium*, likely resulting from recent and nonfunctional IGT events. In addition, there has been an ancestral, functional transfer of mitochondrial *rps1* to the nuclear genome. a notable finding from the mitogenome of *Trifolium* was a novel gene fission of *ccmFn*. Analyses of *ccmF* genes in selected land plants provided further insights into the fission events. Although the current study is based on limited sampling of the three genomic compartments, our findings expand the understanding of how these genomes evolved in *Trifolium* The underlying evolutionary and molecular mechanisms should be examined in future comparisons that incorporate broader taxonomic sampling for all three genomic compartments.

**Supplementary Materials:** Supplementary Materials can be found at Figure S1. Nucleotide alignment showing deletion of mitochondrial *rps1* in *Trifolium* species. Figure S2. Gene and cis-spliced intron content across six Papilionoideae genera. Figure S3. Circoletto map showing similar sequences between mitogenome of *T. meduseum* (left arc) and a continuous region of nuclear genome (right arc) of *T. repens* (chromosome 4; NCBI accession: VCDJ01010667; position: 72,476,623-72,825,180). Figure S4. Sequence variation of *ccmFn* in *Trifolium* species. Figure S5. Alignment of the mitochondrial region containing *ccmFn* and *ccmFc* genes from five species of Marchantiales. Table S1. Information on nuclear genomes, used for comparative study. Table S2. List of taxa for *ccmF* analysis with information about sequence sources and status of the genes.

**Author Contributions:** Conceptualization, I.-S.C. and R.K.J.; Validation, R.K.J. and T.A.R.; Formal Analysis, I.-S.C.; Investigation, I.-S.C.; Resources, R.K.J. and T.A.R.; Data Curation, R.K.J. and T.A.R.; Writing—Original Draft Preparation, I.-S.C.; Writing—Review & Editing, I.-S.C., T.A.R. and R.K.J.; Visualization, I.-S.C.; Supervision, R.K.J.; Funding Acquisition, R.K.J. and T.A.R. All authors have read and agreed to the published version of the manuscript.

## Abbreviations

| | |
|---|---|
| IGT | Intracellular gene transfer |
| NUMT | Nuclear mitochondrial DNA sequences |
| NUPT | Nuclear plastid DNA sequences |
| IR | Inverted repeat |
| IRLC | Inverted repeat lacking clade |
| ORF | Open reading frame |
| NGS | Next-generation sequencing |
| DNA-RRR | DNA replication, recombination and repair |
| ML | Maximum likelihood |
| DSB | Double strand break |
| NCBI | National Center for Biotechnology Information |

## References


1. Jansen, R.K.; Ruhlman, T.A. Plastid Genomes of Seed Plants. In *Genomics of Chloroplasts and Mitochondria*; Bock, R., Knoop, V., Eds.; Springer: Dordrecht, The Netherlands, 2012; pp. 103–126.
2. Ruhlman, T.A.; Jansen, R.K. The plastid genomes of flowering plants. In *Chloroplast Biotechnology: Methods and Protocols*; Maliga, P., Ed.; Spring: New York, NY, USA, 2014; pp. 3–38.
3. Knoop, V. The mitochondrial DNA of land plants: Peculiarities in phylogenetic perspective. *Curr. Genet.* **2004**, *46*, 123–139. [CrossRef] [PubMed]
4. Wynn, E.L.; Christensen, A.C. Repeats of unusual size in plant mitochondrial genomes: Identification, incidence and evolution. *G3 Genes Genomes Genet.* **2019**, *9*, 549–559. [CrossRef] [PubMed]
5. Adams, K.L.; Qiu, Y.-L.; Stoutemyer, M.; Palmer, J.D. Punctuated evolution of mitochondrial gene content: High and variable rates of mitochondrial gene loss and transfer to the nucleus during angiosperm evolution. *Proc. Natl. Acad. Sci. USA* **2002**, *99*, 9905–9912. [CrossRef] [PubMed]
6. Kubo, T.; Mikami, T. Organization and variation of angiosperm mitochondrial genome. *Physiol. Plant.* **2007**, *129*, 6–13. [CrossRef]
7. Gualberto, J.M.; Newton, K.J. Plant mitochondrial genomes: Dynamics and mechanisms of mutation. *Annu. Rev. Plant Biol.* **2017**, *68*, 225–252. [CrossRef] [PubMed]
8. Choi, I.-S.; Schwarz, E.N.; Ruhlman, T.A.; Khiyami, M.A.; Sabir, J.S.; Hajarah, N.H.; Sabir, M.J.; Rabah, S.O.; Jansen, R.K. Fluctuations in Fabaceae mitochondrial genome size and content are both ancient and recent. *BMC Plant Biol.* **2019**, *19*, 448. [CrossRef] [PubMed]
9. Wolfe, K.H.; Li, W.-H.; Sharp, P.M. Rates of nucleotide substitution vary greatly among plant mitochondrial, chloroplast, and nuclear DNAs. *Proc. Natl. Acad. Sci. USA* **1987**, *84*, 9054–9058. [CrossRef]
10. Drouin, G.; Daoud, H.; Xia, J. Relative rates of synonymous substitutions in the mitochondrial, chloroplast and nuclear genomes of seed plants. *Mol. Phylogenet. Evol.* **2008**, *49*, 827–831. [CrossRef]
11. Selosse, M.-A.; Albert, B.; Godelle, B. Reducing the genome size of organelles favours gene transfer to the nucleus. *Trends Ecol. Evol.* **2001**, *16*, 135–141. [CrossRef]
12. Lopez, J.V.; Yuhki, N.; Masuda, R.; Modi, W.; O'Brien, S.J. Numt, a recent transfer and tandem amplification of mitochondrial DNA to the nuclear genome of the domestic cat. *J. Mol. Evol.* **1994**, *39*, 174–190.
13. Timmis, J.N.; Ayliffe, M.A.; Huang, C.Y.; Martin, W. Endosymbiotic gene transfer: Organelle genomes forge eukaryotic chromosomes. *Nat. Rev. Genet.* **2004**, *5*, 123–135. [CrossRef] [PubMed]
14. Richardson, A.O.; Palmer, J.D. Horizontal gene transfer in plants. *J. Exp. Bot.* **2006**, *58*, 1–9. [CrossRef] [PubMed]
15. Bonen, L. Mitochondrial genes leave home. *New Phytol.* **2006**, *172*, 379–381. [CrossRef] [PubMed]
16. Huang, C.Y.; Grünheit, N.; Ahmadinejad, N.; Timmis, J.N.; Martin, W. Mutational decay and age of chloroplast and mitochondrial genomes transferred recently to angiosperm nuclear chromosomes. *Plant Physiol.* **2005**, *138*, 1723–1733. [CrossRef]
17. Adams, K.L.; Song, K.; Roessler, P.G.; Nugent, J.M.; Doyle, J.L.; Doyle, J.J.; Palmer, J.D. Intracellular gene transfer in action: Dual transcription and multiple silencings of nuclear and mitochondrial *cox2* genes in legumes. *Proc. Natl. Acad. Sci. USA* **1999**, *96*, 13863–13868. [CrossRef]

18. Laroche, J.; Li, P.; Maggia, L.; Bousquet, J. Molecular evolution of angiosperm mitochondrial introns and exons. *Proc. Natl. Acad. Sci. USA* **1997**, *94*, 5722–5727. [CrossRef]
19. Adams, K.L.; Daley, D.O.; Qiu, Y.-L.; Whelan, J.; Palmer, J.D. Repeated, recent and diverse transfers of a mitochondrial gene to the nucleus in flowering plants. *Nature* **2000**, *408*, 354–357. [CrossRef]
20. Adams, K.L.; Ong, H.C.; Palmer, J.D. Mitochondrial gene transfer in pieces: Fission of the ribosomal protein gene *rpl2* and partial or complete gene transfer to the nucleus. *Mol. Biol. Evol.* **2001**, *18*, 2289–2297. [CrossRef]
21. Szafranski, P. Intercompartmental piecewise gene transfer. *Genes* **2017**, *8*, 260. [CrossRef]
22. Oda, K.; Yamato, K.; Ohta, E.; Nakamura, Y.; Takemura, M.; Nozato, N.; Akashi, K.; Kanegae, T.; Ogura, Y.; Kohchi, T. Gene organization deduced from the complete sequence of liverwort *Marchantia polymorpha* mitochondrial DNA: a primitive form of plant mitochondrial genome. *J. Mol. Biol.* **1992**, *223*, 1–7. [CrossRef]
23. Handa, H.; Bonnard, G.R.; Grienenberger, J.-M. The rapeseed mitochondrial gene encoding a homologue of the bacterial protein *Ccl1* is divided into two independently transcribed reading frames. *Mol. Gen. Genet.* **1996**, *252*, 292–302. [CrossRef]
24. Unseld, M.; Marienfeld, J.R.; Brandt, P.; Brennicke, A. The mitochondrial genome of *Arabidopsis thaliana* contains 57 genes in 366,924 nucleotides. *Nat. Genet.* **1997**, *15*, 57–61. [CrossRef]
25. Kim, B.; Kim, K.; Yang, T.-J.; Kim, S. Completion of the mitochondrial genome sequence of onion (*Allium cepa* L.) containing the CMS-S male-sterile cytoplasm and identification of an independent event of the *ccmFN* gene split. *Curr. Genet.* **2016**, *62*, 873–885. [CrossRef]
26. Thöny-Meyer, L.; Fischer, F.; Künzler, P.; Ritz, D.; Hennecke, H. *Escherichia coli* genes required for cytochrome c maturation. *J. Bacteriol.* **1995**, *177*, 4321–4326. [CrossRef]
27. Giegé, P.; Grienenberger, J.; Bonnard, G. Cytochrome c biogenesis in mitochondria. *Mitochondrion* **2008**, *8*, 61–73. [CrossRef]
28. LPWG, Legume Phylogeny Working Group. a new subfamily classification of the Leguminosae based on a taxonomically comprehensive phylogeny. *Taxon* **2017**, *66*, 44–77. [CrossRef]
29. Lewis, G.P.; Schrire, B.; Mackinder, B.; Lock, M. *Legumes of the World*; Royal Botanic Gardens: Kew, UK, 2005.
30. Wojciechowski, M.F.; Sanderson, M.J.; Steele, K.P.; Liston, A. Molecular phylogeny of the "temperate herbaceous tribes" of papilionoid legumes: a supertree approach. In *Advances in Legume Systematics*; Herendeen, P., Bruneau, A., Eds.; Royal Botanic Gardens: Kew, UK, 2000; Volume 9, pp. 277–298.
31. Sabir, J.; Schwarz, E.; Ellison, N.; Zhang, J.; Baeshen, N.A.; Mutwakil, M.; Jansen, R.; Ruhlman, T. Evolutionary and biotechnology implications of plastid genome variation in the inverted-repeat-lacking clade of legumes. *Plant Biotechnol. J.* **2014**, *12*, 743–754. [CrossRef]
32. Magee, A.M.; Aspinall, S.; Rice, D.W.; Cusack, B.P.; Sémon, M.; Perry, A.S.; Stefanović, S.; Milbourne, D.; Barth, S.; Palmer, J.D. Localized hypermutation and associated gene losses in legume chloroplast genomes. *Genome Res.* **2010**, *20*, 1700–1710. [CrossRef]
33. Cai, Z.; Guisinger, M.; Kim, H.-G.; Ruck, E.; Blazier, J.C.; McMurtry, V.; Kuehl, J.V.; Boore, J.; Jansen, R.K. Extensive reorganization of the plastid genome of *Trifolium subterraneum* (Fabaceae) is associated with numerous repeated sequences and novel DNA insertions. *J. Mol. Evol.* **2008**, *67*, 696–704. [CrossRef]
34. Sveinsson, S.; Cronk, Q. Evolutionary origin of highly repetitive plastid genomes within the clover genus (*Trifolium*). *BMC Evol. Biol.* **2014**, *14*, 228. [CrossRef]
35. Milligan, B.G.; Hampton, J.N.; Palmer, J.D. Dispersed repeats and structural reorganization in subclover chloroplast DNA. *Mol. Biol. Evol.* **1989**, *6*, 355–368.
36. Choi, I.S.; Jansen, R.; Ruhlman, T. Lost and Found: Return of the inverted repeat in the legume clade defined by its absence. *Genome Biol. Evol.* **2019**, *11*, 1321–1333. [CrossRef]
37. Negruk, V. Mitochondrial genome sequence of the legume *Vicia faba*. *Front. Plant Sci.* **2013**, *4*, 128. [CrossRef]
38. Bi, C.; Wang, X.; Xu, Y.; Wei, S.; Shi, Y.; Dai, X.; Yin, T.; Ye, N. The complete mitochondrial genome of *Medicago truncatula*. *Mitochondrial DNA B* **2016**, *1*, 122–123. [CrossRef]
39. Hazle, T.; Bonen, L. Status of genes encoding the mitochondrial S1 ribosomal protein in closely-related legumes. *Gene* **2007**, *405*, 108–116. [CrossRef]
40. Kazakoff, S.H.; Imelfort, M.; Edwards, D.; Koehorst, J.; Biswas, B.; Batley, J.; Scott, P.T.; Gresshoff, P.M. Capturing the biofuel wellhead and powerhouse: The chloroplast and mitochondrial genomes of the leguminous feedstock tree *Pongamia pinnata*. *PLoS ONE* **2012**, *7*, e51687. [CrossRef]
41. Lock, J. Trifolieae. In *Legumes of the World*; Lewis, G.P., Schrire, B., Mackinder, B., Lock, M., Eds.; Royal Botanic Gardens: Kew, UK, 2005; pp. 499–504.

42. Ellison, N.W.; Liston, A.; Steiner, J.J.; Williams, W.M.; Taylor, N.L. Molecular phylogenetics of the clover genus (*Trifolium*—Leguminosae). *Mol. Phylogenet. Evol.* **2006**, *39*, 688–705. [CrossRef]
43. De Vega, J.J.; Ayling, S.; Hegarty, M.; Kudrna, D.; Goicoechea, J.L.; Ergon, Å.; Rognli, O.A.; Jones, C.; Swain, M.; Geurts, R. Red clover (*Trifolium pratense* L.) draft genome provides a platform for trait improvement. *Sci. Rep.* **2015**, *5*, 17394. [CrossRef]
44. Hirakawa, H.; Kaur, P.; Shirasawa, K.; Nichols, P.; Nagano, S.; Appels, R.; Erskine, W.; Isobe, S.N. Draft genome sequence of subterranean clover, a reference for genus *Trifolium*. *Sci. Rep.* **2016**, *6*, 30358. [CrossRef]
45. Dudchenko, O.; Pham, M.; Lui, C.; Batra, S.S.; Hoeger, M.; Nyquist, S.K.; Durand, N.C.; Shamim, M.S.; Machol, I.; Erskine, W. Hi-C yields chromosome-length scaffolds for a legume genome, *Trifolium subterraneum*. *bioRxiv* **2018**, 473553. [CrossRef]
46. Griffiths, A.G.; Moraga, R.; Tausen, M.; Gupta, V.; Bilton, T.P.; Campbell, M.A.; Ashby, R.; Nagy, I.; Khan, A.; Larking, A. Breaking free: The genomics of allopolyploidy-facilitated niche expansion in white clover. *Plant Cell* **2019**, *31*, 1466–1487. [CrossRef] [PubMed]
47. Huang, C.-H.; Sun, R.; Hu, Y.; Zeng, L.; Zhang, N.; Cai, L.; Zhang, Q.; Koch, M.A.; Al-Shehbaz, I.; Edger, P.P. Resolution of Brassicaceae phylogeny using nuclear genes uncovers nested radiations and supports convergent morphological evolution. *Mol. Biol. Evol.* **2015**, *33*, 394–412. [CrossRef] [PubMed]
48. Christensen, A.C. Plant mitochondrial genome evolution can be explained by DNA repair mechanisms. *Genome Biol. Evol.* **2013**, *5*, 1079–1086. [CrossRef] [PubMed]
49. Christensen, A.C. Genes and junk in plant mitochondria—Repair mechanisms and selection. *Genome Biol. Evol.* **2014**, *6*, 1448–1453. [CrossRef]
50. Christensen, A.C. Mitochondrial DNA repair and genome evolution. In *Annual Plant Reviews*; Logan, D.C., Ed.; Wiley-Blackwell: New York, NY, USA, 2018; Volume 50, pp. 11–32.
51. Zhang, J.; Ruhlman, T.A.; Sabir, J.S.; Blazier, J.C.; Weng, M.-L.; Park, S.; Jansen, R.K. Coevolution between nuclear-encoded DNA replication, recombination, and repair genes and plastid genome complexity. *Genome Biol. Evol.* **2016**, *8*, 622–634. [CrossRef]
52. Bergsten, J. a review of long-branch attraction. *Cladistics* **2005**, *21*, 163–193. [CrossRef]
53. Ong, H.C.; Palmer, J.D. Pervasive survival of expressed mitochondrial *rps14* pseudogenes in grasses and their relatives for 80 million years following three functional transfers to the nucleus. *BMC Evol. Biol.* **2006**, *6*, 55. [CrossRef]
54. Noutsos, C.; Richly, E.; Leister, D. Generation and evolutionary fate of insertions of organelle DNA in the nuclear genomes of flowering plants. *Genome Res.* **2005**, *15*, 616–628. [CrossRef]
55. Lin, X.; Kaul, S.; Rounsley, S.; Shea, T.P.; Benito, M.-I.; Town, C.D.; Fujii, C.Y.; Mason, T.; Bowman, C.L.; Barnstead, M. Sequence and analysis of chromosome 2 of the plant *Arabidopsis thaliana*. *Nature* **1999**, *402*, 761–768. [CrossRef]
56. Stupar, R.M.; Lilly, J.W.; Town, C.D.; Cheng, Z.; Kaul, S.; Buell, C.R.; Jiang, J. Complex mtDNA constitutes an approximate 620-kb insertion on *Arabidopsis thaliana* chromosome 2: Implication of potential sequencing errors caused by large-unit repeats. *Proc. Natl. Acad. Sci. USA* **2001**, *98*, 5099–5103. [CrossRef]
57. Cousins, G.; Woodfield, D. Effect of inbreeding on growth of white clover. In Proceedings of the 13th Australasian Plant Breeding Conference, Christchurch, New Zealand, 18–21 April 2006; pp. 568–572.
58. Venkatesh, B.; Dandona, N.; Brenner, S. Fugu genome does not contain mitochondrial pseudogenes. *Genomics* **2006**, *87*, 307–310. [CrossRef]
59. Shi, H.; Xing, Y.; Mao, X. The little brown bat nuclear genome contains an entire mitochondrial genome: Real or artifact? *Gene* **2017**, *629*, 64–67. [CrossRef]
60. Rayapuram, N.; Hagenmuller, J.; Grienenberger, J.M.; Bonnard, G.; Giegé, P. The three mitochondrial encoded CcmF proteins form a complex that interacts with CCMH and c-type apocytochromes in Arabidopsis. *J. Biol. Chem.* **2008**, *283*, 25200–25208. [CrossRef]
61. Kisiel, K.; Miwa, H.; Odrzykoski, I.J. Taxonomic identification of chloroplast genome of Marchantia polymorpha using DNA barcode sequences. In Proceedings of the Fourth International Barcode of Life Conference, Adelaide, Australia, 28 November–3 December 2011.
62. Grewe, F.; Edger, P.P.; Keren, I.; Sultan, L.; Pires, J.C.; Ostersetzer-Biran, O.; Mower, J.P. Comparative analysis of 11 Brassicales mitochondrial genomes and the mitochondrial transcriptome of *Brassica oleracea*. *Mitochondrion* **2014**, *19*, 135–143. [CrossRef]
63. Altschul, S.F.; Madden, T.L.; Schäffer, A.A.; Zhang, J.; Zhang, Z.; Miller, W.; Lipman, D.J. Gapped BLAST and PSI-BLAST: a new generation of protein database search programs. *Nucleic Acids Res.* **1997**, *25*, 3389–3402. [CrossRef]

64. Tillich, M.; Lehwark, P.; Pellizzer, T.; Ulbricht-Jones, E.S.; Fischer, A.; Bock, R.; Greiner, S. GeSeq–versatile and accurate annotation of organelle genomes. *Nucleic Acids Res.* **2017**, *45*, W6–W11. [CrossRef]
65. Lowe, T.M.; Chan, P.P. tRNAscan-SE On-line: Integrating search and context for analysis of transfer RNA genes. *Nucleic Acids Res.* **2016**, *44*, W54–W57. [CrossRef]
66. Benson, G. Tandem repeats finder: a program to analyze DNA sequences. *Nucleic Acids Res.* **1999**, *27*, 573–580. [CrossRef]
67. Guo, W.; Grewe, F.; Fan, W.; Young, G.J.; Knoop, V.; Palmer, J.D.; Mower, J.P. *Ginkgo* and *Welwitschia* mitogenomes reveal extreme contrasts in gymnosperm mitochondrial evolution. *Mol. Biol. Evol.* **2016**, *33*, 1448–1460. [CrossRef]
68. Darzentas, N. Circoletto: Visualizing sequence similarity with Circos. *Bioinformatics* **2010**, *26*, 2620–2621. [CrossRef]
69. Katoh, K.; Misawa, K.; Kuma, K.i.; Miyata, T. MAFFT: a novel method for rapid multiple sequence alignment based on fast Fourier transform. *Nucleic Acids Res.* **2002**, *30*, 3059–3066. [CrossRef]
70. Darriba, D.; Taboada, G.L.; Doallo, R.; Posada, D. jModelTest 2: More models, new heuristics and parallel computing. *Nat. Methods* **2012**, *9*, 772. [CrossRef]
71. Stamatakis, A. RAxML version 8: a tool for phylogenetic analysis and post-analysis of large phylogenies. *Bioinformatics* **2014**, *30*, 1312–1313. [CrossRef]
72. Miller, M.A.; Pfeiffer, W.; Schwartz, T. Creating the CIPRES Science Gateway for inference of large phylogenetic trees. In Proceedings of the Gateway Computing Environments Workshop (GCE), New Orleans, LA, USA, 14 November 2010; pp. 1–8.
73. Bowman, J.L.; Kohchi, T.; Yamato, K.T.; Jenkins, J.; Shu, S.; Ishizaki, K.; Yamaoka, S.; Nishihama, R.; Nakamura, Y.; Berger, F. Insights into land plant evolution garnered from the *Marchantia polymorpha* genome. *Cell* **2017**, *171*, 287–304. [CrossRef]
74. Marks, R.A.; Smith, J.J.; Cronk, Q.; Grassa, C.J.; McLetchie, D.N. Genome of the tropical plant *Marchantia inflexa*: Implications for sex chromosome evolution and dehydration tolerance. *Sci. Rep.* **2019**, *9*, 8722. [CrossRef]
75. Kwon, W.; Kim, Y.; Park, J. The complete chloroplast genome sequence of *Dumortiera hirsuta* (Sw.) Nees (Marchantiophyta, Dumortieraceae). *Mitochondrial DNA B* **2019**, *4*, 318–319. [CrossRef]
76. Myszczyński, K.; Ślipiko, M.; Sawicki, J. Potential of transcript editing across mitogenomes of early land plants shows novel and familiar trends. *Int. J. Mol. Sci.* **2019**, *20*, 2963. [CrossRef]
77. Pagni, M.; Ioannidis, V.; Cerutti, L.; Zahn-Zabal, M.; Jongeneel, C.V.; Hau, J.; Martin, O.; Kuznetsov, D.; Falquet, L. MyHits: Improvements to an interactive resource for analyzing protein sequences. *Nucleic Acids Res.* **2007**, *35*, W433–W437. [CrossRef]
78. Sigrist, C.J.; Cerutti, L.; De Castro, E.; Langendijk-Genevaux, P.S.; Bulliard, V.; Bairoch, A.; Hulo, N. PROSITE, a protein domain database for functional characterization and annotation. *Nucleic Acids Res.* **2009**, *38*, D161–D166. [CrossRef]

# 13

# Differential RNA Editing and Intron Splicing in Soybean Mitochondria during Nodulation

**Yuzhe Sun [1], Min Xie [2,3], Zhou Xu [1], Koon Chuen Chan [1], Jia Yi Zhong [1], Kejing Fan [2,3], Johanna Wong-Bajracharya [2,3], Hon-Ming Lam [2,3,*] and Boon Leong Lim [1,2,*]**

[1] School of Biological Sciences, University of Hong Kong, Pokfulam, Hong Kong, China; yzsun@connect.hku.hk (Y.S.); JodieXu85@hotmail.com (Z.X.); mikchankc@gmail.com (K.C.C.); zjiaer@gmail.com (J.Y.Z.)

[2] Center for Soybean Research of the State Key Laboratory of Agrobiotechnology, The Chinese University of Hong Kong, Shatin, Hong Kong, China; feixue1039@gmail.com (M.X.); kejing68164614@gmail.com (K.F.); johannawh.wong@gmail.com (J.W.-B.)

[3] School of Life Sciences, The Chinese University of Hong Kong, Shatin, Hong Kong, China

* Correspondence: honming@cuhk.edu.hk (H.-M.L.); bllim@hku.hk (B.L.L.)

**Abstract:** Nitrogen fixation in soybean consumes a tremendous amount of energy, leading to substantial differences in energy metabolism and mitochondrial activities between nodules and uninoculated roots. While C-to-U RNA editing and intron splicing of mitochondrial transcripts are common in plant species, their roles in relation to nodule functions are still elusive. In this study, we performed RNA-seq to compare transcript profiles and RNA editing of mitochondrial genes in soybean nodules and roots. A total of 631 RNA editing sites were identified on mitochondrial transcripts, with 12% or 74 sites differentially edited among the transcripts isolated from nodules, stripped roots, and uninoculated roots. Eight out of these 74 differentially edited sites are located on the *matR* transcript, of which the degrees of RNA editing were the highest in the nodule sample. The degree of mitochondrial intron splicing was also examined. The splicing efficiencies of several introns in nodules and stripped roots were higher than in uninoculated roots. These include *nad1* introns 2/3/4, *nad4* intron 3, *nad5* introns 2/3, *cox2* intron 1, and *ccmFc* intron 1. A greater splicing efficiency of *nad4* intron 1, a higher NAD4 protein abundance, and a reduction in supercomplex I + $III_2$ were also observed in nodules, although the causal relationship between these observations requires further investigation.

**Keywords:** complex I; intron splicing; maturase; NDH; RNA editing; mitochondria

## 1. Introduction

Soybean is an important cash crop for protein and edible oil. Its ability to perform symbiotic nitrogen fixation in root nodules makes it a nitrogen-rich food source. Root cells obtain photo-assimilates from source tissues and catabolize these chemical compounds through root mitochondria to generate the ATPs required for various physiological and biochemical processes. In eukaryotic cells, cellular respiration mainly takes place inside mitochondria. The tricarboxylic acid (TCA) cycle and mitochondrial electron transport chain (mETC) convert biochemical energy from nutrients into ATP, which is an important energy currency of the cell.

The energy demand in root nodules is much higher than in uninoculated roots, since nitrogen fixation consumes a tremendous amount of energy [1]. Therefore, substantial differences in carbon metabolism and mitochondrial activities between root nodules and uninoculated root cells are expected. Sucrolytic activities (invertase and sucrose synthase) are found to be 3–4-fold higher in root nodules than in uninoculated roots and these enzymatic activities are restricted to the uninoculated cortical tissue and are absent in the infected central tissue of nodules [2]. Bacteroids of the nodules were shown to exhibit a limited capacity to utilize carbohydrates [3], with limited glycolytic activities and no sucrolytic activities. It is reported that sucrose taken up by root nodule cells is converted into malate and succinate as the main energy sources for bacteroids [1]. Hence, to facilitate the carbohydrate supply to bacteroids, the physiology of soybean mitochondria in nodules is significantly different from the uninoculated roots. Mitochondria in nodules were shown to oxidize malate at a rate 2-fold higher than cotyledon mitochondria. However, the activities of TCA cycle enzymes, except malate and succinate dehydrogenases, were lower in nodule mitochondria [4].

The vast majority of mitochondrial proteins from these crucial metabolic pathways are encoded in the nucleus and, after their translation in the cytoplasm, are imported into the mitochondria [5,6]. There are 110 predicted open reading frames in soybean mtDNA and 36 of them can be translated into proteins with known functions [7,8], including some subunits of the respiratory chain: NADH dehydrogenase (complex I), cytochrome oxidase (complex IV), ATP synthase (complex V), and cytochrome C biogenesis. Coordination of the expression and accumulation of mitochondrial proteins derived from the nuclear and mitochondrial genomes are complex, especially in the post-transcriptional regulation of mtDNA expression [9,10].

Some mitochondrial mRNA transcripts are subjected to C-to-U RNA editing, a house-keeping post-transcriptional process in plants [11]. RNA editing is carried out by the editosome complexes, which are composed of pentatricopeptide repeat (PPR) proteins, Multiple Organellar RNA Editing Factors (MORF), organelle RNA Recognition Motif-containing proteins (ORRM), and organelle zinc finger editing factor family (OZ) proteins [12]. RNA editing is complex and energy-consuming. The biological function and evolutionary significance of RNA editing in plants still remain unclear [13,14]. RNA editing may lead to amino acid substitutions and affect the function of the translated proteins [15]. Editing sites, which are edited with different efficiencies in different tissues, have also been identified in the plastids of diverse species [16,17].

RNA editing has been shown to be crucial for *nad1* splicing in *Oenothera* [18] and *nad7* splicing in maize [19]. The genomes of plant mitochondria house about 20 group-II introns and their splicing requires maturases [20–22]. In *Arabidopsis thaliana*, there is one maturase gene (*matR*) in the mitochondrial genome and four nuclear maturase genes (*nMat*1-4) in the nuclear genome [23]. The four nuclear-encoded maturases are imported into mitochondria after translation [24,25]. It has been experimentally demonstrated that nuclear-encoded maturases [20,23,26] and MatR [27] are required for the splicing of various group-II introns in Arabidopsis mitochondria and the assembly of functional complex I. Here, we compare the degrees of RNA editing and intron splicing on mitochondrial transcripts in soybean nodules and roots.

## 2. Results

### *2.1. RNA Sequencing and Differentially Expressed Mitochondrial Genes*

To study the differences in mitochondrial transcripts between soybean nodules and uninoculated roots, we collected nodule (N) samples 28 days after rhizobium inoculation. Stripped roots (SR), main roots after the removal of nodules, and uninoculated roots (UR) of the same age were also harvested. We extracted total RNA from N, SR, and UR for RNA sequencing (RNA-seq) analysis (Table 1). The BioProject accession numbers of the RNA-seq data are PRJNA627909 (UR) and PRJNA626514 (N and SR), respectively. Here, we focused on mitochondrial gene expression and identified 93 gene transcripts (Supplementary Table S1). Besides transcripts encoding for hypothetical proteins, we identified 14 transcripts differentially expressed between N and UR, SR and UR, or SR and N (fold change $> 1.5$, $p$-value $< 0.05$ using Student's $t$-test). These include transcripts of cytochrome c biogenesis C/$F_N$ (*ccmC*, *ccmF*$_N$), cytochrome c oxidase subunit III (*cox3*), NADH dehydrogenase subunit 1/2/4L/5 (*nad1*, *nad2*, *nad4L*, *nad5*), and some ribosomal proteins (Table 2). Compared to UR, eight transcripts, including six *nad* transcripts and two ribosomal protein transcripts, were significantly upregulated in SR, while the abundance of six transcripts including *cox3*, *nad1*, *nad5*, and three ribosomal protein transcripts was significantly higher in N. On the contrary, *ccmF*$_N$ was significantly down-regulated in N compared with UR. The abundance of four transcripts, including *cytochrome c biogenesis C/F*$_N$ (*ccmC*, *ccmF*$_N$) and *ribosomal proteins L5* and *S14*, was lower in N than in SR, whereas the abundance of *cox3* and *rps12* transcripts was higher in N when compared with SR and UR.

**Table 1.** Summary of RNA-seq sequencing data.

| Sample | | Read Length (bp) | Read Count | Total Length (bp) |
|---|---|---|---|---|
| Stripped root | D3 | 125 | 66,851,390 | 8,356,423,750 |
| | E3 | 125 | 75,008,428 | 9,376,053,500 |
| | F3 | 125 | 70,072,218 | 8,759,027,250 |
| Nodule | D4 | 125 | 68,175,300 | 8,521,912,500 |
| | E4 | 125 | 63,858,336 | 7,982,292,000 |
| | F4 | 125 | 62,615,136 | 7,826,892,000 |
| Control Root | C08-Root1 | 150 | 111,230,354 | 16,684,553,100 |
| | C08-Root2 | 150 | 106,852,728 | 16,027,909,200 |
| | C08-Root3 | 150 | 115,102,338 | 17,265,350,700 |

**Table 2.** Differentially expressed mitochondrial transcripts.

| Gene ID | Gene | Function Description | UR FPKM | SR FPKM | N FPKM | N/UR | *p*-Value | N/SR | *p*-Value | SR/UR | *p*-Value |
|---|---|---|---|---|---|---|---|---|---|---|---|
| GlmaxMp01 | *ccmC* | cytochrome c biogenesis C | 15,863 | 20,620 | 10,842 | 0.68 | 0.000 | **0.53** | 0.000 | 1.30 | 0.001 |
| GlmaxMp51 | *ccmFn* | cytochrome c biogenesis FN | 4048 | 2982 | 1586 | **0.39** | 0.000 | **0.53** | 0.000 | 0.74 | 0.006 |
| GlmaxMp04 | *cox3* | cytochrome c oxidase subunit III | 31,643 | 31,486 | 60,520 | **1.91** | 0.000 | **1.92** | 0.000 | 1.00 | 0.943 |
| GlmaxMp20-2 | *nad1* | NADH dehydrogenase subunit 1 | 12,430 | 21,952 | 27,133 | **2.18** | 0.000 | 1.24 | 0.002 | **1.77** | 0.000 |
| GlmaxMp31-1 | *nad2* | NADH dehydrogenase subunit 2 | 12,565 | 19,324 | 16,626 | 1.32 | 0.048 | 0.86 | 0.001 | **1.54** | 0.009 |
| GlmaxMp32 | *nad4L-1* | NADH dehydrogenase subunit 4L | 30,659 | 51,843 | 39,578 | 1.29 | 0.257 | 0.76 | 0.007 | **1.69** | 0.035 |
| GlmaxMp46 | *nad4L-2* | NADH dehydrogenase subunit 4L | 29,685 | 47,502 | 36,863 | 1.24 | 0.353 | 0.78 | 0.009 | **1.60** | 0.057 |
| GlmaxMp05-1 | *nad5* | NADH dehydrogenase subunit 5 | 17,739 | 26,994 | 25,170 | 1.42 | 0.025 | 0.93 | 0.036 | **1.52** | 0.014 |
| GlmaxMp05-2 | *nad5* | NADH dehydrogenase subunit 5 | 26,303 | 42,023 | 55,654 | **2.12** | 0.000 | 1.32 | 0.003 | **1.60** | 0.004 |
| GlmaxMp07 | *rp15* | ribosomal protein subunit L5 | 15,181 | 19,217 | 10,240 | 0.67 | 0.145 | **0.53** | 0.000 | 1.27 | 0.213 |
| GlmaxMp53 | *rps1* | ribosomal protein S1 | 9695 | 21,173 | 17,874 | **1.84** | 0.013 | 0.84 | 0.005 | **2.18** | 0.004 |
| GlmaxMp16 | *rps12* | ribosomal protein S12 | 10,421 | 11,110 | 19,813 | **1.90** | 0.001 | **1.78** | 0.000 | 1.07 | 0.572 |
| GlmaxMp08 | *rps14* | ribosomal protein subunit S14 | 24,486 | 25,459 | 14,704 | 0.60 | 0.077 | **0.58** | 0.000 | 1.04 | 0.826 |
| GlmaxMp52 | *rps4* | ribosomal protein S4 | 6737 | 15,174 | 16,296 | **2.42** | 0.004 | 1.07 | 0.012 | **2.25** | 0.006 |

Three samples of each tissue were sequenced and the average FPKM is presented in this table. Student's t-tests were carried out between tissues. Ratios that are significantly increased or decreased (fold change > 1.5 or < 0.067, *p*-value < 0.05) are shown in bold.

### *2.2. Differential RNA Editing in Root Nodules*

RNA-seq reads were mapped to the mitochondrial genome. A total of 631 RNA editing sites, with at least 15% edited reads in all three biological replicates of any one of the three samples, were identified (Supplementary Table S2). The average editing degrees of the three groups (N, SR, and UR) were compared, and 74 sites with ≥ 15% differences in the editing degrees between any two of the three tissues were identified (Supplementary Table S3). Out of these 74 editing sites, 12 sites were intronic, 23 sites were synonymous, and 39 sites were non-synonymous. There were a few observations: (i) UR showed higher editing degrees in *atp1-1*, *atp1-2*, *atp1-3*, and *atp1-4* than the other two samples, whereas all these sites are synonymous; (ii) There were 29 differentially edited sites in *nad* transcripts (*nad1*, *nad2*, *nad4*, *nad4L*, *nad5* and *nad7*). Most of these sites were intronic (11 sites) or synonymous (nine sites). Only six and three sites in *nad1* and *nad4*, respectively, could lead to amino acid substitutions; (iii) There were 18 differentially edited sites in ribosomal protein transcripts (*rpl5*, *rps1*, *rps4*, *rps10* and *rps12*). Only one site was intronic (*rps10*), but all the other sites could lead to amino acid substitutions. Comparing N to UR, 16 sites in *rps 1*, *4*, *10*, and *12* had higher editing degrees, while two sites in *rpl5* had lower editing degrees.

### *2.3. matR Transcripts Underwent Extensive RNA Editing in Root Nodules*

We identified 17 editing sites on *matR*, of which eight sites had been identified in *A. thaliana* [15] (Table 3) and 11 sites had been reported previously in uninoculated soybean by comparing genomic DNA and cDNA sequences [28]. Out of these 17 editing sites, the degrees of RNA editing of eight sites were higher in the N than UR and SR, of which seven could lead to amino acid substitutions. To confirm that these sites were actually edited, RT-PCR and Sanger sequencing were performed on three biological replicates of each sample (Supplementary Figure S1). In general, the Sanger sequencing data confirmed the next-generation sequencing (NGS) data, and both showed that N had a higher degree of editing than UR and SR in multiple editing sites (Figure 1 and Table 2).

**Table 3.** RNA editing sites in *matR* transcripts.

| Genome Position | Transcript Position | A.A. Position | Codon Change | A.A. Change | Nodule Average Editing Degree (%) | Stripped Root Average Editing Degree (%) | Root Average Editing Degree (%) | Changes (N-UR) (%) | Changes (N-SR) (%) | Changes (SR-UR) (%) | Editing Sites in Arabidopsis (Y/N) * | Editing Degree (%) * | Editing Sites in Soybean (Y/N) # |
|---|---|---|---|---|---|---|---|---|---|---|---|---|---|
| 340,005 | 32 | 11 | TCC>-<TTC | S>-<F | $88 \pm 3^{a}$ | $72 \pm 3^{b}$ | $67 \pm 4^{b}$ | 21 | 16 | 5 | No | - | Yes |
| 339,890 | 147 | 49 | TTC>-<TTT | No (F) | $26 \pm 1^{a}$ | $12 \pm 2^{b}$ | $13 \pm 4^{b}$ | 13 | 14 | −1 | No | - | No |
| 339,844 | 193 | 65 | CCA>-<TCA | P>-<S | $92 \pm 3^{a}$ | $79 \pm 2^{b}$ | $77 \pm 3^{b}$ | 15 | 13 | 2 | No | - | Yes |
| 339,801 | 236 | 79 | TCC>-<TTT | S>-<F | $93 \pm 2^{a}$ | $81 \pm 2^{b}$ | $69 \pm 2^{c}$ | 24 | 12 | 12 | No | - | Yes |
| 339,711 | 326 | 109 | CCA>-<CTA | P>-<L | $75 \pm 4^{a}$ | $69 \pm 1^{b}$ | $74 \pm 2^{a}$ | 0 | 6 | −6 | Yes | 87 | Yes |
| 339,633 | 404 | 135 | TCG>-<TTG | S>-<L | $84 \pm 0^{a}$ | $48 \pm 3^{b}$ | $63 \pm 2^{c}$ | 20 | 35 | −15 | Yes | 90 | No |
| 339,126 | 911 | 304 | GCC>-<GTC | A>-<V | $17 \pm 2^{a}$ | $13 \pm 0^{b}$ | $19 \pm 5^{ab}$ | −2 | 4 | −6 | No | - | No |
| 338,976 | 1061 | 354 | CCC>-<CTC | P>-< | $77 \pm 2^{a}$ | $47 \pm 6^{b}$ | $61 \pm 2^{c}$ | 15 | 30 | −15 | No | - | Yes |
| 338,495 | 1542 | 514 | CCC>-<CCT | No (P) | $42 \pm 1^{a}$ | $22 \pm 2^{b}$ | $20 \pm 4^{b}$ | 22 | 20 | 2 | Yes | 43 | No |
| 338,361 | 1676 | 559 | TCT>-<TTT | S>-<F | $91 \pm 1^{a}$ | $79 \pm 2^{b}$ | $87 \pm 1^{c}$ | 4 | 13 | −9 | Yes | 88 | Yes |
| 338,340 | 1697 | 566 | CCT>-<CTT | P>-<L | $91 \pm 1^{a}$ | $68 \pm 2^{b}$ | $84 \pm 1^{c}$ | 7 | 23 | −16 | Yes | 84 | Yes |
| 338,320 | 1717 | 573 | CGC>-<TGT | R>-<C | $80 \pm 2^{a}$ | $55 \pm 2^{b}$ | $69 \pm 2^{c}$ | 11 | 25 | −14 | Yes | 83 | Yes |
| 338,306 | 1731 | 577 | TAC>-<TAT | No (Y) | $97 \pm 1^{a}$ | $93 \pm 1^{b}$ | $95 \pm 0^{c}$ | 2 | 4 | −2 | No | - | No |
| 338,284 | 1753 | 585 | CAC>-<TAC | H>-<Y | $93 \pm 0^{a}$ | $78 \pm 2^{b}$ | $88 \pm 1^{c}$ | 5 | 15 | −10 | Yes | 79 | Yes |
| 338,214 | 1823 | 608 | CCC>-<CTC | P>-<L | $83 \pm 1^{a}$ | $72 \pm 1^{b}$ | $71 \pm 1^{b}$ | 12 | 11 | 2 | No | - | Yes |
| 338,196 | 1841 | 614 | TCA>-<TTA | S>-<L | $97 \pm 1^{a}$ | $91 \pm 1^{b}$ | $96 \pm 1^{a}$ | 1 | 7 | −5 | Yes | 93 | Yes |
| 338,186 | 1851 | 617 | GTC>-<GTT | No (V) | $42 \pm 2^{a}$ | $33 \pm 1^{b}$ | $39 \pm 2^{a}$ | 3 | 9 | −6 | No | - | No |

The *MatR* gene is located on the negative strand. All the RNA editing sites are C-< U changes. Sites validated by RT-PCR in Figure 1 were underlined. * Editing sites identified in *Arabidopsis thaliana* (Sun et al., 2017). # Editing sites previously reported in soybean *matR* (Thomson et al., 1994). [abc] Sites with significant differences (Student *t*-test, $p < 0.05$) in the degree of editing between the samples were labelled by different letters.

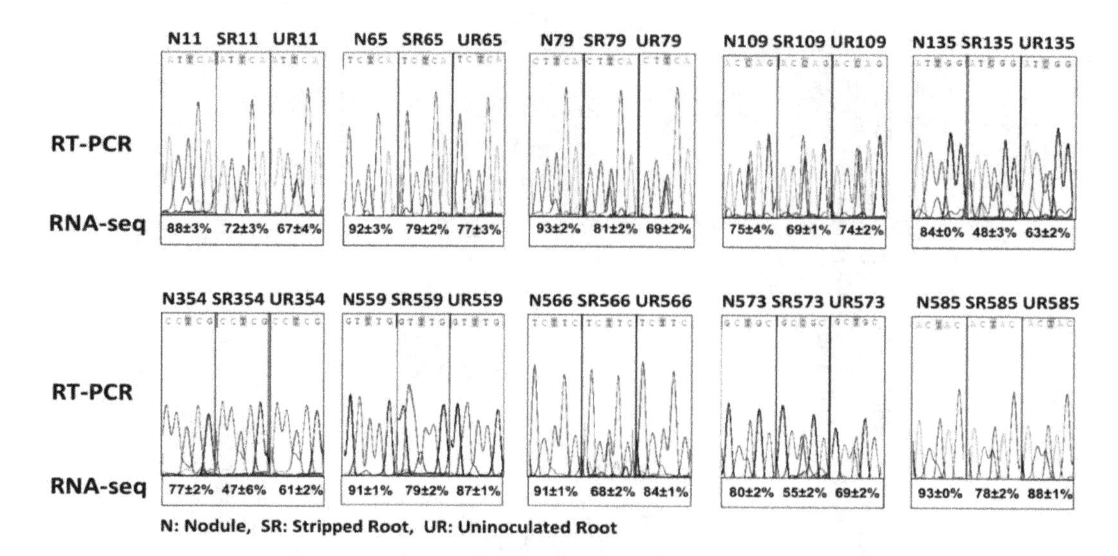


**Figure 1.** Validation of *matR* RNA editing by RT-PCR and Sanger sequencing. N, nodule, SR, stripped root, UR, uninoculated root. The numbers (e.g., N11) represented the editing site locates in the 11th amino acid codon of the *GlmaxMp73* (*matR*) transcript. The average percentages (with SD values) of RNA editing of three biological replicates obtained by RNA-seq are shown below the Sanger sequences.

## 2.4. Identification of Nuclear-Encoded Mitochondrial Intron Maturases in the Soybean Genome

By using the protein sequences of the four Arabidopsis nuclear maturases as baits to search the soybean genome, six homologous nuclear maturases were identified. Among these transcripts, two are likely to have arisen from gene duplication (Figure 2). The expression levels of these six transcripts are presented in Supplementary Table S4. Among these six transcripts, only a transcript homologous to *AtnMAT4* had a significantly higher (2X) expression in N than in UR. By contrast, four transcripts homologous to *AtnMAT1-4* expressed significantly higher in SR than in UR (Supplementary Table S4).

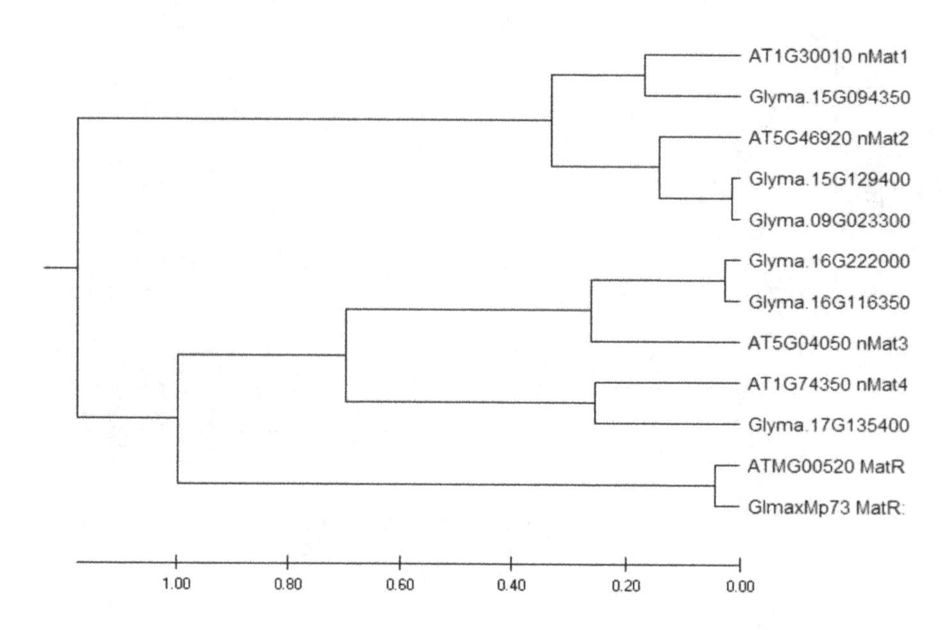


**Figure 2.** Phylogenetic tree of soybean intron maturases. Evolutionary analysis was conducted with MEGA X using the Unweighted Pair Group Method with Arithmetic Mean (UPGMA) method [29].

## 2.5. Intron Splicing of Mitochondrial Transcripts

Next, the splicing efficiencies of the 20 mitochondrial introns were analyzed among the samples by qRT-PCR (Figure 3). The splicing efficiencies of several introns were higher in both N and SR than UR, such as *nad1* introns 2/3/4, *nad4* intron 3, *nad5* introns 2/3, *cox2* intron 1, and *ccmFc* intron 1. While the nuclear intron maturases responsible for the splicing of some of these introns were identified in *A. thaliana* [20,23,26], the scenario in soybean is complicated by the presence of six nuclear intron maturases in the soybean genome (Supplementary Table S4). It should be noted that the splicing efficiency of the intron 1 of *nad4* was greatly enhanced in N than in UR and SR in the following order (N > UR > SR).

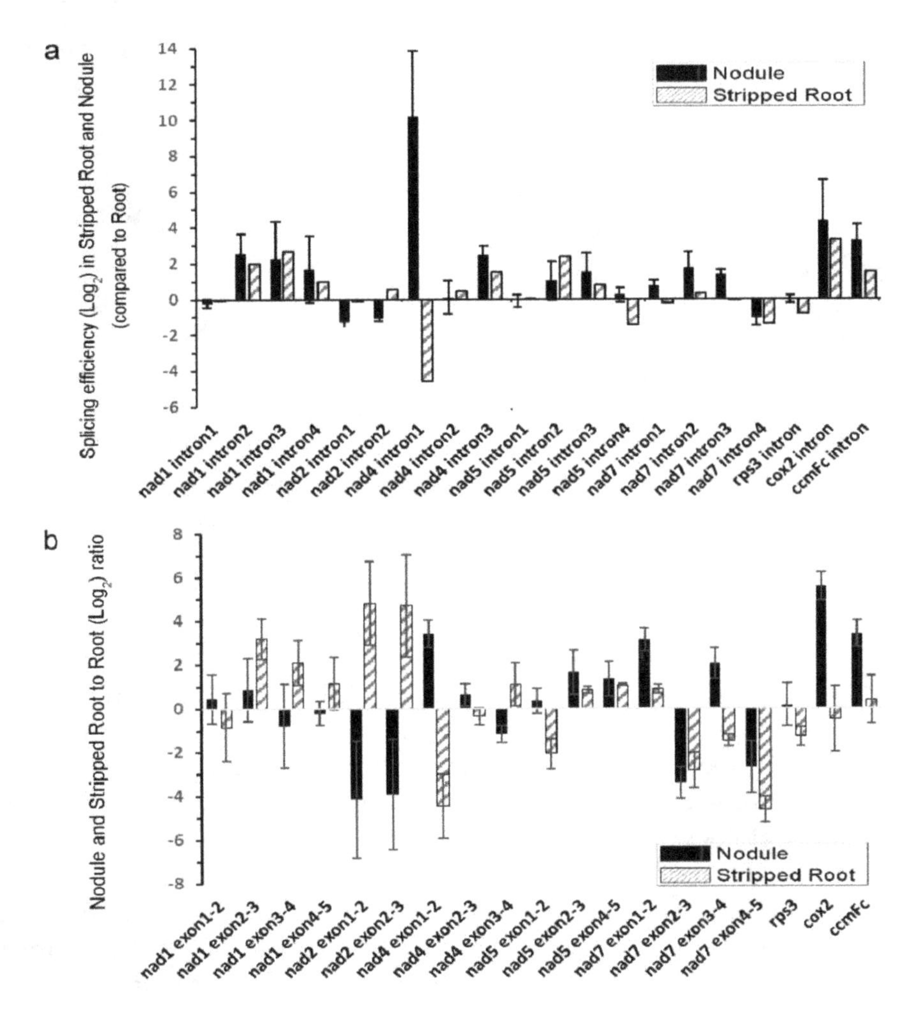


**Figure 3.** Quantitative RT-PCR analysis of splicing efficiency of mitochondrial transcripts. (**a**) Splicing efficiencies of mitochondrial transcripts in nodule and stripped root samples were compared with that of the root sample. The ratio of spliced mRNA to unspliced mRNA is designated as the splicing efficiency. The splicing efficiencies of nodules or stripped roots were divided by the splicing efficiency of roots. (**b**) The relative abundance of spliced mitochondrial mRNAs of nodule and stripped root samples were compared with that of roots. The histogram shows the $\log_2$ ratios of spliced transcripts. Values are means SD of three biological replicates.

### 2.6. In-Gel Activity Assay and Western Blotting

Since qRT-PCR showed that there was more splicing of *nad4* intron 1 in N, we examined the abundance of NAD4 protein in our samples by Western blotting and in-gel activity assays (Figure 4). In-gel activity assay showed that the activity of protein complex II was similar among the three samples, and the complex I activity assay showed that while the activity of the monomeric complex I was similar among the three samples (Figure 4a,b), the activities of the supercomplex composed of complex I and dimeric complex III were lower in N (Figure 4a). In terms of protein abundance, while the level of NAD9 protein was similar among the three samples, a higher protein level of NAD4 was observed in the N sample in comparison to the other two samples. The N sample also had a higher protein abundance of COXIII, which could be due to a 2× increase in its mRNA abundance (Table 2).

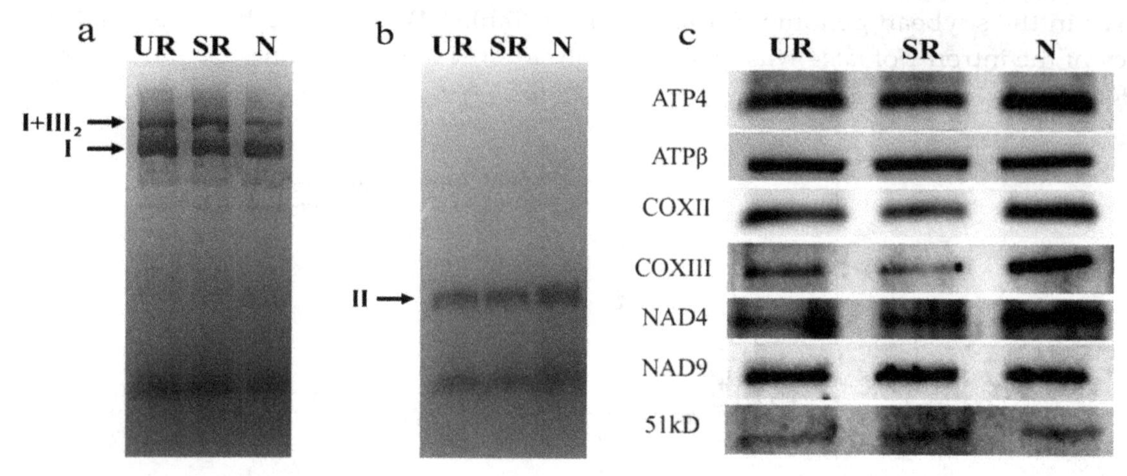


**Figure 4.** In-gel activity of the mitochondrial respiratory complexes separated by BN-PAGE. Complex I activity (**a**), Complex II activity (**b**), and Western blotting (**c**). UR, uninoculated roots; SR, stripped roots; N, nodules. I, complex I; I + $III_2$, supercomplex composed of complex I and dimeric complex III; II, complex II. For Western blotting, 10 µg mitochondrial proteins were separated by SDS-PAGE and blotted with antibodies. Three biological replicates were run and the results were reproducible. The representative images are shown in (**a**–**c**).

## 3. Discussion

RNA editing is a common biological process in the plastids and mitochondria of land plants [30]. It is believed to be a post-transcriptional correction mechanism to circumvent mutations in the organellar genomes [11]. Since RNA editing is an energy-consuming and complex process, it must serve important biological functions, or otherwise it should not be maintained in higher plants during evolution. One possible advantage of RNA editing is to provide an additional regulatory mechanism to organelle physiology. RNA editing was shown to play important roles in organellar tRNA maturation [31] and intron splicing [18,19]. Several hundreds of RNA editing sites on mitochondrial transcripts and differential RNA editing were observed in our previous study on *A. thaliana* [15]. We showed that overexpression of AtPAP2, a phosphatase dually targets the outer membranes of chloroplasts and mitochondria [32,33] and plays a role in the import of pMORF3 into mitochondria [5], might influence cyt *c* biogenesis by modulating RNA editing of *ccm* transcripts through its interaction with the MORF proteins [15].

Here, the objective is to examine how nodule formation affects the degrees of RNA editing in soybean mitochondrial transcripts and investigate its possible biological purpose. We identified 631 RNA editing sites with at least 15% of edited transcripts in all three biological replicates of any one of the samples (Supplementary Table S2). Only 12% of these sites were differentially edited between any two of the three samples. One of the mitochondrial transcripts that underwent extensive RNA editing was the *matR* transcript (Table 3), which encodes an intron maturase that mediates group-II intron splicing. In a previous study, 510 RNA editing sites were identified in Arabidopsis mitochondrial transcripts and 124 sites were differentially edited between the wild-type and a high-energy, fast-growing, transgenic line [15]. There are 12 RNA editing sites in Arabidopsis *matR* transcripts, but none of them were differentially edited between the two lines [15]. Hence, the differential editing of *matR* transcripts is specific in our soybean samples and is likely to play a role in nodule function (Table 2). While plant mitochondria have lost most of their intron-encoded ORFs, a single intron maturase gene (*matR*) is retained in the fourth intron of *nad1* in seed plants [21] (Figure 2). In addition to the mitochondrial *matR* gene, phylogenetic analysis shows that the nuclear maturase genes are also conserved between *A. thaliana* and soybean, since the homologous sequences of all four Arabidopsis *nMat* homologs could be found in the soybean genome (Figure 2). Among the six soybean nuclear maturases, only a gene homologous to Arabidopsis *nMat4* exhibited a 2X higher expression in N than in UR (Supplementary Table S4).

By suppressing the expression of *matR* in Arabidopsis, the splicing efficiencies of *nad1* introns 3 and 4, *nad4* intron 3, *cox2* intron 1, and *ccmFc* intron 1 were downregulated, indicating that AtMatR carries out the splicing of these introns [27]. In soybean, the splicing efficiencies of these introns were higher in N and SR than in UR (Figure 3a). It should be noted that the abundance of *matR* transcripts were 27% and 38% higher in the SR and the N samples than in the UR sample ($p < 0.005$), respectively (Supplementary Table S1). Hence, the changes in the splicing efficiencies of these introns could be due to an increase in *matR* transcript abundance and/or due to the changes in RNA editing. The *nad4* intron 1 was not identified as a *matR*-mediated splicing site in Arabidopsis by *matR* knockdown experiments. However, it was identified in the MatR-ribonucleoprotein complex in an RNA co-immunoprecipitation experiment [27]. Hence, it is possible that RNA editing of *MatR* and the subsequent changes in its amino acid sequence might affect the composition of its associated ribonucleoprotein complex, thus affecting the splicing efficiency of *nad4* intron 1 by the other maturases. Alternatively, MatR might have an additional splicing activity toward *nad4* intron 1 in soybean. A P-type PPR protein, MISF68, was recently shown to be essential for the splicing of *nad4* intron 1 in *A. thaliana*, but the maturase responsible for its splicing has not yet been identified [34]. Nevertheless, several nuclear-encoded splicing factors can independently regulate the intron splicing in plant mitochondria, such as *nMat1-4*, mitochondrial transcription termination factors (*mTERFs*), and some PPR proteins. They are responsible for the splicing regulation of multiple introns in the mitochondrial genomes [35,36]. However, the most significant change between SR and N came from *nad4* intron 1 in the qRT-PCR results, suggesting that the splicing factor which targets the *nad4* intron 1 may play a role in nodule function.

Several *nad* genes are retained in the mitochondrial genome but are not transferred to the nucleus during evolution, probably because they encode the hydrophobic membrane subunits (NAD1-6) of the complex I [37]. Some *nad* introns are subjected to RNA editing (Supplementary Table S3). The latest research reveals that some RNA editing sites on the intronic sequences of maize mitochondrial *nad7* transcript can affect its intron splicing [19]. For example, in maize, it was previously shown that splicing would be abolished without the C-to-U editing at a specific position in two *nad7* introns [19]. In this study, 12 intronic RNA editing sites were identified, 11 of which were situated in *nad1/2/4/5/7* introns, including a site located in *nad4* intron 1 (base 102195, Supplementary Table S3). RNA editing of these intronic sequences could potentially affect intron splicing in soybean.

We observed the highest splicing efficiency of the *nad4* intron 1 in N compared to UR and SR (Figure 3), which might, in turn, affect the NAD4 protein abundance and supercomplex formation (Figure 4). In mammalian cells, spliced mRNA yielded more proteins than identical mRNA not made by splicing, possibly due to an enhanced association of spliced mRNA with polyribosomes [38]. This phenomenon is also observed in viruses [39]. Likewise, a higher splicing efficiency of the first intron of *nad4* in nodule (Figure 3a) might enhance the efficiency of translation. Evidentially, we detected a higher NAD4 protein abundance in nodule comparing to roots (Figure 4c) despite a similar *nad4* transcript abundance in all samples (Supplementary Table S1). NAD4 is a membrane component of complex I. In a maize *nad4* mutant (NCS2), while the missing NAD4 results in the destabilization of the NDH complex, a smaller complex can still exhibit complex I activity in the in-gel enzyme assay [40]. In BN-gel, complex I has two forms, either alone or forming a supercomplex ($I + III_2$) with dimeric complex III, and their ratios are 40%/60% in plants, mammals, and fungi [41]. Our result showed that the mitochondria in nodules contain less $I + III_2$ supercomplex, but contain a similar amount of free complex I to that of the root samples (Figure 4). Factors such as differential intron splicing, differential RNA editing of various *nad* transcripts and a change in NAD4 abundance could collectively affect the formation of the $I + III_2$ supercomplex during nodulation. It was suggested that the $I + III_2$ supercomplex may enable a more efficient electron transfer from complex I to complex III, as the close proximity of the two complexes enables efficient channeling of reduced quinol [42]. The lower abundance of supercomplex I + $III_2$ in nodule mitochondria might affect electron transport rate in the mETC [41].

In nodules, nitrogen fixation by *S. fredii* is an extensive energy-consuming process. To fix one molecule of $N_2$, 16 ATP molecules and eight electrons (and eight $H^+$) are required [43]. C4-dicarboxylates (succinate, fumarate, and malate) were suggested to be the primary carbon source provided to the rhizobia by the root cells [1], and a recent study showed that malate, but not succinate or fumarate, is the essential dicarboxylate for bacteroid growth and symbiosis [44]. Normal root mitochondria do not exhibit substantial glycine decarboxylase (GDC) as there is no photorespiration in roots [45]. However, a proteomic study showed that nodule mitochondria contain a significantly higher amount of GDC, but a lower amount of ATP synthase $\alpha$ and $\beta$ chains than uninoculated roots [46]. A high glycine flux is present in nodule mitochondria due to nitrogen fixation and ureide biosynthesis [47]. The presence of GDC in nodule mitochondria will generate a large amount of NADH. If the electron transport in the mETC is tuned down by the downregulation of the formation of supercomplex I + $III_2$ [42], more NADH generated from GDC can be exported from the nodule mitochondria in the form of malate, which can then be consumed by the bacterioid for energy production. In this study, we reported the changes in RNA editing and intron splicing of the mitochondrial genome in soybean nodules. The relationship among RNA editing, intron splicing, and supercomplex I+$III_2$ formation is complicated and requires further investigation.

## 4. Materials and Methods

### 4.1. Plant Materials and Sinorhizobium Fredii Inoculation

Nodules were harvested from a cultivated soybean C08 (*Glycine max*), which is closely related to Williams 82 from Illinois, USA [48]. Seeds of C08 were surface-sterilized with chlorine gas for 16 h and germinated in the dark in sterilized vermiculite with de-ionized water in the greenhouse. At three days post-germination, seedlings were transferred to individual containers with sterilized vermiculite with 1X low nitrogen nutrient solution (6.35 μM $Ca(NO_3)_2$, 133.59 μM $CaSO_4$, 50.30 μM KCl, 12.17 μM $MgSO_4{\bullet}7H_2O$, 39.04 μM $K_2HPO_4$, 15.31 μM ferric citrate, 2.31 μM $H_3BO_3$, 0.6 μM $MnSO_4$, 0.07 μM $ZnSO_4$, 0.16 μM $CuSO_4{\bullet}5H_2O$, 0.01 μM $H_2MoO_4$) in a 16h/8h light/dark cycle at 25–30 °C [49] and inoculated with *S. fredii* strain *CCBAU45436* [50]. A set of uninoculated soybean control was also prepared alongside this. The *Rhizobium* strain was cultured on TY medium [51] at 28 °C with shaking at 180 rpm for 40 h. The cells were then pelleted and diluted in saline (0.9% *w/v* NaCl) to a final concentration of $10^{20}$ cells $mL^{-1}$ ($OD_{600} = 0.2$) for inoculation. Ten-day-old seedlings were inoculated with 1 mL inoculum per plant. On the 28th day after inoculation, UR, N, and SR samples were collected separately and frozen immediately in liquid nitrogen and stored at −80 °C for RNA extraction. For mitochondria isolation, uninoculated roots, nodules, and the stripped roots were harvested and kept on ice for immediate isolation [5].

### 4.2. RNA Extraction and Sequencing

RNA was extracted using TRIzol (Life Technologies, Carlsbad, CA, USA) following the manufacturer's protocol. Three biological replicates were prepared for each of the three conditions to produce a total of nine samples. Nine strand-specific RNA-seq libraries were generated using TruSeq RNA Sample Preparation Kit (Illumina, San Diego, CA, USA). Messenger RNA (mRNA) was enriched by depleting ribosomal protein RNA using Ribo-Zero Plant kit, rather than poly-A enrichment. These libraries were sequenced on Illumina Hiseq series platforms (sequencing service provided by Groken Bioscience, Hong Kong, China).

### 4.3. Bioinformatics Analysis

*G. max* Williams 82 v275 reference genome was downloaded from the Phytozome database (v9.0: https://phytozome.jgi.doe.gov/). The complete soybean mitochondrion genome was downloaded from the NCBI database (ID: NC_020455.1) [8]. The sequences were combined into one reference genome for subsequent analysis. Genomic variations (SNPs and Indels) between C08 and the reference genome were identified using the GATK pipeline (version 4.0.5.2) [52] according to GATK best practice workflow for germline short variant discovery (https://software.broadinstitute.org/gatk/best-practices/workflow?id=11145). Briefly, C08 DNA sequencing reads were mapped to the reference genome using BWA-MEM (version 0.7.15) implementation with default parameters. Duplicated reads were marked with GATK MarkDuplicates implementation. Then, base quality score recalibration was performed with known variation sites downloaded from the Phytozome database (v12). Then, SNPs and Indels between C08 and the reference genome were called based on mapped and quality score recalibrated reads using GATK HaplotypeCaller implementation.

RNA-seq reads were mapped to reference genome using Tophat2 (version 2.1.1) [53]. Properly mapped read pairs were assigned to each annotated gene by featureCounts from the subread package [54]. Read pair count of each mitochondrial gene was normalized by transcript length and total read count to calculate FPKM value.

To identify RNA editing sites, RNA-seq reads that could be mapped to the mitochondrial genome were extracted. SNPs were identified, and allele frequency was calculated for each polymorphic site using samtools mpileup (version 1.7) [55] and varScan2 (version 2.4.3) [56]. SNPs that were also identified by GATK with DNA reads were assumed to be germline variations and filtered. The remaining SNPs were considered as candidate RNA-editing sites.

### 4.4. RT-PCR and Quantitative RT-PCR

cDNAs were synthesized with M-MLV RT (200 U/μL) (Invitrogen, Hong Kong, China) and random hexamers (Invitrogen, Hong Kong, China), according to the manufacturer's instructions. To verify the differential editing of mitochondrial *matR* transcript observed in RNA-seq data, specific primers were designed to amplify regions that contain each *matR* editing site by RT-PCR (Supplementary Table S5). PCR products were sent out to BGI-Shenzhen for Sanger sequencing.

Quantitative reverse transcription PCR (qRT-PCR) analysis was carried out using the same batch of RNA samples. Primers used in qRT-PCR were derived from a previous *A. thaliana* study in which specific oligonucleotides were designed to target intron-exon and exon-exon regions [23]. New primers were designed based on the homology between *A. thaliana* and soybean mitochondrial genomes (Supplementary Table S5). SYBR Green Master Mix (ABIsystems, Hong Kong) was used in a 10 μL volume PCR reactions. Tubulin gene (Gene ID: Glyma20g27280) was used as the internal house-keeping control. The assessment of relative expression levels was calculated using the Ct comparative threshold method [15,57]. The expression levels of spliced mRNA and unspliced mRNA were first calculated and the ratio was defined as the splicing efficiency. To compare the splicing efficiencies between samples, the splicing efficiencies of SR and N were divided by that of UR (Figure 3a).

### 4.5. Isolation of Soybean Mitochondria

Soybean mitochondria were isolated as previously described with modifications [58,59]. All procedures were done at 4 °C including sample harvest and centrifugation. A total of 20 g of the UR, 20 g of SR, and 10 g of the N were sampled and ground in 50 mL grinding buffer (pH 7.5) containing 0.3 M sucrose, 25 mM Tetrasodiumpyrophosphate, 2 mM EDTA, 10 mM $KH_2PO_4$, 1.0% (*w/v*) PVP-40, 1% (*w/v*) BSA, 20 mM ascorbate and L-cysteine. After 2 min of grinding, the homogenates were filtered through a double layer of Miracloth and rinsed again with 50 mL grinding buffer and centrifuged at 4000 *g* for 5 min. The supernatant was transferred to a tube and centrifuged at 10,000 *g* for 15 min. The pellet was resuspended in a wash buffer (0.3 M sucrose, 10 mM TES, 0.1% (*w/v*) BSA, pH 7.5) and layered on 30 mL of wash buffer containing 45% (*v/v*) Percoll in a tube and centrifuged at 40,000 *g* for 30 min. The crude mitochondria located in a tight brown band near the top of the tube were transferred and diluted at least 5-fold with the wash buffer and concentrated by centrifuging at 15,000 *g* for 10 min. The pellet was resuspended in around 5 mL of wash buffer before loading to a continuous gradient solution containing 0 to 4.4% (*v/v*) PVP-40 and 28% (*v/v*) Percoll in the wash buffer. After centrifugation at 40,000 *g* for 30 min, the mitochondria were concentrated in a pale-yellow band located near the bottom of the tube. This layer was then transferred to a new polycarbonate centrifuge tube with the BSA-free wash buffer and centrifuged at 2450 *g* for 15 min. After 3-4 wash steps, the mitochondria pellet was resuspended in the wash buffer without BSA. After the protein concentration was determined by the Bradford protein assay (BIO-RAD, Hercules, CA, USA), the mitochondria were stored in aliquots at −80 °C.

### *4.6. Blue Native-Polyacrylamide Gel Electrophoresis (BN-PAGE) and In-Gel Enzyme Activity Staining*

The mitochondrial protein complex extraction and BN-PAGE were carried out as previously published with modifications [60]. An equal amount of mitochondria was collected by centrifugation at 14,300 g for 10 min at 4 °C and resuspended in 5% (*w/v*) digitonin extraction buffer to a final ratio of 10:1 (*w/v*) of protein to detergent and incubated on ice for 20 min. The solubilized proteins were then centrifuged at 18,300 g for 20 min at 4 °C. The supernatant (100 µg per sample) was transferred to a new tube supplemented with 5% (*v/v*) Serva blue G250 solution by a final ratio of 100:1 (*w/v*) of protein to dye and was loaded to a standard 1.0 mm × 10 well NativePAGE$^{TM}$ 3–12% Bis-Tris Gel (Invitrogen, Hong Kong, China). The cathode buffer (50 mM Tricine, 15 mM bis-Tris, 0.02% (*w/v*) Serva Blue G250, pH 7.0) and the anode buffer (50 mM Bis-Tris, pH 7.0 with HCl) were freshly prepared. The electrophoresis was carried out at 4 °C at 75 V for 30 min, followed by 100 V for 30 min, 125 V for 30 min, 150 V for 1 h, 175 V for 30 min, and then set to a constant voltage 200 V until the sample reached the bottom of the gel.

After electrophoresis, the gels were washed twice with MiliQ water for 10 min. Then, the gels were equilibrated in the appropriate reaction buffer without reagents for 10 min. The gel was then incubated in a fresh reaction buffer of complex I (0.1 M Tris, 0.2 mM NADH, 0.2% (*w/v*) nitro-blue tetrazolium, pH 7.4) for 30 min. The other gel was incubated in a fresh reaction buffer of complex II (50 mM $KH_2PO_4$, 0.1 mM ATP, 0.2 mM Phenazine methosulphate, 10 mM succinate, 0.2% (*w/v*) nitro-blue tetrazolium) for 2 h. The reactions were terminated by fixing the gels in 40% (*v/v*) methanol and 10% (*v/v*) acetic acid for at least 1 h. The gels were destained overnight in 20% (*v/v*) methanol to remove residual Serva Blue G. For Western blot analysis, 10 µg mitochondrial proteins were run into SDS-PAGE and the proteins were transferred to Hybond-P nitrocellulose membranes (GE Healthcare, Hong Kong, China). The following antibodies were used: anti-beta subunit of ATP synthase (ATP4, PhytoAB PHY0587S, 1:1000; ATPβ, Agrisera AS05 085, 1:4000); anti-cytochrome oxidase subunit II (COXII, Agrisera AS04 053A, 1:2000); anti-cytochrome oxidase subunit III (COXIII, PhytoAB PHY0580S, 1:1000); anti-NAD4 (PhytoAB PHY0511S, 1:1000); anti-NAD9 (from Dr. G. Bonnard, 1:50,000); anti-51kDa (PhytoAB PHY0525S, 1:1000). The signals were developed by the Enhanced Chemiluminescence method (ECL; GE Healthcare, Hong Kong, China).

**Supplementary Materials:** Figure S1: Sanger sequencing of three biological replicates of each sample. Table S1: Normalized read count for each mitochondrial transcript in the three tissues. Table S2. RNA edited sites in the mitochondrial genome identified in this study. Table S3. Differentially edited RNA editing sites. Table S4: Expression of nuclear intron maturases in the three tissues. Table S5. Primers used in this study.

**Author Contributions:** B.L.L. and H.-M.L. designed and coordinated the study; Y.S. led the experimental works related to Sanger sequencing intron splicing, isolation of mitochondria, in-gel activity assays, and Western blotting; M.X. performed bioinformatic analysis of the RNA and DNA sequencing data; Z.X. participated in the in-gel activity assays; K.C.C. and J.Y.Z. participated in the isolation of mitochondria and Western blotting; K.F. and J.W.-B. grew the soybean, performed inoculation, and prepared the RNA samples; B.L.L., H.-M.L. and Y.S. wrote the manuscript. All authors have read and agreed to the published version of the manuscript.

**Acknowledgments:** We thank Zeta Mui for her help in sample preparation. Man-Wah Li and Qianwen Wang assisted with the analysis of genomic and transcriptome data.

## References

1. Liu, A.; Contador, C.; Fan, K.; Lam, H.-M. Interaction and regulation of carbon, nitrogen, and phosphorus metabolisms in root nodules of legumes. *Front. Plant Sci.* **2018**, *9*, 1860. [CrossRef] [PubMed]
2. Copeland, L.; Vella, J.; Hong, Z. Enzymes of carbohydrate metabolism in soybean nodules. *Phytochemistry* **1989**, *28*, 57–61. [CrossRef]
3. Dunn, M.F. Tricarboxylic acid cycle and anaplerotic enzymes in rhizobia. *FEMS Microbiol. Rev.* **1998**, *22*, 105–123. [CrossRef] [PubMed]
4. Bryce, J.H.; Day, D.A. Tricarboxylic acid cycle activity in mitochondria from soybean nodules and cotyledons. *J. Exp. Bot.* **1990**, *41*, 961–967. [CrossRef]
5. Law, Y.S.; Zhang, R.; Guan, X.; Cheng, S.; Sun, F.; Duncan, O.; Murcha, M.; Whelan, J.; Lim, B.L. Phosphorylation and dephosphorylation of the presequence of pMORF3 during import into mitochondria from *Arabidopsis thaliana*. *Plant Physiol.* **2015**, *169*, 1–12. [CrossRef]
6. Law, Y.S.; Ngan, L.; Yan, J.; Kwok, L.Y.; Sun, Y.; Cheng, S.; Schwenkert, S.; Lim, B.L. Multiple kinases can phosphorylate the N-terminal sequences of mitochondrial proteins in *Arabidopsis thaliana*. *Front. Plant Sci.* **2018**, *9*, 982. [CrossRef]
7. Mower, J.P. Variation in protein gene and intron content among land plant mitogenomes. *Mitochondrion* **2020**, *53*, 203. [CrossRef]
8. Chang, S.; Wang, Y.; Lu, J.; Gai, J.; Li, J.; Chu, P.; Guan, R.; Zhao, T. The mitochondrial genome of soybean reveals complex genome structures and gene evolution at intercellular and phylogenetic levels. *PLoS ONE* **2013**, *8*, e56502.
9. Takenaka, M.; Verbitskiy, D.; Merwe, J.A.V.D.; Zehrmann, A.; Brennicke, A. The process of RNA editing in plant mitochondria. *Mitochondrion* **2008**, *8*, 35–46. [CrossRef]
10. Giegé, P.; Sweetlove, L.J.; Cognat, V.; Leaver, C.J. Coordination of nuclear and mitochondrial genome expression during mitochondrial biogenesis in Arabidopsis. *Plant Cell* **2005**, *17*, 1497–1512. [CrossRef]
11. Ichinose, M.; Sugita, M. RNA editing and its molecular mechanism in plant organelles. *Genes* **2016**, *8*, 5. [CrossRef] [PubMed]
12. Sun, T.; Shi, X.; Friso, G.; Wijk, K.V.; Bentolila, S.; Hanson, M.R. A zinc ginger motif-containing protein is essential for chloroplast RNA editing. *PLoS Genet.* **2015**, *11*, e1005028. [CrossRef] [PubMed]
13. Shikanai, T. RNA editing in plants: Machinery and flexibility of site recognition. *Biochim. Biophys. Acta BBA Bioenerg.* **2015**, *1847*, 779–785. [CrossRef] [PubMed]
14. Castandet, B.; Araya, A. RNA editing in plant organelles. Why make it easy? *Biochem. Mosc.* **2011**, *76*, 924–931. [CrossRef]
15. Sun, Y.; Law, Y.S.; Cheng, S.; Lim, B.L. RNA editing of cytochrome c maturation transcripts is responsive to the energy status of leaf cells in *Arabidopsis thaliana*. *Mitochondrion* **2017**, *35*, 23. [CrossRef]
16. Tseng, C.-C.; Lee, C.-J.; Chung, Y.-T.; Sung, T.-Y.; Hsieh, M.-H. Differential regulation of Arabidopsis plastid gene expression and RNA editing in non-photosynthetic tissues. *Plant Mol. Biol.* **2013**, *82*, 375–392. [CrossRef]
17. Miyata, Y.; Sugita, M. Tissue- and stage-specific RNA editing of *rps14* transcripts in moss (*Physcomitrella patens*) chloroplasts. *J. Plant Physiol.* **2004**, *161*, 113–115. [CrossRef]
18. B#xF6;rner, G.V.; Mörl, M.; Wissinger, B.; Brennicke, A.; Schmelzer, C. RNA editing of a group II intron in Oenothera as a prerequisite for splicing. *Mol. Gen. Genet.* **1995**, *246*, 739–744. [CrossRef]
19. Xu, C.; Song, S.; Yang, Y.; Lu, F.; Zhang, M.; Sun, F.; Jia, R.; Song, R.; Tan, B.-C. DEK46 performs C-to-U editing of a specific site in mitochondrial *nad7* introns that is critical for intron splicing and seed development in maize. *Plant J.* **2020**, *5*, 1767. [CrossRef]
20. Keren, I.; Bezawork-Geleta, A.; Kolton, M.; Maayan, I.; Belausov, E.; Levy, M.; Mett, A.; Gidoni, D.; Shaya, F.; Ostersetzer-Biran, O. AtnMat2, a nuclear-encoded maturase required for splicing of group-II introns in Arabidopsis mitochondria. *RNA* **2009**, *15*, 2299–2311. [CrossRef]
21. Guo, W.; Mower, J.P. Evolution of plant mitochondrial intron-encoded maturases: Frequent lineage-specific loss and recurrent intracellular transfer to the nucleus. *J. Mol. Evol.* **2013**, *77*, 43–54. [CrossRef] [PubMed]
22. Brown, G.G.; Francs-Small, C.C.D.; Ostersetzer-Biran, O. Group II intron splicing factors in plant mitochondria. *Front. Plant Sci.* **2014**, *5*, 35. [CrossRef] [PubMed]
23. Keren, I.; Tal, L.; Francs-Small, C.C.D.; Araújo, W.L.; Shevtsov, S.; Shaya, F.; Fernie, A.R.; Small, I.; Ostersetzer-Biran, O. nMAT1, a nuclear-encoded maturase involved in the trans-splicing of *nad1* intron 1, is essential for mitochondrial complex I assembly and function. *Plant J.* **2012**, *71*, 413–426. [CrossRef] [PubMed]

24. Bonen, L. Cis- and trans-splicing of group II introns in plant mitochondria. *Mitochondrion* **2008**, *8*, 26–34. [CrossRef] [PubMed]
25. Verbitskiy, D.; Härtel, B.; Zehrmann, A.; Brennicke, A.; Takenaka, M. The DYW-E-PPR protein MEF14 is required for RNA editing at site *matR* -1895 in mitochondria of *Arabidopsis thaliana*. *FEBS Lett.* **2011**, *585*, 700–704. [CrossRef]
26. Cohen, S.; Zmudjak, M.; Francs-Small, C.C.d.; Malik, S.; Shaya, F.; Keren, I.; Belausov, E.; Many, Y.; Brown, G.G.; Small, I.; et al. nMAT4, a maturase factor required for *nad1* pre-mRNA processing and maturation, is essential for holocomplex I biogenesis in Arabidopsis mitochondria. *Plant J.* **2014**, *78*, 253–268. [CrossRef]
27. Sultan, L.D.; Mileshina, D.; Grewe, F.; Rolle, K.; Abudraham, S.; G#x142;odowicz, P.; Niazi, A.K.; Keren, I.; Shevtsov, S.; Klipcan, L.; et al. The reverse transcriptase/RNA maturase protein MatR is required for the splicing of various Group II introns in Brassicaceae mitochondria. *Plant Cell* **2016**, *28*, 2805. [CrossRef]
28. Thomson, M.C.; Macfarlane, J.L.; Beagley, C.T.; Wolstenholme, D.R. RNA editing of *matR* transcripts in maize and soybean increases similarity of the encoded protein to fungal and bryophyte group II intron maturases: Evidence that *matR* encodes a functional protein. *Nucleic Acids Res.* **1994**, *22*, 5745–5752. [CrossRef]
29. Kumar, S.; Stecher, G.; Li, M.; Knyaz, C.; Tamura, K. MEGA X: Molecular Evolutionary Genetics Analysis across computing platforms. *Mol. Biol. Evol.* **2018**, *35*, 1547. [CrossRef]
30. Small, I.; Schallenberg-Rüdinger, M.; Takenaka, M.; Mireau, H.; Ostersetzer-Biran, O. Plant organellar RNA editing: What 30 years of research has revealed. *Plant J.* **2019**, *101*, 1040. [CrossRef]
31. Mareéchal-Drouard, L.; Ramamonjisoa, D.; Cosset, A.; Weil, J.; Dietrich, A. Editing corrects mispairing in the acceptor stem of bean and potato mitochondrial phenylalanine transfer RNAs. *Nucleic Acids Res.* **1993**, *21*, 4909–4914. [CrossRef]
32. Sun, F.; Carrie, C.; Law, S.; Murcha, M.W.; Zhang, R.; Law, Y.S.; Suen, P.K.; Whelan, J.; Lim, B.L. AtPAP2 is a tail-anchored protein in the outer membrane of chloroplasts and mitochondria. *Plant Signal. Behav.* **2012**, *7*, 927–932. [CrossRef] [PubMed]
33. Sun, F.; Suen, P.K.; Zhang, Y.; Liang, C.; Carrie, C.; Whelan, J.; Ward, J.L.; Hawkins, N.D.; Jiang, L.; Lim, B.L. A dual-targeted purple acid phosphatase in *Arabidopsis thaliana* moderates carbon metabolism and its overexpression leads to faster plant growth and higher seed yield. *New Phytol.* **2012**, *194*, 206–219. [CrossRef] [PubMed]
34. Wang, C.; Aubé, F.; Quadrado, M.; Dargel-Graffin, C.; Mireau, H. Three new pentatricopeptide repeat proteins facilitate the splicing of mitochondrial transcripts and complex I biogenesis in Arabidopsis. *J. Exp. Bot.* **2018**, *69*, 5131. [CrossRef] [PubMed]
35. Hsu-Liang, H.; Wang, H.-J.; Hsieh, M.-H.; Hsieh, H.-L.; Jauh, G.-Y. Arabidopsis mTERF15 is required for mitochondrial *nad2* intron 3 splicing and functional complex I activity. *PLoS ONE* **2014**, *9*, e112360. [CrossRef]
36. Zhao, P.; Wang, F.; Li, N.; Shi, D.-Q.; Yang, W.-C. Pentatricopeptide repeat protein MID1 modulates *nad2* intron 1 splicing and Arabidopsis development. *Sci. Rep.* **2020**, *10*, 2008. [CrossRef] [PubMed]
37. Braun, H.-P.; Binder, S.; Brennicke, A.; Eubel, H.; Fernie, A.R.; Finkemeier, I.; Klodmann, J.; König, A.-C.; Kühn, >K.; Meyer, E.; et al. The life of plant mitochondrial complex I. *Mitochondrion* **2014**, *19*, 295–313. [CrossRef]
38. Long, R. Faculty Opinions recommendation of Splicing enhances translation in mammalian cells: An additional function of the exon junction complex. *Genes Dev.* **2004**, *18*, 210–222. [CrossRef]
39. Machinaga, A.; Ishihara, S.; Shirai, A.; Takase-Yoden, S. Splicing of friend murine leukemia virus env-mRNA enhances its ability to form polysomes. *Front. Microbiol.* **2016**, *7*, 160. [CrossRef]
40. Karpova, O.V.; Newton, K.J. A partially assembled complex I in NAD4-deficient mitochondria of maize. *Plant J.* **1999**, *17*, 511–521. [CrossRef]
41. Davies, K.M.; Blum, T.B.; Kuhlbrandt, W. Conserved in situ arrangement of complex I and $III_2$ in mitochondrial respiratory chain supercomplexes of mammals, yeast, and plants. *Proc. Natl. Acad. Sci. USA* **2018**, *115*, 3024. [CrossRef] [PubMed]
42. Hunte, C. Faculty Opinions recommendation of structure of a mitochondrial supercomplex formed by respiratory-chain complexes I and III. *Proc. Natl. Acad. Sci. USA* **2005**, *102*, 3225–3229. [CrossRef]
43. Seefeldt, L.C.; Hoffman, B.M.; Dean, D.R. Mechanism of Mo-dependent nitrogenase. *Annu. Rev. Biochem.* **2009**, *78*, 701–722. [CrossRef] [PubMed]
44. Mitsch, M.J.; DiCenzo, G.C.; Cowie, A.; Finan, T.M. Succinate transport is not essential for symbiotic nitrogen fixation by *Sinorhizobium meliloti* or *Rhizobium leguminosarum*. *Appl. Environ. Microbiol.* **2017**, *84*, 1. [CrossRef] [PubMed]

45. Gardeström, P.; Bergman, A.; Ericson, I. Oxidation of glycine via the respiratory chain in mitochondria prepared from different parts of Spinach. *Plant Physiol.* **1980**, *65*, 389–391. [CrossRef]
46. Hoa, L.T.-P.; Nomura, M.; Kajiwara, H.; Day, D.A.; Tajima, S. Proteomic analysis on symbiotic differentiation of mitochondria in soybean nodules. *Plant Cell Physiol.* **2004**, *45*, 300–308. [CrossRef]
47. Tajima, S. Ureide biosynthesis in legume nodules. *Front. Biosci.* **2004**, *9*, 1374–1381. [CrossRef]
48. Qi, X.; Li, M.-W.; Xie, M.; Liu, X.; Ni, M.; Shao, G.; Song, C.; Yim, A.K.-Y.; Tao, Y.; Wong, F.-L.; et al. Identification of a novel salt tolerance gene in wild soybean by whole-genome sequencing. *Nat. Commun.* **2014**, *5*, 4340. [CrossRef]
49. Callow, J.A.; Vincent, J.M. A manual for the practical study of root-nodule bacteria. *J. Appl. Ecol.* **1971**, *8*, 977. [CrossRef]
50. Rehman, H.M.; Cheung, W.-L.; Wong, K.-S.; Xie, M.; Luk, C.-Y.; Wong, F.-L.; Li, M.-W.; Tsai, S.-N.; To, W.-T.; Chan, L.-Y.; et al. High-throughput mass spectrometric analysis of the whole proteome and secretome from *Sinorhizobium fredii* strains CCBAU25509 and CCBAU45436. *Front. Microbiol.* **2019**, *10*, 2569. [CrossRef]
51. Beringer, J.E. R factor transfer in *Rhizobium leguminosarum*. *J. Gen. Microbiol.* **1974**, *84*, 188–198. [CrossRef] [PubMed]
52. Auwera, G.A.V.D.; Carneiro, M.O.; Hartl, C.; Poplin, R.; Del Angel, G.; Levy-Moonshine, A.; Jordan, T.; Shakir, K.; Roazen, D.; Thibault, J.; et al. From FastQ data to high-confidence variant calls: The genome analysis toolkit best practices pipeline. *Curr. Protoc. Bioinform.* **2013**, *43*, 10–11. [CrossRef] [PubMed]
53. Kim, D.; Pertea, G.; Trapnell, C.; Pimentel, H.; Kelley, R.; Salzberg, S.L. TopHat2: Accurate alignment of transcriptomes in the presence of insertions, deletions and gene fusions. *Genome Biol.* **2013**, *14*, R36. [CrossRef] [PubMed]
54. Liao, Y.; Smyth, G.K.; Shi, W. Feature Counts: An efficient general purpose program for assigning sequence reads to genomic features. *Bioinformatics* **2013**, *30*, 923–930. [CrossRef] [PubMed]
55. Li, H. A statistical framework for SNP calling, mutation discovery, association mapping and population genetical parameter estimation from sequencing data. *Bioinformatics* **2011**, *27*, 2987–2993. [CrossRef]
56. Koboldt, D.C.; Zhang, Q.; Larson, D.E.; Shen, D.; McLellan, M.D.; Lin, L.; Miller, C.A.; Mardis, E.R.; Ding, L.; Wilson, R.K. VarScan 2: Somatic mutation and copy number alteration discovery in cancer by exome sequencing. *Genome Res.* **2012**, *22*, 568–576. [CrossRef]
57. Liang, C.; Zhang, Y.; Cheng, S.; Osorio, S.; Sun, Y.; Fernie, A.R.; Cheung, C.Y.M.; Lim, B.L. Impacts of high ATP supply from chloroplasts and mitochondria on the leaf metabolism of *Arabidopsis thaliana*. *Front. Plant Sci.* **2015**, *6*, 922. [CrossRef]
58. Day, D.; Neuburger, M.; Douce, R. Biochemical characterization of chlorophyll-free mitochondria from pea leaves. *Funct. Plant Biol.* **1985**, *12*, 219–228. [CrossRef]
59. Lister, R.; Carrie, C.; Duncan, O.; Ho, L.H.; Howell, K.A.; Murcha, M.W.; Whelan, J. Functional definition of outer membrane proteins involved in preprotein import into mitochondria. *Plant Cell* **2007**, *19*, 3739–3759. [CrossRef]
60. Eubel, H.; Braun, H.-P.; Millar, A.H. Blue-native PAGE in plants: A tool in analysis of protein-protein interactions. *Plant Methods* **2005**, *1*, 11. [CrossRef]

# 14

# Intraspecific Variation within the *Utricularia amethystina* Species Morphotypes based on Chloroplast Genomes

**Saura R. Silva [1,*], Daniel G. Pinheiro [1], Helen A. Penha [1], Bartosz J. Płachno [2], Todd P. Michael [3], Elliott J. Meer [4], Vitor F. O. Miranda [5,*] and Alessandro M. Varani [1,*]**

1. Departamento de Tecnologia, Faculdade de Ciências Agrárias e Veterinárias, Jaboticabal, Universidade Estadual Paulista (Unesp), Sao Paulo 14884-900, Brazil; daniel.pinheiro@unesp.br (D.G.P.); helen.penha@gmail.com (H.A.P.)
2. Department of Plant Cytology and Embryology, Institute of Botany, Faculty of Biology, Jagiellonian University in Kraków, 30-387 Krakow, Poland; bartosz.plachno@uj.edu.pl
3. J. Craig Venter Institute, La Jolla, CA 92037, USA; tmichael@jcvi.org
4. 10X Genomics, Pleasanton, CA 94566, USA; ejmeer@gmail.com
5. Departamento de Biologia Aplicada à Agropecuária, Faculdade de Ciências Agrárias e Veterinárias, Jaboticabal, Universidade Estadual Paulista (Unesp), Sao Paulo 14884-900, Brazil

* Correspondence: saura.silva@gmail.com (S.R.S.); vitor.miranda@unesp.br (V.F.O.M.); alessandro.varani@unesp.br (A.M.V.)

**Abstract:** *Utricularia amethystina* Salzm. ex A.St.-Hil. & Girard (Lentibulariace*ae*) is a highly polymorphic carnivorous plant taxonomically rearranged many times throughout history. Herein, the complete chloroplast genomes (cpDNA) of three *U. amethystina* morphotypes: purple-, white-, and yellow-flowered, were sequenced, compared, and putative markers for systematic, populations, and evolutionary studies were uncovered. In addition, RNA-Seq and RNA-editing analysis were employed for functional cpDNA evaluation. The cpDNA of three *U. amethystina* morphotypes exhibits typical quadripartite structure. Fine-grained sequence comparison revealed a high degree of intraspecific genetic variability in all morphotypes, including an exclusive inversion in the *psb*M and *pet*N genes in *U. amethystina* yellow. Phylogenetic analyses indicate that *U. amethystina* morphotypes are monophyletic. Furthermore, in contrast to the terrestrial *Utricularia reniformis* cpDNA, the *U. amethystina* morphotypes retain all the plastid NAD(P)H-dehydrogenase (*ndh*) complex genes. This observation supports the hypothesis that the *ndh*s in terrestrial *Utricularia* were independently lost and regained, also suggesting that different habitats (aquatic and terrestrial) are not related to the absence of *Utricularia ndh*s gene repertoire as previously assumed. Moreover, RNA-Seq analyses recovered similar patterns, including nonsynonymous RNA-editing sites (e.g., *rps14* and *petB*). Collectively, our results bring new insights into the chloroplast genome architecture and evolution of the photosynthesis machinery in the Lentibulariaceae.

**Keywords:** *Utricularia amethystina*; Lentibulariaceae; chloroplast phylogenomics; organelle genome; carnivorous plants; polymorphic species; intraspecific variation; *ndh* genes

---

## 1. Introduction

The species of the carnivorous plant family Lentibulariaceae are grouped in three genera: *Pinguicula* L., *Genlisea* A.St-Hil., and *Utricularia* L. [1,2], and are increasingly becoming important plant models mainly due to their alternative nutrient uptake system, their morphological non-orthodox body structure, characterized by Fuzzy Arberian Morphology [3,4], and particular genomic characteristics, such as high mutational levels with nuclear genome shrinkage and expansion in some lineages [5,6].

*Utricularia* is the biggest genus and most widespread group of carnivorous plants and is very diverse regarding its distribution and habit (e.g., terrestrial, aquatic, lithophytes, epiphytes, and reophytes) [7]. Moreover, several species are polymorphic, which may lead to controversial taxonomic classification. For instance, *Utricularia amethystina* Salzm. ex A.St.-Hil. & Girard is a terrestrial herb, with petiolate and rosetted leaves. The species is broadly distributed in about 18 different countries of the tropical and subtropical America [7], commonly found in different altitudes (from inselbergs in the Guianas [8] to the coast of Brazil [9]) and habitats such as humid sandy soil of the savannas, swamps, and soil between rocks usually near streams, rivers, and waterfalls.

The species is classified in the *Utricularia* sect. *Foliosa* Kamieński, showing common morphological characteristics making it easy to distinguish from other *Utricularia* sections due to its connate bracts and bracteoles, which is the singular morphology of utricles (carnivorous traps), and the capsule dorsoventrally and bivalvate dehiscence. However, the distinction between the conspecific species is not a trivial task, as *U. amethystina* shows high intraspecific variation, mainly between reproductive characters, such as the corolla shapes and colors, which can vary from shades of purple (Figure 1A), white (Figure 1B) to yellow (Figure 1C) [7].

**Figure 1.** *Utricularia amethystina* species morphotypes are differentiated mainly by corolla color. (**A**) *U. amethystina* purple morphotype; (**B**) *U. amethystina* white morphotype; (**C**) *U. amethystina* yellow morphotype.

This intraspecific morphological variation resulted in several taxonomic rearrangements since the earliest descriptions at nineteenth century [10,11] and even now there is much controversy about if the species is one or more [12,13]. Taylor (1989) [7] struggled to separate the species based on reproductive characters, such as corolla shape, pedicel sizes, palynological characters, and calyx indumentum, but he couldn't find traits for enough taxonomical circumscriptions to split the different morphotypes. Indeed, in his *Utricularia* taxonomic monograph, he synonymized 31 taxa under the binomial "*Utricularia amethystina*" and he wrote "*U. amethystina* is a most 'difficult' and excessively polymorphic species..." (Taylor, 1989 [7], p. 291). Therefore, he assumed one name for the species, as he was unable to find discontinuities to support taxa separation due to the high degree of polymorphism between populations. However, to date, there is no proper taxonomic treatment to solve this question. In addition, only a few genetic differences have been explored [13], such as the chloroplast regions *rps*16, *trn*L-F, *trn*D-T, and nuclear ITS, but these markers were not able to give enough resolution to distinguish them all. In this context, chloroplast genomes are a valuable resource for phylogenies, and the study of their structure and content can provide clues for improving inter- and intraspecific studies, such as population biology [14], and even the discovery of new species [15].

The chloroplast genomes of most angiosperms have conserved quadripartite structure separated in Large and Small Single Copy regions (LSC and SSC, respectively) and two inverted repetitive regions (IRs) [16]. However, comparative analyses indicate that some plants, such as parasitic [17], mycoheterotrophic (e.g., in [18,19]), and species of carnivorous plants from the order Caryophyllales [20,21], have suffered substantial rearrangement and gene losses throughout plant evolution. For example, across diverse lineages of plants, chloroplast genomes lack NAD(P)H-dehydrogenase (*ndh*) complex genes, genes that could have been involved in the transition from aquatic to terrestrial habit thought plant evolutionary history [22,23].

Within Lentibulariaceae family, there are published chloroplast genomes (cpDNA) of *Pinguicula ehlersiae* [24] and seven *Genlisea* species [24,25]. There are cpDNA genomes available for four *Utricularia* species: *U. gibba* [26], *U. macrorhiza* [24], *U. reniformis* [27], and *U. foliosa* [28].

All published *Utricularia* cpDNAs have the typical quadripartite structure and the same genes as most angiosperms. However, some species exhibit variation, such as two complete copies of *ycf*1 and *ndh*F in *U. gibba*, and *U. reniformis*, which has reduced chloroplast size due to several losses in all *ndh*s genes repertoire that could not be integrally found either in the mitochondrial genome [29] nor in the nuclear DNA (unpublished data). Indeed, *U. reniformis* is a terrestrial species and other assessed *Utricularia* are aquatic, and based on this observation, Silva et al. (2016) proposed that the loss of *ndh*s could be related to terrestrial habit [27]. Nonetheless, other studies are still needed for a better understanding of genes especially the evolution of the *ndh* genes in the genus.

Herein, we present the chloroplast genomes of three *Utricularia amethystina* morphotypes to assess inter- and intraspecific sequence variability and polymorphic regions that could be used for further phylogenetic studies. In addition, we employed chloroplast transcriptome to assess gene expression and identify the RNA editing sites in each chloroplast. We also compare *ndh* gene gains and losses across the sequenced *Utricularia* cpDNA species, to examine the variation of structural changes across the genus.

## 2. Results

### 2.1. Structure of Chloroplasts in Utricularia amethystina

A total of 2,873,574 million paired-end reads were generated of all *Utricularia amethystina* morphotypes. Approximately 7.57%, 9.10%, and 9.56% represents cpDNA-derived reads and were used for the de novo assembly of *U. amethystina* purple, white, and yellow, respectively. For each morphotype, the assembly using SPAdes resulted in a contig with the entire LSC region, followed by two contigs containing the IR and SSC regions. The cpDNA contigs were joined in a supercontig and circularized using MITOBim iterative read mapping. The three *U. amethystina* cpDNAs have a consistent quadripartite structure similar to the majority of other angiosperms, and slightly varying in size (Figure 2) (Table 1) (Supplementary Table S1).

**Table 1.** The summary of characteristics in *Utricularia amethystina* chloroplast genomes. Between parentheses the percentage that represents each part in comparison to the cpDNA genome.

| U. amethystina Morphotypes | Purple | White | Yellow |
|---|---|---|---|
| Genbank accession number | MN223721 | MN223722 | MN223720 |
| Genome size (bp) | 150,115 | 150,332 | 150,020 |
| Large single copy (LSC) length (bp) | 82,388 (54.9%) | 82,561 (54.9%) | 82,256 (54.8%) |
| Small single copy (SSC) length (bp) | 16,969 (11.3%) | 17,070 (11.4%) | 16,870 (11.3%) |
| Inverted repeats ($IR_a$+$IR_b$) length (bp) | 25,399 (34.7%) | 25,350 (33.8%) | 25,447 (34.0%) |
| Noncoding regions (bp) | 40,199 | 41,901 | 41,259 |
| Intronic regions(bp) | 18,052 | 19,260 | 20,280 |
| %GC | 37.5 | 37.5 | 37.7 |
| Coverage | 84× | 119× | 123× |

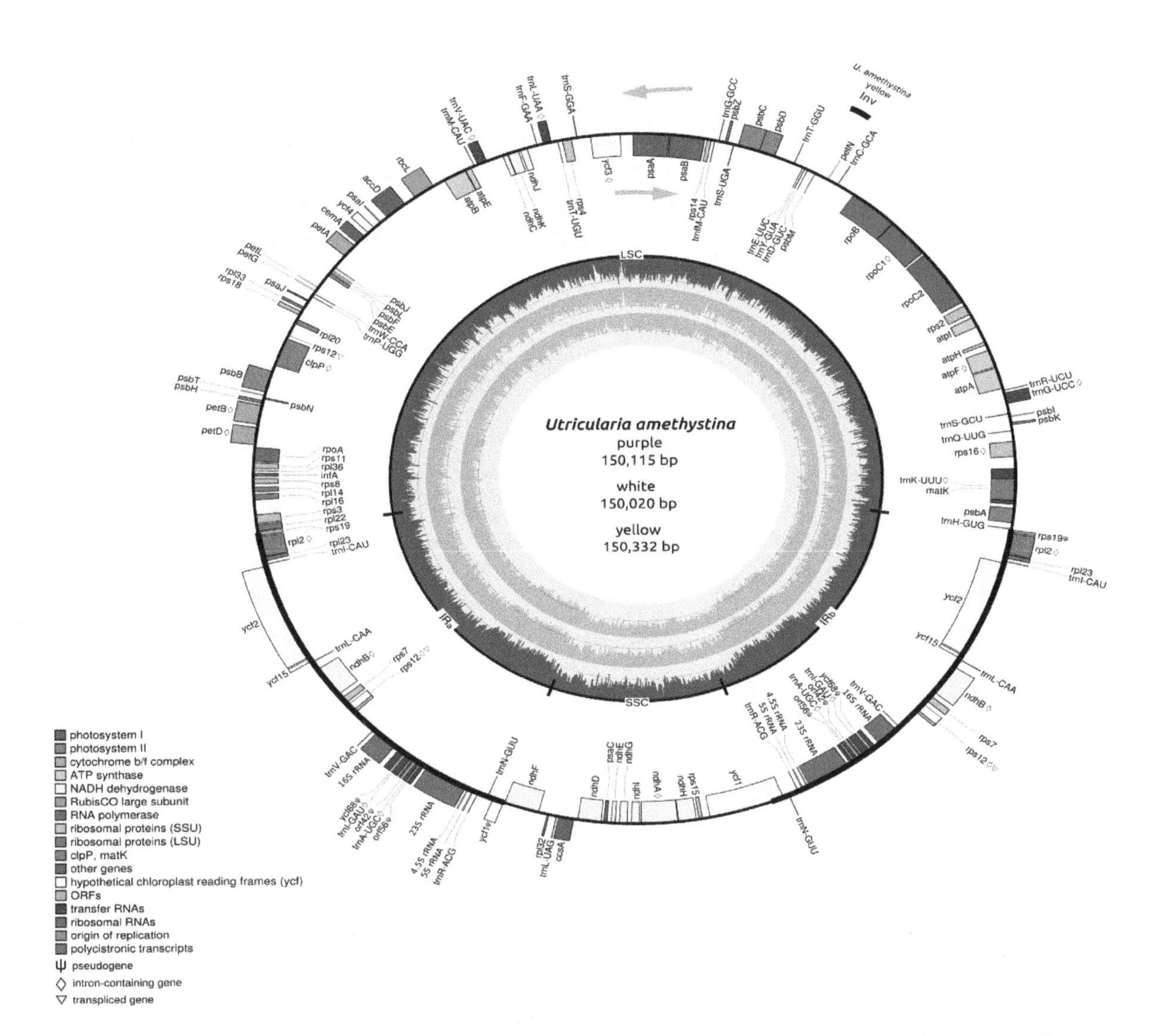


**Figure 2.** Chloroplast genome map for *Utricularia amethystina* purple, white, and yellow. The map represents all three cpDNAs. Gene order and number are the same, except that yellow has an inversion in the *pet*N and *psb*M genes (see at 2 o'clock in the map). Black thick lines of the outer circle indicate the extension of the inverted repeats. The direction of the arrows denotes the transcription direction. Genes are colored according to their functional groups. The inner graph corresponds to the GC content for each cpDNA region in the chloroplast of each species morphotype. Purple, yellow, and gray bars denote *U. amethystina* purple, yellow, and white morphotypes, respectively.

The three cpDNA exhibited in total of 137 annotated genes, including 39 unique protein-coding genes, 30 tRNA, and four rRNA. Eighteen genes (*pet*B, *pet*D, *atp*F, *rpo*C1, *rps*12, *rps*16, *rpl*2, *ndh*A, *ndh*B (2×), *trn*K-UUU, *trn*A-UGC (2×), *trn*I-GAC (2×), *trn*G-UCC, *trn*L-UAA, *trn*V-UAC) contain one intron, and two genes (*clp*P, *ycf*3) have two introns; 18 genes (*rpl*2, *rpl*23, *ycf*2, *ycf*15, *ndh*B, *rps*7, *rps*12, *trn*I-CAU, *trn*L-CAA, *trn*V-GAC, 16S rRNA, *trn*I-GAU, *trn*A-UGC, *trn*R-ACG, *trn*N-GUU, 23S rRNA, 4.5S rRNA, 5S rRNA) have duplicates and five (*ycf*68, *orf*42, *orf*56, *ycf*1, *rps*19) partial genes (putative pseudogenes) in the IR regions. All assessed cpDNAs have collinear gene content and arrangement. The main difference among the three morphotypes is the inversion of *pet*N and *psb*M genes position in *U. amethystina* yellow in comparison to other *Utricularia* (Figure 2).

### 2.2. *Repeats and Chloroplast Microsatellites (cpSSR)*

REPuter identified 22, 29, and 27 repeats in *U. amethystina* purple, white, and yellow, respectively (Figure 3A). Most were characterized as forwarding and palindromic repeats. The repeats were mainly distributed among the *ycf*2, *rpo*C1 and *trn*S-GCU, *ycf*3, *ndh*A genes, and *rbc*L-*acc*D, and *rps*12-*trn*V-GAC intergenic region (Supplementary Tables S2–S4). The identified microsatellite (cpSSR), vary from 7 to 369-bp for *U. amethystina* purple, 7 to 264-bp for *U. amethystina* yellow, and 7 to 245-bp for *U. amethystina* white (Figure 3B). The amount of identified cpSSR repeats are similar within all morphotypes, with the mono- (346, 350, 343), and dinucleotide repeats as the most abundant (42, 43, 40), followed by tri- (4, 3, 4) and tetra- repeats (2 for *U. amethystina* purple and 10 for white and 7 for yellow). Interestingly, penta- nucleotides were only found in *U. amethystina* yellow, and hexa- repeats were not found in any morphotypes (Supplementary Table S5).

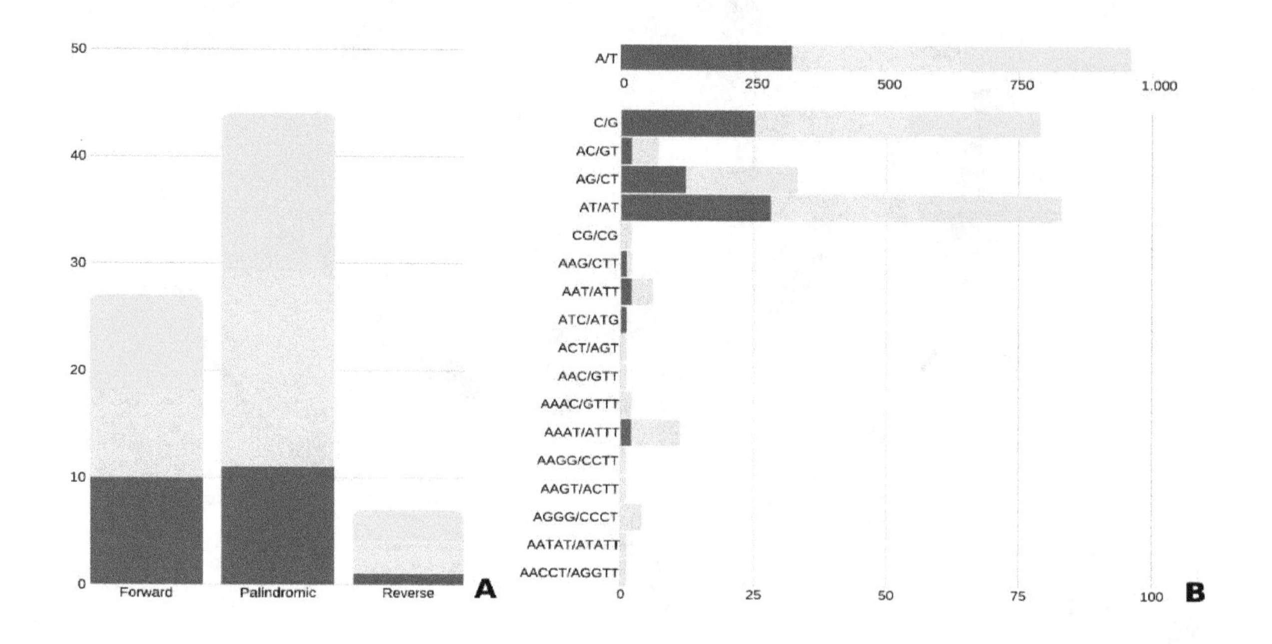


**Figure 3.** Quantity of repeats in *Utricularia amethystina*. (**A**) Long repeats. (**B**) Simple sequence repeats (cpSSRs). Purple, yellow, and gray bars denote *U. amethystina* purple, yellow, and white morphotypes, respectively. (Additional information can be found in Supplementary Table S5).

### *2.3. Interspecific Comparison*

The interspecific chloroplast genome divergence among all available *Utricularia* cpDNAs, using *U. amethystina* purple as a reference, showed that *U. amethystina* specimens have a high degree of synteny; however, for *ycf*1 and non-coding sequences, they are highly divergent (<50% identity; Figure 4). Furthermore, the coding regions *ccs*A, *mat*K, *rpo*C2, *rpo*C1, *rps*19, *ycf*1, *ycf*2; the introns *atp*F and *rps*16; and most intergenic spacers, such as *trn*K-*rps*16, *rps*16-*trn*Q, *psb*K-*psb*I, *trn*L-*trn*F, and *trn*H-*psb*A, showed high levels of variation that can be used for phylogenetic and DNA-barcoding studies (Figure 4).

The SSC regions of *Utricularia amethystina* morphotypes are similar to most *Utricularia*, except for *U. reniformis*, due to the deletion of *ndh*s genes (Figure 5), and *U. gibba*, which have an extra copy of *ycf*1 and *ndh*F.

### *2.4. Intraspecific Comparison: Species Polymorphism*

The nucleotide diversity ($\pi$) analyses indicated that the IRs exhibited lower variability than LSC and SSC regions (Figure 6). There are twelve spots of chloroplast genome regions that showed remarkably higher $\pi$ values (>0.02), including nine gene (*trn*H, *psb*A, *trn*C, *pet*N, *psb*M, *trn*D, *psa*A, *ycf*4 and *ndh*F) and 12 intergenic regions (*trn*H-*psb*A, *rps*16-*trn*Q, *trn*C-*pet*N, *rpo*B-*trn*C, *pet*N-*psb*M, *trn*C-*pet*N, *psb*M-*trn*D, *trn*D-*trn*Y, *psa*A-*ycf*3, *ycf*3-*trn*S, *psa*I-*ycf*4, *ndh*F-*rpl*32). (More details in Supplementary Table S6.)

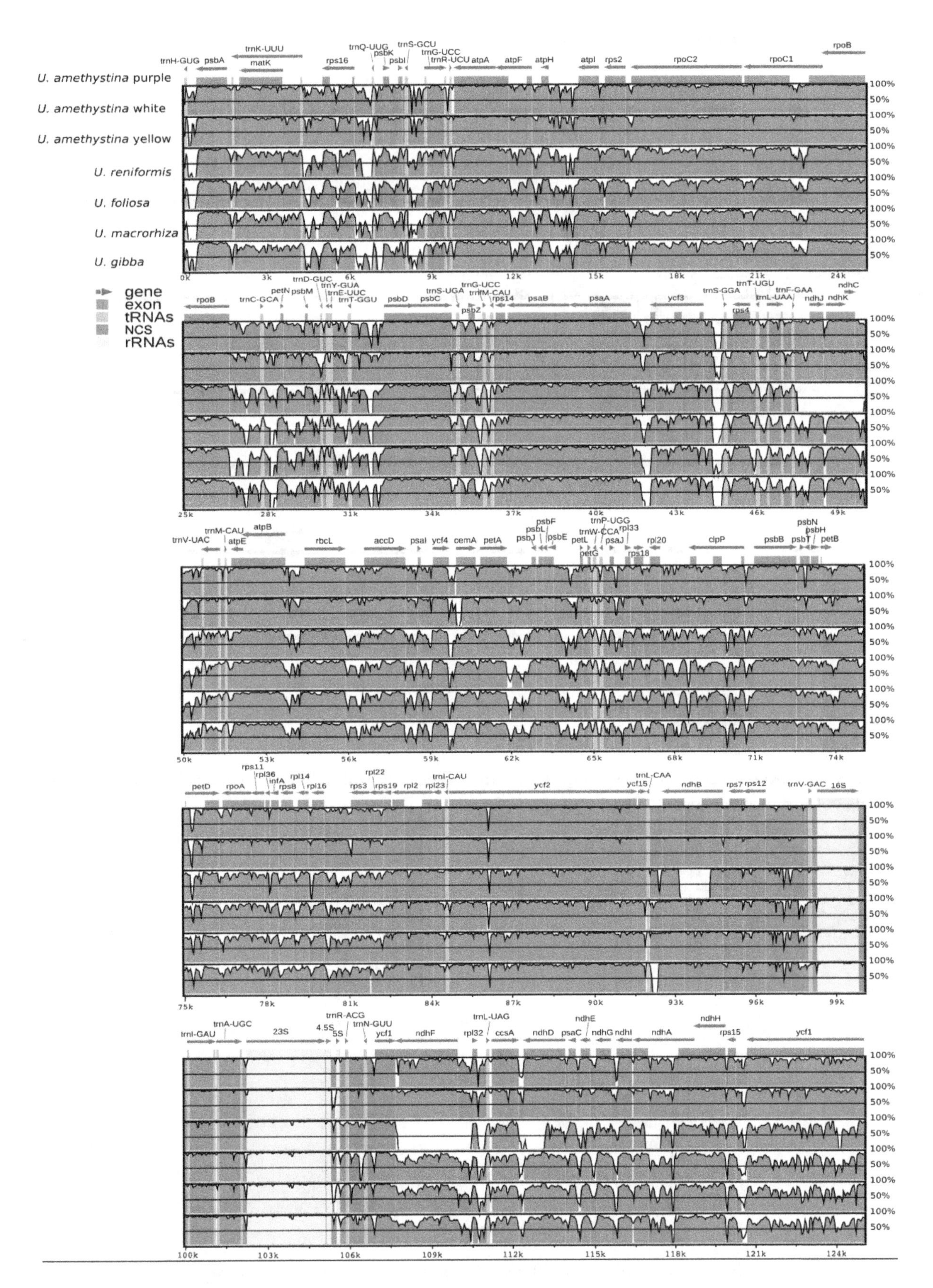


**Figure 4.** mVISTA identity plot based in Shuffle-LAGAN alignment for *Utricularia amethystina* morphotypes and previously reported chloroplasts of other *Utricularia* species. *U. amethystina* purple was used as a reference. NCS denotes non-coding sequence.

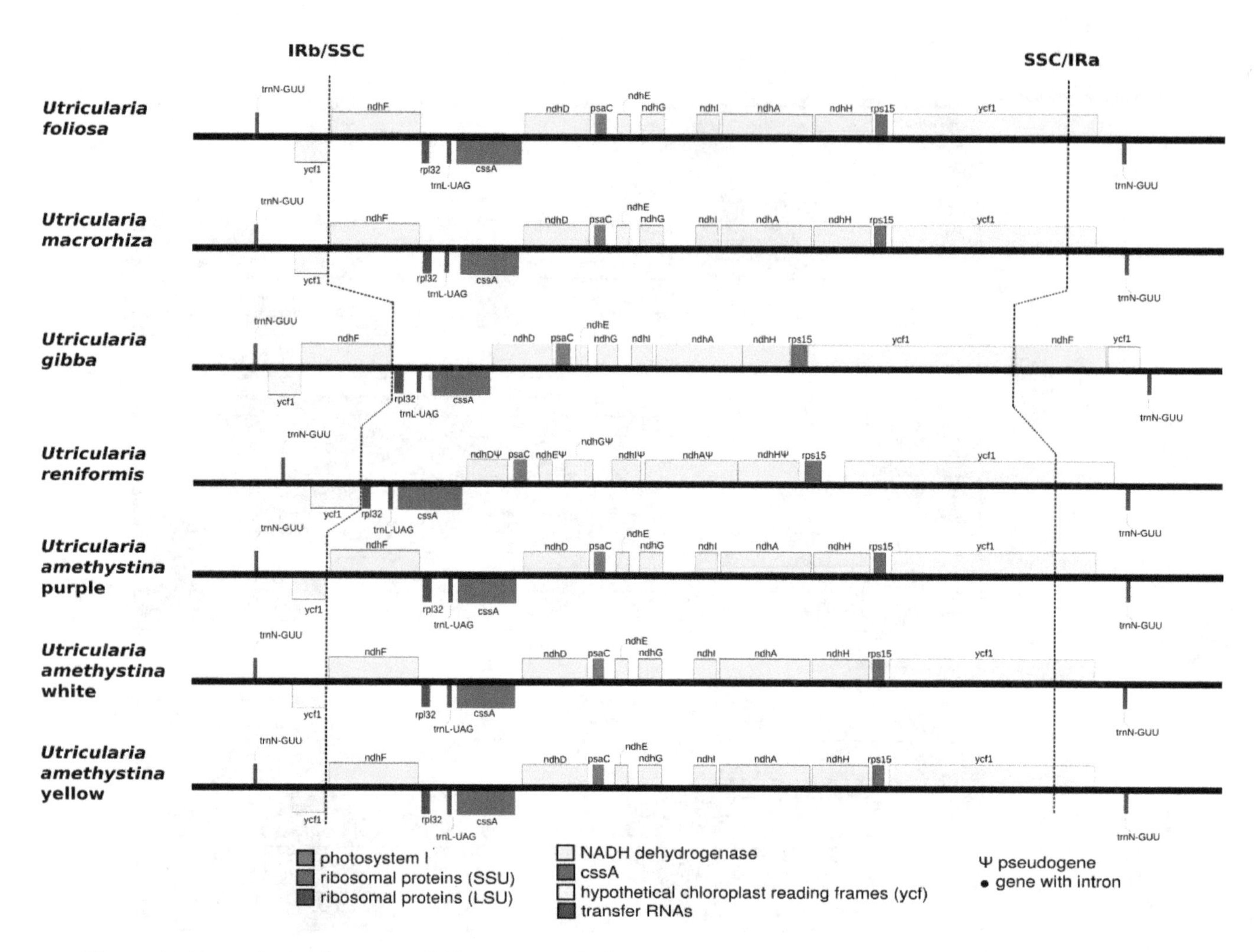


**Figure 5.** Boundaries between SSC regions of all previously published *Utricularia* species. cpDNA regions color denote different chloroplast gene families.

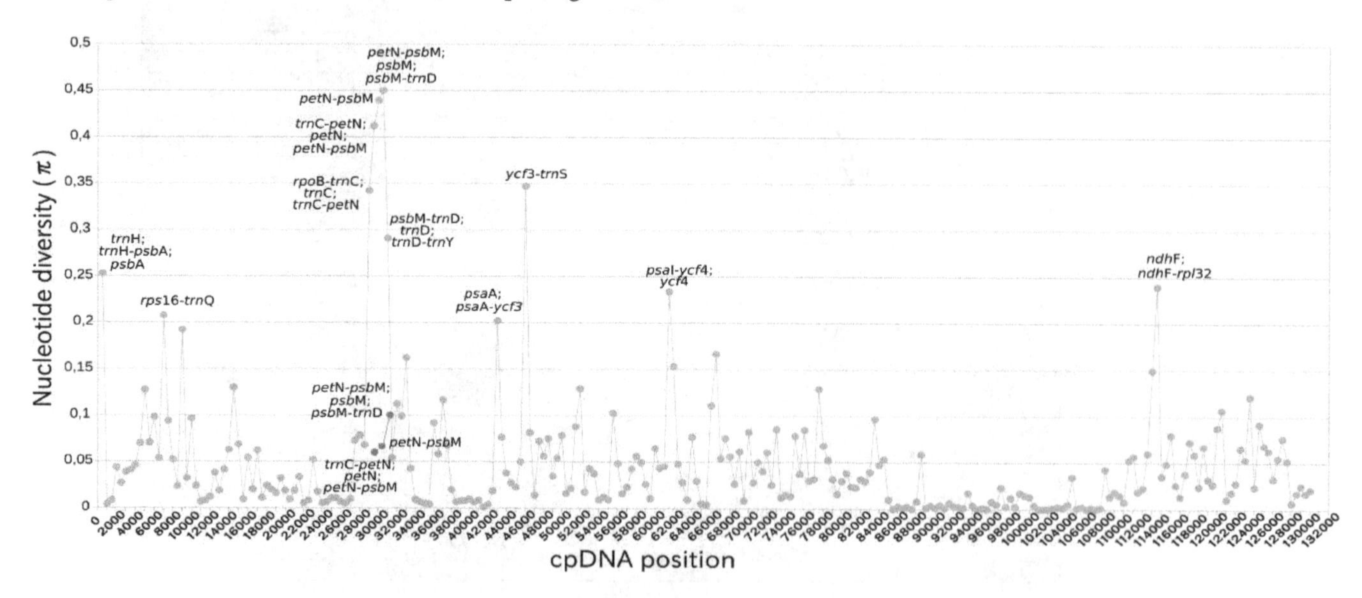


**Figure 6.** Nucleotide diversity (π) for the three *Utricularia amethystina* morphotypes. Each blue dot represents the nucleotide diversity per 500 bp. Orange dots denote the nucleotide diversity for the un-inverted region in *U. amethystina* yellow (*psb*M-*pet*N).

*2.5. Chloroplast Expression*

RNAseq clustering analyses indicated distinct expression profiles for each *Utricularia amethystina* morphotype. The *psb*A and *rbc*L genes showed much higher levels of expression, followed by *psa*A, *psa*B, *psb*B, *psb*C, and *psb*D, in comparison to other genes in all samples (Figure 7; Supplementary Table S7).

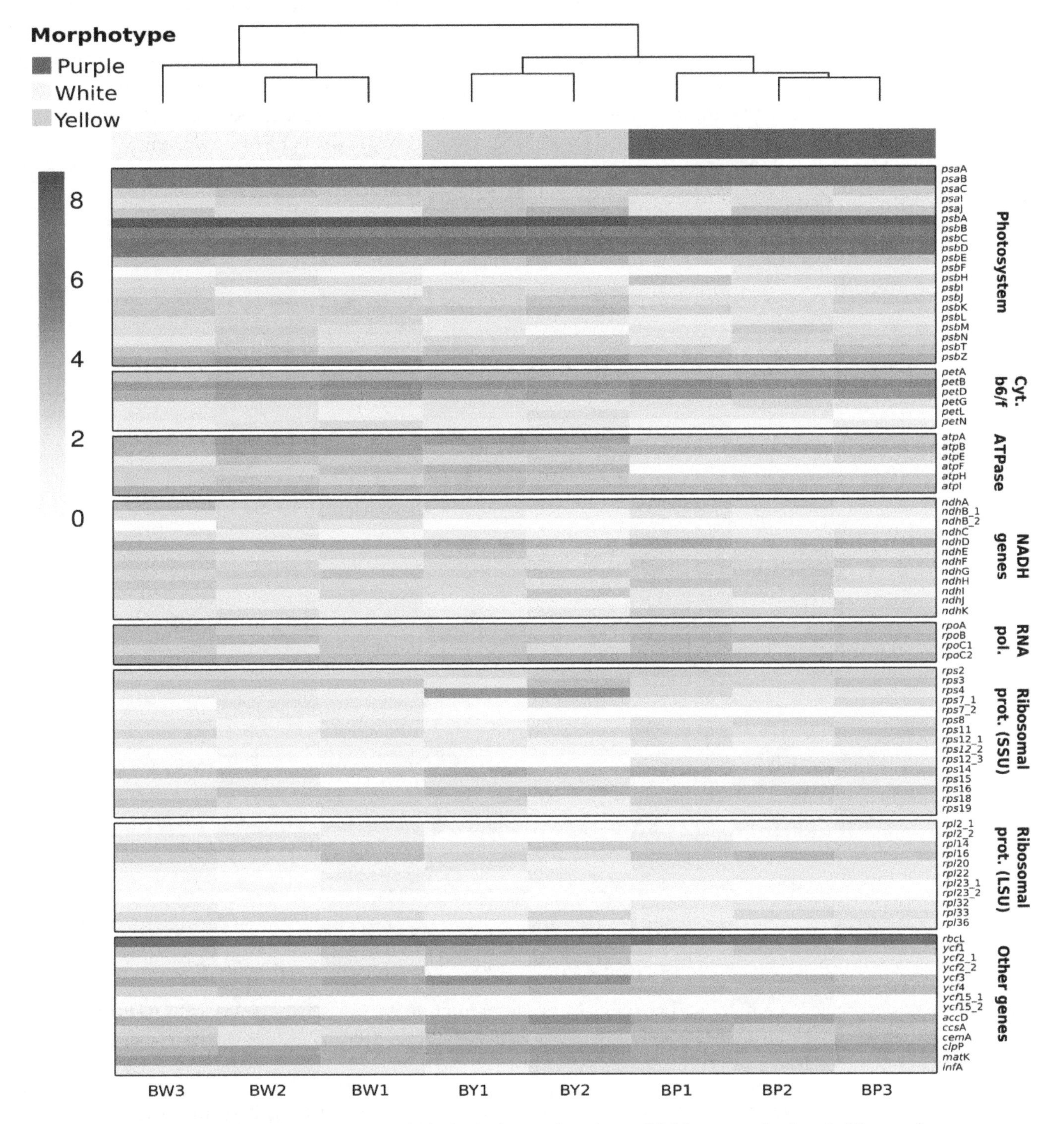


**Figure 7.** Heat map representation of *Utricularia amethystina* cpDNA transcript level. The underscore "_1" and "_2" denotes each gene duplicate. The *rps*12 is a duplicated trans-spliced gene, therefore it was analyzed in three parts.

*2.6. RNA Edit*

The RNA editing analyses were carried out using the PREPACT3 tool, and in-house script to search for validation of editing sites using RNA-Seq mapped reads. The PREPACT3 predicted 154 sites for *U. amethystina* purple, 140 for *U. amethystina* white, and 146 for *U. amethystina* yellow. Most amino acid changes are shared between populations, and comparison between *Utricularia amethystina* morphotypes showed 22 genes with the same editing sites and 15 genes with differences in the quantity and amino acid composition changes (for more information see Supplementary Tables S8–S10). According to the results, there are 13 types of amino acid transitions in the three *U. amethystina*. The most prevalent nonsynonymous substitutional changes occurred between Alanine to Valine and Serine to Leucine, followed by Leucine to Phenylalanine, Proline to Serine, Threonine to Isoleucine, Serine to Phenylalanine, Proline to Leucine, Histidine to Tyrosine, Threonine to Methionine, Proline to Phenylalanine, Arginine to Cysteine, Glutamine to Stop codon, and Arginine to Tryptophan (Figure 8). For RNA-Seq-based results, only three nonsynonymous amino acid substitutions were found (Figure 8; Table 2).

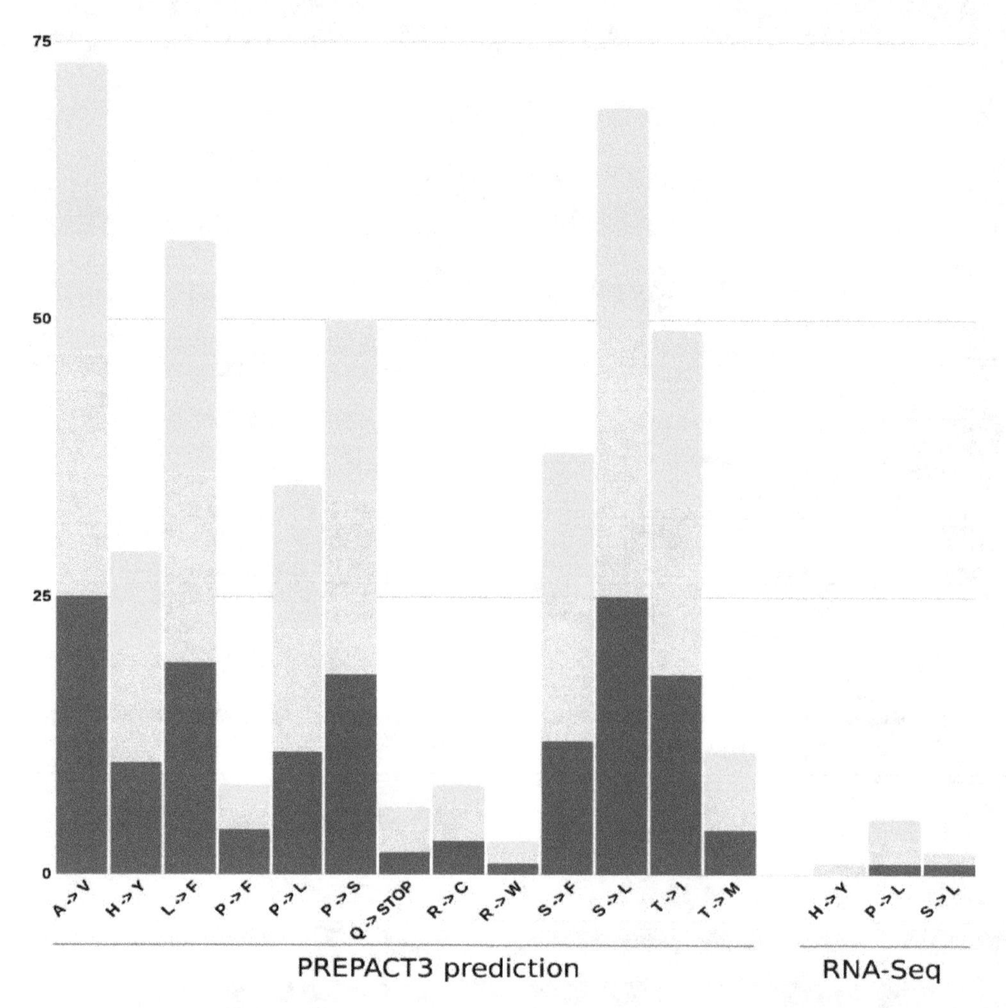


**Figure 8.** Quantities of amino acid changes from the prediction and RNA-Seq data of RNA-editing sites for each *Utricularia amethystina* morphotype. Purple, yellow, and gray bars denote *U. amethystina* purple, yellow, and white morphotypes, respectively.

The results from RNA-Seq data showed eight nonsynonymous and eight synonymous sites for the three *U. amethystina* morphotypes. Regarding nonsynonymous sites, *U. amethystina* purple and yellow have *rps*14 and *pet*B genes edition, and for *U. amethystina* white, edition sites were two in *rps*14, *pet*B, and *ndh*B (Table 2). The same edited sites in the same gene position were found in *pet*B for all *U. amethystina* samples. In addition, the *rps*14 gene from purple is the same as *rps*14 in position 36,572 for white, and *rps*14 gene from yellow is the same as the *rps*14 in position 36,497 for white.

**Table 2.** RNA editing sites predicted using RNAseq data in *Utricularia amethystina* cpDNA. The editing level is given in percentage and is showed for each biological replicate. * denotes amino acids with a change in physicochemical composition.

| Morphotype | Gene | Genome (cpDNA) | | Codon | | Codon Position | Amino Acid | | Editing Level (Each Biorep) | | |
|---|---|---|---|---|---|---|---|---|---|---|---|
| | | Position | Strand | from | to | | from | to | | | |
| **Purple** | | | | | | | | | **P1** | **P2** | **P3** |
| | *rps*14 * | 36,640 | - | UCA | UUA | 2 | S | L | 96 | 100 | 100 |
| | *pet*B | 74,797 | + | CCA | CUA | 2 | P | L | 100 | 100 | 97 |
| **White** | | | | | | | | | **W1** | **W2** | **W3** |
| | *ndh*B | 138,239 | + | CCA | CUA | 2 | H | Y | 92 | 82 | 0 |
| | *ndh*D | 113,305 | - | CUA | UUA | 1 | Synonym | | 100 | 0 | 100 |
| | *rbc*L | 55,651 | + | GCC | GCU | 3 | Synonym | | 100 | 100 | 0 |
| | | 55,777 | + | AUC | AUU | 3 | Synonym | | 100 | 100 | 0 |
| | *rps*14 * | 36,572 | - | UCA | UUA | 2 | S | L | 38 | 14 | 75 |
| | | 36,497 | - | CCA | CUA | 2 | P | L | 100 | 0 | 72 |
| | | 36,553 | - | AAC | AAU | 3 | Synonym | | 100 | 100 | 0 |
| | *psb*B | 72,140 | + | GGC | GGU | 3 | Synonym | | 100 | 100 | 0 |
| | | 71,372 | + | UAC | UAU | 3 | Synonym | | 100 | 0 | 100 |
| | *psa*A | 39,370 | - | AUC | AUU | 3 | Synonym | | 100 | 100 | 100 |
| | *pet*B | 74,816 | + | CCA | CUA | 2 | P | L | 100 | 0 | 97 |
| **Yellow** | | | | | | | | | **Y1** | **Y2** | - |
| | *rps*14 | 36,439 | - | CCA | CUA | 2 | P | L | 69 | 71 | - |
| | *ccs*A | 111,950 | + | CUA | UUA | 1 | Synonym | | 18 | 16 | - |
| | *pet*B | 74,948 | + | CCA | CUA | 2 | P | L | 100 | 100 | - |

## 2.7. Phylogeny

The phylogenetic tree based on available chloroplast genomes of 15 Lentibulariaceae specimens is shown in Figure 9. Both maximum likelihood (ML) and Bayesian inference (BI) trees exhibited identical phylogenetic topologies, and support values (bootstrap for ML and posterior probabilities for BI) are 100% for all clades. The Lentibulariaceae is known to be monophyletic, and *Pinguicula* is a sister clade to *Utricularia-Genlisea*. The results show that the *Utricularia* genus is monophyletic and the three *U. amethystina* are closely related to *U. reniformis*. Also, the *U. amethystina* yellow is sister to purple, with white as having the same common ancestor (Figure 9).

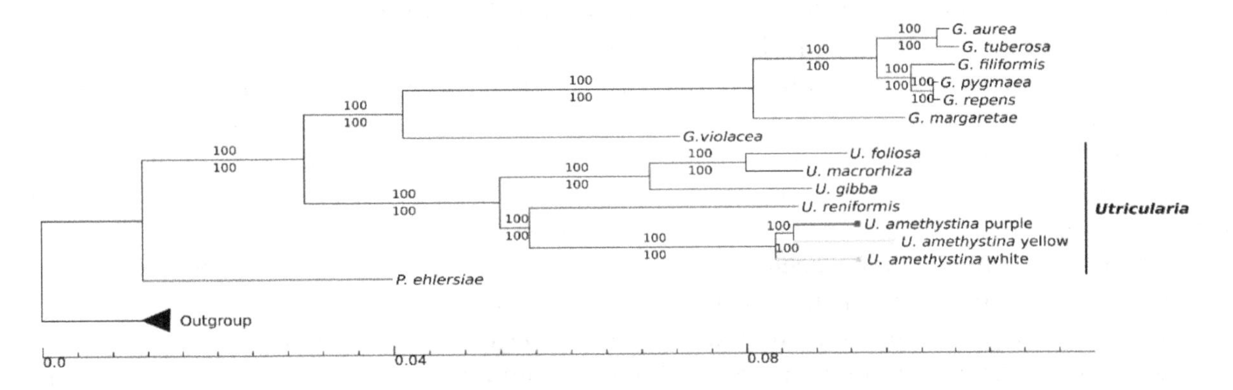


**Figure 9.** Bayesian inference tree for available Lentibulariaceae chloroplast genomes and studied *Utricularia amethystina*. The values above branches denote the posterior probabilities and below maximum likelihood bootstrap. The branch colors correspond to each *U. amethystina* morphotype flower color. (*G. = Genlisea; P. = Pinguicula; U. = Utricularia*). The scale represents the expected changes per site.

## 3. Discussion

Due to the plasticity in corolla shape and color, *Utricularia amethystina* is one of the most polymorphic species within the *Utricularia* genus. This polymorphism resulted in a historically taxonomic complicated group with its systematics only partially resolved to date. During the last decades, several efforts attempted to separate the different *U. amethystina* morphotypes into different species, yet without much success [7,12,13].

In this study, we analyzed the cpDNAs of three morphologically distinct *Utricularia amethystina* from different populations: the purple, white, and yellow morphotypes, aiming to detect intra- and interspecific variations and phylogenetic signals and provide new cpDNA regions for evolutionary studies. In addition, we evaluated the transcription and RNA editing sites for *U. amethystina* populations.

In an attempt to diminish the environmental conditions bias, we have collected the specimens from close populations (~2.8 km between purple and white, 0.2 km between white and yellow, and 2.82 km between purple and yellow. The specimens of *U. amethystina* cpDNA have a typical quadripartite structure present in most land plants and have a similar organization and GC content to other *Utricularia* [24,27].

Among the three *Utricularia amethystina* morphotypes, we found an inversion between the *pet*N and *psb*M genes in *U. amethystina* yellow, representing the first known gene inversion in LSC region identified in Lentibulariaceae chloroplast genomes. Indeed, the same inversion was detected in the chloroplast genome of species of Cannabaceae [30], and microstructural short inversions of 10 bp were also found in the *pet*N-*psb*M region in *Solanus* species [31]. Some comparative cpDNA studies have also identified structural mutations in monilophyte chloroplast genomes, including as many as six inversions and some gene losses (e.g., in [32–34]).

In general, chloroplast deletions/losses are observed among Lentibulariaceae. Indeed, *Utricularia reniformis* suffered a major SSC region retraction due to the losses of NAD(P)H-dehydrogenase (*ndh*) complex genes [27]. In contrast, all other sequenced *Utricularia* cpDNAs have complete *ndh*s gene complexes. These chlororespiratory genes are *ndh*A, B, C, D, E, F, G, H, I, J, and K, and encode subunits of the NADH dehydrogenase complex in plant chloroplast genomes that play a role in plant signaling in the photosynthesis reaction [35] and the reduction and oxidation of plastoquinones [36]. As *U. reniformis* is a terrestrial species, it has been proposed that possibly all terrestrial species of *Utricularia* may lack members from the *ndh* genes complex [27,28]. However, *U. amethystina* is terrestrial, and all three morphotypes retain all plastid *ndhs* complex genes. Therefore, our results now suggest that the *ndh*s in terrestrial *Utricularia* were independently lost and regained, thus refuting the hypothesis (at least for *Utricularia*) that terrestrial species have experienced the loss of *ndh*s genes.

Chloroplast repeats are important regions for replication and DNA stability [37]. Microsatellites or SSRs are tandem repeats of 1–6 base pairs units long that can be used as genetic markers [38]. They are most commonly found in plants and due to genetic variation in the number of tandem repeats units. Therefore, as they produce polymorphism detectable with PCR-based methods banding pattern and genotyping, the SSRs are widely used in population genetics and evolutionary studies [39]. *Utricularia amethystina* has high amounts of mononucleotide repeats in the cpSSR, which is similar to other angiosperms, such as *Arabidopsis thaliana* [40], and other Lentibulariaceae [24,25,27]. Previous results for *Utricularia* indicated that most of SSR were found in coding regions for *U. gibba*, *U. macrorhiza*, *Genlisea margaretae*, and *Pinguicula ehlersiae*. However, for *U. reniformis*, more cpSSR were found in non-coding regions. In *U. amethystina*, long repeats have similar quantities between populations, and as seen in other *Utricularia*, most of them are in coding regions [24,27], an uncommon fact for other angiosperms chloroplast genomes (e.g., see [41]), which could indicate high rates of recombination and rearrangement, as discussed in Silva et al. (2016) [27]. Although long repeats could be the cause for gene rearrangements, we could not find repeats in flanking regions of the genes *psb*M and *pet*N in *U. amethystina* yellow. Therefore, this indicates that other evolutionary forces were involved with the observed inversion of these genes in this species.

For some *Utricularia*, DNA barcoding approaches have been considered a difficult task to perform. For instance, the DNA-barcoding markers, such as ITS, *rbc*L, and *mat*K, could not discriminate all *Utricularia* accessions at the species and population level due to their low level of polymorphism (e.g., *Utricularia* sect. *Utricularia* in Astuti et al., 2019 [42]). Furthermore, *rps*16, *trn*L-F, and *trn*D-T markers cannot discriminate *U. amethystina* populations [13]. Therefore, it is important to explore regions with high variability at inter- and intraspecies levels that represent potentially useful markers for future studies. Using mVISTA results for the interspecific divergence analysis, it is noticeable that the LSC and SSC regions are more variable than IR regions, corroborating with the results found for identity analyses with *Genlisea* species [25] and other angiosperms [41]. The results showed highly variable regions between the different species, mostly represented by intergenic spacers, such as *trn*H-*psb*A, *trn*K-*rps*16, and *rps*16-*trn*Q, which could be used for interspecies identification.

It is previously proposed that populations from closely related environments should be less divergent if they are of the same species. However, we observed high intraspecific chloroplast sequence variability, although geographical sampling covered a restricted area. Among the regions with high nucleotide diversity and intraspecific variations, there is the intergenic spacer, *trn*H-*psb*A, which is already being used as DNA barcoding in many studies [43]. This study also revealed spots that can be used for populations and phylogenetic analyses due to high variability, such as the genes *trn*H, *psb*A, and intergenic regions, such as *trn*H-*psb*A, *ycf*3-*trn*S, and *rps*16-*trn*Q (see more in the Results section and Supplementary Material S6). Nevertheless, the spots of diversity near the genes *pet*N and *psb*M should be avoided due to low primer annealing considering that the region could be inverted, as seen in *U. amethystina* yellow.

The preparation of paired-end libraries was enriched for polyadenylated transcripts which causes the instability of organelle transcripts, therefore there is probably underrepresentation of transcripts [44]. However, we were able to observe that almost all chloroplast protein-coding genes are expressed in all sampled flower tissue of *U. amethystina*, except for *ycf*15, both *rpl*23 duplicated genes in *U. amethystina* purple and yellow, and the *atp*F gene in *U. amethystina* purple (Supplementary Table S7).

The expression profile is similar between samples of the same morphotypes and even expression profile clustering corresponds to the phylogenetic hypothesis proposed in this research. The *rbc*L gene was one of the most highly expressed genes and encodes for one of the most abundant enzymes in nature, the large subunit of ribulose-1-5-biphosphate carboxylase [45]. This protein is involved in fixing $CO_2$ and photorespiration [46]. Moreover, high levels of gene expression were found in Photosystem I (PSI) and II genes (PSII), such as *psa*A and *psa*B, and *psb*A, *psb*B, *psb*C, and *psb*D, these proteins are involved in photosynthesis [47]. Studies of barley leaf activities showed that dark-grown plants were deficient in PSI and PSII proteins [48]. Moreover, Klein et al. (1988) showed that the elongation of translation in *psa*A, *psa*B, *psb*A, and *rbc*L are regulated by light [49,50]. Therefore, considering that the corollas were collected during the day, our results are congruent with the hypothesis of a protein exhibiting light-induced translation.

Interestingly, the *pet*N and *psb*M genes are expressed in all *Utricularia amethystina* biological samples, indicating that, the inversion observed in *U. amethystina* yellow did not affect the expression of these genes.

RNA editing sites are common features of a plant chloroplast. These mutations usually occur from C-to-U in mRNA molecules, and thus have an important role in the differential amino acid generation that can lead to different proteins originated from the same gene [51]. RNA-Seq-based results showed that there is a sum of eight editing sites for all *U. amethystina* morphotypes (Table 2).

The PREPACT3's prediction showed that most nonsynonymous substitutions were characterized as Alanine to Valine and Serine to Leucine. Both lead to protein variations (Table 2, Tables S8–S10), whereas amino acid changes from Alanine to Valine, Histidine to Tyrosine, Leucine to Phenylalanine, Proline to Phenylalanine, Proline to Leucine, Proline to Serine, Arginine to Tryptophan, Threonine to Isoleucine, and Threonine to Methionine result in no physicochemical properties changes in protein. In addition, the Arginine to Cysteine, Serine to Leucine, and Serine to Phenylalanine mutations can modify protein formation due to hydrophilic (Serine and Arginine) to hydrophobic (Leucine, Phenylalanine, and Cysteine) molecule changes [52,53]. Moreover, PREPACT3 has predicted that the genes *rps*2 and *rpl*32 can be edited from Glutamine into a Stop codon, and despite they could be polycistronic genes as in other plants [54], these genes are still transcribed according to RNA-Seq data.

The presented evolutionary history, based on whole chloroplast DNA genomes, and reconstructed by ML and BI approaches supported the same relationship within the Lentibulariaceae when compared with one or few loci–loci analyses (Figure 8) [2,55]. These analyses and many other studies indicated that *Utricularia amethystina* can be paraphyletic [13]. However, in this study, despite the differences in the specimen, they are still a monophyletic taxon. This indicated that *U. amethystina* morphotypes have a common ancestry, but the sampling of other species from sect. *Foliosa* (*U. tricolor* and *U. tridentata*) and species from the close phylogenetically related sect. *Psyllosperma* would be necessary to shine this issue.

Our results support that the sampling based on three different morphotypes proved to be insufficient to allow firm conclusions on the *U. amethystina* species separation, considering we sampled one individual per morphotype. However, the scenario presented here based on chloroplast genomes suggests that *U. amethystina* morphotypes may be different species as previous studies based on morphometric approach [12] and phylogeny with few loci [13], but with more populations, have suggested.

Moreover, the comparative and functional analyses provided by this study bring new insights into the *Utricularia* chloroplast genome architecture, in particular, the evolutionary history of *ndh* complex genes and other important photosynthesis-related genes. Taken together, these results prove that we are just in the beginning for the understanding of the evolution of chloroplast photosynthesis machinery in the Lentibulariaceae.

## 4. Materials and Methods

### *4.1. Sampling and DNA Extraction*

The three *Utricularia amethystina* morphotypes were collected from natural populations geographically close to each other (~2.8 km between purple and white; 0.2 km between white and yellow; and 2.82 km between purple and yellow). The samples were preserved in silica gel and stored at room temperature. Vouchers were deposited at Universidade Estadual Paulista (UNESP) in the Herbarium JABU (Table S1). The total genomic DNA was extracted from approximately 0.1 μg of flowers with Qiagen DNeasy Plant Mini extraction kit (Qiagen, Hilden, Germany) following manufacturer's protocol. The quality and quantity of DNA were estimated with Nanodrop 2000 (Thermo Scientific, MA, USA) and Qubit fluorometer (Life Technologies, CA, USA), respectively.

### *4.2. Organellar Genome Sequencing, Assembly and Annotation*

Sequence libraries were quantified using Bioanalyzer 2100 (Agilent, CA, USA). The paired-end libraries were prepared using Illumina library preparation manufacturer's protocol, and genomic DNA of 2 × 100 bp and insert size of ~200 bp was sequenced using Illumina MiSeq platform (Illumina, San Diego, CA, USA)

The produced paired-end reads filtered for adapters, low-quality bases (Phred score Q > 24) and size (length cutoff for 50 bp), and possible contaminants using Trimmomatic v. 0.38 [56]. The resulting paired-end reads were mapped in the search for discarding mitochondria and nuclear reads with Bowtie2 v.2.2.3 [57] using very sensitive local and -N 1 parameters and *Utricularia* spp. (NC_021449, KY025562, and KT336489) chloroplasts as reference genomes. The resulting reads were assembled using SPAdes v. 3.7.1 [58] and regions with assembly uncertainties were extended using iterative read mapping performed using MITObim v.1.8 [59].

The organelles genomes were primarily annotated using DOGMA [60], cross checked with GeSeq [61], and start and stop codons were adjusted manually for annotation refinements. The tRNAs were annotated using tRNA-scan [62], implemented in DOGMA and Aragorn [63]. The rRNAs were annotated using RNAmmer and BLASTn searches with available *Utricularia* cpDNA genomes. The cp genome map was constructed using the Organellar Genome Draw program [64].

### *4.3. Repeats and SSR Analyses*

To avoid redundant results, only one IR of each *Utricularia amethystina* cpDNA was used and direct, forward, reverse, and palindromic repeats were identified using REPuter [65] with a minimal size of 30 pb and Hamming distance of 3. Simple sequence repeats (SSR) were detected using MISA-web [66] by setting the minimum number of repeats to 7, 4, 4, 3, 3, and 3, for mono-, di-, tri-, tetra-, penta-, and hexanucleotides, respectively.

*4.4. Phylogenetic Reconstruction*

Sequence alignment for all published and available Lentibulariaceae cpDNA genomes (Species and accession number, respectively: *Utricularia foliosa*, KY025562; *U. gibba*, KC997777; *U. reniformis*, KT336489; *U. macrorhiza*, HG803177; *Pinguicula ehlersiae*, HG803178, *Genlisea aurea*, MF593121; *G. violacea*, MF593126; *G. tuberosa*, NC_037082; *G. filiformis*, MF593122; *G. pygmaea*, MF593123; *G. repens*, MF593124; *G. margaretae*, HG530134) were performed with online MAFFT v.7 [67] with default parameters. The phylogenetic tree reconstruction was performed using the Bayesian inference (BI) and maximum likelihood (ML) approaches under the best-of-fit model GTR+G+I, in accordance with Akaike Information Criterion (AIC), assessed by Mr.ModelTest v.2.4 [68]. For BI, set for the substitution model accordingly, we employed Mr.Bayes v.3.2.6 software with $5 \times 10^7$ generations sampled for each 1000 generations using two runs and four chains, until the average standard deviation of split frequencies became less than 0.01, beginning with random trees. The initial trees were discarded after reaching stationary (~25%). For ML we used the software RAxML v.8.2.10 [69] with 1000 bootstrap pseudoreplicates. The analyses were performed using implementations within CIPRES Science Gateway v.3.3 (https://www.phylo.org/) and cladograms were edited using TreeGraph v.2.15 beta [70]. The species *Tectona grandis* (NC020098), *Sesamum indicum* (NC016433), and *Tanaecium tetragonolobum* (NC027955) were used as outgroup.

*4.5. Intraspecific Polymorphism Analyses*

For polymorphism analyses, the *Utricularia amethystina* chloroplasts were aligned using MAFFT v.7, with default parameters. Based on the cpDNA multiple alignments, polymorphism analysis was conducted for coding genes, introns, and intergenic spacers. The nucleotide diversity was calculated using Tassel v.5.2.54 [71] with a sliding window of 500 bp length.

*4.6. Interspecific Comparison*

Using the *Utricularia amethystina* purple as a reference and previously published *Utricularia (Utricularia foliosa*, KY025562; *U. gibba*, KC997777; *U. reniformis*, KT336489; *U. macrorhiza*, HG80317), the identity of cpDNA was assessed using mVISTA online software (http://genome.lbl.gov/vista/mvista/submit.shtml) with Shuffle-LAGAN Mode.

*4.7. RNA Extraction, Sequencing and RNA Editing Site Analyses*

The corollas of *Utricularia amethystina* were stored in RNAlater® (Thermo Fisher Scientific, MA, USA) from each analyzed population and were used as plant tissues for RNA-Seq. The corollas (~5 per specimen) were pooled in three replicates for *U. amethystina* white and purple and two for *U. amethystina* yellow, and total RNA was extracted using PureLink RNA MiniKit (Thermo Fisher Scientific, MA, USA), according to manufacturer's protocol. The extracted RNA was analyzed with Agilent 2100 Bioanalyzer and Qubit 2.0 Fluorometer for quality and quantity assessment, and only samples with RNA Integrity Number (RIN) > 7.0 were used for the sequencing.

The eight libraries (3 libraries for each *U. amethystina* purple and white and 2 for *U. amethystina* yellow) were constructed following the TruSeq Stranded mRNA LS Protocol sample preparation protocol. The paired-end ($2 \times 100$ pb) sequencing was performed in one lane on an Illumina platform (HiSeq 2500) following supplier-provided protocols (Illumina, San Diego, CA, USA).

The raw sequencing data, was preprocessing with high stringency using the following steps. (1) For the 3′ end, the adapter and low-quality reads were removed using Scythe (https://github.com/vsbuffalo/scythe; default parameters, except for -n 5 and -M 15); (2) for the 5′ end, the removal of adapter and low quality reads were performed with Cutadapt [72]; default parameters, except for –overlap 5; –minimum-length = 15; –times = 2); (3) to filter reads with more than 30% of unknown base (Ns), polyA/T tails we used the software Prinseq [73].

Filtered RNA-seq reads were mapped against the assembled chloroplast genome using STAR version 2.7.2a [74], using default parameters except for adjusted parameters to perform an end-to-end mapping, diminish multiple mapping of the same reads, minimum and maximum size of introns and the number of allowed mismatches (–outFilterMultiMapMax = 3; –outFilterMismatchNmax = 2; –outFilterMismatchNoverLmax = 0.1; –outSJfilterReads = Unique; –alignEndsType = EndtoEnd; –alignIntronMin = 70; –alignIntronMax = 2500). To estimate differential transcripts abundance between biological replicates, normalized count data was obtained using relative log expression (RLE) method in DESeq2 version 3.9 [75] and results were showed following with log2(norm. counts+1). The *rps*12 is a duplicated trans-spliced gene, therefore it was analyzed in three parts and "_2" and "_3" represent the duplicated regions. The putative RNA edit sites were predicted following PREPACT3 software [76] with BlastX searches (using default parameters) against the *Nicotiana tabacum*, as reference. The prediction results were compared with the results obtained with an in-house script that counts the number of editing sites according to the previous STAR mapped reads, except for the number of mapped reads, which was set to 1 (only uniquely mapped reads). All of the sites were inspected for C to U nucleotide substitutions by a custom Perl script, with the use of the following parameters; presence in at least two of the biological replicates, editing set with a minimum coverage of 10×.

**Author Contributions:** Conceptualization, S.R.S., V.F.O.M., and A.M.V.; methodology, S.R.S., A.M.V., and D.G.P., V.F.O.M., T.P.M., H.A.P., and E.J.M.; software, S.R.S., A.M.V., V.F.O.M., and D.G.P.; validation, S.R.S., D.G.P., V.F.O.M., and A.M.V.; formal analysis, S.R.S., A.M.V., V.F.O.M., and D.G.P.; investigation, S.R.S., A.M.V., V.F.O.M., and D.G.P.; resources, S.R.S., V.F.O.M., T.P.M., E.J.M., and A.M.V.; data curation, S.R.S., V.F.O.M. and A.M.V.; writing—original draft preparation, S.R.S., A.M.V., and V.F.O.M.; writing—review and editing, S.R.S., A.M.V., D.G.P., T.P.M., H.A.P., B.J.P., and E.J.M.; visualization, S.R.S., V.F.O.M., A.M.V., and T.P.M.; supervision, V.F.O.M. and A.M.V.; project administration, S.R.S., V.F.O.M., and A.M.V.; funding acquisition, V.F.O.M., A.M.V., and T.P.M.

**Acknowledgments:** We thank Cristine Gobbo Menezes and all colleagues from the Laboratory of Plant Systematics (Unesp/FCAV) for collaboration in searching for *Utricularia* natural populations and fruitful discussions. We also thank Tiago Santana Balbuena (Unesp/FCAV) for providing the opportunity to use the server to support the analyses.

## References

1. Müller, K.F.; Borsch, T.; Legendre, L.; Porembski, S.; Barthlott, W. Recent progress in understanding the evolution of carnivorous Lentibulariaceae (Lamiales). *Plant Biol. (Stuttg)* **2006**, *8*, 748–757. [CrossRef]
2. Silva, S.R.; Gibson, R.; Adamec, L.; Domínguez, Y.; Miranda, V.F.O. Molecular phylogeny of bladderworts: A wide approach of *Utricularia* (Lentibulariaceae) species relationships based on six plastidial and nuclear DNA sequences. *Mol. Phylogenet. Evol.* **2018**, *118*, 244–264. [CrossRef]
3. Rutishauser, R.; Isler, B. Developmental genetics and morphological evolution of flowering plants, especially bladderworts (*Utricularia*): Fuzzy Arberian Morphology complements Classical Morphology. *Ann. Bot.* **2001**, *88*, 1173–1202. [CrossRef]
4. Reut, M.S.; Płachno, B.J. Unusual developmental morphology and anatomy of vegetative organs in *Utricularia dichotoma* – leaf, shoot and root dynamics. *Protoplasma* **2019**. [CrossRef]
5. Albert, V.A.; Jobson, R.W.; Michael, T.P.; Taylor, D.J. The carnivorous bladderwort (*Utricularia*, Lentibulariaceae): A system inflates. *J. Exp. Bot.* **2010**, *61*, 5–9. [CrossRef]
6. Veleba, A.; Bureš, P.; Adamec, L.; Šmarda, P.; Lipnerová, I.; Horová, L. Genome size and genomic GC content evolution in the miniature genome-sized family Lentibulariaceae. *New Phytol.* **2014**, *203*, 22–28. [CrossRef]
7. Taylor, P. *The Genus Utricularia—A Taxonomic Monograph*; The Royal Botanic Gardens, Kew: London, UK, 1989.
8. Seine, R.; Porembski, S.; Barthlott, W. A neglected habitat of carnivorous plants: Inselbergs. *Feddes Repert.* **1996**, *106*, 555–562. [CrossRef]
9. BFG Brazilian Flora 2020: Innovation and collaboration to meet Target 1 of the Global Strategy for Plant Conservation (GSPC). *Rodriguésia* **2018**, *69*, 1513–1527. [CrossRef]
10. Vellozo, J.M.C. 19. Utricularia. In *Florae Fluminensis, Seu, Descriptionum Plantarum Praefectura Fluminensi Sponte Mascentium Liber Primus ad Systema Sexuale Concinnatus*; Typographia National: Rio de Janeiro, Brazil, 1825; p. 18.
11. Saint-Hilaire, A.F.C.P.; Girard, F. *Comptes Rendus Hebdomadaires des Séances de L'académie des Sciences*; French Academy of Sciences: Paris, France, 1838; Volume 7, ISBN 0001-4036.

12. Baleeiro, P.C.; Jobson, R.W.; Sano, P.T. Morphometric approach to address taxonomic problems: The case of *Utricularia* sect. *Foliosa (Lentibulariaceae). J. Syst. Evol.* **2016**, *54*, 175–186. [CrossRef]
13. Baleeiro, P.C.; Sano, P.T.; Jobson, R.W. Molecular Phylogeny of the *Utricularia amethystina* Complex (*Utricularia* sect. *Foliosa) Assessed Using Plastid and Nuclear Sequence Data. Syst. Bot.* **2019**, *44*, 398–404.
14. Zhang, Y.; Iaffaldano, B.J.; Zhuang, X.; Cardina, J.; Cornish, K. Chloroplast genome resources and molecular markers differentiate rubber dandelion species from weedy relatives. *BMC Plant Biol.* **2017**, *17*, 1–14. [CrossRef]
15. Oh, S.H.; Suh, H.J.; Park, J.; Kim, Y.; Kim, S. The complete chloroplast genome sequence of a morphotype of *Goodyera schlechtendaliana* (Orchidaceae) with the column appendages. *Mitochondrial DNA Part B Resour.* **2019**, *4*, 626–627. [CrossRef]
16. Jansen, R.K.; Ruhlman, T.A. Plastid Genomes of Seed Plants. In *Genomics of Chloroplasts and Mitochondria. Advances in Photosynthesis and Respiration (Including Bioenergy and Related Processes)*; Bock, R., Knoop, V., Eds.; Springer: Dordrecht, The Netherlands, 2012; ISBN 978-94-007-2920-9.
17. Westwood, J.H.; Yoder, J.I.; Timko, M.P.; DePamphilis, C.W. The evolution of parasitism in plants. *Trends Plant Sci.* **2010**, *15*, 227–235. [CrossRef]
18. Delannoy, E.; Fujii, S.; Colas Des Francs-Small, C.; Brundrett, M.; Small, I. Rampant Gene loss in the underground orchid *Rhizanthella* Gardneri highlights evolutionary constraints on plastid genomes. *Mol. Biol. Evol.* **2011**, *28*, 2077–2086. [CrossRef]
19. Braukmann, T.; Stefanović, S. Plastid genome evolution in mycoheterotrophic Ericaceae. *Plant Mol. Biol.* **2012**, *79*, 5–20. [CrossRef]
20. Gruzdev, E.V.; Kadnikov, V.V.; Beletsky, A.V.; Kochieva, E.Z.; Mardanov, A.V.; Skryabin, K.G.; Ravin, N.V. Plastid Genomes of Carnivorous Plants *Drosera rotundifolia* and *Nepenthes × ventrata* Reveal Evolutionary Patterns Resembling Those Observed in Parasitic Plants. *Int. J. Mol. Sci.* **2019**, *20*, 4107. [CrossRef]
21. Nevill, P.G.; Howell, K.A.; Cross, A.T.; Williams, A.V.; Zhong, X.; Tonti-Filippini, J.; Boykin, L.M.; Dixon, K.W.; Small, I. Plastome-wide rearrangements and gene losses in carnivorous droseraceae. *Genome Biol. Evol.* **2019**, *11*, 472–485. [CrossRef]
22. Martín, M.; Sabater, B. Plastid ndh genes in plant evolution. *Plant Physiol. Biochem.* **2010**, *48*, 636–645. [CrossRef]
23. Ruhlman, T.A.; Chang, W.-J.; Chen, J.J.W.; Huang, Y.-T.; Chan, M.-T.; Zhang, J.; Liao, D.-C.; Blazier, J.C.; Jin, X.; Shih, M.-C.; et al. NDH expression marks major transitions in plant evolution and reveals coordinate intracellular gene loss. *BMC Plant Biol.* **2015**, *15*, 100. [CrossRef]
24. Wicke, S.; Schäferhoff, B.; Depamphilis, C.W.; Müller, K.F. Disproportional plastome-wide increase of substitution rates and relaxed purifying selection in genes of carnivorous Lentibulariaceae. *Mol. Biol. Evol.* **2014**, *31*, 529–545. [CrossRef]
25. Silva, S.R.; Michael, T.P.; Meer, E.J.; Pinheiro, D.G.; Varani, A.M.; Miranda, V.F.O. Comparative genomic analysis of *Genlisea* (corkscrew plants—Lentibulariaceae) chloroplast genomes reveals an increasing loss of the ndh genes. *PLoS ONE* **2018**, *13*, e0190321. [CrossRef]
26. Ibarra-Laclette, E.; Lyons, E.; Hernández-Guzmán, G.; Pérez-Torres, C.A.; Carretero-Paulet, L.; Chang, T.H.; Lan, T.; Welch, A.J.; Juárez, M.J.A.; Simpson, J.; et al. Architecture and evolution of a minute plant genome. *Nature* **2013**, *498*, 94–98. [CrossRef]
27. Silva, S.R.; Diaz, Y.C.A.; Penha, H.A.; Pinheiro, D.G.; Fernandes, C.C.; Miranda, V.F.O.; Michael, T.P.; Varani, A.M. The chloroplast genome of *Utricularia reniformis* sheds light on the evolution of the ndh gene complex of terrestrial carnivorous plants from the Lentibulariaceae family. *PLoS ONE* **2016**, *11*, e0165176. [CrossRef]
28. Silva, S.R.; Pinheiro, D.G.; Meer, E.J.; Michael, T.P.; Varani, A.M.; Miranda, V.F.O. The complete chloroplast genome sequence of the leafy bladderwort, *Utricularia foliosa* L. (Lentibulariaceae). *Conserv. Genet. Resour.* **2017**, *9*, 213–216. [CrossRef]
29. Silva, S.R.; Alvarenga, D.O.; Aranguren, Y.; Penha, H.A.; Fernandes, C.C.; Pinheiro, D.G.; Oliveira, M.T.; Michael, T.P.; Miranda, V.F.O.; Varani, A.M. The mitochondrial genome of the terrestrial carnivorous plant *Utricularia reniformis* (Lentibulariaceae): Structure, comparative analysis and evolutionary landmarks. *PLoS ONE* **2017**, *12*, e0180484. [CrossRef]
30. Zhang, H.; Jin, J.; Moore, M.J.; Yi, T.; Li, D. Plastome characteristics of Cannabaceae. *Plant Divers.* **2018**, *40*, 127–137. [CrossRef]
31. Gargano, D.; Scotti, N.; Vezzi, A.; Bilardi, A.; Valle, G.; Grillo, S.; Cozzolino, S.; Cardi, T. Genome-wide analysis of plastome sequence variation and development of plastidial CAPS markers in common potato and related Solanum species. *Genet. Resour. Crop Evol.* **2012**, *59*, 419–430. [CrossRef]

32. Gao, L.; Zhou, Y.; Wang, Z.W.; Su, Y.J.; Wang, T. Evolution of the rpoB-psbZ region in fern plastid genomes: Notable structural rearrangements and highly variable intergenic spacers. *BMC Plant Biol.* **2011**, *11*, 1–13. [CrossRef]
33. Gao, L.; Yi, X.; Yang, Y.X.; Su, Y.J.; Wang, T. Complete chloroplast genome sequence of a tree fern *Alsophila spinulosa*: Insights into evolutionary changes in fern chloroplast genomes. *BMC Evol. Biol.* **2009**, *9*, 1–14. [CrossRef]
34. Wolf, P.; Roper, J.; Duffy, A. The evolution of chloroplast genome structure in ferns. *Genome* **2010**, *53*, 731–738. [CrossRef]
35. Shikanai, T. The NAD(P)H Dehydrogenase Complex in Photosynthetic Organisms: Subunit Composition and Physiological Function. *Funct. Plant Sci. Biotechnol.* **2007**, *1*, 129–137.
36. Sazanov, L.A.; Burrows, P.A.; Nixon, P.J. The plastid ndh genes code for an NADH-specific dehydrogenase: Isolation of a complex I analogue from pea thylakoid membranes. *Proc. Natl. Acad. Sci. USA* **2002**, *95*, 1319–1324. [CrossRef]
37. Pearson, C.E.; Zorbas, H.; Price, G.B.; Zannis-Hadjopoulos, M. Inverted repeats, stem-loops, and cruciforms: Significance for initiation of DNA replication. *J. Cell. Biochem.* **1996**, *63*, 1–22. [CrossRef]
38. Gupta, P.K.; Balyan, H.S.; Sharma, P.C.; Ramesh, B. Microsatellites in plants: A new class of molecular markers. *Curr. Sci.* **1996**, *70*, 45–54.
39. Gao, C.; Ren, X.; Mason, A.S.; Li, J.; Wang, W.; Xiao, M.; Fu, D. Revisiting an important component of plant genomes: Microsatellites. *Funct. Plant Biol.* **2013**, *40*, 645–661. [CrossRef]
40. Lawson, M.J.; Zhang, L. Distinct patterns of SSR distribution in the *Arabidopsis thaliana* and rice genomes. *Genome Biol.* **2006**, *7*, R14. [CrossRef]
41. Yao, X.; Tang, P.; Li, Z.; Li, D.; Liu, Y.; Huang, H. The first complete chloroplast genome sequences in Actinidiaceae: Genome structure and comparative analysis. *PLoS ONE* **2015**, *10*, e0129347. [CrossRef]
42. Astuti, G.; Petroni, G.; Adamec, L.; Miranda, V.F.O.; Peruzzi, L. DNA barcoding approach fails to discriminate Central European bladderworts (*Utricularia*, Lentibulariaceae), but provides insights concerning their evolution. *Plant Biosyst.* **2019**. [CrossRef]
43. Kress, W.J.; Wurdack, K.J.; Zimmer, E.A.; Weigt, L.A.; Janzen, D.H. Use of DNA barcodes to identify flowering plants. *Proc. Natl. Acad. Sci. USA* **2005**, *102*, 8369–8374. [CrossRef]
44. Rorbach, J.; Bobrowicz, A.; Pearce, S.; Minczuk, M. Polyadenylation in bacteria and organelles. In *Methods in Molecular Biology*; Rorbach, J., Bobrowicz, A., Eds.; Humana Press: Totowa, NJ, USA, 2014; Volume 1125, pp. 211–227.
45. Saschenbrecker, S.; Bracher, A.; Rao, K.V.; Rao, B.V.; Hartl, F.U.; Hayer-Hartl, M. Structure and Function of RbcX, an Assembly Chaperone for Hexadecameric Rubisco. *Cell* **2007**, *129*, 1189–1200. [CrossRef]
46. Andrews, T.J.; Lorimer, G.H. Rubisco: Structure, Mechanisms, and Prospects for Improvement. In *Photosynthesis*; Hatch, M.D., Boardman, N.K., Eds.; Academic Press: Cambridge, MA, USA, 1987; pp. 131–218. ISBN 9780126754100.
47. Nelson, N.; Yocum, C.F. Structure and function of photosystems I and II. *Annu. Rev. Plant Biol.* **2006**, *57*, 521–565. [CrossRef] [PubMed]
48. Plesnicar, M.; Bendall, D.S. The photochemical activities and electron carriers of developing barley leaves. *Biochem. J.* **1973**, *136*, 803–812. [CrossRef] [PubMed]
49. Klein, R.R.; Mason, H.S.; Mullet, J.E. Light-regulated translation of chloroplast proteins. I. Transcripts of psaA-psaB, psbA, and rbcL are associated with polysomes in dark-grown and illuminated barley seedlings. *J. Cell Biol.* **1988**, *106*, 289–301. [CrossRef] [PubMed]
50. Zoschke, R.; Bock, R. Chloroplast translation: Structural and functional organization, operational control, and regulation. *Plant Cell* **2018**, *30*, 745–770. [CrossRef]
51. Wolf, P.G.; Rowe, C.A.; Hasebe, M. High levels of RNA editing in a vascular plant chloroplast genome: Analysis of transcripts from the fern *Adiantum capillus-veneris*. *Gene* **2004**, *339*, 89–97. [CrossRef]
52. Betts, M.J.; Russell, R.B. Amino-Acid Properties and Consequences of Substitutions. In *Bioinformatics for Geneticists: A Bioinformatics Primer for the Analysis of Genetic Data*, 2nd ed.; Barnes, M.R., Gray, I.C., Eds.; John Wiley & Sons, Ltd.: Hoboken, NJ, USA, 2007; ISBN 9780470026199.
53. Wang, M.; Liu, H.; Ge, L.; Xing, G.; Wang, M.; Weining, S.; Nie, X. Identification and analysis of RNA editing sites in the chloroplast transcripts of *Aegilops tauschii* L. *Genes (Basel)* **2017**, *8*, 13. [CrossRef]

54. Sugita, M.; Sugiura, M. Regulation of gene expression in chloroplasts of higher plants. *Plant Mol. Biol.* **1996**, *32*, 315–326. [CrossRef]
55. Müller, K.; Borsch, T. Phylogenetics of *Utricularia* (Lentibulariaceae) and molecular evolution of the trnK intron in a lineage with high substitutional rates. *Plant Syst. Evol.* **2005**, *250*, 39–67. [CrossRef]
56. Bolger, A.M.; Lohse, M.; Usadel, B. Trimmomatic: A flexible trimmer for Illumina sequence data. *Bioinformatics* **2014**, *30*, 2114–2120. [CrossRef]
57. Langmead, B.; Salzberg, S.L. Fast gapped-read alignment with Bowtie 2. *Nat. Methods* **2012**, *9*, 357–359. [CrossRef]
58. Bankevich, A.; Nurk, S.; Antipov, D.; Gurevich, A.A.; Dvorkin, M.; Kulikov, A.S.; Lesin, V.M.; Nikolenko, S.I.; Pham, S.; Prjibelski, A.D.; et al. SPAdes: A New Genome Assembly Algorithm and Its Applications to Single-Cell Sequencing. *J. Comput. Biol.* **2012**, *19*, 455–477. [CrossRef] [PubMed]
59. Hahn, C.; Bachmann, L.; Chevreux, B. Reconstructing mitochondrial genomes directly from genomic next-generation sequencing reads—A baiting and iterative mapping approach. *Nucleic Acids Res.* **2013**, *41*, e129. [CrossRef] [PubMed]
60. Wyman, S.K.; Jansen, R.K.; Boore, J.L. Automatic annotation of organellar genomes with DOGMA. *Bioinformatics* **2004**, *20*, 3252–3255. [CrossRef] [PubMed]
61. Tillich, M.; Lehwark, P.; Pellizzer, T.; Ulbricht-Jones, E.S.; Fischer, A.; Bock, R.; Greiner, S. GeSeq—Versatile and accurate annotation of organelle genomes. *Nucleic Acids Res.* **2017**, *45*, W6–W11. [CrossRef] [PubMed]
62. Lowe, T.M.; Eddy, S.R. tRNAscan-SE: A program for improved detection of transfer RNA genes in genomic sequence. *Nucleic Acids Res.* **1997**, *25*, 955–964. [CrossRef]
63. Laslett, D.; Canback, B. ARAGORN, a program to detect tRNA genes and tmRNA genes in nucleotide sequences. *Nucleic Acids Res.* **2004**, *32*, 11–16. [CrossRef]
64. Greiner, S.; Lehwark, P.; Bock, R. OrganellarGenomeDRAW (OGDRAW) version 1.3.1: Expanded toolkit for the graphical visualization of organellar genomes. *Nucleic Acids Res.* **2019**, *47*, W59–W64. [CrossRef]
65. Kurtz, S.; Choudhuri, J.V.; Ohlebusch, E.; Scheleiermacher, C.; Stoye, J.; Giegerich, R. REPuter: The manifold applications of repeat analysis on a genomic scale. *Nucleic Acids Res.* **2001**, *29*, 4633–4642. [CrossRef]
66. Beier, S.; Thiel, T.; Münch, T.; Scholz, U.; Mascher, M. MISA-web: A web server for microsatellite prediction. *Bioinformatics* **2017**, *33*, 2583–2585. [CrossRef]
67. Katoh, K. MAFFT: A novel method for rapid multiple sequence alignment based on fast Fourier transform. *Nucleic Acids Res.* **2002**, *30*, 3059–3066. [CrossRef]
68. Nylander, J.A.A. *MrModeltest v2.3. Program Distributed by the Author*; Evolutionary Biology Centre, Uppsala University: Uppsala, Sweden, 2004.
69. Stamatakis, A. RAxML version 8: A tool for phylogenetic analysis and post-analysis of large phylogenies. *Bioinformatics* **2014**, *30*, 1312–1313. [CrossRef] [PubMed]
70. Stöver, B.C.; Müller, K.F. TreeGraph 2: Combining and visualizing evidence from different phylogenetic analyses. *BMC Bioinform.* **2010**, *11*, 1–9. [CrossRef] [PubMed]
71. Bradbury, P.J.; Zhang, Z.; Kroon, D.E.; Casstevens, T.M.; Ramdoss, Y.; Buckler, E.S. TASSEL: Software for association mapping of complex traits in diverse samples. *Bioinformatics* **2007**, *23*, 2633–2635. [CrossRef] [PubMed]
72. Martin, M. Cutadapt removes adapter sequences from high-throughput sequencing reads. *EMBnet. J.* **2011**, *17*, 10. [CrossRef]
73. Schmieder, R.; Edwards, R. Quality control and preprocessing of metagenomic datasets. *Bioinformatics* **2011**, *27*, 863–864. [CrossRef]
74. Dobin, A.; Davis, C.A.; Schlesinger, F.; Drenkow, J.; Zaleski, C.; Jha, S.; Batut, P.; Chaisson, M.; Gingeras, T.R. STAR: Ultrafast universal RNA-seq aligner. *Bioinformatics* **2013**, *29*, 15–21. [CrossRef]
75. Love, M.I.; Huber, W.; Anders, S. Moderated estimation of fold change and dispersion for RNA-seq data with DESeq2. *Genome Biol.* **2014**, *15*, 550. [CrossRef]
76. Lenz, H.; Hein, A.; Knoop, V. Plant organelle RNA editing and its specificity factors: Enhancements of analyses and new database features in PREPACT 3.0. *BMC Bioinform.* **2018**, *19*, 255. [CrossRef]

# PERMISSIONS

All chapters in this book were first published by MDPI; hereby published with permission under the Creative Commons Attribution License or equivalent. Every chapter published in this book has been scrutinized by our experts. Their significance has been extensively debated. The topics covered herein carry significant findings which will fuel the growth of the discipline. They may even be implemented as practical applications or may be referred to as a beginning point for another development.

The contributors of this book come from diverse backgrounds, making this book a truly international effort. This book will bring forth new frontiers with its revolutionizing research information and detailed analysis of the nascent developments around the world.

We would like to thank all the contributing authors for lending their expertise to make the book truly unique. They have played a crucial role in the development of this book. Without their invaluable contributions this book wouldn't have been possible. They have made vital efforts to compile up to date information on the varied aspects of this subject to make this book a valuable addition to the collection of many professionals and students.

This book was conceptualized with the vision of imparting up-to-date information and advanced data in this field. To ensure the same, a matchless editorial board was set up. Every individual on the board went through rigorous rounds of assessment to prove their worth. After which they invested a large part of their time researching and compiling the most relevant data for our readers.

The editorial board has been involved in producing this book since its inception. They have spent rigorous hours researching and exploring the diverse topics which have resulted in the successful publishing of this book. They have passed on their knowledge of decades through this book. To expedite this challenging task, the publisher supported the team at every step. A small team of assistant editors was also appointed to further simplify the editing procedure and attain best results for the readers.

Apart from the editorial board, the designing team has also invested a significant amount of their time in understanding the subject and creating the most relevant covers. They scrutinized every image to scout for the most suitable representation of the subject and create an appropriate cover for the book.

The publishing team has been an ardent support to the editorial, designing and production team. Their endless efforts to recruit the best for this project, has resulted in the accomplishment of this book. They are a veteran in the field of academics and their pool of knowledge is as vast as their experience in printing. Their expertise and guidance has proved useful at every step. Their uncompromising quality standards have made this book an exceptional effort. Their encouragement from time to time has been an inspiration for everyone.

The publisher and the editorial board hope that this book will prove to be a valuable piece of knowledge for researchers, students, practitioners and scholars across the globe.

# LIST OF CONTRIBUTORS

**Jinwei Suo**
Alkali Soil Natural Environmental Science Center, Northeast Forestry University, Key Laboratory of Saline-alkali Vegetation Ecology Restoration in Oil Field, Ministry of Education, Harbin 150040, China

**Qi Zhao**
Development Center of Plant Germplasm Resources, College of Life and Environmental Sciences, Shanghai Normal University, Shanghai 200234, China

**Shaojun Dai**
Alkali Soil Natural Environmental Science Center, Northeast Forestry University, Key Laboratory of Saline-alkali Vegetation Ecology Restoration in Oil Field, Ministry of Education, Harbin 150040, China
Development Center of Plant Germplasm Resources, College of Life and Environmental Sciences, Shanghai Normal University, Shanghai 200234, China

**Lisa David and Sixue Chen**
Department of Biology, Genetics Institute, Plant Molecular and Cellular Biology Program, Interdisciplinary Center for Biotechnology Research, University of Florida, Gainesville, FL 32610, USA

**Rui Liu, Shi-Kai Cao, Aqib Sayyed, Chunhui Xu, Feng Sun and Bao-Cai Tan**
Key Laboratory of Plant Development and Environment Adaptation Biology, Ministry of Education, School of Life Sciences, Shandong University, Qingdao 266237, China

**Xiaomin Wang**
Key Laboratory of Cell Activities and Stress Adaptations, Ministry of Education, School of Life Sciences, Lanzhou University, Lanzhou 730000, China

**Rana Khalid Iqbal, Daria Mileshina, Romain Val, Frédérique Weber-Lotfi, José Manuel Gualberto and André Dietrich**
Institute of Plant Molecular Biology (IBMP), CNRS and University of Strasbourg, 12 rue du Général Zimmer, 67084 Strasbourg, France

**Adnan Khan Niazi**
Institute of Plant Molecular Biology (IBMP), CNRS and University of Strasbourg, 12 rue du Général Zimmer, 67084 Strasbourg, France
Centre ofAgricultural Biochemistry and Biotechnology (CABB), University ofAgriculture, Faisalabad 38000, Pakistan

**Ludivine Soubigou-Taconnat and Etienne Delannoy**
Institute of Plant Sciences Paris-Saclay IPS2, CNRS, INRA, Université Paris-Sud, Université Evry, Université Paris-Saclay, Paris Diderot, Sorbonne Paris-Cité, 91405 Orsay, France

**Marta Gabryelska and Eliza Wyszko**
Institute of Bioorganic Chemistry, Polish Academy of Sciences, Ul. Z. Noskowskiego 12/14, 61-704 Poznan, Poland

**Maciej Szymanski**
Department of Computational Biology, Institute of Molecular Biology and Biotechnology, A. Mickiewicz University Poznan, Ul. Umultowska 89, 61-614 Poznan, Poland

**Jan Barciszewski**
Institute of Bioorganic Chemistry, Polish Academy of Sciences, Ul. Z. Noskowskiego 12/14, 61-704 Poznan, Poland
NanoBioMedical Centre of the Adam Mickiewicz University, Umultowska 85, 61614 Poznan, Poland

**Liping Nie, Yingxian Cui, Liwei Wu, Jianguo Zhou, Zhichao Xu, Yu Wang and Hui Yao**
Key Lab of Chinese Medicine Resources Conservation, State Administration of Traditional Chinese Medicine of the People's Republic of China, Institute of Medicinal Plant Development, Chinese Academy of Medical Sciences and Peking Union Medical College, Beijing 100193, China
Engineering Research Center of Chinese Medicine Resources, Ministry of Education, Beijing 100193, China

**Yonghua Li**
College of Pharmacy, Guangxi University of Traditional Chinese Medicine, Nanning 530200, China

**Xiwen Li**
Institute of Chinese Materia Medica, China Academy of Chinese Medical Sciences, Beijing 100700, China

**Changwei Bi, Na Lu, Chunpeng He and Zuhong Lu**
State Key Laboratory of Bioelectronics, School of Biological Science and Medical Engineering, Southeast University, Nanjing 210096, Jiangsu, China

**Yiqing Xu**
School of Information Science and Technology, Nanjing Forestry University, Nanjing 210037, Jiangsu, China

**Yi Zhang, Aihong Zhang and Congming Lu**
State Key Laboratory of Crop Biology, College of Life Sciences, Shandong Agricultural University, Taian 271018, China

**Xiuming Li**
State Key Laboratory of Crop Biology, College of Horticulture Science and Engineering, Shandong Agricultural University, Taian 271018, China

**Yi Pan, Zhi Liu, Yucheng Liu, Zongbiao Duan, Baoge Zhu and Guoan Zhou**
State Key Laboratory of Plant Cell and Chromosome Engineering, Institute of Genetics and Developmental Biology, Innovative Academy of Seed Design, Chinese Academy of Sciences, Beijing 100101, China

**Xiaowei Zhu**
State Key Laboratory of Plant Cell and Chromosome Engineering, Institute of Genetics and Developmental Biology, Innovative Academy of Seed Design, Chinese Academy of Sciences, Beijing 100101, China
Horticulture Research Institute, Shanghai Academy of Agricultural Sciences, Shanghai 201403, China

**Deyi Zhong**
State Key Laboratory of Plant Cell and Chromosome Engineering, Institute of Genetics and Developmental Biology, Innovative Academy of Seed Design, Chinese Academy of Sciences, Beijing 100101, China
Agricultural Genomics Institute, Chinese Academy of Agricultural Sciences, Shenzhen 518120, China

**Zhixi Tian**
State Key Laboratory of Plant Cell and Chromosome Engineering, Institute of Genetics and Developmental Biology, Innovative Academy of Seed Design, Chinese Academy of Sciences, Beijing 100101, China
College of Advanced Agricultural Sciences, University of Chinese Academy of Sciences, Beijing 100049, China

**Huan Zhu, Yuxin Hu and Guoxiang Liu**
Key Laboratory of Algal Biology, Institute of Hydrobiology, Chinese Academy of Sciences, Wuhan 430072, China

**Feng Liu**
Key Laboratory of Marine Ecology and Environmental Sciences, Institute of Oceanology, Chinese Academy of Sciences, Qingdao 266071, China

**Zhengyu Hu**
State Key Laboratory of Freshwater Ecology and Biotechnology, Institute of Hydrobiology, Chinese Academy of Sciences, Wuhan 430072, China

**Pedro Robles and Víctor Quesada**
Instituto de Bioingeniería, Universidad Miguel Hernández, Campus de Elche, 03202 Elche, Spain

**Yihe Yu and Keke Yu**
College of Forestry, Henan University of Science and Technology, Luoyang 471023, China

**Po-Cheng Yu**
Department of Chemistry, Temple University, Philadelphia, PA 19122, USA

**Wan-Jung Chang**
Perelman School of Medicine, University of Pennsylvania, Philadelphia, PA 19104, USA

**Choun-Sea Lin**
Agricultural Biotechnology Research Center, Academia Sinica, Taipei 115, Taiwan

**Ran Dong, Li-Na Lan, Shu-Fen Li, Wu-Jun Gao and Hong-Xing Niu**
College of Life Sciences, Henan Normal University, Xinxiang 453007, China

**Guo-Jun Zhang**
College of Life Sciences, Henan Normal University, Xinxiang 453007, China
School of Basic Medical Sciences, Xinxiang Medical University, Xinxiang 453003, China

**In-Su Choi and Tracey A. Ruhlman**
Department of Integrative Biology, University of Texas at Austin, Austin, TX 78712, USA

**Robert K. Jansen**
Department of Integrative Biology, University of Texas at Austin, Austin, TX 78712, USA
Centre of Excellence in Bionanoscience Research, Department of Biological Sciences, Faculty of Science, King Abdulaziz University, Jeddah 21589, Saudi Arabia

**Yuzhe Sun, Zhou Xu, Koon Chuen Chan and Jia Yi Zhong**
School of Biological Sciences, University of Hong Kong, Pokfulam, Hong Kong, China

**Boon Leong Lim**
School of Biological Sciences, University of Hong Kong, Pokfulam, Hong Kong, China
Center for Soybean Research of the State Key Laboratory of Agrobiotechnology, The Chinese University of Hong Kong, Shatin, Hong Kong, China

**Min Xie, Kejing Fan, Johanna Wong-Bajracharya and Hon-Ming Lam**
Center for Soybean Research of the State Key Laboratory of Agrobiotechnology, The Chinese University of Hong Kong, Shatin, Hong Kong, China
School of Life Sciences, The Chinese University of Hong Kong, Shatin, Hong Kong, China

**Saura R. Silva, Daniel G. Pinheiro, Helen A. Penha and Alessandro M. Varani**
Departamento de Tecnologia, Faculdade de Ciências Agrárias e Veterinárias, Jaboticabal, Universidade Estadual Paulista (Unesp), Sao Paulo 14884-900, Brazil

**Bartosz J. Płachno**
Department of Plant Cytology and Embryology, Institute of Botany, Faculty of Biology, Jagiellonian University in Kraków, 30-387 Krakow, Poland

**Todd P. Michael**
J. Craig Venter Institute, La Jolla, CA 92037, USA

**Elliott J. Meer**
10X Genomics, Pleasanton, CA 94566, USA

**Vitor F. O. Miranda**
Departamento de Biologia Aplicada à Agropecuária, Faculdade de Ciências Agrárias e Veterinárias, Jaboticabal, Universidade Estadual Paulista (Unesp), Sao Paulo 14884-900, Brazil

# Index

Printed in the USA
CPSIA information can be obtained
at www.ICGtesting.com
JSHW052312231023
50683JS00006BA/79